MW01070161

VARIATIONAL METHODS WITH APPLICATIONS IN SCIENCE AND ENGINEERING

There is an ongoing resurgence of applications in which the calculus of variations has direct relevance. *Variational Methods with Applications in Science and Engineering* reflects the strong connection between calculus of variations and the applications for which variational methods form the fundamental foundation. The material is presented in a manner that promotes development of an intuition about the concepts and methods with an emphasis on applications, and the priority of the application chapters is to provide a brief introduction to a variety of physical phenomena and optimization principles from a unified variational point of view. The first part of the book provides a modern treatment of the calculus of variations suitable for advanced undergraduate students and graduate students in applied mathematics, physical sciences, and engineering. The second part gives an account of several physical applications from a variational point of view, such as classical mechanics, optics and electromagnetics, modern physics, and fluid mechanics. A unique feature of this part of the text is derivation of the ubiquitous Hamilton's principle directly from the first law of thermodynamics, which enforces conservation of total energy, and the subsequent derivation of the governing equations of many discrete and continuous phenomena from Hamilton's principle. In this way, the reader will see how the traditional variational treatments of statics and dynamics are unified with the physics of fluids, electromagnetic fields, relativistic mechanics, and quantum mechanics through Hamilton's principle. The third part covers applications of variational methods to optimization and control of discrete and continuous systems, including image and data processing as well as numerical grid generation. The application chapters in parts two and three are largely independent of each other so that the instructor or reader can choose a path through the topics that aligns with their interests.

A graduate of Lehigh University and former National Research Council Postdoctoral Fellow at the National Institute of Standards and Technology, Kevin W. Cassel is Associate Professor of Mechanical and Aerospace Engineering at the Illinois Institute of Technology (IIT). He has been a visiting researcher at the University of Manchester, UK, and University College London, UK, and is a visiting professor at the University of Palermo, Italy. Professor Cassel's research utilizes computational fluid dynamics in conjunction with advanced analytical methods to address problems in bio-fluids, unsteady aerodynamics, multiphase flow, and cryogenic fluid flow and heat transfer. Dr. Cassel is an Associate Fellow of the American Institute of Aeronautics and Astronautics, and his honors include IIT's University Excellence in Teaching Award (2008) and Ralph L. Barnett Excellence in Teaching Award (2007, 2001), the 2002 Alfred Noble Prize, and the Army Research Office Young Investigator Award (1998–2001).

Variational Methods with Applications in Science and Engineering

Kevin W. Cassel

Mechanical, Materials, and
Aerospace Engineering Department
Illinois Institute of Technology

CAMBRIDGE UNIVERSITY PRESS
Cambridge, New York, Melbourne, Madrid, Cape Town,
Singapore, São Paulo, Delhi, Mexico City

Cambridge University Press
32 Avenue of the Americas, New York, NY 10013-2473, USA

www.cambridge.org
Information on this title: www.cambridge.org/9781107022584

First published 2013

Printed in the United States of America

A catalog record for this publication is available from the British Library.

Library of Congress Cataloging in Publication Data
Cassel, Kevin W., 1966–
Variational methods with applications in science and engineering / Kevin W. Cassel.
pages cm.
Includes bibliographical references and index.
ISBN 978-1-107-02258-4 (hardback)
1. Variational principles. 2. Science—Methodology. 3. Engineering—Methodology.
I. Title.
Q172.5.V37C27 2013
515′.64–dc23 2013002762

ISBN 978-1-107-02258-4 Hardback

To my father, Professor D. Wayne Cassel, for imparting a love of mathematics.

To Professor J. David A. Walker, for imparting a love of the application of mathematics to fluid mechanics.

To my wife Adrienne, and children Ryan, Nathan, and Sarah, for their unconditional love and support.

Contents

Preface

In a review of the book *Mathematics for Physics: A Guided Tour for Graduate Students* by Michael Stone and Paul Goldbart (2009), David Khmelnitskii perceptively writes:

> Without textbooks, the education of scientists is unthinkable. Textbook authors rearrange, repackage, and present established facts and discoveries – along the way straightening logic, excluding unnecessary details, and, finally, shrinking the volume of preparatory reading for the next generation. Writing them is therefore one of the most important collective tasks of the academic community, and an often underrated one at that. Textbooks are not easy to create, but once they are, the good ones become cornerstones, often advancing and redefining common knowledge.[1]

There is perhaps no other branch of applied mathematics that is more in need of such a "repackaging" than the calculus of variations.

A glance through the reference list at the end of the book will reveal that there are a number of now-classic texts on the calculus of variations from up through the mid-1960s, such as Weinstock (1952) and Gelfand and Fomin (1963), with a sharp drop subsequent to that period. The classic texts that emphasize applications, such as Morse and Feshback (1953) and Courant and Hilbert (1953), typically focus the majority of their discussion of variational methods on classical mechanics, for example, statics, dynamics, elasticity, and vibrations. Since that time, it has been more common to simply include the necessary elements of variational calculus in books dedicated to specific topics, such as analytical dynamics, dynamical systems, mechanical vibrations, elasticity, finite-element methods, and optimal control theory. More recently, the trend has been to avoid treating these subjects from a variational point of view altogether. Until very recently, therefore, modern stand-alone texts on the calculus of variations have been nearly nonexistent, and a small subset of modern texts in the above-mentioned subject areas include a limited treatment of variational calculus.

Not having had a course on the subject myself, I learned calculus of variations along with my students in a first-year graduate engineering analysis course that I first taught at the Illinois Institute of Technology (IIT) in 2001. The course covers matrices, eigenfunction theory, complex variable theory, and calculus

[1] *Physics Today*, October 2009.

of variations. It required several semesters of teaching the course for me to transition from the – unstated – attitude of "we all need to suffer through the calculus of variations for the sake of those students in solid mechanics, dynamics, and controls" to the – stated – attitude that "calculus of variations is one of the most widely applicable branches of mathematics for scientists and engineers!"

There has been a remarkable resurgence of applications in the last half century in which the calculus of variations has direct relevance. In addition to its traditional application to mechanics of solids and dynamics, it is now being applied in a variety of numerical methods, numerical grid generation, modern physics, various optimization settings, and fluid dynamics. Many of these applications, such as nonlinear optimal control theory applied to continuous systems, have only recently become tractable computationally with the advent of advanced algorithms and large computer systems. In my area of fluid mechanics, which has not traditionally been considered an area ripe for utilization of calculus of variations, there is increasing interest in applying optimal control theory to fluid mechanics and applying variational calculus to numerical methods and grid generation in computational fluid dynamics. With the growing interest in flow control, whereby small actuators are used to excite mechanisms within a flow to produce large changes in the overall flow structure, optimal control theory is making significant headway in providing a formal framework in which to consider various flow control techniques. In addition, numerical grid generation increasingly is being based on formal optimization principles and variational calculus rather than ad hoc techniques, such as elliptic grid generation. Variational calculus also impacts fluid mechanics through its applications to shape optimization and recent advances in hydrodynamic stability theory via transient-growth analysis.

Unfortunately, the majority of modern texts on applied mathematics do not reflect this revival of interest in variational methods. This is demonstrated by the fact that the genre of modern *advanced engineering mathematics* textbooks in wide use today typically do not include calculus of variations. See, for example, Kreyszig (2011), which is now in its 10th edition; O'Neil (2012); Zill and Wright (2014); Jeffrey (2002); and Greenberg (1998). Regrettably, calculus of variations has become too closely linked with certain application areas, such as mechanics of solids, analytical dynamics, and control theory. Such connections are reflected in many of the classic books on these subjects. This gives the impression that it is not a subject that has wide applicability in other areas of science and engineering. The paucity of current stand-alone textbooks on the calculus of variations reinforces the impression that the subject has matured and that there is little modern interest in the field.

Owing to the renaissance of applications in which variational methods have direct relevance and development of the computational resources required to treat them, there is need for a modern treatment of the subject from a general point of view that highlights the breadth of applications demonstrating the widespread applicability of variational methods. The present text is my humble attempt at providing such a treatment of variational methods. The objectives are twofold and include assisting the reader in *learning* and *applying* variational methods:

1. *Learn variational methods:* Part I provides a concise treatment of the fundamentals of the calculus of variations that are accessible to the typical student having a background in differential calculus and ordinary differential equations, but none in variational calculus.
2. *Apply variational methods:* Parts II and III provide a bridge between the fundamental material in Part I that is common to many areas of application and the area-specific applications. They provide an introduction to applications of calculus of variations in various fields and a link to dedicated texts and research literature in these subjects.

The content of the text reflects the strong connection between calculus of variations and the applications for which variational methods form the fundamental foundation. Most readers will be pleased to note that the mathematical fundamentals of calculus of variations (at least those necessary to pursue applications) are rather compact and are contained in a single chapter of the book. Therefore, the majority of the text consists of applications of variational calculus in various fields. The priority of these application chapters is to provide a brief introduction to a variety of physical phenomena and optimization principles from a unified variational point of view. The emphasis is on illustrating the wide range of applications of the calculus of variations, and the reader is referred to dedicated texts for more complete treatments of each topic. The centerpiece of these disparate subjects is *Hamilton's principle*, which provides a compact form of the dynamical equations of motion (its traditional area of application) and the governing equations for many other physical phenomena as illustrated throughout the text.

Given the emphasis on breadth of applications addressed using variational methods, it is necessary to sacrifice depth of treatment of each topic. This is the case not merely for space considerations but, more importantly, for clarity and unification of presentation. The objective in the application chapters is to provide a clear and concise introduction to the variational underpinnings of each field as the basis for further study and investigation. For example, optimization and control are being applied so broadly, including financial optimization, shape optimization, control of dynamical systems, grid generation, image processing, and so on, that it is instructive to view these topics within a common optimal control theory framework. It is also felt that there is value in readers with a background in a given field to be exposed to other related and complementary topics in order to see potential analogies in approach or opportunities for application of one topic in another.

A unique feature of the present text is the derivation of the ubiquitous Hamilton's principle directly from the first law of thermodynamics, which enforces conservation of total energy, and the subsequent derivation of the governing equations of many discrete and continuous phenomena from Hamilton's principle. In this way, the reader will see how the traditional variational treatments of statics and dynamics of discrete and continuous systems are unified with the physics of fluids, electromagnetic fields, relativistic mechanics, quantum mechanics, and so on through Hamilton's principle. I hope to impart to the reader some of the joy of discovering the interrelationship between these seemingly disparate physical principles that submit to the variational framework in general and the first law of thermodynamics through Hamilton's principle specifically.

In writing and arranging the text, I have made the following pedagogical assumptions:

1. Most readers are handicapped in learning a difficult topic unless they are persuaded of its relevance to their academic and/or professional development. That is, engineers and scientists must be convinced of the need for a subject or topic before learning it. It is insufficient to say (or imply), "Trust me; you need to know this." Although this should be a primary concern of instructors on the front lines of pedagogy, there is much that can be done in selecting and arranging the content of a textbook to assist both the instructor and the reader. For example, to motivate the engineering, scientific, and/or mathematical need for a subject or topic, one could include both broad and specific applications early in the development of each topic that students are, or will become, familiar with and/or provide historical motivations for development of a topic.
2. Engineers and scientists learn details better when they have an overall intuition or understanding on which to hang the details. This is why we typically feel like we have learned more about a topic when encountering it a second (or third, ...) time as compared to the first. Therefore, one must seek to provide and develop a physical and/or mathematical intuition that will inform theoretical developments, derivations, or proofs.
3. It is better to err on the side of too much detail in derivations and worked examples than too little. It is easier for the reader to skip a familiar step than to fill in an unfamiliar one. This is also done in order to clearly illustrate the concepts and methods developed in the text and to promote good problem-solving techniques with a minimum of "shortcuts."

This treatment is aimed at the advanced undergraduate or graduate engineering, physical sciences, or applied mathematics student. In addition, it could serve as a reference for researchers and practitioners in fields that are based on, or make use of, variational methods. The text assumes that readers are proficient with differential calculus, including vector calculus, and ordinary differential equations. Familiarity with basic matrix operations and partial differential equations is helpful, but not essential. Moreover, it is not necessary to have any background in the application areas treated in Chapters 4–12; however, such a background would certainly provide a useful perspective.

The book provides sufficient material to form the basis for a one-semester course on calculus of variations or a portion of an applied mathematics, mathematical physics, or engineering analysis course that includes such topics. The latter is the case at IIT, where this material is part of the engineering analysis course taken by many first-year graduate students in mechanical, aerospace, materials, chemical, civil, structural, electrical, and biomedical engineering. In addition, it could serve as a supplement or reference for courses in analytical dynamics, elasticity, mechanical vibrations, modern physics, fluid mechanics, optimal control theory, image processing, and so on that are taught from a variational point of view. Exercises are provided at the end of selected "fundamentals," or nonapplication, chapters, including each chapter in Part I and the first chapter in each of Parts II and III. Solutions are available to instructors at the book's website.

Because of the intended audience of the book, there is little emphasis on mathematical proofs. Instead, the material is presented in a manner that promotes development of an intuition about the concepts and methods with an emphasis on applications. Although primarily couched in terms of optimization theory, Luenberger (1969) does an excellent job of putting calculus of variations in the context of the underlying branches of mathematics known as linear vector spaces and functional analysis and is replete with useful geometric interpretations. Therefore, it provides an excellent complement to the present text for those with a more mathematical orientation than the present text typifies.

I would like to acknowledge certain individuals who have contributed to this book project. Professors Sudhakar Nair and Xiaoping Qian of IIT provided helpful comments and insights throughout development of the book. In particular, Professor Qian directed me to several useful references, and Professor Nair laid the foundation for the course from which this book had its start. Moreover, Professor Nair is largely responsible for keeping the mechanics tradition alive at IIT. I would also like to thank the reviewers who provided a host of useful comments that helped shape the emphasis and content of the book. The numerous students over the years whose questions have challenged me to continually improve the core content of the book are owed a special thank you; you are the ones that I had pictured in my mind as I penned each word and equation. In particular, I would like to acknowledge Mr. Jiacheng Wu for his insightful comments, probing questions, and thorough reading of the text. Finally, Mr. Peter Gordon of Cambridge University Press is to be thanked for his encouragement, advice, and editorial assistance throughout this book project. His "old-school" approach is refreshing in an age of increasingly detached publishers.

I would welcome any comments on the text from instructors and readers. I can be reached at cassel@iit.edu.

PART I

VARIATIONAL METHODS

1 Preliminaries

The vessels, heavy laden, put to sea
With prosp'rous winds; a woman leads the way.
I know not, if by stress of weather driv'n,
Or was their fatal course dispos'd by Heav'n;
At last they landed, where from far your eyes
May view the turrets of new Carthage rise;
There bought a space of ground, which (Byrsa call'd,
From the bull's hide) they first inclos'd, and wall'd.

(Virgil, *The Aeneid*)

The above excerpt cryptically recounts the legend of Dido, the Queen of Carthage, from the ninth century BC. After being exiled from Tyre in Lebanon, Dido purportedly sailed to the shores of North Africa (now Tunisia) and requested that the local inhabitants give her and her party the land that could be enclosed by the hide of an ox. Not thinking that an oxhide could encompass a large portion of land, they granted her wish. She then proceeded to have the hide cut into narrow strips and extended end-to-end to form a semicircle bounding the shoreline and encompassing a nearby hill. This area became known as Carthage.

Certain branches of mathematics have arisen out of consideration of the theoretical consequences of known mathematical theorems. More often than not, however, new branches of mathematics have been developed to provide the means to address certain types of practical problems that initially are often very specific. For example, how did Dido know to arrange the oxhide in a circular shape in order to enclose the largest possible area with a given perimeter length?

One of the best examples of a branch of mathematics motivated by applications, and the one required to solve the problem of Dido, is the *calculus of variations*, which today encompasses numerous, far-reaching applications. Its application-driven roots live on in several fundamental variational principles that form the foundation of a number of important fields in the physical sciences and engineering. A glance through the table of contents of this text will hint at the variety of topics that can be treated using variational methods. This will become more evident in Section 1.3 as we motivate the need for variational calculus through consideration of three practical problems and by elucidation of a variety of physical phenomena and optimization principles through a unified variational framework throughout the remainder of the text.

1.1 A Bit of History

To fully appreciate our treatment of the calculus of variations, a bit of history is in order.[1] Doing so will assist us in placing the principles, methods, and applications to be encountered in a useful framework. In addition, our discussions will be replete with some of the greatest names in mathematics and physics, so it is hoped that a brief history will enrich our study of these topics and the people behind them.

Each era of human existence is marked to some extent by the scientific and technological problems of the day, and it is always the case that some of the most fertile minds are drawn to addressing the greatest challenges and grandest questions. After seminal contributions by the Greek philosophers in the sixth through fourth centuries BC and the Roman emphasis on infrastructure and engineering for the following 800 years, high-level intellectual pursuits were largely abandoned for a number of centuries while Western society was dominated by instability and decentralized power during the Middle Ages. Whereas this period was marked by the need to accurately launch projectiles into neighboring feudal castles, the subsequent periods of the Renaissance and Age of Enlightenment began a scientific revolution that we are still experiencing today. This period led to mathematical contributions, scientific progress, artistic expression, architectural prowess, and musical genius that have stood the test of time well into the modern – and now postmodern – era.

Coming out of the Middle Ages at the dawn of the sixteenth century, the Church provided needed stability, but it also wielded excessive power over the largely uneducated and illiterate populace throughout Europe. While the spiritual authority of the Church was being challenged by the reformers, such as Martin Luther, its scientific authority was being challenged by the increasingly rigorous methods of Nicolaus Copernicus and Galileo Galilei based on observation and experimentation. The Renaissance also saw a return to classical Greek philosophy and the Greek emphasis on realism in art and expression. Moreover, it was a time in which integration and harmonization of broad endeavors were encouraged, giving rise to the "Renaissance man." While Leonardo da Vinci was the quintessential Renaissance man of Italy, the center of gravity of the Renaissance, Christopher Wren exemplified the British variant. Best known for being the architect of St. Paul's Cathedral in London (and approximately fifty other churches after the Great Fire of London in 1666), he was also a noted astronomer, physicist, and mathematician, as well as the founder of the Royal Society of London.

Much of the most celebrated art, architecture, and music of the time was commissioned by or directed toward the Church, reflecting an overarching emphasis on unifying and integrating various human intellectual, spiritual, and artistic pursuits. There was a developing sense that an intrinsic order, beauty, and simplicity governs both the universe and the highest human endeavors. The Renaissance produced artists, musicians, scientists, and mathematicians emboldened by the potential contributions that they as individuals and as humans collectively could make in all human undertaking.

[1] Our brief history is admittedly European-centric, as this region provided the immediate historical context for the development of variational methods.

Released from the patriarchal constraints on scientific thought rendered by the Church and emboldened with a strong sense of realism and drive toward harmonization of all aspects of life by the Renaissance artists and thinkers, the Age of Enlightenment was born. While the Renaissance up through the seventeenth century was marked primarily by artistic and musical innovation, a revolution in science and mathematics began that laid the foundation for the Age of Enlightenment, which spanned the eighteenth century and led to unprecedented mathematical and scientific progress. Both the Renaissance and Age of Enlightenment shared the ideals of realism in which objects, thought, and indeed the entire universe were viewed, rendered, and modeled as they really were.

The Age of Enlightenment brought a renewed emphasis on how to explain the existence and operation of the universe, and this required a new set of mathematical tools. An increased priority was placed on providing explanations and mathematical models to elucidate observed phenomena. Together these led to an explosion of development and discovery in science and mathematics, the two feeding off of each other's questions and contributions. Much of today's mathematical arsenal had its origins during the Age of Enlightenment, including complex variables, differential equations, and, most notably, differential calculus.

The stage was set for tremendous advances during the Age of Enlightenment by the work of Isaac Newton and the formation of scientific societies in the latter part of the seventeenth century. These societies provided a means for interaction among mathematicians and scientists and the broader communication of findings through printed journals, and Newton's work inspired a renewed search for universal physical principles and the mathematics to describe them. In particular, Newton's *Principia* (*Mathematical Principles of Natural Philosophy*) (1687) had several important influences beyond the actual articulation of his laws of motion and gravity:

1. The accurate prediction of the movement of the celestial bodies must have seemed almost magical to a world accustomed to its scientists primarily providing observations, with only occasional explanation.
2. The realization that the same physical principles that govern the falling of an apple also govern the movement of those celestial bodies throughout the universe accelerated the human pursuit of a unified theory that explains all observations in our universe.
3. A recognition of the power of mathematics to provide the unifying language for such models and laws.
4. Newton's methods formed the basis of the scientific method.

These four influences of Newton's *Principia* emboldened scientists and mathematicians for centuries to pursue a complete understanding of the universe. As concluded by Greer (2005),[2] "[Newton] demonstrated that, on the basis of experiments conducted in a tiny corner of the universe, and aided by the lever of mathematics, he had discovered the nature of gravitation *everywhere*."

In large part, Newton inaugurated the Age of Enlightenment with his physical laws articulated in the *Principia* and his – and Leibniz's – calculus. Calculus provided the mathematical tools, and his physical laws provided the framework and impetus

[2] *A Brief History of the Western World* by Thomas H. Greer.

for further developments in the ensuing years. The attitudes and successes of the Age of Enlightenment fostered a search for the ultimate truths that govern the physical world in which we live and the reasons for its existence. Permeating many aspects of life and thought was a renewed sense of order, idealism, and perfection. Scientists concluded that because the universe is ordered and behaves in a rational manner, it is amenable to explanation via universal laws expressed mathematically. It is into this world seeking order and beauty in its many forms that the contributions of Pierre Louis Moreau de Maupertuis, Leonhard Euler, and Joseph Louis Lagrange, among others, were made.

Coupled with Newton's influence on our understanding of the mechanics of moving bodies, and the ancillary philosophical contributions that his laws supplied, was Pierre de Fermat's principle asserting that light optimizes its path according to a minimum principle. These realizations inaugurated a search for other minimum principles that may govern the universe and gave rise to what we now call *optimization*. Today, we would classify Dido's problem as an optimization problem, in which we seek the "best" or "optimal" means of accomplishing a task (maximizing the land grab to form Carthage). Problems such as this occupied the efforts of some of the greatest minds of the late seventeenth and early eighteenth centuries. The differential calculus of Isaac Newton and Gottfried Wilhelm Leibniz succeeded in solving many such problems; however, some of them demanded a new calculus, later to be labeled *calculus of variations*. Some of the early optimization problems that motivated the development of the calculus of variations include:

- The shape of an object moving through a fluid to minimize resistance (Isaac Newton in 1687). See Section 10.2.
- The brachistochrone problem to determine the shape of a wire to minimize travel time of a sliding mass (Johann Bernoulli, Gottfried Wilhelm Leibniz, Isaac Newton, Jacob Bernoulli, and Guillaume de l'Hôpital in 1696). See Section 2.2.
- Minimal surface shapes (Leonhard Euler in 1744). See Section 2.5.2.

Comprehensive treatment of optimization problems had to await the arrival of the digital computer in the latter half of the twentieth century. In the intervening two hundred years, however, the calculus of variations revolutionized mechanics and physics through the development and application of de Maupertuis's least action principle, later refined to become Hamilton's principle, which will occupy our considerations in Part II of this text. It is remarkable that this mathematical framework, marked by an enduring elegance, is capable of encapsulating many physical and optimization principles of which its inventors could have never conceived. This is the "miracle of the appropriateness of the language of mathematics to the formulation of the laws of physics" that Physics Nobel Laureate Eugene Wigner spoke of in 1960. The "miracle" of variational methods is stronger than ever today, in particular as formal optimization techniques permeate more and more areas of science, engineering, and economics. As such, the calculus of variations has come full circle; originally conceived to formally address a limited set of early optimization problems, variational methods now provide the mathematical foundation and framework to treat modern, large-scale optimization problems. Every day we optimize our decisions, activities, processes, and designs; therefore, variational methods are proving to be an indispensable tool in providing a formal basis for optimization along with providing the foundation for an impressive array of physical principles that encompass classical mechanics and modern physics.

1.2 Introduction

With respect to its strong ties to applications, this treatment of the calculus of variations will emphasize three themes. First, formulating many problems in the physical sciences and engineering from first principles[3] leads naturally to variational forms. Second, these variational forms often provide additional physical insight that is not readily apparent from the equivalent, typically more familiar, differential formulation. Third, the variational approach to optimization and control supplies a general and formal framework within which to apply such principles to a broad spectrum of diverse fields. Variational methods furnish the mathematical tools to both encapsulate a wide variety of physical systems and processes under a unified principle, known as Hamilton's principle, and to provide a framework for their optimization and control. This is because the fundamental principles underlying so much of classical and modern physics, as well as optimization and control, are based on extremum principles to which the calculus of variations supplies an unrivaled mathematical framework that is tailor-made for their treatment.

This text can be divided into three distinct parts: Part I addresses the variational methods themselves and Parts II and III treat the applications of such methods. In the remainder of Chapter 1, three examples are formulated to instill the need for the calculus of variations to the reader who is new to the subject. The first two examples are based on physical principles, and the third is an optimization problem. In addition, a review of material from differential calculus related to finding extrema of functions with and without constraints is provided. This material serves as the basis for developing similar techniques for functionals in the calculus of variations. There is also a review of integration by parts and a proof of the fundamental lemma of the calculus of variations along with a review of self-adjoint differential operators. Chapter 2 encompasses all of the mathematical techniques necessary to solve these examples and treat numerous other problems that arise in the calculus of variations. The applications-minded reader will be pleased to note that the relatively compact Chapter 2 contains all of the necessary material required for a scientist or engineer to become proficient in applying these variational methods to practical and research areas that utilize variational calculus. Part I closes with a brief account of approximate and numerical methods as applied to variational methods in Chapter 3. In Part II, encompassing Chapters 4 through 9, we take a rather long journey through the application of variational methods to physical principles. Chapter 4 provides the physical centerpiece, Hamilton's principle, that forms the foundation for the physical applications treated throughout Part II. The remaining chapters in this part address classical mechanics of nondeformable and deformable bodies, stability of dynamical systems, optics and electromagnetics, modern physics, and fluid mechanics. At the risk of overwhelming the reader, the breadth of these

[3] When we use the phrase "first principles" in physical contexts, as is the case here, we mean starting directly from established physical laws, that is, without assumptions, approximations, or empirical modeling.

topics is intended to illustrate the range and variety of fields that yield to variational methods. Part III of the text is devoted to applying the calculus of variations to a variety of applications in optimization. These applications include optimization and control, image processing, data analysis, and grid generation for numerical methods. Whereas Chapter 2 contains a cohesive treatment of the fundamental mathematical methods required, which is of use to all readers, the subsequent applications-oriented material is designed such that each chapter largely stands on its own and does not depend on material other than that in Chapter 2 (and Chapter 4 for Chapters 5 through 9, and Chapter 10 for Chapters 11 and 12). Interestingly, the ordering of the topics of this text, which is intended to reflect a logical progression through the material, also roughly follows their chronological development historically.

1.3 Motivation

Before developing the mathematical techniques required to solve variational problems, it is instructive to formulate a few such problems. In doing so, we will illustrate that formulating these problems from first principles leads naturally to variational forms, thereby motivating the need for an ability to solve variational problems. We consider two physical scenarios and one optimization problem that will illustrate what a functional is and that will highlight some of the issues requiring attention in the remainder of the text.

1.3.1 Optics

Optics is an excellent example in which to illustrate the difference between differential and variational calculus. We utilize *differential calculus* when finding *extrema*, that is, minimums and maximums, *of functions*. In this case we differentiate the function, set it equal to zero, and seek the values of the independent variable(s) where the function is an extremum. As you will see, *variational calculus* is utilized when finding *extrema* of *functionals*, where a functional is a definite integral involving an unknown function and its derivatives. In the case of calculus of variations, we seek the function for which the functional is an extremum.

It will be shown that when we apply Fermat's principle of optics, which states that the path of light between two points is the one that requires the minimum travel time,[4] results in a problem in *differential calculus* if the media through which the light is traveling are homogeneous and a *variational calculus* problem if the medium is nonhomogeneous.

To apply Fermat's principle, we seek to minimize the travel time $T[u(x)]$ of light along all the possible paths $u(x)$ that the light could take through a medium between the points x_0 and x_1 at times t_0 and t_1, respectively. This requires us to minimize

$$T[u(x)] = \int_{t_0}^{t_1} dt = \int_{x_0}^{x_1} \frac{ds}{v(x,u)} = \int_{x_0}^{x_1} \frac{n(x,u)}{c} ds, \tag{1.1}$$

[4] In fact, the travel time is not always a minimum, which is an important subtlety that we will return to in due course.

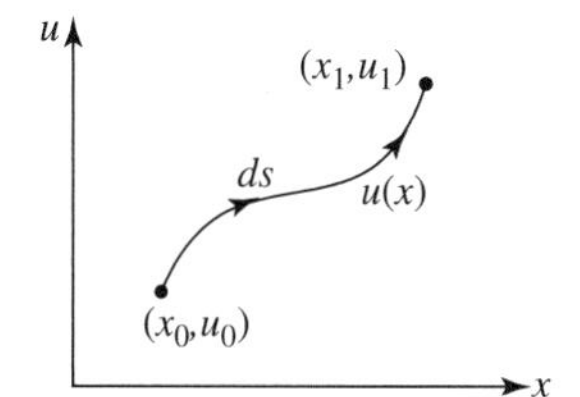

Figure 1.1. Path of light through a medium with variable index of refraction.

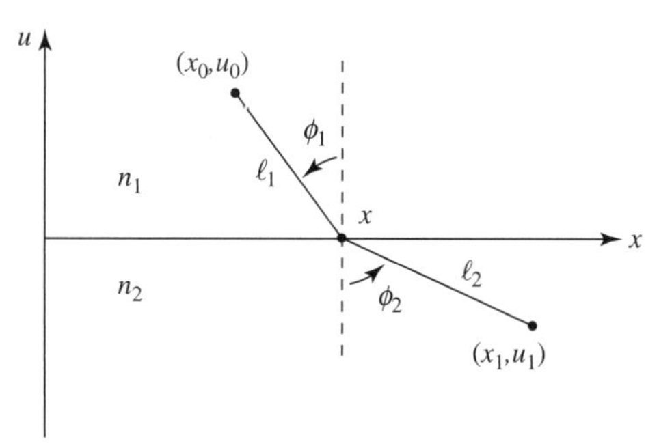

Figure 1.2. Path of light through two homogeneous media.

where $v(x,u)$ is the speed of light in the medium, c is the speed of light in a vacuum, $n(x,u)$ is the index of refraction in the medium, and ds is a differential element along the path $u(x)$ as illustrated in Figure 1.1. Equation (1.1) sums (through integration) the time it takes light to travel along each infinitesimally short element ds of the path $u(x)$ to obtain the total travel time $T[u(x)]$.

First, let us consider light traveling through two homogeneous media that each have constant index of refraction with their interface coinciding with the x-axis as shown in Figure 1.2. Because the indices of refraction n_1 and n_2 are constant in their respective media, the light travels along a straight path through each medium. That is, the path having the shortest travel time is the same as that having the shortest length in a homogeneous medium. For fixed end points (x_0,u_0) and (x_1,u_1), therefore, we seek to find the value of x where the actual path of the light intersects with the interface between the two media. With constant indices of refraction, the travel time (1.1) is

$$\begin{aligned}
T(x) &= \frac{1}{c}\left[n_1\int_{x_0}^{x} ds + n_2\int_{x}^{x_1} ds\right] \\
&= \frac{1}{c}\left[n_1\ell_1 + n_2\ell_2\right] \\
T(x) &= \frac{1}{c}\left[n_1\sqrt{(x-x_0)^2+u_0^2} + n_2\sqrt{(x_1-x)^2+u_1^2}\right].
\end{aligned}$$

Accordingly, the travel time $T(x)$ is an algebraic function of x, and we use *differential calculus* to minimize it. We seek the value of x for which $T(x)$ is a minimum corresponding to the value of x where $dT/dx = 0$. Evaluating $dT/dx = 0$ leads to

$$\frac{1}{2}n_1\left[(x-x_0)^2+u_0^2\right]^{-1/2}[2(x-x_0)] + \frac{1}{2}n_2\left[(x_1-x)^2+u_1^2\right]^{-1/2}[-2(x_1-x)] = 0.$$

Note that the square root factors are the lengths ℓ_1 and ℓ_2, and given that

$$x - x_0 = \ell_1 \sin\phi_1, \quad x_1 - x = \ell_2 \sin\phi_2$$

from the geometry, we have

$$\frac{n_1}{\ell_1}(\ell_1 \sin\phi_1) = \frac{n_2}{\ell_2}(\ell_2 \sin\phi_2).$$

This leads to *Snell's law* for two homogenous media

$$n_1 \sin\phi_1 = n_2 \sin\phi_2,$$

which indicates that the ratio of the sines of the angles of incidence and refraction is the same as the ratio of the indices of refraction in the two homogeneous media.

Now consider the general case with a variable index of refraction $n(x,u)$ throughout the medium, in which case Fermat's principle (1.1) becomes

$$T[u(x)] = \frac{1}{c}\int_{x_0}^{x_1} n(x,u)ds = \frac{1}{c}\int_{x_0}^{x_1} n(x,u)\sqrt{1+\left(\frac{du}{dx}\right)^2}\,dx.$$

The last expression follows from the fact that $ds = \sqrt{dx^2 + du^2}$ from the geometry, and because $u = u(x)$, the total differential of $u(x)$ is $du = \frac{du}{dx}dx$. This leads to the result that $ds = \sqrt{1+[u'(x)]^2}dx$. Consequently, in the case of nonhomogeneous media, we seek the path represented by the function $u(x)$ that results in the minimum travel time $T[u(x)]$. Note that evaluation of the definite integral for each possible path $u(x)$ produces a unique scalar value for the travel time $T[u(x)]$. We seek the path that produces the least travel time according to Fermat's principle. In other words, the *functional* $T[u(x)]$, which is a definite integral involving an unknown function $u(x)$, is to be minimized. This requires *calculus of variations*.

Observe that the variational form of the governing equation arises naturally from formulating the problem from first principles. That is, applying the physical law (Fermat's principle) directly leads to the variational form. In summary, we see that finding extrema of an *algebraic function* is a problem in differential calculus, while finding extrema of a *functional* (definite integral) is a problem in variational calculus.

1.3.2 Shape of a Liquid Drop

Second, let us consider the shape that a liquid drop takes when placed on a smooth horizontal surface as shown in Figure 1.3. The shape of the liquid drop will be axisymmetric about the vertical u-axis and be such that the total energy of the drop is a minimum.

The two forms of energy of the liquid drop are that due to its potential energy under a gravitational field and the surface energy due to surface tension at the liquid–gas interface. First, let us consider the potential energy of the drop. The potential energy per unit volume of a horizontal portion of the drop with differential thickness du is $\rho g u(r)$, and the volume of this infinitesimally thin portion is $\pi r^2 du$. Here, ρ is the density of the liquid, g is acceleration due to gravity, and $u(r)$ is the vertical

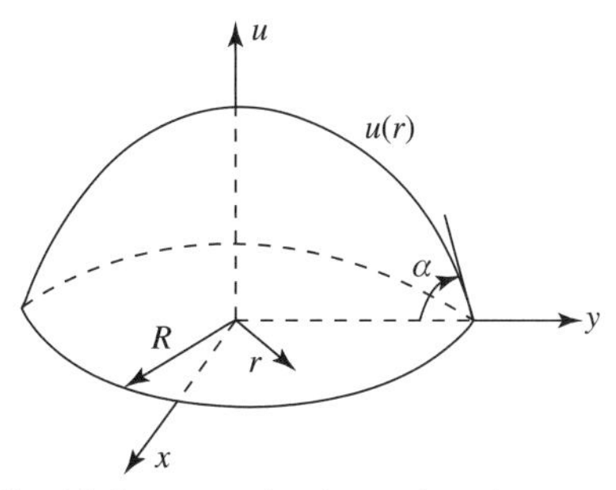

Figure 1.3. Schematic of a liquid drop on a horizontal surface.

distance to the interface from the surface. Therefore, the total potential energy of the drop is

$$E_p = \int_{r=R}^{r=0} \rho g u(r)\left(\pi r^2 du\right) = \pi \rho g \int_{r=R}^{r=0} r^2 u(r) \frac{du}{dr} dr,$$

where R is the base radius of the drop.

The surface energy within the liquid–gas interface of the horizontal portion of the drop with vertical thickness du is the product of the surface energy per unit area σ and the surface area $2\pi r ds$, where ds is along the interface. As a result, the total surface energy is

$$E_s = \int_{r=R}^{r=0} \sigma (2\pi r ds) = 2\pi\sigma \int_{r=R}^{r=0} r ds = 2\pi\sigma \int_{r=R}^{r=0} r\sqrt{1+[u'(r)]^2} dr,$$

where we have used the fact that the differential arc length is $ds = \sqrt{dr^2 + du^2} = \sqrt{1+[u'(r)]^2} dr$ because $du = \frac{du}{dr} dr = u'(r) dr$, similar to the optics example.

Combining the above contributions, the total energy of the liquid drop is

$$E[u(r)] = E_p + E_s = \pi \int_{r=R}^{r=0} \left\{\rho g r^2 u(r) u'(r) + 2\sigma r\sqrt{1+[u'(r)]^2}\right\} dr,$$

and we seek the axisymmetric drop shape $u(r)$ that results in minimization of the total energy $E[u(r)]$ of the drop subject to a constraint. The constraint is that the total volume $\bar{V}$ of the liquid drop is fixed and given by

$$\bar{V} = \pi \int_{r=R}^{r=0} r^2 \frac{du}{dr} dr.$$

In addition to the constraint on total volume, we have boundary conditions on $u(r)$. The boundary condition at $r = R$ involves the contact angle where the gas, liquid, and solid phases intersect. The contact angle α is defined in Figure 1.4 and can be determined using Young's equation (see Butt, Graf, and Kappl 2004). From the definition of the contact angle, we have that $du = \sin\alpha \, ds$ and $dr = -\cos\alpha \, ds$. Hence,

$$\left.\frac{du}{dr}\right|_{r=R} = u'(R) = -\tan\alpha.$$

We also have that $u(R) = 0$. Along the vertical centerline of the bubble, we have

$$\left.\frac{du}{dr}\right|_{r=0} = 0,$$

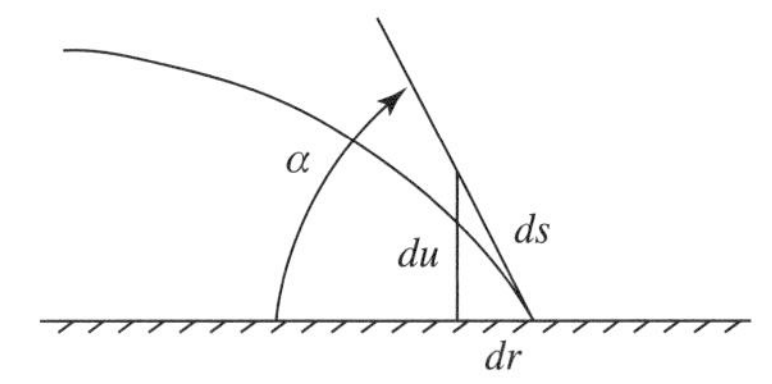

Figure 1.4. Contact angle at the solid-liquid-gas interface.

or $u'(0) = 0$ owing to the axisymmetry of the smooth bubble. Note that we only require two boundary conditions; however, we have three. The extra boundary condition is used to determine the unknown base radius R of the bubble.

Again, we have a definite integral, in this case for the total energy $E[u(r)]$, that is to be minimized to obtain the function $u(r)$ that specifies the shape of the liquid–gas interface of the drop. Unlike in the case of optics, however, we seek to minimize this *functional* subject to the integral constraint on the total volume of the bubble.

Physically, the functional $E[u(r)]$ represents the fact that there is a competition between two physical effects. Minimization of the total potential energy of the drop alone would lead to a large base radius R and small maximum height $u(0)$ in order to minimize u, and thus the potential energy of the liquid. On the other hand, minimization of the surface energy alone would lead to a spherical drop, which minimizes the surface area and, therefore, the total surface energy for a drop of a given volume. The actual shape of the drop is somewhere between these two extremes depending upon the physical parameters g, ρ, σ, and α.

Because we seek to minimize the definite integral $E[u(r)]$ involving an unknown function $u(r)$, this is a problem in calculus of variations. Again, observe that the variational form of the problem arises naturally from a consideration of the physical problem from first principles.

1.3.3 Optimization of a River-Crossing Trajectory

As a final example to motivate the need for variational calculus, the problem of determining the trajectory that should be taken by a boat crossing a river such that the crossing time is a minimum is considered. This is often referred to as Zermelo's problem, and it is an example of optimization, which will be considered in more detail in Chapter 10. Whereas the previous two examples are physical scenarios in which the functional to be minimized results from consideration of the relevant physical principles, in the case of optimization, the functional is devised in order to meet some objective; accordingly, it is often called an *objective functional*. Alternatively, it may be formulated in such a way as to minimize a cost, in which case it is referred to as a *cost functional*.

The speed of the river current through which the boat is navigating is given by $v(x)$ as illustrated in Figure 1.5. The boat maintains a constant speed V relative to the moving water with heading angle $\theta(x)$ relative to the fixed coordinate system. We seek the path $u(x)$ along which the boat should travel in order to cross the river from the point (x_0, u_0) on the left bank to the point (x_1, u_1) on the right bank in minimum time.

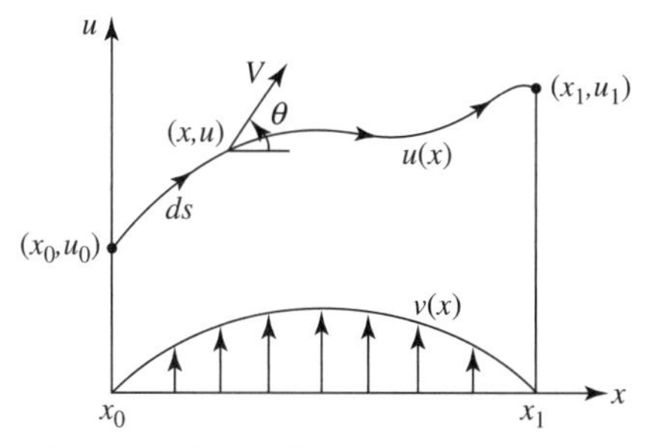

Figure 1.5. Schematic of a river-crossing trajectory.

Similar to Fermat's principle, in order to minimize the time of travel, we want to minimize the functional

$$T[u(x),\theta(x)] = \int_{t_0}^{t_1} dt = \int_{(x_0,u_0)}^{(x_1,u_1)} \frac{ds}{v_B(x,u)}, \tag{1.2}$$

where $v_B(x,u)$ is the speed of the boat relative to the fixed coordinate system. That is, $v_B(x,u)$ is the magnitude of the resultant velocity vector of the boat owing to the combination of the river's current $v(x)$ and the boat's relative speed V. The velocity components of the boat are

$$\dot{x} = V\cos[\theta(x)], \quad \dot{u} = v(x) + V\sin[\theta(x)]. \tag{1.3}$$

Therefore, with the velocity vector $\mathbf{v}_B = \dot{x}\mathbf{i} + \dot{u}\mathbf{j}$, the speed is

$$\begin{aligned} v_B(x,u) &= \sqrt{\dot{x}^2 + \dot{u}^2} \\ &= \left(V^2\cos^2\theta + v^2 + 2vV\sin\theta + V^2\sin^2\theta\right)^{1/2} \\ v_B(x,u) &= \left(V^2 + v^2 + 2vV\sin\theta\right)^{1/2}. \end{aligned}$$

As in the previous examples, the differential arc length is

$$ds = \sqrt{dx^2 + du^2} = \sqrt{1 + [u'(x)]^2}dx.$$

Substituting into the functional (1.2) gives

$$T[u(x),\theta(x)] = \int_{x_0}^{x_1} \frac{\left[1 + (u')^2\right]^{1/2}}{\left(V^2 + v^2 + 2vV\sin\theta\right)^{1/2}} dx. \tag{1.4}$$

We seek the path $u(x)$ and heading angle $\theta(x)$ that minimizes this objective functional for the time of transit $T[u(x),\theta(x)]$. Observe that once again we have a definite integral to be minimized involving, in this case, two unknown functions.

Given the river current speed $v(x)$ and the relative boat speed V, a relationship can be developed between the path $u(x)$ and heading angle $\theta(x)$ that reduces the number of unknown functions in equation (1.4) to one. The slope of the trajectory is given by the ratio of velocity components according to

$$\frac{du}{dx} = \frac{\dot{u}}{\dot{x}},$$

Table 1.1. *Sample applications of variational calculus with associated variational principles.*

Application	Variational Principle
Optics	Fermat's principle
Static equilibrium	Minimum potential energy
Dynamical systems	Hamilton's principle (variational form of Newton's Second Law)
Optimization and control	Minimize objective or cost functional

or from equation (1.3)

$$\frac{du}{dx} = \frac{v(x) + V\sin[\theta(x)]}{V\cos[\theta(x)]}. \tag{1.5}$$

This could be substituted for u' in the functional (1.4) to produce a functional for the heading angle $\theta(x)$. Alternatively, this relationship can be imposed as a constraint on the minimization of the functional (1.4). That is, we seek to minimize (1.4) subject to the constraint (1.5) being satisfied. Once the heading angle $\theta(x)$ is obtained, equation (1.5) could then be integrated across the river to obtain the corresponding optimal path of the boat $u(x)$.

1.3.4 Summary

Observe that in each of the examples, the *calculus of variations* involves determining extrema of a definite integral containing an unknown function. In other words, we seek the function for which the functional is an extremum. In the case of optics, the unknown function is the path of the light, and the functional to be minimized is the time of travel. For the liquid drop shape, the unknown function is the shape, and the functional is the total energy of the drop. In the river crossing trajectory, the unknown function is the heading angle, and the functional is the total travel time.

Some of the applications of calculus of variations are summarized in Table 1.1. In addition to the listed applications, applications to modern physics, fluid mechanics, and numerical methods will be considered in Chapters 4 through 12.

1.4 Extrema of Functions

The motivational examples provided in the previous section suggest that the calculus of variations is concerned with finding extrema, or, more generally, stationary functions, of *functionals*. In anticipation of a formal treatment of such extrema of functionals, let us review extrema of algebraic *functions* in the context of differential calculus.

First, let us consider a one-dimensional function $f(x)$ as shown in Figure 1.6. The point x_0 is a *stationary* (or *critical*) *point* of $f(x)$ if

$$\frac{df}{dx} = 0 \quad \text{at} \quad x = x_0,$$

or equivalently

$$df = \frac{df}{dx}dx = 0 \quad \text{at} \quad x = x_0,$$

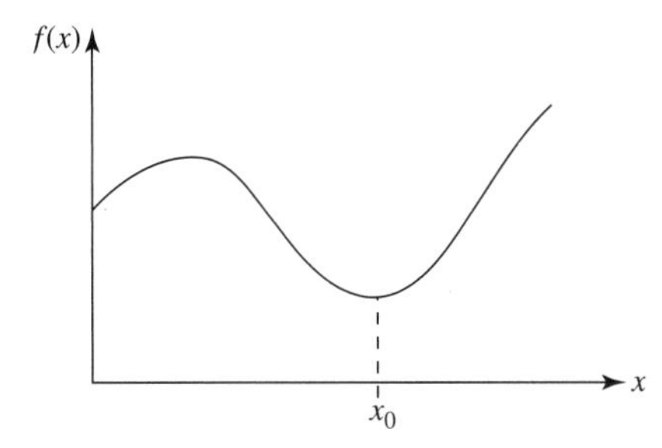

Figure 1.6. One-dimensional function $f(x)$ with a local minimum at $x = x_0$.

where df is the *total differential* of $f(x)$. That is, the slope of $f(x)$ is zero at $x = x_0$. The following possibilities exist at a stationary point:

1. If $d^2f/dx^2 < 0$ at x_0, the function $f(x)$ has a *local maximum* at $x = x_0$.
2. If $d^2f/dx^2 > 0$ at x_0, the function $f(x)$ has a *local minimum* at $x = x_0$.
3. If $d^2f/dx^2 = 0$, then $f(x)$ may still have a local minimum (e.g., $f(x) = x^4$ at $x = 0$) or a local maximum (e.g., $f(x) = -x^4$ at $x = 0$), or it may have neither (e.g., $f(x) = x^3$ at $x = 0$).

The important point here is that the requirement that $f'(x_0) = 0$, in which case x_0 is a stationary point, provides a *necessary* condition for a local extremum, while possibilities (1) and (2) provide additional *sufficient* conditions for a local extremum at x_0. We have seen an example of this case in Section 1.3.1.

Now consider the two-dimensional function $f(x,y)$. For an extremum to occur at (x_0, y_0), it is necessary (but not sufficient) that

$$\frac{\partial f}{\partial x} = \frac{\partial f}{\partial y} = 0 \quad \text{at} \quad x = x_0, y = y_0,$$

or equivalently

$$df = \frac{\partial f}{\partial x}dx + \frac{\partial f}{\partial y}dy = 0 \quad \text{at} \quad x = x_0, y = y_0,$$

where df is the *total differential* of $f(x,y)$. The point (x_0, y_0) is a *stationary point* of $f(x,y)$ if $df = 0$ at (x_0, y_0), and the rate of change of $f(x,y)$ at (x_0, y_0) in all directions is zero. At a stationary point (x_0, y_0), the following possibilities exist, where subscripts denote partial differentiation with respect to the indicated variable:

1. If $f_{xx}f_{yy} - f_{xy}^2 > 0$ and $f_{xx} < 0$ at (x_0, y_0), then it is a *local maximum*.
2. If $f_{xx}f_{yy} - f_{xy}^2 > 0$ and $f_{xx} > 0$ at (x_0, y_0), then it is a *local minimum*.
3. If $f_{xx}f_{yy} - f_{xy}^2 < 0$ at (x_0, y_0), then it is a *saddle point*.
4. If $f_{xx}f_{yy} - f_{xy}^2 = 0$ at (x_0, y_0), then (1), (2), or (3) is possible.

Observe that the evaluated expression can be expressed as the determinant

$$\begin{vmatrix} f_{xx} & f_{xy} \\ f_{yx} & f_{yy} \end{vmatrix} = f_{xx}f_{yy} - f_{xy}^2.$$

Possibilities (1) and (2) above provide sufficient conditions, along with the necessary condition that $df = 0$ at (x_0, y_0), for the existence of an extremum. See Hildebrand (1976) for a derivation of these conditions.

The above criteria can be generalized to n-dimensions. That is, we have a *stationary point* of $f(x_1,x_2,\ldots,x_n)$ if

$$df = \frac{\partial f}{\partial x_1}dx_1 + \frac{\partial f}{\partial x_2}dx_2 + \cdots + \frac{\partial f}{\partial x_n}dx_n = 0.$$

EXAMPLE 1.1 Obtain the location (x_1,x_2,x_3) at which a minimum value of the function

$$f(x_1,x_2,x_3) = \frac{f_1(x_1,x_2,x_3)}{f_2(x_1,x_2,x_3)} = \frac{x_1^2 + 6x_1x_2 + 4x_2^2 + x_3^2}{x_1^2 + x_2^2 + x_3^2}$$

occurs.

Solution: At a point (x_1,x_2,x_3) where f has zero slope, we have

$$\frac{\partial f}{\partial x_i} = 0, \quad i = 1,2,3.$$

With $f = f_1/f_2$, this requires that

$$\frac{1}{f_2}\left(\frac{\partial f_1}{\partial x_i} - \frac{f_1}{f_2}\frac{\partial f_2}{\partial x_i}\right) = 0, \quad i = 1,2,3,$$

or, letting $\lambda = f = f_1/f_2$,

$$\frac{\partial f_1}{\partial x_i} - \lambda\frac{\partial f_2}{\partial x_i} = 0, \quad i = 1,2,3.$$

Substituting

$$f_1 = x_1^2 + 6x_1x_2 + 4x_2^2 + x_3^2, \quad f_2 = x_1^2 + x_2^2 + x_3^2,$$

and evaluating the partial derivatives produces three equations for the three unknown coordinate values. In matrix form, these equations are

$$\begin{bmatrix} 1-\lambda & 3 & 0 \\ 3 & 4-\lambda & 0 \\ 0 & 0 & 1-\lambda \end{bmatrix}\begin{bmatrix} x_1 \\ x_2 \\ x_3 \end{bmatrix} = \begin{bmatrix} 0 \\ 0 \\ 0 \end{bmatrix}.$$

For a nontrivial solution, the required eigenvalues are

$$\lambda_1 = 1, \quad \lambda_2 = \frac{1}{2}(5 + 3\sqrt{5}), \quad \lambda_3 = \frac{1}{2}(5 - 3\sqrt{5}),$$

and the corresponding eigenvectors are

$$\mathbf{u}_1 = \begin{bmatrix} 0 \\ 0 \\ 1 \end{bmatrix}, \quad \mathbf{u}_2 = \begin{bmatrix} -\frac{4}{3} + \frac{1}{6}(5 + 3\sqrt{5}) \\ 1 \\ 0 \end{bmatrix}, \quad \mathbf{u}_3 = \begin{bmatrix} -\frac{4}{3} + \frac{1}{6}(5 - 3\sqrt{5}) \\ 1 \\ 0 \end{bmatrix}.$$

Of the three eigenvalues, $f = \lambda_3 = \frac{1}{2}(5 - 3\sqrt{5})$ has the minimum value, which occurs at

$$\begin{bmatrix} x_1 \\ x_2 \\ x_3 \end{bmatrix} = \mathbf{u}_3 = \begin{bmatrix} -\frac{4}{3} + \frac{1}{6}(5 - 3\sqrt{5}) \\ 1 \\ 0 \end{bmatrix}.$$

1.5 Constrained Extrema and Lagrange Multipliers

It is sometimes necessary to find an extremum of a functional subject to constraints as in the liquid-drop shape and river-crossing trajectory problems. Again, in anticipation of applying a similar approach to variational problems, let us consider the procedure for determining extrema of algebraic functions subject to a constraint.

Recall from the previous section that a necessary condition for an extremum of a function $f(x,y,z)$ at (x_0,y_0,z_0) without constraints is

$$df = f_x dx + f_y dy + f_z dz = 0. \tag{1.6}$$

Because dx, dy, and dz are arbitrary (x, y, and z are independent variables), this requires that

$$f_x = 0, \quad f_y = 0, \quad f_z = 0,$$

which is solved to find x_0, y_0, and z_0.

Now let us determine the stationary point(s) of a function $f(x,y,z)$ subject to a constraint, say

$$g(x,y,z) = c,$$

where c is a specified constant. The constraint provides a geometric relationship among the coordinates x, y, and z. In addition to equation (1.6), we have that the total differential of $g(x,y)$ is

$$dg = g_x dx + g_y dy + g_z dz = 0, \tag{1.7}$$

which is zero because g is equal to a constant. Because both (1.6) and (1.7) equal zero, it follows that we can add them according to

$$df + \Lambda dg = (f_x + \Lambda g_x)dx + (f_y + \Lambda g_y)dy + (f_z + \Lambda g_z)dz = 0, \tag{1.8}$$

where Λ is an arbitrary constant, which we call the *Lagrange multiplier*.[5]

Note that because of the constraint $g = c$, the variables x, y, and z are no longer independent. With one constraint, for example, we can only have two of the three variables being independent. As a result, we cannot regard dx, dy, and dz as all being arbitrary and set the three coefficients equal to zero as in (1.6). Instead, suppose that $g_z \neq 0$ at (x_0, y_0, z_0). Then the last term in (1.8) may be eliminated by specifying the arbitrary Lagrange multiplier to be $\Lambda = -f_z/g_z$ giving

$$(f_x + \Lambda g_x)dx + (f_y + \Lambda g_y)dy = 0.$$

The remaining variables, x and y, may now be regarded as independent, and the coefficients of dx and dy must each vanish. This results in four equations for the four unknowns x_0, y_0, z_0, and Λ as follows

$$f_x + \Lambda g_x = 0, \quad f_y + \Lambda g_y = 0, \quad f_z + \Lambda g_z = 0, \quad g = c.$$

[5] Most authors use λ to denote Lagrange multipliers. Throughout this text, however, we use λ to denote eigenvalues (as is common) and Λ for Lagrange multipliers.

Consequently, finding the stationary point of the function $f(x,y,z)$ subject to the constraint $g(x,y,z) = c$ is equivalent to finding the stationary point of the *augmented function*

$$\tilde{f}(x,y,z) = f(x,y,z) + \Lambda g(x,y,z)$$

subject to no constraint. Thus, we seek the point (x_0, y_0, z_0) where

$$\tilde{f}_x = \tilde{f}_y = \tilde{f}_z = 0.$$

REMARKS:

1. Observe that because c is a constant, we may write the augmented function as $\tilde{f} = f + \Lambda g$, as above, or as $\tilde{f} = f + \Lambda(g - c)$.
2. Additional constraints may be imposed in the same manner, with each constraint having its own Lagrange multiplier.

EXAMPLE 1.2 Find the semi-major and semi-minor axes of the ellipse shown in Figure 1.7 and defined by

$$(x_1 + x_2)^2 + 2(x_1 - x_2)^2 = 8,$$

which may be written

$$3x_1^2 - 2x_1x_2 + 3x_2^2 = 8.$$

Solution: To determine the semi-major (minor) axis, calculate the farthest (nearest) point on the ellipse from the origin. Therefore, we maximize (minimize) $d = x_1^2 +$

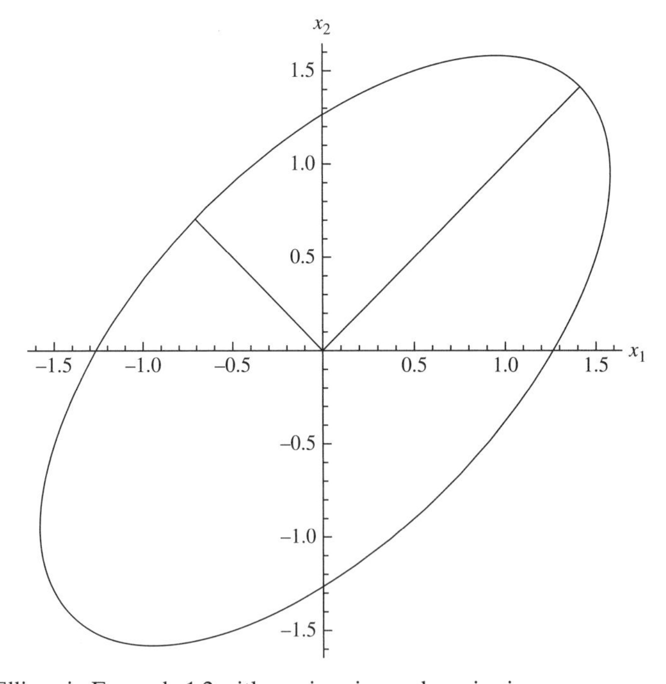

Figure 1.7. Ellipse in Example 1.2 with semi-major and semi-minor axes.

x_2^2, the square of the distance from the origin, subject to the constraint that the coordinates (x_1,x_2) be on the ellipse. Define an augmented function as follows

$$\tilde{d} = d + \Lambda\left(3x_1^2 - 2x_1x_2 + 3x_2^2 - 8\right),$$

where Λ is the *Lagrange multiplier* and is multiplied by the *constraint* that the extrema be on the ellipse. To determine the extrema of the algebraic function $\tilde{d}$, we evaluate

$$\frac{\partial \tilde{d}}{\partial x_1} = 0, \quad \frac{\partial \tilde{d}}{\partial x_2} = 0,$$

with

$$\tilde{d} = x_1^2 + x_2^2 + \Lambda\left(3x_1^2 - 2x_1x_2 + 3x_2^2 - 8\right).$$

Evaluating the partial derivatives gives

$$\frac{\partial \tilde{d}}{\partial x_1} = 2x_1 + \Lambda\left(6x_1 - 2x_2\right) = 0,$$

$$\frac{\partial \tilde{d}}{\partial x_2} = 2x_2 + \Lambda\left(-2x_1 + 6x_2\right) = 0.$$

Thus, we have two equations for x_1 and x_2 given by

$$3x_1 - x_2 = \lambda x_1,$$

$$-x_1 + 3x_2 = \lambda x_2,$$

where $\lambda = -\frac{1}{\Lambda}$. This is an eigenproblem of the form $\mathbf{A}\mathbf{x} = \lambda\mathbf{x}$, where

$$\mathbf{A} = \begin{bmatrix} 3 & -1 \\ -1 & 3 \end{bmatrix}.$$

The eigenvalues of the symmetric matrix $\mathbf{A}$ are $\lambda_1 = 2$ and $\lambda_2 = 4$ with the corresponding eigenvectors

$$\mathbf{u}_1 = \begin{bmatrix} 1 \\ 1 \end{bmatrix}, \quad \mathbf{u}_2 = \begin{bmatrix} -1 \\ 1 \end{bmatrix}.$$

The two eigenvectors $\mathbf{u}_1$ and $\mathbf{u}_2$ (along with $-\mathbf{u}_1$ and $-\mathbf{u}_2$) give the directions of the semi-major and semi-minor axes, which are along lines that bisect the first and third quadrants and second and fourth quadrants, respectively. Note that because $\mathbf{A}$ is real and symmetric with distinct eigenvalues, the eigenvectors are mutually orthogonal.

In order to determine which eigenvectors correspond to the semi-major and semi-minor axes, we recognize that a point on the ellipse must satisfy

$$(x_1 + x_2)^2 + 2(x_1 - x_2)^2 = 8.$$

Considering $\mathbf{u}_1^T = [1,\ 1]$, let us set $x_1 = c_1$ and $x_2 = c_1$. Substituting into the equation for the ellipse yields

$$4c_1^2 + 0 = 8,$$

in which case $c_1 = \pm\sqrt{2}$. Therefore, $x_1 = \sqrt{2}$ and $x_2 = \sqrt{2}$ or ($x_1 = -\sqrt{2}$ and $x_2 = -\sqrt{2}$), and the length of the corresponding axis is $\sqrt{x_1^2 + x_2^2} = 2$. Similarly, considering $\mathbf{u}_2^T = [-1, 1]$, let us set $x_1 = -c_2$ and $x_2 = c_2$. Substituting into the equation for the ellipse yields

$$0 + 8c_2^2 = 8,$$

in which case $c_2 = \pm 1$. Therefore, $x_1 = -1$ and $x_2 = 1$ (or $x_1 = 1$ and $x_2 = -1$), and the length of the corresponding axis is $\sqrt{x_1^2 + x_2^2} = \sqrt{2}$. As a result, the eigenvector $\mathbf{u}_1$ corresponds to the semi-major axis, and $\mathbf{u}_2$ corresponds to the semi-minor axis.

The Lagrange multiplier method is a general approach to determining extrema of a function or set of functions subject to constraints. It will also be used in calculus of variations, where it is applied to functionals.

1.6 Integration by Parts

Because integration by parts plays such an important role in the calculus of variations, we derive the integration by parts formula using two approaches. First, consider the product rule of differential calculus for the derivative of the product of two one-dimensional functions $p(x)$ and $q(x)$, which is

$$\frac{d}{dx}(pq) = p\frac{dq}{dx} + q\frac{dp}{dx}.$$

Integrating with respect to x gives

$$pq = \int p\frac{dq}{dx}dx + \int q\frac{dp}{dx}dx,$$

or

$$pq = \int pdq + \int qdp.$$

Rearranging yields the usual formula for integration by parts as follows

$$\int pdq = pq - \int qdp.$$

Alternatively, for a definite integral over the interval $x_0 \leq x \leq x_1$, this takes the form

$$\int_{x_0}^{x_1} pdq = \Big[pq\Big]_{x_0}^{x_1} - \int_{x_0}^{x_1} qdp, \tag{1.9}$$

where the term in brackets $[\cdot]$ is evaluated only at the end points.

In the two-dimensional context, integration by parts is replaced by the *divergence theorem* of vector calculus. Therefore, it is also possible to derive the one-dimensional integration by parts formula (1.9) directly from the two-dimensional divergence (Gauss's) theorem

$$\iint_A \nabla \cdot \mathbf{v} dA = \oint_C \mathbf{v} \cdot \mathbf{n} ds, \tag{1.10}$$

where $\mathbf{v}$ is any vector, and $\mathbf{n}$ is the outward facing normal to the curve C bounding the area A. The divergence theorem relates the area integral of the divergence of a vector field to the line integral of the component of the vector normal to the bounding curve C.

Let us apply the divergence theorem to the vector

$$\mathbf{v} = \psi \nabla \phi,$$

where $\psi = \psi(x,y)$ and $\phi = \phi(x,y)$ are arbitrary two-dimensional scalar functions. Then

$$\nabla \cdot \mathbf{v} = \nabla \cdot (\psi \nabla \phi) = \psi \nabla^2 \phi + \nabla \psi \cdot \nabla \phi,$$

$$\mathbf{v} \cdot \mathbf{n} = \psi (\nabla \phi \cdot \mathbf{n}).$$

Substituting these results into the divergence theorem (1.10) gives *Greeen's first theorem*

$$\iint_A \nabla \psi \cdot \nabla \phi dA = \oint_C \psi (\nabla \phi \cdot \mathbf{n}) ds - \iint_A \psi \nabla^2 \phi dA. \tag{1.11}$$

In one dimension, with $\psi = \psi(x)$ and $\phi = \phi(x)$, we have $\nabla = \frac{d}{dx}\mathbf{i}$, $\mathbf{n}_0 = -\mathbf{i}$, $\mathbf{n}_1 = \mathbf{i}$, and $dA = dx$. Therefore, substituting into equation (1.11) leads to

$$\int_{x_0}^{x_1} \frac{d\psi}{dx}\frac{d\phi}{dx} dx = \left[\psi \frac{d\phi}{dx}\right]_{x=x_1} - \left[\psi \frac{d\phi}{dx}\right]_{x=x_0} - \int_{x_0}^{x_1} \psi \frac{d^2\phi}{dx^2} dx.$$

Let $\psi(x) = q(x)$ and $\phi(x) = \int p(x)dx$, in which case

$$\frac{d\phi}{dx} = p(x).$$

Then we have

$$\int_{x_0}^{x_1} pdq = \Big[pq\Big]_{x_0}^{x_1} - \int_{x_0}^{x_1} qdp,$$

which again is the formula for integration by parts in one dimension. Therefore, we see that indeed integration by parts follows from applying the divergence theorem in one dimension.

1.7 Fundamental Lemma of the Calculus of Variations

Along with integration by parts, the *fundamental lemma of the calculus of variations* will prove essential to our developments in Chapter 2. The lemma[6] says the following:

LEMMA 1.1. If $g(x)$ is a continuous function in $x_0 \le x \le x_1$, and if

$$\int_{x_0}^{x_1} g(x)h(x)dx = 0,$$

where $h(x)$ is an arbitrary function in the same interval with $h(x_0) = h(x_1) = 0$, then $g(x) = 0$ at every point in the interval $x_0 \le x \le x_1$.

The proof of this lemma by contradiction is very straightforward and hinges on the fact that the function $h(x)$ is arbitrary. Let us suppose that the function $g(x)$ is

[6] A *lemma* is a short theorem that is used in the proof of a more comprehensive theorem.

nonzero and positive in some subinterval $a \le x \le b$ within the interval $x_0 \le x \le x_1$. Because $h(x)$ is arbitrary, we can set it to be any function of our choice, such as

$$h(x) = \begin{cases} 0, & x < a \\ (x-a)(b-x), & a \le x \le b \\ 0, & x > b \end{cases}.$$

Observe that $h(x)$ satisfies the required conditions at the end points. Noting that $(x-a) > 0$ and $(b-x) > 0$ for $a \le x \le b$, let us consider the integral

$$\int_{x_0}^{x_1} g(x)h(x)dx = \int_a^b g(x)(x-a)(b-x)dx > 0,$$

which is positive because the integrand is positive throughout the subinterval except at $x = a$ and $x = b$, where it is zero. Thus, the lemma is proven by contradiction, as the only way for the integral over the entire interval to be zero is if the function $g(x) = 0$ everywhere in the interval.

1.8 Adjoint and Self-Adjoint Differential Operators

Finally, it will prove beneficial to review some of the properties of differential operators. Consider an $n \times n$ matrix $\mathbf{A}$ and arbitrary $n \times 1$ vectors $\mathbf{u}$ and $\mathbf{v}$. Let us premultiply $\mathbf{A}$ by $\mathbf{v}^T$ and postmultiply by $\mathbf{u}$ and take the transpose

$$\left(\mathbf{v}^T\mathbf{A}\mathbf{u}\right)^T = \mathbf{u}^T\mathbf{A}^T\mathbf{v}.$$

Observe from the matrix multiplication that the left-hand side is a scalar, which is a 1×1 matrix; therefore, we may remove the transpose and write in terms of inner products

$$\langle \mathbf{v}, \mathbf{A}\mathbf{u} \rangle = \left\langle \mathbf{u}, \mathbf{A}^T\mathbf{v} \right\rangle.$$

Similarly, a differential operator $\mathcal{L}$ has an *adjoint* operator $\mathcal{L}^*$, which is analogous to taking the transpose, that satisfies

$$\langle v, \mathcal{L}u \rangle = \left\langle u, \mathcal{L}^* v \right\rangle, \tag{1.12}$$

where $u(x)$ and $v(x)$ are arbitrary functions with homogeneous boundary conditions.

In order to illustrate the approach for determining the adjoint of a differential operator, consider the second-order linear differential equation with variable coefficients

$$\mathcal{L}u = \frac{1}{r(x)}\left[a_0(x)u'' + a_1(x)u' + a_2(x)u\right] = 0, \quad a \le x \le b, \tag{1.13}$$

where $r(x)$ is a weight function. To obtain the adjoint operator, consider an arbitrary function $v(x)$, and take the inner product with $\mathcal{L}u$ to obtain the left-hand side of (1.12) as follows

$$\langle v, \mathcal{L}u \rangle = \int_a^b r(x)v(x)\left\{\frac{1}{r(x)}\left[a_0(x)u'' + a_1(x)u' + a_2(x)u\right]\right\}dx, \tag{1.14}$$

where the inner product is taken with respect to the weight function $r(x)$. We want to switch the roles of $u(x)$ and $v(x)$ in the inner product, that is, interchange derivatives on $u(x)$ for derivatives on $v(x)$, in order to obtain the right-hand side of (1.12). This is accomplished using integration by parts as described in Section 1.6. Integrating the second term by parts gives

$$\int_a^b a_1 v u' dx = a_1 v u \Big|_a^b - \int_a^b u(a_1 v)' dx,$$

where

$$\int p dq = pq - \int q dp$$

with

$$p = va_1, \qquad q = u,$$
$$dp = (va_1)' dx, \qquad dq = u' dx.$$

Similarly, integrating the first term of (1.14) by parts twice results in

$$\int_a^b a_0 v u'' dx = \underbrace{a_0 v u' \Big|_a^b - \int_a^b u'(a_0 v)' dx}_{(1)} = \underbrace{\Big[a_0 v u' - (a_0 v)' u \Big]_a^b + \int_a^b u(a_0 v)'' dx}_{(2)},$$

(1)

$$p = va_0, \qquad q = u',$$
$$dp = (va_0)' dx, \qquad dq = u'' dx,$$

(2)

$$p = (va_0)', \qquad q = u,$$
$$dp = (va_0)'' dx, \qquad dq = u' dx.$$

Substituting into (1.14) yields

$$\langle v, \mathcal{L}u \rangle = \Big[a_0 v u' - (a_0 v)' u + a_1 v u \Big]_a^b + \int_a^b r(x) u(x) \left\{ \frac{1}{r(x)} \left[(a_0 v)'' - (a_1 v)' + a_2 v \right] \right\} dx,$$

where the expression in {} is $\mathcal{L}^* v$, and the integral is $\langle u, \mathcal{L}^* v \rangle$. Therefore, if the boundary conditions on $u(x)$ and $v(x)$ are homogeneous, in which case the terms evaluated at $x = a$ and $x = b$ vanish, we have

$$\langle v, \mathcal{L}u \rangle = \langle u, \mathcal{L}^* v \rangle,$$

where the *adjoint* operator $\mathcal{L}^*$ of the differential operator $\mathcal{L}$ is

$$\mathcal{L}^* v = \frac{1}{r(x)} \left\{ [a_0(x) v]'' - [a_1(x) v]' + a_2(x) v \right\}. \tag{1.15}$$

Note that the variable coefficients move inside the derivatives, and the odd-order derivatives change sign as compared to $\mathcal{L}u$. This is the case for higher-order derivatives as well.

EXAMPLE 1.3 Find the adjoint operator for

$$\mathcal{L} = \frac{d^2}{dx^2} + x\frac{d}{dx}, \quad 0 \leq x \leq 1,$$

with homogeneous boundary conditions.

Solution: From equation (1.13)

$$a_0(x) = 1, \quad a_1(x) = x, \quad a_2(x) = 0, \quad r(x) = 1.$$

Then from (1.15), the adjoint operator is

$$\begin{aligned} \mathcal{L}^*v &= [a_0 v]'' - [a_1 v]' + a_2 v, \\ &= v'' - (xv)', \\ \mathcal{L}^*v &= v'' - xv' - v. \end{aligned}$$

Thus, the adjoint operator of $\mathcal{L}$ is

$$\mathcal{L}^* = \frac{d^2}{dx^2} - x\frac{d}{dx} - 1.$$

Note that $\mathcal{L}^* \neq \mathcal{L}$ in this example.

Recall that a matrix that is equal to its transpose is said to be *symmetric*, or *Hermitian*, and has the property that its eigenvectors are mutually orthogonal if its eigenvalues are distinct. Likewise, if the differential operator and its adjoint are the same, in which case $\mathcal{L} = \mathcal{L}^*$, then the differential operator $\mathcal{L}$ is said to be *self-adjoint*, or *Hermitian*, and distinct eigenvalues produce orthogonal eigenfunctions.

To show this, consider the case when $\mathcal{L} = \mathcal{L}^*$ given above and let $u(x)$ and $v(x)$ be two eigenfunctions of the differential operator, such that $\mathcal{L}u = \lambda_1 u$, and $\mathcal{L}v = \lambda_2 v$. Evaluating (1.12)

$$\begin{aligned} \langle u, \mathcal{L}v \rangle &= \langle v, \mathcal{L}u \rangle \\ \langle u, \lambda_2 v \rangle &= \langle v, \lambda_1 u \rangle \\ (\lambda_1 - \lambda_2) \langle u, v \rangle &= 0. \end{aligned}$$

Thus, if $\lambda_1 \neq \lambda_2$, the corresponding eigenfunctions must be orthogonal in order for their inner product to be zero.

As illustrated in the previous example, not all second-order linear differential equations of the form (1.13) have differential operators that are self-adjoint, that is, for arbitrary $a_0(x), a_1(x)$, and $a_2(x)$ coefficients. Let us determine the subset of such equations that are self-adjoint.

We rewrite the adjoint $\mathcal{L}^*v$, given by (1.15), of $\mathcal{L}u$ by carrying out the differentiations and collecting terms as follows:

$$\mathcal{L}^*v = \frac{1}{r(x)} \left\{ [a_0 v]'' - [a_1 v]' + a_2 v \right\},$$

or from the product rule

$$\mathcal{L}^*v = \frac{1}{r(x)}\left\{a_0v'' + \left[2a_0' - a_1\right]v' + \left[a_0'' - a_1' + a_2\right]v\right\}. \tag{1.16}$$

For $\mathcal{L}$ to be self-adjoint, the operators $\mathcal{L}$ and $\mathcal{L}^*$ in equations (1.13) and (1.16), respectively, must be the same. Given that the first term is already identical, consider the second and third terms. Equivalence of the second terms requires that

$$a_1(x) = 2a_0'(x) - a_1(x),$$

or

$$a_1(x) = a_0'(x). \tag{1.17}$$

Equivalence of the third terms requires that

$$a_2(x) = a_0''(x) - a_1'(x) + a_2(x),$$

but this is always the case if (1.17) is satisfied. Therefore, substitution of the condition (1.17) into (1.13) requires that

$$\mathcal{L}u = \frac{1}{r(x)}\left\{a_0u'' + a_0'u' + a_2u\right\} = 0.$$

The differential operator in the above expression may be written in the form

$$\mathcal{L} = \frac{1}{r(x)}\left\{\frac{d}{dx}\left[a_0(x)\frac{d}{dx}\right] + a_2(x)\right\},$$

which is called the *Sturm-Liouville differential operator.*

Therefore, a second-order linear differential operator is self-adjoint *if and only if* it is of the Sturm-Liouville form

$$\mathcal{L} = \frac{1}{r(x)}\left\{\frac{d}{dx}\left[p(x)\frac{d}{dx}\right] + q(x)\right\}, \tag{1.18}$$

and the boundary conditions are homogeneous. It follows that the corresponding eigenfunctions of the Sturm-Liouville differential operator are orthogonal with respect to the weight function $r(x)$.

Similarly, consider the fourth-order Sturm-Liouville differential operator

$$\mathcal{L} = \frac{1}{r(x)}\left\{\frac{d^2}{dx^2}\left[s(x)\frac{d^2}{dx^2}\right] + \frac{d}{dx}\left[p(x)\frac{d}{dx}\right] + q(x)\right\}.$$

This operator can also be shown to be self-adjoint if the boundary conditions are homogeneous of the form

$$u = 0, \quad u' = 0,$$

or

$$u = 0, \quad s(x)u'' = 0,$$

or

$$u' = 0, \quad \left[s(x)u''\right]' = 0,$$

Table 1.2. *Notational conventions.*

Quantity/Operation	Convention	Examples
Vectors	Bold lowercase letters and square brackets	$\mathbf{a} = [\cdot]$
Matrices	Bold uppercase letters and square brackets	$\mathbf{A} = [\cdot]$
Determinant		$\lvert\mathbf{A}\rvert$
Vector norm		$\lvert\lvert\mathbf{a}\rvert\rvert$
Matrix transpose	Superscript T	$\mathbf{A}^T$
Scalar/dot/inner product		$\mathbf{a} \cdot \mathbf{b} = \mathbf{a}^T\mathbf{b} = \langle \mathbf{a}, \mathbf{b} \rangle$
Inner product of functions		$\langle g, h \rangle = \int_a^b g(x)h(x)dx$
Functions	Lower case with independent variables in parentheses	$f(x,t)$
Functionals	Upper case with dependent functions in square brackets	$I[u(x), v(x)]$
Spatial ordinary derivative	Prime	$u'(x) = du/dx,\, u^{(n)}(x) = d^n u/dx^n$
Temporal ordinary derivative	Dot	$\dot{u}(t) = du/dt, \ddot{u}(t) = d^2u/dt^2$
Partial derivatives	Subscripts	$u_{xx} = \partial^2 u/\partial x^2$
Eigenvalues	Lowercase lambda	λ
Lagrange multipliers	Uppercase lambda	Λ
Gradient operator	Del (nabla)	∇
Laplacian operator	Del (nabla) squared	∇^2
General differential operator		$\mathcal{L}$
Adjoint differential operator		$\mathcal{L}^*$

at $x = a$ or $x = b$. Then the corresponding eigenfunctions are orthogonal on the interval $a \leq x \leq b$ with respect to $r(x)$.

We will encounter adjoint and/or self-adjoint, that is, Sturm-Liouville, differential operators in our discussions of functionals with integral constraints (Section 2.7.1), the Rayleigh-Ritz method (Section 3.1.2), stability of continuous systems (Section 6.6), and control of continuous systems (Section 10.5). There are analogous criteria for partial differential equations to be self-adjoint. The notational conventions introduced in Chapter 1 and used throughout the text are summarized in Table 1.2.

EXERCISES

1.1 Determine the point on the plane

$$c_1x_1 + c_2x_2 + c_3x_3 = c_4$$

that is closest to the origin. What is this distance?

1.2 Determine the points on the sphere

$$x_1^2 + x_2^2 + x_3^2 = 4$$

that are closest and farthest from the point $(x_1,x_2,x_3) = (1,-1,1)$.

1.3 Determine the maximum value of the function $f(x_1,x_2,x_3) = x_1^2 + 2x_2 - x_3^2$ on the line along which the planes $2x_1 - x_2 = 0$ and $x_2 + x_3 = 0$ intersect.

1.4 Perform the integration by parts necessary to determine the adjoint operator of

$$\mathcal{L}u = \frac{d^2u}{dx^2} + u,$$

with homogeneous boundary conditions. Check your result using equation (1.15). Is the differential operator $\mathcal{L}$ self-adjoint?

1.5 Perform the integration by parts necessary to determine the adjoint operator of

$$\mathcal{L}u = \frac{d^2u}{dx^2} - 2\frac{du}{dx} + u,$$

with homogeneous boundary conditions. Check your result using equation (1.15). Is the differential operator $\mathcal{L}$ self-adjoint?

1.6 Perform the integration by parts necessary to determine the adjoint operator of

$$\mathcal{L}u = \frac{d^2u}{dx^2} + \frac{1}{2}x^{-1/2}\frac{du}{dx} + x^2u,$$

with homogeneous boundary conditions. Check your result using equation (1.15). Is the differential operator $\mathcal{L}$ self-adjoint?

2 Calculus of Variations

> To those who do not know mathematics it is difficult to get across a real feeling as to the beauty, the deepest beauty, of nature ... If you want to learn about nature, to appreciate nature, it is necessary to understand the language that she speaks in.
>
> (Richard Feynman)

Calculus is the mathematics of change. Differential calculus addresses the change of a *function* as one moves from point to point. A function $u(x)$ is such that the dependent variable u takes on a unique value for each value of the independent variable x, and the derivative of the function at a point indicates the rate of change, or slope, at that point. In calculus of variations, we deal with the *functional*, which may be regarded as a "function of functions," that is, a function that depends on other functions. More specifically, a functional is a definite integral whose integrand contains a function that is yet to be determined. A functional $I[u(x)]$ is such that I takes on a unique scalar value for each function $u(x)$.[1] In Section 1.3, the travel time $T[u(x)]$ and total energy $E[u(z)]$ are functionals, which are functions of the path of light or the boat $u(x)$ and bubble shape $u(z)$, respectively. Variational calculus addresses the change in a functional as one moves from function to function. Accordingly, whereas differential calculus is the calculus of functions, variational calculus is the calculus of functionals.

In this chapter, we consider the mathematical aspects of obtaining stationary functions (possibly also extrema) of functionals. Armed with these techniques, we will be prepared to solve the examples formulated in Chapter 1 and address the applications in subsequent chapters. This chapter will provide us with the capability to formulate and solve problems in fields as diverse as dynamics of discrete and continuous bodies, optics, electromagnetics, modern physics, fluid mechanics, and optimization and control. The breadth and scope of the applications of variational methods are truly remarkable and reflect its fundamental connection with extremum principles, which are ubiquitous in science and engineering.

[1] Whereas a function takes a scalar value as its argument and returns another scalar value, a functional takes a function and returns a scalar value. That is, a function is a mapping from points to points, and a functional is a mapping from functions to points. Although there are other operations that are functionals, we focus our attention in calculus of variations on functionals in the form of definite integrals.

We begin by treating a functional of one function, which itself is a function of a single independent variable. In subsequent sections, the approach is generalized to include additional independent and dependent variables, more general boundary conditions, and incorporation of constraints.

2.1 Functionals of One Independent Variable

Consider the functional $I[u(x)]$ of the function $u(x)$ of the form

$$I[u(x)] = \int_{x_0}^{x_1} F\left[x, u(x), u'(x)\right] dx, \tag{2.1}$$

where $u(x)$ is smooth (differentiable) and is the family of curves passing through the specified end points

$$u(x_0) = u_0, \quad u(x_1) = u_1. \tag{2.2}$$

Boundary conditions in which $u(x)$ is specified at the end points are sometimes referred to as *essential* or *Dirichlet*. Of all the smooth functions that satisfy these end conditions, we seek the function $u(x)$ for which the functional $I[u(x)]$ is an extremum, or more generally, for which $u(x)$ is a *stationary function* of $I[u(x)]$.

In order to use arguments from differential calculus to determine the stationary function of equation (2.1) as in Section 1.4, let us expand a function $\hat{u}(x)$ "close" to the actual, as yet unknown, stationary function $u(x)$ in the small parameter ϵ about $\epsilon = 0$ as follows

$$\hat{u}(x) = u(x) + \left.\frac{\partial u}{\partial \epsilon}\right|_{\epsilon=0} \epsilon + O(\epsilon^2). \tag{2.3}$$

The functions $u(x)$ and $\hat{u}(x)$ are "close" in the sense that their norms are "close". The two functions $u(x)$ and $\hat{u}(x)$ are illustrated in Figure 2.1. We call the $O(\epsilon)$ term the "variation of u" and denote it by

$$\delta u = \left.\frac{\partial u}{\partial \epsilon}\right|_{\epsilon=0} \epsilon; \tag{2.4}$$

it is the variation between the actual stationary function $u(x)$ and another function $\hat{u}(x)$ close to it. Letting $\eta(x) = \partial u / \partial \epsilon|_{\epsilon=0}$, equation (2.3) becomes

$$\hat{u}(x) = u(x) + \epsilon \eta(x) + O(\epsilon^2),$$

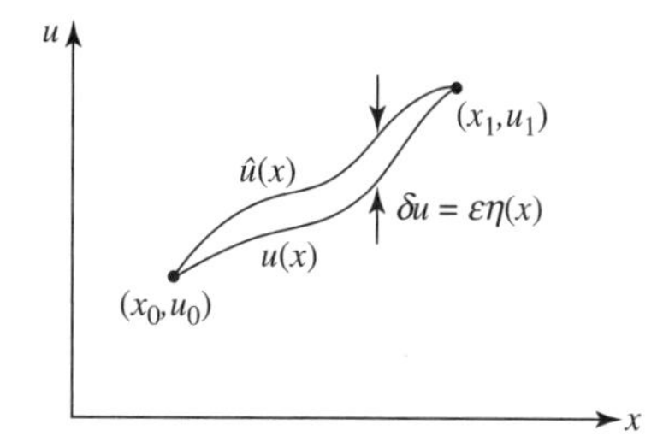

Figure 2.1. The stationary function $u(x)$ and a function $\hat{u}(x)$ close to it.

where ϵ is a small parameter that is a constant for a given function, but it varies from function to function. Therefore, the variation δ is the differential with respect to ϵ as reflected in (2.4), instead of x. The function $\eta(x)$ has

$$\eta(x_0) = \eta(x_1) = 0,$$

such that $\hat{u}(x)$ satisfies the end conditions (2.2). The difference between the two functions is

$$\hat{u}(x) - u(x) = \epsilon\eta(x) = \delta u,$$

such that δu is the *variation* of $u(x)$ from curve to curve. Compare this to differential calculus, where dx is the *differential* of x from point to point.

2.1.1 Functional Derivative

In differential calculus, the partial derivative indicates the rate of change of a function with respect to one of the coordinate directions. For example, given a function $f(x,y,z)$, the rate of change of the function f in the x-direction is given by the partial derivative $\partial f/\partial x$. Similarly, we would like to be able to characterize the rate of change of a functional with respect to the function on which it depends. This is done using the *functional*, or *Fréchet, derivative*[2] and is denoted by $\delta I/\delta u$ for the functional $I[u(x)]$.

From the definition of the derivative, let us differentiate the functional $I[u(x)]$ with respect to the small parameter ϵ as follows

$$\left.\frac{dI}{d\epsilon}\right|_{\epsilon=0} = \lim_{\epsilon\to 0} \frac{I[u+\epsilon\eta] - I[u]}{\epsilon}.$$

For a functional $I[u(x)]$ of the form (2.1), this yields

$$\left.\frac{dI}{d\epsilon}\right|_{\epsilon=0} = \lim_{\epsilon\to 0} \frac{1}{\epsilon}\left\{\int_{x_0}^{x_1} F(x,u+\epsilon\eta,u'+\epsilon\eta')dx - \int_{x_0}^{x_1} F(x,u,u')dx\right\}.$$

Let us use a Taylor series expansion for the first of these integrals according to

$$\left.\frac{dI}{d\epsilon}\right|_{\epsilon=0} = \lim_{\epsilon\to 0} \frac{1}{\epsilon}\left\{\int_{x_0}^{x_1} \left[F(x,u,u') + \epsilon\eta\frac{\partial F}{\partial u} + \epsilon\frac{d\eta}{dx}\frac{\partial F}{\partial u'} + O(\epsilon^2)\right]dx \right.$$
$$\left. - \int_{x_0}^{x_1} F(x,u,u')dx\right\}.$$

Canceling the first and last terms results in

$$\left.\frac{dI}{d\epsilon}\right|_{\epsilon=0} = \int_{x_0}^{x_1}\left[\eta\frac{\partial F}{\partial u} + \frac{d\eta}{dx}\frac{\partial F}{\partial u'}\right]dx.$$

Using integration by parts [$\int p dq = pq - \int q dp, p = \partial F/\partial u', dp = (\partial F/\partial u')'dx, q = \eta,$ $dq = \eta' dx$] to remove the derivative from $\eta(x)$ in the second term allows us to write

[2] The Fréchet derivative, which is the variational analog of the partial derivative (or gradient), is a special case of the Gâteaux derivative, which is the variational analog of the directional derivative.

the above result in the form

$$\left.\frac{dI}{d\epsilon}\right|_{\epsilon=0}=\left[\eta\frac{\partial F}{\partial u'}\right]_{x_0}^{x_1}+\int_{x_0}^{x_1}\left[\frac{\partial F}{\partial u}-\frac{d}{dx}\left(\frac{\partial F}{\partial u'}\right)\right]\eta dx.$$

Because $\eta(x_0)=\eta(x_1)=0$, the first term evaluated at the end points vanishes. We define the functional derivative as the expression in $[\cdot]$. Thus, we have

$$\left.\frac{dI}{d\epsilon}\right|_{\epsilon=0}=\int_{x_0}^{x_1}\frac{\delta I}{\delta u}\eta dx,$$

where the functional derivative is given by

$$\frac{\delta I}{\delta u}=\frac{\partial F}{\partial u}-\frac{d}{dx}\left(\frac{\partial F}{\partial u'}\right). \tag{2.5}$$

It is the rate of change of the functional I with respect to the function $u(x)$ on which it depends. Intuitively, we might expect that for $u(x)$ to be a stationary function of I, the variational derivative should vanish according to

$$\frac{\delta I}{\delta u}=\frac{\partial F}{\partial u}-\frac{d}{dx}\left(\frac{\partial F}{\partial u'}\right)=0.$$

This is indeed the case, which will be established more formally in the next section.

2.1.2 Derivation of Euler's Equation

Although we are in a position to address the question of what function $u(x)$ is a stationary function of the functional $I[u(x)]$, let us first consider the nuances of our developments in the previous section in order to provide a more formal proof and introduce some of the arguments to be used later.

Taking ϵ as the independent variable as we vary the functions $\hat{u}(x)$, $I[u(x)]$ has a stationary function when the variation of the functional vanishes according to

$$\delta I=\left.\frac{dI}{d\epsilon}\right|_{\epsilon=0}\epsilon=0.$$

Note that we seek the stationary function of a *functional* rather than the stationary point of a *function*. While we have already evaluated $dI/d\epsilon|_{\epsilon\to 0}$ in the previous section, we will use a slightly different approach here.

Taking the derivative of the functional (2.1) with respect to ϵ and setting it equal to zero gives (x_0 and x_1 do not depend on ϵ)

$$\frac{dI}{d\epsilon}=\frac{d}{d\epsilon}\int_{x_0}^{x_1}Fdx=\int_{x_0}^{x_1}\frac{dF}{d\epsilon}dx=0. \tag{2.6}$$

Treating x, u, and u' as independent variables, the integrand is

$$\frac{dF}{d\epsilon}=\frac{\partial F}{\partial x}\frac{dx}{d\epsilon}+\frac{\partial F}{\partial u}\frac{du}{d\epsilon}+\frac{\partial F}{\partial u'}\frac{du'}{d\epsilon}.$$

Because the independent variable x does not depend on ϵ, the first term vanishes, and with the definition $u(x) = \hat{u}(x) - \epsilon\eta(x)$, we have $du/d\epsilon = -\eta(x)$ and $du'/d\epsilon = -\eta'(x)$. As a result, equation (2.6) requires that

$$\int_{x_0}^{x_1} \left(\frac{\partial F}{\partial u}\eta + \frac{\partial F}{\partial u'}\eta' \right) dx = 0. \tag{2.7}$$

In order to factor η out of the remainder of the integrand, integrate the second term by parts as in the previous section, giving

$$\int_{x_0}^{x_1} \frac{\partial F}{\partial u'}\eta' dx = \left[\frac{\partial F}{\partial u'}\eta \right]_{x_0}^{x_1} - \int_{x_0}^{x_1} \frac{d}{dx}\left(\frac{\partial F}{\partial u'} \right) \eta dx. \tag{2.8}$$

Because $\eta(x_0) = \eta(x_1) = 0$, the term evaluated at the end points vanishes, and from (2.7), we are left with

$$\int_{x_0}^{x_1} \left[\frac{\partial F}{\partial u} - \frac{d}{dx}\left(\frac{\partial F}{\partial u'} \right) \right] \eta(x) dx = 0.$$

According to the *fundamental lemma of the calculus of variations* (see Section 1.7), for the definite integral of the product of $[\cdot]$ and $\eta(x)$ to vanish, where $\eta(x)$ is an arbitrary function, the expression in $[\cdot]$ must vanish everywhere in the interval $x_0 \leq x \leq x_1$, that is,

$$\frac{\partial F}{\partial u} - \frac{d}{dx}\left(\frac{\partial F}{\partial u'} \right) = 0. \tag{2.9}$$

This is the *Euler*, or *Euler-Lagrange, equation* for the case with $F = F(x,u,u')$. It is a necessary, but not sufficient, condition for an extremum of $I[u(x)]$. The solution $u(x)$ of (2.9) is a *stationary function* of the functional $I[u(x)]$ and requires that the functional derivative vanishes for a stationary function.

Euler's equation may be regarded as an equivalent differential form to the variational form (2.1). It is invariant in the sense that it does not change when transformed to alternative curvilinear coordinate systems, which is shown in Section 4.4. As we will see, this is one of the significant advantages of the variational approach. In order to more clearly reveal the form of Euler's differential equation, it may also be written in terms of the sought after stationary function $u(x)$ in the form

$$\left(\frac{\partial^2 F}{\partial u'^2} \right) \frac{d^2 u}{dx^2} + \left(\frac{\partial^2 F}{\partial u' \partial u} \right) \frac{du}{dx} + \left(\frac{\partial^2 F}{\partial u' \partial x} - \frac{\partial F}{\partial u} \right) = 0, \tag{2.10}$$

which is a second-order ordinary differential equation for $u(x)$. The Euler equation may be linear or nonlinear depending upon the nature of the integrand $F = F(x,u,u')$.

REMARKS:

1. We will use the more compact form (2.9) in practice; the form (2.10) is provided to more explicitly illustrate the final form of the Euler equation to be solved.
2. This, the simplest form of the Euler equation that we will encounter, turns out to encompass a surprising array of mathematical and physical problems.
3. When evaluating the partial derivatives in Euler's equation (2.9), u and u' are treated as independent variables.

4. The set of *admissible* (allowable) *functions* $u(x)$ that may be a stationary function of a functional is all the continuously differentiable functions. In some cases, the actual solution is discontinuous (or piecewise continuous), and such solutions cannot be obtained using variational methods. See Chapter 10 for an important class of such solutions that arise in optimal control.
5. Isaac Newton and the Bernoulli brothers provided the seminal ideas that would become the calculus of variations in the late 1600s, and Euler and Lagrange formalized this new branch of mathematics one-half century later. The Euler equation was first derived by Leonhard Euler in 1744 and later by Joseph Louis Lagrange in 1755 using the above derivation (Lagrange was 19 years old!).

2.1.3 Variational Notation

It is convenient to fully utilize *variational notation* – with δ – in order to take advantage of its similarities to differentials – with d or ∂ – and to avoid the need to appeal to unnecessary mathematical constructs involving ϵ and η. The primary difference is that d (or ∂) is a change of a function from point to point, whereas δ is a change from function to function. Recall that δu is the difference between the actual stationary function and a function close to it. That is, the variation of a quantity is a virtual displacement from function to function, and we imagine considering all possible functions $u(x)$ in order to find that which is stationary. The analog of equation (2.6) in variational notation is

$$\delta I = \delta \int_{x_0}^{x_1} F\left(x,u,u'\right) dx = \int_{x_0}^{x_1} \delta F\left(x,u,u'\right) dx = 0.$$

We want to find the function $u(x)$ for which the functional $I[u(x)]$ is stationary, in which case the variation of I is zero, that is, $\delta I = 0$. This is analogous to taking a total differential of a function and setting it equal to zero to find its stationary points as in Section 1.4. In taking the variation inside the integral, it has been assumed that the end points x_0 and x_1 do not vary with x. We say that a quantity varies when it changes from function to function. Therefore, $I[u(x)], u(x)$, and $u'(x)$ vary, but x does not.

Using this variational notation, note the analogies between differentials and variations of functions and functionals:

- The total *differential* of a *function* $f(x,y,z)$ is

$$df = \frac{\partial f}{\partial x}dx + \frac{\partial f}{\partial y}dy + \frac{\partial f}{\partial z}dz,$$

  which is the change in $f(x,y,z)$ along the curve from point to point.
- The total *differential* of a *function* $F(x,u,u')$ is

$$dF = \frac{\partial F}{\partial x}dx + \frac{\partial F}{\partial u}du + \frac{\partial F}{\partial u'}du'.$$

- The *variation* of a *function* $F(x,u,u')$ is

$$\delta F = \frac{\partial F}{\partial x}\delta x + \frac{\partial F}{\partial u}\delta u + \frac{\partial F}{\partial u'}\delta u'.$$

However, because x is an independent variable, which does not vary, $\delta x = 0$ and

$$\delta F = \frac{\partial F}{\partial u}\delta u + \frac{\partial F}{\partial u'}\delta u',$$

which is the variation of F from function to function.

- The *stationary point* of a *function* $f(x,y,z)$ is the point (x,y,z) where the *differential* vanishes, that is, $df = 0$.
- The *stationary function* of a *functional* $I[u]$ is the function $u(x)$ that causes the *variation* to vanish, that is, $\delta I = 0$.
- The laws of variations of sums and products are analogous to those for differentials. For example, the product rule is

$$\delta(F_1 F_2) = F_1 \delta F_2 + F_2 \delta F_1.$$

In order to illustrate the use of variational notation, let us re-derive Euler's equation. We seek the stationary function $u(x)$ of the functional

$$I[u] = \int_{x_0}^{x_1} F\left[x, u(x), u'(x)\right] dx.$$

Taking the variation of I and setting it equal to zero gives

$$\delta I = \delta \int_{x_0}^{x_1} F\left[x,u,u'\right] dx = \int_{x_0}^{x_1} \delta F\left[x,u,u'\right] dx = 0.$$

Evaluating the variation of the integrand F as before yields

$$\int_{x_0}^{x_1} \left(\frac{\partial F}{\partial x}\delta x + \frac{\partial F}{\partial u}\delta u + \frac{\partial F}{\partial u'}\delta u'\right) dx = 0. \tag{2.11}$$

However, x is an independent variable and does not vary; therefore, $\delta x = 0$ [we are considering the variation of I with respect to $u(x)$]. Also

$$\delta u' = \delta\left(\frac{du}{dx}\right) = \frac{d}{dx}(\delta u),$$

that is, the variational and differential operators δ and d/dx are commutative if differentiation is with respect to an independent variable. Then (2.11) becomes

$$\int_{x_0}^{x_1} \left[\frac{\partial F}{\partial u}\delta u + \frac{\partial F}{\partial u'}(\delta u)'\right] dx = 0.$$

To factor δu out of the remainder of the integrand, integrate the last term by parts as in Section 2.1.2 giving

$$\delta I = \left[\frac{\partial F}{\partial u'}\delta u\right]_{x_0}^{x_1} + \int_{x_0}^{x_1} \left[\frac{\partial F}{\partial u} - \frac{d}{dx}\left(\frac{\partial F}{\partial u'}\right)\right] \delta u dx = 0. \tag{2.12}$$

All possible functions $u(x)$ must satisfy the fixed end conditions; therefore, the term evaluated at x_0 and x_1 vanishes. That is, although $u(x)$ varies throughout the domain, it does not vary at the endpoints, and $\delta u(x_0) = \delta u(x_1) = 0$. Furthermore, the function

$u(x)$, and thus its variation δu, are arbitrary; therefore, in order for the variation of the functional to vanish, that is, $\delta I = 0$, we must have

$$\frac{\partial F}{\partial u} - \frac{d}{dx}\left(\frac{\partial F}{\partial u'}\right) = 0,$$

which is Euler's equation (2.9) for $F = F(x,u,u')$.

While the derivation of Euler's equation in Section 2.1.2 provides a clearer understanding of the underlying principles through use of arguments from differential calculus, the derivation given in this section using variational notation is more concise. Moreover, it highlights the parallels with differential calculus of extrema of functions as follows:

- Taking the variation of the functional and setting it equal to zero, that is, $\delta I = 0$, is equivalent to differentiating $I[u]$ with respect to ϵ and setting equal to zero.
- The variation of $u(x)$, δu, plays the same role as $\eta(x)$ accounting for the variation from curve to curve.
- Taking $\delta I = 0$ is analogous to setting the total differential of a function equal to zero, that is, $df = 0$.

Mathematically, the function $u(x)$ for which the first variation is zero, that is, $\delta I[u] = 0$, is a *stationary function*. Evaluation of the second variation δ^2 (analogous to evaluating the second derivative, or curvature, of a function) would be required to determine whether it is also a minimum or maximum. Formally, we say that Euler's equation is a *necessary*, but not *sufficient*, condition for $u(x)$ to be an extremum of $I[u(x)]$. See Gelfand and Fomin (1963) and MacCluer (2005) for more on sufficiency.

It is typically not necessary to evaluate second variations in applications for one of two reasons. First, in certain applications the stationary function truly is a minimum, and it is not necessary to confirm that this is the case. Second, many applications are only concerned with finding stationary functions, and not necessarily extrema. This is the case, for example, in the application of Hamilton's principle to dynamical systems, which will be discussed in Section 4.1. Accordingly, many texts use the terms minimum, extremum, and stationary function loosely.

EXAMPLE 2.1 Determine the stationary function $u(x)$ for the functional

$$I[u] = \int_0^1 \left[(xu')^2 + 2u^2\right] dx \tag{2.13}$$

with the specified boundary conditions

$$u(0) = 0, \quad u(1) = 2.$$

Solution: Note that the functional $I[u]$ is of the form (2.1) with $F(x,u,u') = (xu')^2 + 2u^2$. Evaluating the partial derivatives in the Euler equation, we have

$$\frac{\partial F}{\partial u} = 4u, \quad \frac{\partial F}{\partial u'} = 2x^2u'.$$

Substituting into the Euler equation (2.9) yields

$$\frac{1}{2}\frac{\partial F}{\partial u} - \frac{d}{dx}\left(\frac{\partial F}{\partial u'}\right) = 0$$

$$4u - \frac{d}{dx}\left(2x^2u'\right) = 0$$

$$2u - 2xu' - x^2u'' = 0$$

$$x^2u'' + 2xu' - 2u = 0.$$

This is a *Cauchy-Euler* ordinary differential equation[3], which has general solutions of the form $u(x) = x^m$. Substituting and solving the resulting algebraic equation for m gives

$$x^2m(m-1)x^{m-2} + 2xmx^{m-1} - 2x^m = 0$$

$$m(m-1) + 2m - 2 = 0$$

$$m^2 + m - 2 = 0$$

$$(m+2)(m-1) = 0$$

$$m = -2, 1.$$

As a result, the general solution of the Euler equation is

$$u(x) = c_1x + \frac{c_2}{x^2},$$

where c_1 and c_2 are constants of integration to be found using the boundary conditions. The boundary conditions require that $c_1 = 2$ and $c_2 = 0$, in which case the solution to the Euler equation becomes

$$u(x) = 2x. \tag{2.14}$$

This is the stationary function of the functional (2.13). Recall that for each function $u(x)$, the functional (2.13) takes on a constant value resulting from evaluation of the definite integral. For the stationary function (2.14), the value of the functional is

$$\begin{aligned} I[u] &= \int_0^1 \left[(xu')^2 + 2u^2\right] dx \\ &= \int_0^1 \left[(2x)^2 + 8x^2\right] dx \\ &= 12\int_0^1 x^2 dx \\ &= 4x^3\Big|_0^1 \\ I[u] &= 4. \end{aligned}$$

[3] Various authors refer to such ordinary differential equations as *Euler, Cauchy-Euler, Euler-Cauchy*, or *equidimensional* equations.

If (2.14) is truly a minimum, then all other functions $u(x)$ would produce a larger value for the functional $I[u]$. For example, if we take $u = 2x^2$, which satisfies the boundary conditions, the functional results in $I[u] = 24/5$, which is larger than the value obtained for the stationary function.

The previous example illustrates the typical procedure for solving variational problems:

1. Formulate the functional (variational) form of the problem (provided in the previous example).
2. Obtain the equivalent differential Euler equation.
3. Solve the differential equation to obtain the stationary function of the functional.

2.1.4 Special Cases of Euler's Equation

In this section we consider two common special cases of Euler's equation, in which the process of obtaining the stationary function sometimes can be streamlined as compared to the general procedure illustrated in the previous example. Recall that the Euler equation for $F(x,u,u')$ is

$$\frac{\partial F}{\partial u} - \frac{d}{dx}\left(\frac{\partial F}{\partial u'}\right) = 0.$$

Special Cases:

I) If the integrand does not explicitly depend on the *dependent variable* $u(x)$, in which case $F = F(x,u')$, then $\partial F/\partial u = 0$, and integration with respect to x yields

$$\frac{\partial F}{\partial u'} = c, \tag{2.15}$$

where c is a constant of integration.

II) If the integrand does not explicitly depend on the *independent variable* x, in which case $F = F(u,u')$, then we can interchange the roles of the independent and dependent variables as follows

$$\begin{aligned} I[u(x)] &= \int_{x_0}^{x_1} F(u,u')dx \\ &= \int_{u_0}^{u_1} F\left[u, \left(\frac{dx}{du}\right)^{-1}\right]\frac{dx}{du}du \\ &= \int_{u_0}^{u_1} F\left[u, (x')^{-1}\right]x'du \\ I[x(u)] &= \int_{u_0}^{u_1} \bar{F}(u,x')du, \end{aligned}$$

where

$$\bar{F}(u,x') = x'F\left[u, (x')^{-1}\right]. \tag{2.16}$$

Because $x(u)$ is missing from $\bar{F}$, we now have a case similar to special case I for which the Euler equation is $(\partial\bar{F}/\partial x = 0)$

$$-\frac{d}{du}\left(\frac{\partial\bar{F}}{\partial x'}\right) = 0.$$

Integrating with respect to u gives

$$\frac{\partial\bar{F}}{\partial x'} = c. \tag{2.17}$$

Observe that in the integrand F, we interchange du/dx with $(dx/du)^{-1}$.

These special cases are sometimes called *first integrals* and correspond to the situations in which *reduction of order* may be used to convert the second-order, ordinary differential Euler equation to first order. They are applied directly to the general form of the Euler equation, however, rather than the second-order form. See the soap film problem in Example 2.3 for an illustration of both methods carried out for the same problem. Interestingly, this mathematical "trick" that sometimes simplifies solutions of these special cases can provide clues into the physics of a problem by exposing conservation laws. This will be taken up in Section 4.7 within the context of mechanics.

Alternatively, we can treat special case II by considering the Euler equation directly. Observe that for $F(x,u,u')$

$$\frac{dF}{dx} = \frac{\partial F}{\partial x} + \frac{\partial F}{\partial u}\frac{du}{dx} + \frac{\partial F}{\partial u'}\frac{d^2u}{dx^2}.$$

Thus,

$$\frac{\partial F}{\partial u}\frac{du}{dx} + \frac{\partial F}{\partial u'}\frac{d^2u}{dx^2} = \frac{dF}{dx} - \frac{\partial F}{\partial x}. \tag{2.18}$$

Let us take the Euler equation, which is

$$\frac{\partial F}{\partial u} - \frac{d}{dx}\left(\frac{\partial F}{\partial u'}\right) = 0,$$

and multiply by du/dx to obtain

$$\frac{\partial F}{\partial u}\frac{du}{dx} - \frac{d}{dx}\left(\frac{\partial F}{\partial u'}\frac{du}{dx}\right) + \frac{\partial F}{\partial u'}\frac{d^2u}{dx^2} = 0.$$

Substituting equation (2.18) for the first and third terms leads to the following form of the Euler equation

$$\frac{dF}{dx} - \frac{\partial F}{\partial x} - \frac{d}{dx}\left(\frac{\partial F}{\partial u'}\frac{du}{dx}\right) = 0.$$

In special case II, in which $F(u,u')$ does not explicitly depend on the independent variable x, $\partial F/\partial x = 0$, and we have

$$\frac{dF}{dx} - \frac{d}{dx}\left(\frac{\partial F}{\partial u'}\frac{du}{dx}\right) = 0.$$

Integrating with respect to x leads to the first integral

$$F - \frac{\partial F}{\partial u'}\frac{du}{dx} = c, \qquad (2.19)$$

where c is the constant of integration. It can readily be shown that the resulting form of the Euler equation (2.19) is equivalent to (2.17) with (2.16). While this approach is conceptually easier to apply as it does not require interchanging the roles of the independent and dependent variables, the first approach is more concise and more straightforward to remember. Moreover, it more clearly follows the reduction of order technique applied to differential equations.

We know intuitively that the shortest distance between two points is a straight line. For example, we used this fact in the optics example in Section 1.3.1. However, the reader has likely never seen a formal mathematical proof of this universally accepted fact. The first example provides such a proof and illustrates application of special case I. The subsequent soap film problem illustrates special case II.

EXAMPLE 2.2 What plane curve connecting the two points (x_0, u_0) and (x_1, u_1) has the shortest length?

Solution: We want to find the curve $u(x)$ shown in Figure 2.2 with $u(x_0) = u_0, u(x_1) = u_1$, for which the length is a minimum. As in Chapter 1, the arc length of a differential portion of the curve (see Figure 2.3) is

$$ds = \sqrt{dx^2 + du^2}, \quad du = \frac{du}{dx}dx;$$

therefore,

$$ds = \sqrt{1 + u'(x)^2}dx.$$

We define a functional $I[u]$ to be the length of the curve $u(x)$, and integrating from x_0 to x_1 gives

$$I[u] = \int_{x_0}^{x_1} ds = \int_{x_0}^{x_1} \left[1 + (u')^2\right]^{1/2} dx; \qquad (2.20)$$

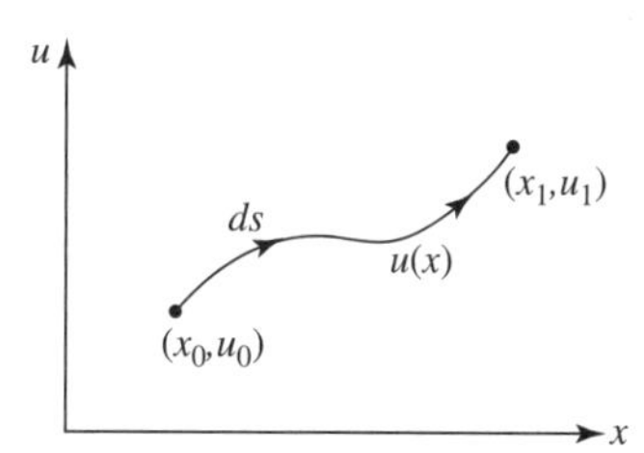

Figure 2.2. Plane curve.

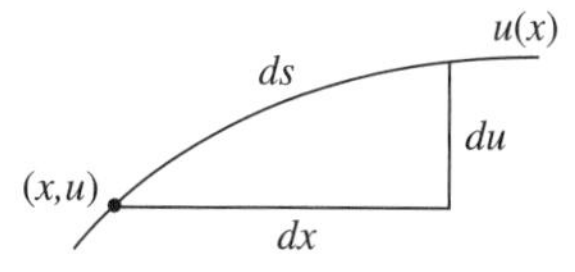

Figure 2.3. Differential arc length.

therefore, the integrand is

$$F(u') = \left[1 + (u')^2\right]^{1/2}.$$

We seek the curve $u(x)$ that minimizes the length functional $I[u]$. Observe that the integrand does not involve the dependent variable $u(x)$ explicitly (nor does it involve the independent variable x); therefore, this corresponds to special case I.

To evaluate the Euler equation, we need

$$\frac{\partial F}{\partial u} = 0, \quad \frac{\partial F}{\partial u'} = \frac{1}{2}\left[1 + (u')^2\right]^{-1/2}(2u').$$

Substituting into the Euler equation

$$\frac{\partial F}{\partial u} - \frac{d}{dx}\left(\frac{\partial F}{\partial u'}\right) = 0,$$

results in

$$-\frac{d}{dx}\left\{u'\left[1 + (u')^2\right]^{-1/2}\right\} = 0.$$

Integrating and solving for u' leads to

$$\begin{aligned}
u'\left[1 + (u')^2\right]^{-1/2} &= \bar{c} \\
u' &= \bar{c}\left[1 + (u')^2\right]^{1/2} \\
(u')^2 &= \bar{c}^2\left[1 + (u')^2\right] \\
(1 - \bar{c}^2)(u')^2 &= \bar{c}^2 \\
(u')^2 &= \hat{c} \\
u' &= c_1.
\end{aligned}$$

Integrating again, we have

$$u(x) = c_1 x + c_2,$$

which as expected is a straight line. Applying the boundary conditions $u(x_0) = u_0$ and $u(x_1) = u_1$ leads to the solution

$$u(x) = \frac{u_1 - u_0}{x_1 - x_0}(x - x_0) + u_0.$$

We could obtain the length of this curve by substituting $u(x)$ into the functional (2.20) and evaluating the definite integral.

Although it may be stretching the point somewhat, we can go so far as to say that the fundamental building blocks of Euclidean geometry are based on variational principles:

- straight line → shortest distance between two points.
- circle → shortest curve enclosing a given area.
- sphere → smallest area enclosing a given volume.
- isosceles triangle → a triangle having the shortest perimeter with a given base line and area.

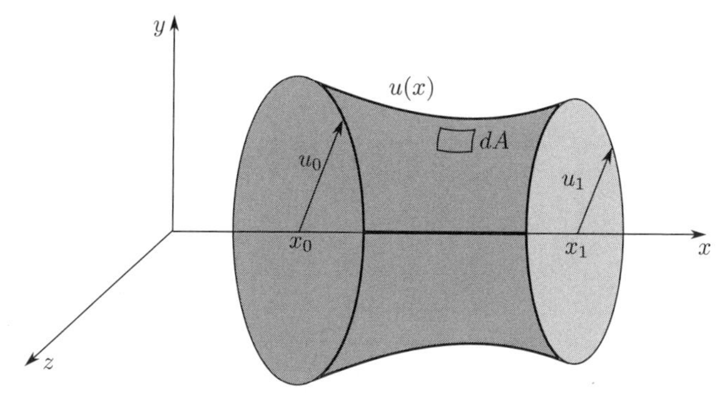

Figure 2.4. Soap film between two coaxial circular rings.

The above result can be extended to the case of finding the shortest paths on curved surfaces, which are called *geodesics*. First used to find the shortest path between two points on the spherical Earth, geodesics are an important topic in *differential geometry* along with *minimal surfaces* to be considered in Section 2.5.2 and the following example. Geodesics also play a role in general relativity, where they are determined along the curved spacetime relating matter and energy (see Section 8.1.2).

EXAMPLE 2.3 Determine the shape of the soap film spanning the two coaxial circular rings shown in Figure 2.4.

Solution: The soap film takes on the shape that minimizes its total surface energy, which is directly proportional to its surface area. Because the surface of revolution defining the soap film shape is axisymmetric for coaxial circular rings, the surface area to be minimized is given by

$$I[u] = \iint_A dA = \int_{x_0}^{x_1} 2\pi u(x) ds = 2\pi \int_{x_0}^{x_1} u\sqrt{1+(u')^2}dx, \tag{2.21}$$

as for the liquid drop considered in Section 1.3.2. Here, $u(x)$ is the radial distance from the x-axis to the soap film surface. The integrand is

$$F(u,u') = 2\pi u\sqrt{1+(u')^2},$$

in which the independent variable x does not appear explicitly; therefore, we have special case II. We form the new integrand

$$\bar{F}(u,x') = x'F\left[u,(x')^{-1}\right] = 2\pi x'u\sqrt{1+(x')^{-2}} = 2\pi u\sqrt{(x')^2+1},$$

in which the role of the independent and dependent variables have been interchanged. The Euler equation (2.17) for this case is

$$\frac{\partial \bar{F}}{\partial x'} = c,$$

which yields

$$2\pi u\left\{\frac{1}{2}\left[(x')^2+1\right]^{-1/2}(2x')\right\} = 2\pi c_1.$$

Solving for x' yields

$$\frac{ux'}{\sqrt{(x')^2+1}} = c_1$$
$$u^2(x')^2 = c_1^2\left[(x')^2+1\right]$$
$$(u^2-c_1^2)(x')^2 = c_1^2$$
$$x' = \frac{dx}{du} = \frac{c_1}{\sqrt{u^2-c_1^2}},$$

or

$$\frac{du}{dx} = \frac{1}{c_1}\sqrt{u^2-c_1^2}. \tag{2.22}$$

Separating variables and integrating gives

$$c_1\int\frac{du}{\sqrt{u^2-c_1^2}} = \int dx = x+c_2.$$

This integral may be evaluated using the substitution

$$u = c_1\cosh\theta, \tag{2.23}$$

for which

$$du = c_1\sinh\theta\, d\theta.$$

Then the integral becomes

$$c_1\int\frac{\sinh\theta}{\sqrt{\cosh^2\theta-1}}d\theta = x+c_2$$
$$\int d\theta = \frac{x+c_2}{c_1}$$
$$\theta = \frac{x+c_2}{c_1}.$$

Substituting for θ in (2.23) yields

$$u(x) = c_1\cosh\left(\frac{x+c_2}{c_1}\right),$$

which is known as a *catenary*.

The constants c_1 and c_2 are obtained from the end conditions $u(x_0) = u_0$ and $u(x_1) = u_1$. For example, consider the case when $(x_0,u_0) = (-1,2)$ and $(x_1,u_1) = (1,2)$. Then

$$2 = c_1\cosh\left(\frac{-1+c_2}{c_1}\right), \quad 2 = c_1\cosh\left(\frac{1+c_2}{c_1}\right). \tag{2.24}$$

Equating requires that

$$\cosh\left(\frac{-1+c_2}{c_1}\right) = \cosh\left(\frac{1+c_2}{c_1}\right),$$

which is only true if $c_2 = 0$. To obtain the constant c_1, consider the second equation of (2.24)

$$c_1 \cosh\left(\frac{1}{c_1}\right) = 2.$$

This equation cannot be solved exactly for c_1; therefore, a numerical root-finding technique must be used. For ring radius $u_0 < 1.5$ there are no roots, that is, there is no solution to the soap film problem. For $u_0 = 1.5$, there is a single root, and for $u_0 > 1.5$, there are two roots. We are interested in the root corresponding to a minimum in surface area of the soap film. Let us consider the case with ring radius $u_0 = 2$. The first root for this case is $c_1 = 0.47019$, and the corresponding surface area of the catenoid is $A = 27.3826$. The second root is $c_1 = 1.69668$, which has a surface area $A = 23.9672$. Therefore, the catenoid corresponding to the second root has a smaller surface area and represents the global minimum. Thus, the catenoid shape of the soap film is given by the catenary

$$u(x) = c_1 \cosh\left(\frac{x}{c_1}\right), \quad c_1 = 1.69668,$$

revolved around the x-axis.

Physically, we would expect that another possible solution to the soap film problem would simply consist of two separate soap films in the shape of circular disks spanning each ring individually. If this were the case, the total soap film area of the two circular disks for $u_0 = 2$ would be $A = 2\pi u_0^2 = 25.1327$. Observe that this area is larger than the catenoid shape computed previously. For some ring radii, however, the two-disk solution would result in a smaller total soap film area than the corresponding catenoid for that radius. Numerical experimentation shows that this occurs for $u_0 < 1.895$, which is larger than the lower limit of $u_0 = 1.5$ for which catenary solutions exist. Accordingly, there is a range of radii $1.5 < u_0 < 1.895$ for which the solution obtained using the calculus of variations is not the physically observed solution. This is because the physically observed solution is not a function; hence, it is not an admissible solution of the variational problem. The two-disk solution is known as the *Goldschmidt discontinuous solution*.

Alternatively, let us consider the solution without application of special case II for comparison. Recall that the integrand is

$$F(x,u,u') = 2\pi u\sqrt{1+(u')^2}.$$

Evaluating the terms in the general form of the Euler equation gives

$$\begin{aligned}
\frac{\partial F}{\partial u} &= 2\pi\sqrt{1+(u')^2}, \\
\frac{\partial F}{\partial u'} &= 2\pi u\left[1+(u')^2\right]^{-1/2} u'.
\end{aligned}$$

Then Euler's equation is

$$\frac{\partial F}{\partial u} - \frac{d}{dx}\left(\frac{\partial F}{\partial u'}\right) = 0,$$

which yields

$$\left[1+(u')^2\right]^{1/2} - \frac{d}{dx}\left\{u\left[1+(u')^2\right]^{-1/2} u'\right\} = 0.$$

Performing the differentiation and simplifying, which is rather involved, eventually yields

$$uu'' - (u')^2 = 1.$$

Consequently, the Euler equation is a nonlinear, second-order, ordinary differential equation for $u(x)$. Because the independent variable x does not appear explicitly in the differential equation, the ordinary differential equation may be solved using *reduction of order*. Let $p = u'$, then

$$u'' = \frac{dp}{dx} = \frac{dp}{du}\frac{du}{dx} = p\frac{dp}{du},$$

and the Euler equation transforms to

$$up\frac{dp}{du} - p^2 = 1,$$

which is a first-order ordinary differential equation for $p(u)$, in which u is now the independent variable. Separating variables and integrating leads to

$$u\frac{dp}{du} = \frac{p^2+1}{p}$$

$$\int \frac{p}{p^2+1}dp = \int \frac{du}{u}$$

$$\frac{1}{2}\ln(p^2+1) = \ln u + \bar{c}_3$$

$$(p^2+1)^{1/2} = \frac{1}{c_3}u$$

$$p^2 = \frac{1}{c_3^2}(u^2 - c_3^2)$$

$$p = \frac{du}{dx} = \frac{1}{c_3}\sqrt{u^2 - c_3^2},$$

which is the same as equation (2.22) but after significantly more exertion. We will revisit this problem as the “hanging cable problem” in Section 2.7.1.

Note that it is never necessary to apply the special cases when solving problems. In fact, in some cases it is more straightforward to apply the usual approach as in the previous section. However, as illustrated in the soap film problem, a significant amount of effort can sometimes be avoided by applying the special cases. It is recommended that one first try the usual approach. If difficulties arise in obtaining the Euler equation, consider whether one of the special cases is relevant. Observe that using the special cases bypasses obtaining the typical second-order differential form of the Euler equation, which is sometimes desired (e.g., when obtaining the equations of motion of dynamical systems).

2.2 Natural Boundary Conditions

Thus far we have encountered situations in which the values of $u(x)$ at the end points of the sought after stationary function are known. In some applications of calculus

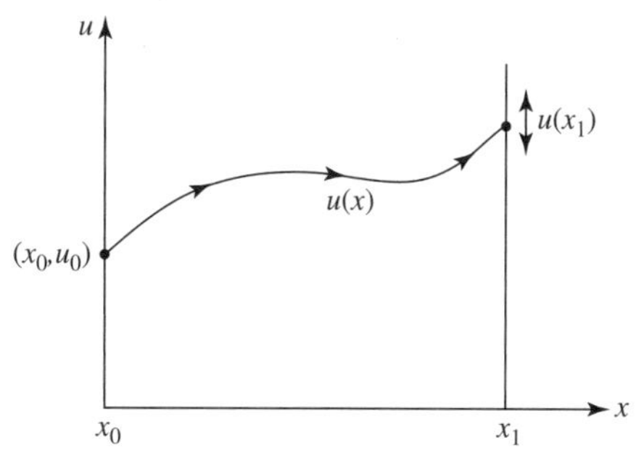

Figure 2.5. River-crossing trajectory with unspecified location at $x = x_1$.

of variations, however, the locations of the end points x_0 and x_1 are specified, but the values of the function at the end points $u(x_0)$ and $u(x_1)$ are not.

Consider, for example, the optimal river-crossing trajectory treated in Section 1.3.3, in which the problem is to determine the minimum time required for a boat to cross a river. In our previous treatment, we imagine that the target point on the right bank of the river is specified. However, what if we do not specify the value $u(x_1)$ of the end point, but simply minimize the time required to cross the river regardless of the end point location as shown in Figure 2.5? The boat leaves the shore at $x = x_0$ at $u(x_0) = u_0$. Given a particular current velocity distribution across the river, what is the path $u(x)$ that minimizes the crossing time to reach $x = x_1$ without regard for the location $u(x_1)$? No longer having a specified boundary condition at $x = x_1$, what should be the boundary condition?

In order to determine the appropriate boundary conditions if $u(x_0)$ and/or $u(x_1)$ are not specified, recall equation (2.12) from the derivation of Euler's equation

$$\left[\frac{\partial F}{\partial u'}\delta u\right]_{x_0}^{x_1} + \int_{x_0}^{x_1}\left[\frac{\partial F}{\partial u} - \frac{d}{dx}\left(\frac{\partial F}{\partial u'}\right)\right]\delta u dx = 0.$$

In the previous case, $u(x)$ was fixed at the end points; consequently, $u(x)$ did not vary at x_0 and x_1 and $\delta u|_{x_0}^{x_1} = 0$, eliminating the first term on the left-hand side. Here, the end points are not fixed and may vary. Because the first term is only evaluated at the end points, however, it cannot in general cancel with the integral term, which is evaluated over the entire domain. Therefore, the integral must still vanish independent of the first term, and we are left with an Euler equation of the same form as in the case with specified boundary conditions. This leaves

$$\left[\frac{\partial F}{\partial u'}\delta u\right]_{x_0}^{x_1} = 0.$$

If the end points are not specified, in which case $\delta u|_{x_0}^{x_1} \neq 0$, this can only occur if

$$\left.\frac{\partial F}{\partial u'}\right|_{x_0} = 0, \quad \text{and} \quad \left.\frac{\partial F}{\partial u'}\right|_{x_1} = 0. \tag{2.25}$$

These are the *natural boundary conditions* for the case with $F = F(x,u,u')$ that replace $u(x_0) = u_0$ and $u(x_1) = u_1$. Observe that the boundary conditions arise "naturally" from the equation if the end points are not specified.

EXAMPLE 2.4 Determine the stationary function $u(x)$ for the functional

$$I[u] = \frac{1}{2}\int_0^{\pi/2}\left[-(u')^2 + u^2 + 2ux\right]dx$$

over the interval $0 \le x \le \pi/2$, where $u(0)$ and $u(\pi/2)$ are not specified.
Solution: Evaluating the terms in Euler's equation for $F = F(x,u,u')$ yields

$$\frac{\partial F}{\partial u} = u + x, \quad \frac{\partial F}{\partial u'} = -u'.$$

Then from Euler's equation

$$\frac{\partial F}{\partial u} - \frac{d}{dx}\left(\frac{\partial F}{\partial u'}\right) = 0$$

$$u + x - \frac{d}{dx}(-u') = 0$$

$$u'' + u = -x,$$

which is a second-order, constant coefficient, nonhomogeneous ordinary differential equation for $u(x)$. Applying the natural boundary conditions gives

$$\left.\frac{\partial F}{\partial u'}\right|_{x=0} = u'(0) = 0,$$

$$\left.\frac{\partial F}{\partial u'}\right|_{x=\pi/2} = u'(\pi/2) = 0,$$

which are the boundary conditions for the ordinary differential equation for $u(x)$.
First consider the homogeneous equation

$$u_H'' + u_H = 0,$$

which has a solution of the form $u_H = e^{rx}$, where $r = \pm i$. Thus, the homogeneous solution is

$$u_H(x) = c_1 \sin x + c_2 \cos x.$$

By inspection (or by the method of undetermined coefficients), a particular solution to

$$u_P'' + u_P = -x$$

is

$$u_P(x) = -x.$$

As a result, the general solution is

$$u(x) = u_H(x) + u_P(x) = c_1 \sin x + c_2 \cos x - x.$$

Applying the natural boundary conditions, where

$$u'(x) = c_1 \cos x - c_2 \sin x - 1,$$

gives $c_1 = 1$ and $c_2 = -1$. Hence, the solution is

$$u(x) = \sin x - \cos x - x.$$

Natural boundary conditions typically produce derivative boundary conditions for Euler's differential equation, as in the previous example. Derivative boundary conditions are often referred to as Neumann, or nonessential, boundary conditions. Accordingly, in various contexts, the terms *natural*, *nonessential*, *derivative*, or *Neumann* boundary conditions may be used interchangeably.

It is often asserted that the *brachistochrone problem* is the one that initiated development of the calculus of variations. Although this fact is disputed,[4] and branches of mathematics rarely have such clearly identifiable beginnings, the brachistochrone problem does have an interesting history dating back to the famous Bernoulli family and involving some of the biggest names in mathematics at the time. The problem was first solved by Johann Bernoulli who published it without solution in 1696 as a challenge to the mathematical community. The first to provide a solution to Bernoulli was Gottfried Wilhelm Leibniz, the inventor of differential calculus, who sent a solution to Bernoulli within a week of Bernoulli's personal letter to Leibniz. A second solution was provided by Isaac Newton, who invented differential calculus independent of Leibniz, allegedly within twelve hours of Newton's delayed discovery of the problem. However, he communicated his solution anonymously to the Royal Society. Johann's older brother, Jacob Bernoulli, and Guillaume de l'Hôpital also provided correct solutions to the problem. These early solutions appealed to geometric arguments but laid the groundwork for later developments in variational calculus by Euler and Lagrange, which we now follow. Newton and the Bernoulli brothers provided the seminal ideas that would become the calculus of variations in the late 1600s, and Euler and Lagrange formalized this new branch of mathematics one-half century later. The brachistochrone problem is illustrated in the following example.

EXAMPLE 2.5 The brachistochrone problem is to determine the path $u(x)$ along which a particle of mass m should fall under the action of gravity such that the time of travel is a minimum (brachistochrone is Greek for "shortest time"). This is shown schematically in Figure 2.6. The mass starts at rest at the origin $(x_0, u_0) = (0, 0)$ and slides along a frictionless wire with shape $u(x)$ until it reaches the vertical line $x_1 = b$ in the shortest time. We seek the wire shape $u(x)$.

Solution: Similar to application of Fermat's principle in Section 1.3.1, the functional to be minimized is

$$T[u(x)] = \int_{x_0}^{x_1} \frac{ds}{v}, \tag{2.26}$$

where $v(x)$ is the velocity of the mass. Physically, there is a competition between two effects: 1) as we saw in Section 2.1.4, the shortest distance between two points is a straight line, and 2) the acceleration due to gravity has its maximum influence along a vertical path. Thus, the solution will be some combination of these two effects.

[4] For example, Isaac Newton solved a minimum drag problem approximately ten years earlier using what would later become known as variational calculus. See Section 10.2.

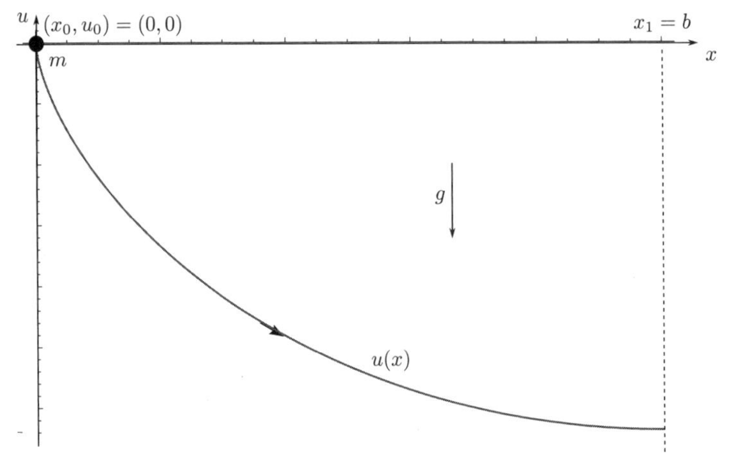

Figure 2.6. Schematic of the brachistochrone problem.

Because there is no friction, the velocity may be found from a balance of kinetic energy, E_k, and potential energy, E_p, as follows

$$E = E_k + E_p = \frac{1}{2}mv^2 + mgu,$$

where E is the total energy, which is a constant. The particle starts at rest ($v = 0$) at the origin ($u = 0$); therefore, the total energy is $E = 0$. Solving for v gives

$$v = \sqrt{-2gu},$$

where $u < 0$. In addition, recall that the differential arc length is

$$ds = \sqrt{1 + (u')^2}dx.$$

Then the functional (2.26) may be written as

$$T[u(x)] = \int_0^b \left[\frac{1 + (u')^2}{-2gu}\right]^{1/2} dx. \tag{2.27}$$

Observe that the integrand of the functional (2.27) is

$$F(u,u') = \frac{1}{\sqrt{2g}}\left[1 + (u')^2\right]^{1/2}(-u)^{-1/2},$$

which does not explicitly include the independent variable x. Therefore, we can apply special case II.

The boundary conditions are

$$u = 0 \quad \text{at} \quad x = 0, \tag{2.28}$$

and a natural boundary condition at $x = b$

$$\frac{\partial F}{\partial u'} = 0 \quad \text{at} \quad x = b.$$

Substituting F into the natural boundary condition yields

$$\frac{1}{2}\frac{1}{\sqrt{2g}}\left[1+(u')^2\right]^{-1/2}(2u')(-u)^{-1/2}=0,$$

or

$$\frac{u'}{\sqrt{-u}\sqrt{1+(u')^2}}=0 \quad \text{at} \quad x=b.$$

For finite u, this requires that

$$u'=0 \quad \text{at} \quad x=b. \tag{2.29}$$

That is, the curve must be normal to its terminal line $x=b$.

Returning to application of special case II, we interchange u' with $(x')^{-1}$ to give

$$\begin{aligned}\bar{F}(u,x') &= x'F\left[u,(x')^{-1}\right]\\ &= \frac{x'}{\sqrt{2g}}\left[1+(x')^{-2}\right]^{1/2}(-u)^{-1/2}\\ \bar{F}(u,x') &= \frac{1}{\sqrt{2g}}\left[(x')^2+1\right]^{1/2}(-u)^{-1/2}.\end{aligned}$$

Recall that u is now the independent variable, and x is the dependent variable. Then

$$\frac{\partial\bar{F}}{\partial x'}=\frac{1}{2\sqrt{2g}}\left[(x')^2+1\right]^{-1/2}\left[2x'\right](-u)^{-1/2}=\frac{1}{\sqrt{2g}}\frac{x'}{\sqrt{-u[(x')^2+1]}},$$

and from (2.17), the Euler equation for this special case is

$$\frac{\partial\bar{F}}{\partial x'}=c,$$

or

$$\frac{x'}{\sqrt{(-u)\left[(x')^2+1\right]}}=c_1.$$

Solving for x' yields (choosing the negative in the square root to obtain the proper slope)

$$x'=\frac{dx}{du}=-\sqrt{\frac{-c_1^2u}{1+c_1^2u}},$$

and integrating results in

$$x(u)=-\int\sqrt{\frac{-c_1^2u}{1+c_1^2u}}du.$$

In order to evaluate this integral, we make the following substitution from the independent variable u to θ

$$u=-\frac{1}{c_1^2}\sin^2\theta, \tag{2.30}$$

for which

$$du=-\frac{2}{c_1^2}\sin\theta\cos\theta d\theta.$$

Upon substituting and integrating we obtain

$$x = \frac{2}{c_1^2}\int\sqrt{\frac{\sin^2\theta}{1-\sin^2\theta}}\sin\theta\cos\theta\,d\theta$$

$$= \frac{2}{c_1^2}\int\sqrt{\frac{\sin^2\theta}{\cos^2\theta}}\sin\theta\cos\theta\,d\theta$$

$$= \frac{2}{c_1^2}\int\sin^2\theta\,d\theta$$

$$= \frac{1}{c_1^2}\int[1-\cos(2\theta)]d\theta$$

$$x = \frac{1}{c_1^2}\left[\theta - \frac{1}{2}\sin(2\theta)\right] + c_2.$$

As in the previous example, we would normally solve this expression for θ in terms of x and substitute into (2.30) to obtain the solution $u(x)$. Because this is not possible in the present case, we instead write the solution in parametric form. Recall the substitution (2.30), which can be written in the form

$$u = -\frac{1}{2c_1^2}[1-\cos(2\theta)]. \tag{2.31}$$

These parametric equations may be simplified slightly by making the substitution $\phi = 2\theta$, which leads to

$$x = \frac{1}{2c_1^2}(\phi - \sin\phi) + c_2, \tag{2.32}$$

$$u = -\frac{1}{2c_1^2}(1-\cos\phi). \tag{2.33}$$

The constants c_1 and c_2 are obtained by applying the boundary condition $(x_0, u_0) = (0,0)$ and the natural boundary condition (2.29)

$$u' = 0 \quad \text{at} \quad x = b.$$

Consider equation (2.33) with $(x_0, u_0) = (0,0)$

$$0 = -\frac{1}{2c_1^2}(1-\cos\phi_0),$$

which is satisfied if $\phi_0 = 0$. Then from equation (2.32) applied at $x_0 = 0$ with $u_0 = 0$ and $\phi_0 = 0$, we find that $c_2 = 0$. Now consider the natural boundary condition at $x_1 = b$. Differentiating (2.33) gives

$$u' = -\frac{1}{2c_1^2}\sin\phi\frac{d\phi}{dx}.$$

Applying this result at $x_1 = b$, we have that $\sin\phi_1 = 0$, which is true for $\phi_1 = \pi$. Then from equation (2.32)

$$b = \frac{1}{2c_1^2}[\pi - \sin\pi],$$

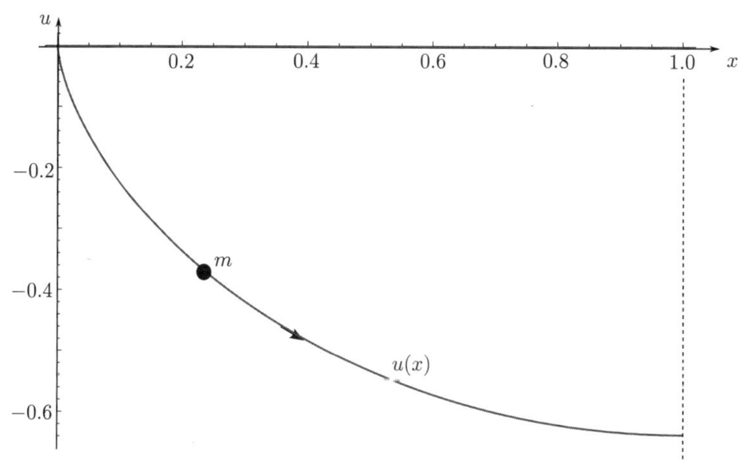

Figure 2.7. Solution of the brachistochrone problem with $b = 1$.

which leads to

$$2c_1^2 = \frac{\pi}{b}.$$

As a result, the solution (2.32) and (2.33) becomes

$$x = \frac{b}{\pi}(\phi - \sin\phi), \tag{2.34}$$

$$u = -\frac{b}{\pi}(1 - \cos\phi), \tag{2.35}$$

where $\phi_0 \leq \phi \leq \phi_1$ or $0 \leq \phi \leq \pi$. Equations (2.34) and (2.35) are the parametric equations for a *cycloid*, which is shown in Figure 2.7. A point on the edge of a circular disk rolling over a flat surface traces out a cycloid. One of the intriguing characteristics of the cycloid is that when two particles are let go from two different points on the curve, they will reach the terminal point of the curve at the same time. This is known as the *tautochrone*, Greek for "same time," problem.

The following is an example of a *dual functional*, in which the functional contains values of the function $u(x)$ evaluated at the end points in addition to the definite integral over the entire domain. We will encounter such dual functionals in Sections 6.5, 9.5.4, and 10.4.1 in the context of transient growth stability analysis and optimal control.

EXAMPLE 2.6 Determine the stationary function $u(x)$ for

$$I[u] = \frac{1}{2}[u(1)]^2 + \frac{1}{2}\int_0^1 (u')^2 dx, \quad u(0) = 1. \tag{2.36}$$

Note that $u(1)$ is not specified, but rather $[u(1)]^2$ is to be minimized along with the definite integral. With respect to the river-crossing example in Section 1.3.3, for example, this would correspond to seeking to minimize the travel time across the river *and* the square of the distance (positive or negative does not matter) to one's cabin on the right bank at $x = 1$.

Solution: In general, we must consider such dual functionals by starting with the variational form as in Section 2.1.3 in order to obtain the natural boundary condition that results. That is, we cannot use the Euler equation along with the result (2.25) for the natural boundary conditions. Taking $\delta I = 0$ gives

$$\delta I = u(1)\delta u(1) + \int_0^1 u'\delta u' dx = 0,$$

where we note that $u(1)$ does vary in this context. Integrating by parts ($p = u', dp = u'' dx, q = \delta u, dq = \delta u' dx$) leads to

$$\Big[u'\delta u\Big]_0^1 + u(1)\delta u(1) - \int_0^1 u''\delta u dx = 0,$$

or

$$u'(1)\delta u(1) - u'(0)\cancelto{0}{\delta u(0)} + u(1)\delta u(1) - \int_0^1 u''\delta u dx = 0,$$

where $\delta u(0) = 0$ because u is specified at $x = 0$.

From the integrand, the Euler equation is

$$u'' = 0. \tag{2.37}$$

To obtain the boundary conditions, consider the coefficients of the variations of $u(x)$ and its derivatives evaluated at the boundaries, which must each vanish:

$$\begin{aligned}
\delta u(0) &: \quad - \\
\delta u(1) &: \quad u'(1) + u(1) = 0 \\
\delta u'(0) &: \quad - \\
\delta u'(1) &: \quad - \\
&\vdots
\end{aligned}$$

As a result, the boundary conditions for (2.37) are

$$u(0) = 1, \quad u(1) + u'(1) = 0,$$

where the first is given, and the second is a natural boundary condition.

The solution to (2.37) is a straight line:

$$u(x) = c_1 x + c_2.$$

Application of the boundary condition $u(0) = 1$ requires that $c_2 = 1$, and the boundary condition

$$u(1) + u'(1) = 0$$

leads to $c_1 = -\frac{1}{2}$. Hence, the stationary function is

$$u(x) = -\frac{1}{2}x + 1.$$

We could substitute $u(x)$ into (2.36) and integrate to find the corresponding stationary value of the functional $I[u]$. Note that although the Euler equation is the same as the case without the $[u(1)]^2$ term in the functional, it is necessary to solve such dual functional problems directly from the variational form in the above manner to obtain any natural boundary conditions.

In summary, we have considered the following two types of end conditions. First, the case in which both the end points of the domain and their boundary values are specified has been considered in Section 2.1. Second, we have considered natural boundary conditions in which the end points are specified but the boundary values are not. Although not common, there is a third possibility in which the end points of the domain are not specified, but rather are taken to be along given curves. This would be the case in the river-crossing trajectory problem, for example, if the banks of the river are curved. The case of variable end points is considered in the next section.

2.3 Variable End Points

Typically, the boundaries of the domain, for example, x_0 and x_1, are known and the stationary function $u(x)$ is sought over the fixed interval $x_0 \le x \le x_1$. In some cases, however, the boundaries of the domain are not fixed and must be determined along with the stationary function. For example, consider determining the shortest distance between the two curves illustrated in Figure 2.8. We know that the stationary function $u(x)$ is a straight line; however, the end points x_0 and x_1 must be determined. An important application involving variable end points is in image processing segmentation in which an operation is to be applied to an object within an image, where the object has curved boundaries. Image processing is considered in Chapter 11.

Consider the functional of the form

$$I[u(x)] = \int_{x_0}^{x_1} F(x,u,u')dx, \tag{2.38}$$

where $u_0 = u(x_0) = f(x_0)$ and $u_1 = u(x_1) = g(x_1)$, such that the end points (x_0,u_0) and (x_1,u_1) lie on the curves $f(x)$ and $g(x)$, respectively. It is shown in Hildebrand (1965) (Section 2.8), and Gelfand and Fomin (1963) (Section 14), that the corresponding

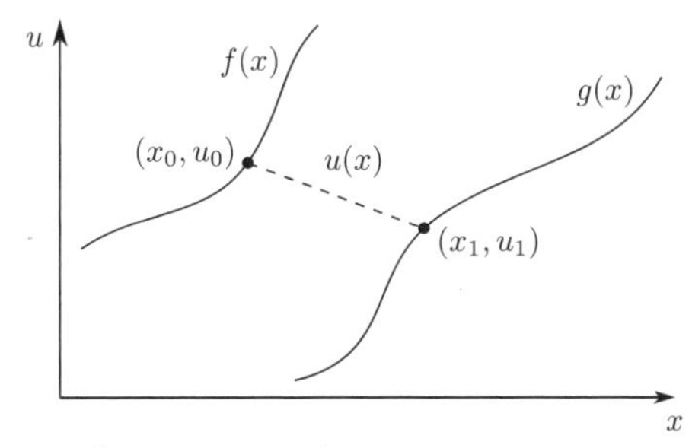

Figure 2.8. Shortest distance between two plane curves.

Euler equation is of the usual form

$$\frac{\partial F}{\partial u} - \frac{d}{dx}\left(\frac{\partial F}{\partial u'}\right) = 0, \tag{2.39}$$

but that the end conditions, called *transversality conditions*, become

$$F + (f' - u')\frac{\partial F}{\partial u'} = 0, \quad \text{at} \quad x = x_0, \tag{2.40}$$

$$F + (g' - u')\frac{\partial F}{\partial u'} = 0, \quad \text{at} \quad x = x_1. \tag{2.41}$$

EXAMPLE 2.7 Find the shortest distance between the line $f(x) = x - 3$ and the curve $g(x) = e^x$.

Solution: The length functional is

$$I[u] = \int_{x_0}^{x_1} \left(1 + u'^2\right)^{1/2} dx, \tag{2.42}$$

and differentiating the end conditions yields

$$f'(x) = 1, \quad g'(x) = e^x. \tag{2.43}$$

As in Example 2.2, the stationary function for (2.42) may be found to be the straight line

$$u(x) = c_1 x + c_2. \tag{2.44}$$

From the transversality condition (2.40) at x_0 with (2.43) and $u' = c_1$, we have

$$\left\{[1 + u'^2]^{1/2} + [1 - u'][(1/2)(1 + u'^2)^{-1/2}(2u')]\right\}_{x=x_0} = 0$$

$$(1 + c_1^2)^{1/2} + (1 - c_1)c_1(1 - c_1^2)^{-1/2} = 0$$

$$(1 + c_1^2) + (c_1 - c_1^2) = 0,$$

which requires that $c_1 = -1$. Similarly, from equation (2.41) at x_1 with (2.43) and $u' = c_1$, we obtain

$$\left\{[1 + u'^2]^{1/2} + [e^{x_1} - u'][u'(1 + u'^2)^{-1/2}]\right\}_{x=x_1} = 0$$

$$(1 + c_1^2)^{1/2} + (e^{x_1} - c_1)c_1(1 - c_1^2)^{-1/2} = 0$$

$$(1 + c_1^2) + (c_1 e^{x_1} - c_1^2) = 0.$$

This requires that $c_1 e^{x_1} = -1$. However, $c_1 = -1$; thus,

$$e^{x_1} = 1,$$

which requires that

$$x_1 = \ln 1 = 0.$$

Then from $g(x) = e^x$

$$u(x_1) = e^{x_1} = e^0 = 1,$$

which results in

$$(x_1, u_1) = (0, 1).$$

To determine the constant c_2, substitute this result into equation (2.44):

$$1 = -1 \cdot 0 + c_2,$$

which yields

$$c_2 = 1.$$

Thus, the shortest distance between the two curves is along the line

$$u(x) = -x + 1.$$

Observing that $(x_0, u_0) = (2, -1)$, the minimum distance between the two curves is

$$\begin{aligned} I[u] &= \int_2^0 \left[1 + u'^2\right]^{1/2} dx \\ &= -\int_0^2 \left[1 + (-1)^2\right]^{1/2} dx \\ &= -\sqrt{2} \int_0^2 dx \\ &= -\sqrt{2}x \Big|_0^2 \\ I[u] &= -2\sqrt{2}, \end{aligned}$$

which is the distance in the negative x-direction. Note that the same result is obtained by finding the distance between the points $(x_0, u_0) = (2, -1)$ and $(x_1, u_1) = (0, 1)$. This result is shown in Figure 2.9.

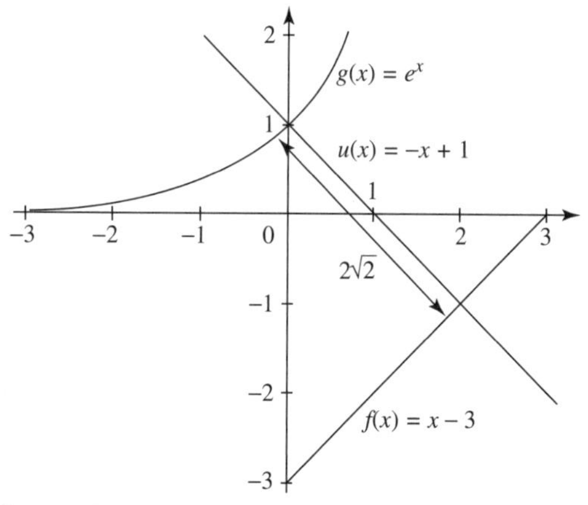

Figure 2.9. Shortest distance between two curves from Example 2.7.

2.4 Higher-Order Derivatives

Thus far, we have limited our considerations to functionals containing a single independent variable x, a single dependent variable $u(x)$, and its first derivative $u'(x)$. We may extend these techniques to include functionals containing more independent variables, more dependent variables, and higher-order derivatives within the integrand. The first two extensions will be considered in subsequent sections. As an example of extension to higher derivatives, consider a functional of the form

$$I[u] = \int_{x_0}^{x_1} F(x,u,u',u'')dx. \tag{2.45}$$

Observe that the integrand $F = F(x,u,u',u'')$ now includes $u''(x)$ (see, for example, Section 5.3.2 for the beam equation).

The derivation of the Euler equation and natural boundary conditions are left as an exercise. The Euler equation corresponding to equation (2.45) is

$$\frac{\partial F}{\partial u} - \frac{d}{dx}\left(\frac{\partial F}{\partial u'}\right) + \frac{d^2}{dx^2}\left(\frac{\partial F}{\partial u''}\right) = 0, \tag{2.46}$$

which results in a fourth-order ordinary differential equation. Specified boundary conditions require that

$$\delta u = 0 \quad \text{and} \quad \delta u' = 0 \quad \text{at} \quad x = x_0, x_1,$$

in which case u and u' are specified at the end points. Alternatively, natural boundary conditions require that

$$\frac{d}{dx}\left(\frac{\partial F}{\partial u''}\right) - \frac{\partial F}{\partial u'} = 0 \quad \text{and} \quad \frac{\partial F}{\partial u''} = 0 \quad \text{at} \quad x = x_0, x_1. \tag{2.47}$$

Note that two boundary conditions are necessary at each boundary for a fourth-order ordinary differential equation.

For the more general case with a functional of the form

$$I[u] = \int_{x_0}^{x_1} F(x,u,u',u'',\ldots,u^{(n)})dx,$$

the Euler equation is

$$\frac{\partial F}{\partial u} - \frac{d}{dx}\left(\frac{\partial F}{\partial u'}\right) + \frac{d^2}{dx^2}\left(\frac{\partial F}{\partial u''}\right) - \ldots + (-1)^n \frac{d^n}{dx^n}\left(\frac{\partial F}{\partial u^{(n)}}\right) = 0.$$

Observe that odd-order derivative terms change sign as a result of the required integration by parts. The ordinary differential equation that results from the Euler equation is of order $2n$.

2.5 Functionals of Two Independent Variables

The single independent variable functionals considered thus far produce ordinary differential equations. In this section, we generalize to the two-dimensional case

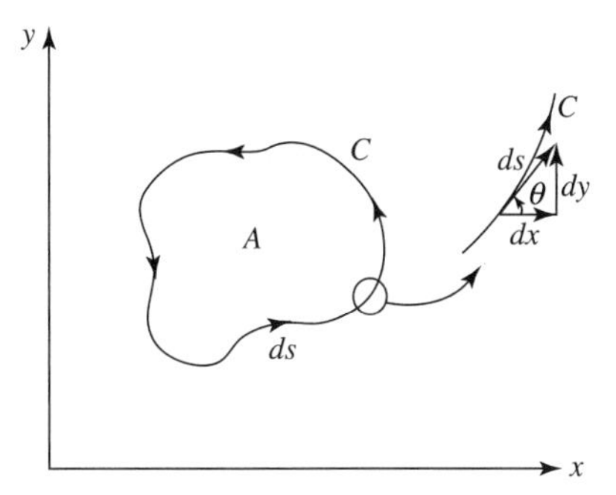

Figure 2.10. Schematic of two-dimensional region A bounded by curve C.

involving two independent variables with obvious extension to three dimensions. Such multidimensional problems produce partial differential Euler equations.

2.5.1 Euler's Equation

Consider a functional in which the function $u(x,y)$ depends upon two independent variables x and y as follows:

$$I[u] = \iint_A F(x,y,u,u_x,u_y)dxdy,$$

where subscripts denote partial differentiation with respect to the indicated variable. A is the region of the domain in the (x,y)-plane, and we define ds to be a differential element along the curve C bounding A proceeding in the counterclockwise direction as shown in Figure 2.10.

We seek the stationary function $u(x,y)$ of the functional $I[u]$ in region A. Setting the variation of the functional equal to zero

$$\delta I[u] = \iint_A \delta F(x,y,u,u_x,u_y)dxdy = 0,$$

and evaluating the variation of F gives

$$\iint_A \left[\frac{\partial F}{\partial x}\delta x + \frac{\partial F}{\partial y}\delta y + \frac{\partial F}{\partial u}\delta u + \frac{\partial F}{\partial u_x}\delta u_x + \frac{\partial F}{\partial u_y}\delta u_y\right] dxdy = 0.$$

The variations of the independent variables are zero; therefore, the first two terms vanish, and we have

$$\iint_A \left[\frac{\partial F}{\partial u}\delta u + \frac{\partial F}{\partial u_x}\delta\left(\frac{\partial u}{\partial x}\right) + \frac{\partial F}{\partial u_y}\delta\left(\frac{\partial u}{\partial y}\right)\right] dxdy = 0,$$

or recalling that variations and differentials are commutative

$$\iint_A \left[\frac{\partial F}{\partial u}\delta u + \frac{\partial F}{\partial u_x}\frac{\partial}{\partial x}(\delta u) + \frac{\partial F}{\partial u_y}\frac{\partial}{\partial y}(\delta u)\right] dxdy = 0.$$

In order to factor δu out of the remainder of the integrand, we write this in the form

$$\iint_A \left[\frac{\partial F}{\partial u}\delta u + \frac{\partial F}{\partial u_x}\frac{\partial}{\partial x}(\delta u) + \frac{\partial}{\partial x}\left(\frac{\partial F}{\partial u_x}\right)\delta u - \frac{\partial}{\partial x}\left(\frac{\partial F}{\partial u_x}\right)\delta u \right.$$
$$\left. + \frac{\partial F}{\partial u_y}\frac{\partial}{\partial y}(\delta u) + \frac{\partial}{\partial y}\left(\frac{\partial F}{\partial u_y}\right)\delta u - \frac{\partial}{\partial y}\left(\frac{\partial F}{\partial u_y}\right)\delta u \right] dxdy = 0,$$

where we have added and subtracted the same terms. Combining the second and third and fifth and sixth terms, respectively, yields

$$\iint_A \left[\frac{\partial F}{\partial u}\delta u + \frac{\partial}{\partial x}\left(\frac{\partial F}{\partial u_x}\delta u\right) - \frac{\partial}{\partial x}\left(\frac{\partial F}{\partial u_x}\right)\delta u \right.$$
$$\left. + \frac{\partial}{\partial y}\left(\frac{\partial F}{\partial u_y}\delta u\right) - \frac{\partial}{\partial y}\left(\frac{\partial F}{\partial u_y}\right)\delta u \right] dxdy = 0. \tag{2.48}$$

Recall the *divergence theorem*[5]

$$\iint_A \nabla \cdot \mathbf{V} dA = \oint_C \mathbf{V} \cdot \mathbf{n}\, ds,$$

where $\mathbf{V}$ is any vector, and $\mathbf{n}$ is the outward facing normal to C. The divergence theorem relates the area integral of the divergence of a vector field to the line integral of the component of the vector normal to the bounding curve C. In two dimensions, we let

$$\mathbf{V} = P(x,y)\mathbf{i} + Q(x,y)\mathbf{j}, \quad \mathbf{n} = \frac{dy}{ds}\mathbf{i} - \frac{dx}{ds}\mathbf{j},$$

where $P(x,y)$ and $Q(x,y)$ are arbitrary functions. With the "del" operator defined in two-dimensional Cartesian coordinates by

$$\nabla = \frac{\partial}{\partial x}\mathbf{i} + \frac{\partial}{\partial y}\mathbf{j},$$

the divergence theorem simplifies to

$$\iint_A \left(\frac{\partial P}{\partial x} + \frac{\partial Q}{\partial y}\right) dxdy = \oint_C \left(P\frac{dy}{ds} - Q\frac{dx}{ds}\right) ds \left[= \oint_C (Pdy - Qdx) \right].$$

We apply the divergence theorem to the second and fourth terms in (2.48) by letting

$$P = \frac{\partial F}{\partial u_x}\delta u, \quad Q = \frac{\partial F}{\partial u_y}\delta u.$$

Then equation (2.48) becomes

$$\iint_A \left[\frac{\partial F}{\partial u} - \frac{\partial}{\partial x}\left(\frac{\partial F}{\partial u_x}\right) - \frac{\partial}{\partial y}\left(\frac{\partial F}{\partial u_y}\right) \right] \delta u dxdy$$
$$+ \oint_C \left[\frac{\partial F}{\partial u_x}\frac{dy}{ds} - \frac{\partial F}{\partial u_y}\frac{dx}{ds} \right] \delta u ds = 0.$$

[5] This is also called *Gauss's theorem*.

In this way, δu is factored out of the remainder of the integrand. Consequently, Euler's equation for the case with $F = F(x,y,u,u_x,u_y)$ is

$$\frac{\partial F}{\partial u} - \frac{\partial}{\partial x}\left(\frac{\partial F}{\partial u_x}\right) - \frac{\partial}{\partial y}\left(\frac{\partial F}{\partial u_y}\right) = 0, \tag{2.49}$$

which leads to a partial differential equation for $u(x,y)$. Along the boundary C, either $u(x,y)$ is specified according to

$$u(x,y) = u_0(x,y) \quad \text{on} \quad C,$$

in which case $\delta u = 0$, or we apply the natural boundary condition

$$\frac{\partial F}{\partial u_x}\frac{dy}{ds} - \frac{\partial F}{\partial u_y}\frac{dx}{ds} = 0 \quad \text{on} \quad C. \tag{2.50}$$

Note that use of the divergence theorem here replaces (or one can think of it as a two-dimensional generalization of) the integration by parts in the one-dimensional cases considered previously. In order to show that this is the case, the one-dimensional integration by parts formula is derived from the two-dimensional divergence theorem in Section 1.6.

In terms of the angle θ between the positive x-axis and the tangent to the curve C (see Figure 2.10), we can write $dx/ds = \cos\theta$ and $dy/ds = \sin\theta$. Alternatively, the natural boundary conditions may be expressed in terms of the angle ν *normal* to the contour C rather than the angle θ *tangent* to it, in which case $\nu = \theta - \pi/2$, $\sin\nu = -\cos\theta$, and $\cos\nu = \sin\theta$.

The Euler equation (2.49) along with the natural boundary condition (2.50) can be written in vector form as well. The general form for an integrand of the form $F = F(x,y,u,\nabla u)$, where $u = u(x,y)$, is

$$\frac{\partial F}{\partial u} - \nabla\cdot\left(\frac{\partial F}{\partial \nabla u}\right) = 0, \tag{2.51}$$

and the natural boundary condition for this general case is

$$\mathbf{n}^T\frac{\partial F}{\partial \nabla u} = 0 \quad \text{on} \quad C, \tag{2.52}$$

where C is the boundary of the domain A, and $\mathbf{n}$ is the outward facing normal to curve C.

EXAMPLE 2.8 Obtain the differential equation for the stationary function $u(x,y)$ of the functional

$$I[u] = \int_{y=0}^{y=1}\int_{x=0}^{x=1}\frac{1}{2}\left(u_x^2 + u_xu_y + u^2 + 2xu\right)dxdy$$

in the square domain $0 \le x \le 1, 0 \le y \le 1$ shown in Figure 2.11. The specified boundary conditions are

$$u(0,y) = 3y, \quad u(x,0) = 2x,$$

and natural boundary conditions are imposed at $x = 1$ and $y = 1$.

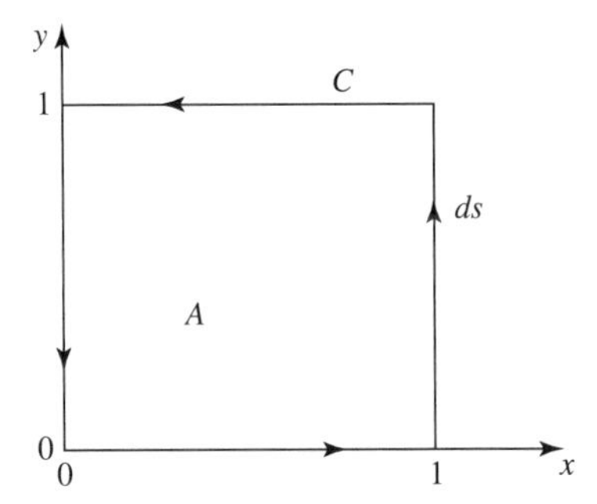

Figure 2.11. Two-dimensional domain for Example 2.8.

Solution: Evaluating the terms in Euler's equation (2.49) with $F = F(x,y,u,u_x,u_y)$ yields

$$\frac{\partial F}{\partial u} = u + x, \quad \frac{\partial F}{\partial u_x} = u_x + \frac{1}{2}u_y, \quad \frac{\partial F}{\partial u_y} = \frac{1}{2}u_x.$$

Then Euler's equation is

$$\frac{\partial F}{\partial u} - \frac{\partial}{\partial x}\left(\frac{\partial F}{\partial u_x}\right) - \frac{\partial}{\partial y}\left(\frac{\partial F}{\partial u_y}\right) = 0$$

$$u + x - \frac{\partial}{\partial x}\left(\frac{\partial u}{\partial x} + \frac{1}{2}\frac{\partial u}{\partial y}\right) - \frac{\partial}{\partial y}\left(\frac{1}{2}\frac{\partial u}{\partial x}\right) = 0$$

$$\frac{\partial^2 u}{\partial x^2} + \frac{\partial^2 u}{\partial x \partial y} - u = x, \tag{2.53}$$

which is a second-order, nonhomogeneous partial differential equation. The boundary conditions are fixed at $x = 0$ and $y = 0$ as follows

$$\begin{aligned} u(0,y) &= 3y \quad \text{at} \quad x = 0, \\ u(x,0) &= 2x \quad \text{at} \quad y = 0. \end{aligned} \tag{2.54}$$

To obtain the natural boundary conditions at $x = 1$ and $y = 1$, we evaluate equation (2.50), which is

$$\frac{\partial F}{\partial u_x}\frac{dy}{ds} - \frac{\partial F}{\partial u_y}\frac{dx}{ds} = 0.$$

Recalling that ds proceeds in the counterclockwise direction along C, for $x = 1$ we have $dx = 0$ and $dy = ds$. Therefore, $dx/ds = 0$ and we require that

$$\frac{\partial F}{\partial u_x} = 0,$$

which leads to

$$\frac{\partial u}{\partial x} + \frac{1}{2}\frac{\partial u}{\partial y} = 0 \quad \text{at} \quad x = 1. \tag{2.55}$$

Similarly, for $y = 1$ we have that $dx = -ds$ and $dy = 0$; therefore, $dy/ds = 0$ requiring that

$$\frac{\partial F}{\partial u_y} = 0.$$

This requires that

$$\frac{\partial u}{\partial x} = 0 \quad \text{at} \quad y = 1. \tag{2.56}$$

In order to determine the stationary function $u(x,y)$, the partial differential Euler equation (2.53) would be solved subject to the four boundary conditions (2.54)–(2.56). This is beyond the scope of this text, so we will typically be content to formulate the partial differential equation with its boundary conditions.

2.5.2 Minimal Surfaces

In Section 2.1.4 we considered the soap film problem, which is one of a series of *minimal surface* problems. The mathematics of minimal surfaces began with Leonhard Euler and Joseph Louis Lagrange in the middle of the eighteenth century. The catenoid, which is the characteristic shape found in the soap film problem, was discovered by Euler in 1744. However, it was the Belgian physicist Joseph Plateau (1801–1883) who first realized that a soap film forms a shape such that the surface area is a minimum in order to minimize energy, thereby connecting the physics of soap films and the mathematics of minimal surfaces. Neglecting gravity (because the film is very thin and nearly massless), the only form of energy in the soap film is that owing to surface tension, which is directly proportional to the surface area. As a result, the general problem of determining the shape of a minimal surface area spanning a given boundary has become known as *Plateau's problem*, and minimal surfaces is a topic in *differential geometry*. This remains an active area of research with applications in structural design, molecular engineering, materials science, and the physics of black holes. Here, we consider a general version of Plateau's problem in which the shape of a soap film stretched across a wire loop of arbitrary shape is governed by a partial differential equation.

EXAMPLE 2.9 Let A be a region in the (x,y)-plane bounded by the simple closed curve C as shown in Figure 2.12. Determine the shape of a soap film stretched across a "wire frame" with the same shape as C when projected on the (x,y)-plane but with height $u(x,y) = g(x,y)$ in the z-direction.

Solution: We seek to minimize the surface area $I[u]$ of $u(x,y)$ with boundary condition

$$u(x,y) = g(x,y) \quad \text{on} \quad C.$$

The surface area functional to be minimized is

$$I[u] = \iint_A \sqrt{1 + \left(\frac{\partial u}{\partial x}\right)^2 + \left(\frac{\partial u}{\partial y}\right)^2}\, dxdy, \tag{2.57}$$

which is a straightforward generalization of the length functional. The Euler equation (2.49) for $F = F(x,y,u,u_x,u_y)$ is

$$\frac{\partial F}{\partial u} - \frac{\partial}{\partial x}\left(\frac{\partial F}{\partial u_x}\right) - \frac{\partial}{\partial y}\left(\frac{\partial F}{\partial u_y}\right) = 0.$$

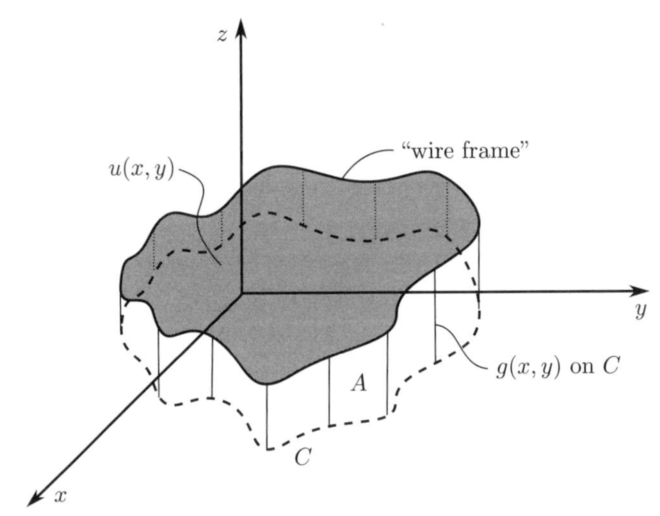

Figure 2.12. Minimal surface schematic.

Upon evaluating the partial derivatives in Euler's equation and substituting, we have

$$0-\frac{\partial}{\partial x}\left[\frac{1}{2}\left(1+u_x^2+u_y^2\right)^{-1/2}(2u_x)\right]-\frac{\partial}{\partial y}\left[\frac{1}{2}\left(1+u_x^2+u_y^2\right)^{-1/2}(2u_y)\right]=0.$$

Differentiating and simplifying yields

$$\left(1+u_x^2+u_y^2\right)(u_{xx}+u_{yy})-\left(u_x^2u_{xx}+2u_xu_yu_{xy}+u_y^2u_{yy}\right)=0,$$

or

$$\left[1+\left(\frac{\partial u}{\partial y}\right)^2\right]\frac{\partial^2 u}{\partial x^2}-2\frac{\partial u}{\partial x}\frac{\partial u}{\partial y}\frac{\partial^2 u}{\partial x\partial y}+\left[1+\left(\frac{\partial u}{\partial x}\right)^2\right]\frac{\partial^2 u}{\partial y^2}=0. \tag{2.58}$$

This is the *minimal surface equation* first derived by Lagrange in 1760. It is a second-order, nonlinear partial differential equation and is equivalent to enforcing that the mean curvature of the surface is zero. Given the base shape C defining the area A, we can determine the shape $u(x,y)$ with the minimal surface area stretched across the wire frame defined by $g(x,y)$ on C.

An important special case occurs when the wire frame defined by $g(x,y)$ on C is such that the film is very nearly a horizontal planar surface. In this case, the slopes are such that $u_x \ll 1$ and $u_y \ll 1$. Even if the slopes are small, the corresponding curvatures u_{xx} and u_{yy} can remain $O(1)$, in which case the minimal surface equation (2.58) reduces to the *Laplace equation*

$$\frac{\partial^2 u}{\partial x^2}+\frac{\partial^2 u}{\partial y^2}=0, \tag{2.59}$$

which is considered further in the next section.

2.5.3 Dirichlet Problem

Previously, we have begun with the variational form and obtained the equivalent differential Euler equation. Alternatively, we may have occasion to convert known differential equations into their equivalent variational form. This is known as the *inverse problem*. For example, let us consider the *Dirichlet problem*.

The Dirichlet problem expressed in two-dimensional Cartesian coordinates is Laplace's equation

$$\mathcal{L}u = \nabla^2 u = \frac{\partial^2 u}{\partial x^2} + \frac{\partial^2 u}{\partial y^2} = 0, \tag{2.60}$$

along with specified boundary conditions around the entire boundary

$$u(x,y) = u_0(x,y) \quad \text{on} \quad C.$$

To obtain the variational form of a differential equation $\mathcal{L}u = 0$ for $u(x,y)$, we multiply the differential equation by the variation of the dependent variable, δu, and integrate over the domain A bounded by C as follows:

$$\iint_A \mathcal{L}u\,\delta u\,dA = 0,$$

or for Laplace's equation

$$\iint_A \left(\frac{\partial^2 u}{\partial x^2} + \frac{\partial^2 u}{\partial y^2} \right) \delta u\,dx\,dy = 0. \tag{2.61}$$

For the sake of illustration, let us consider a rectangular domain with $x_0 \le x \le x_1$. Using integration by parts, the first term becomes

$$\begin{aligned}
\int \left(\int_{x_0}^{x_1} u_{xx}\delta u\,dx \right) dy &= \int \left(\cancelto{0}{\left[u_x \delta u \right]_{x_0}^{x_1}} - \int_{x_0}^{x_1} u_x \delta u_x\,dx \right) dy \\
&= -\iint u_x \delta u_x\,dx\,dy \\
&= -\frac{1}{2}\delta \iint u_x^2 dx\,dy.
\end{aligned}$$

The last step may be generalized as follows:

$$\phi\delta\phi = \frac{1}{2}\delta\phi^2,$$

where in this case $\phi = u_x$. That is, a quantity times its variation is equal to one half the variation of the square of the quantity. Treating the second term in equation (2.61) in a similar manner leads to the variational form

$$\delta I = \delta \iint_A F dA = 0,$$

which for the Dirichlet problem is

$$\delta I[u] = \delta \iint_A \left[\left(\frac{\partial u}{\partial x} \right)^2 + \left(\frac{\partial u}{\partial y} \right)^2 \right] dx\,dy = \delta \iint_A \nabla u \cdot \nabla u\,dA = 0. \tag{2.62}$$

Thus, the Dirichlet problem corresponds to minimizing the sum of the magnitudes of the gradients of $u(x,y)$ in the x- and y-directions throughout the region A. The

Dirichlet problem has many applications involving diffusion, such as heat conduction in a solid with fixed temperatures at the boundaries.

REMARKS:

1. The reader familiar with *finite-element methods* may recognize the similarity between the above procedure and that used to convert the *strong* (differential) *form* to the *weak* (integral) *form* in finite-element methods. This is no coincidence. Although the terminology is somewhat different – for example, one speaks in terms of multiplying by a *weight* (or *test*) *function*, rather than the variation δu – it is the same procedure carried out for the same purpose. See Chapter 3 for more on finite-element methods from a variational point of view.
2. Conversion of differential Euler equations to their variational forms is discussed in more detail in Section 3.1.
3. The inverse problem of determining the variational form from the differential Euler equation is always possible for linear, self-adjoint differential operators (see Section 1.8). This is the case for both ordinary and partial differential equations.
4. Writing differential equations in their equivalent variational form often provides valuable insight into the physical processes that they represent. For example, the variational form of the Dirichlet problem (2.62) clearly indicates that diffusion is a *smoothing* process in which the overall magnitudes of the gradients in the solution are minimized.
5. We may generalize the functional (2.62) as follows

$$I[u] = \iint_A \left[\left(\frac{\partial u}{\partial x} \right)^2 + \left(\frac{\partial u}{\partial y} \right)^2 + 2fu \right] dx\, dy,$$

where $f(x,y)$ is a specified function; that is, only $u(x,y)$ varies. The corresponding Euler equation is

$$\mathcal{L}u = \nabla^2 u = \frac{\partial^2 u}{\partial x^2} + \frac{\partial^2 u}{\partial y^2} = f(x,y),$$

which is known as a *Poisson equation.*

2.6 Functionals of Two Dependent Variables

The derivation in Section 2.5 can easily be extended to what some authors refer to as "the more general case," in which a functional of two functions $u(x,y)$ and $v(x,y)$ is considered of the form

$$I[u,v] = \iint_A F(x,y,u,v,u_x,v_x,u_y,v_y)dxdy.$$

The resulting Euler equations are

$$\frac{\partial F}{\partial u} - \frac{\partial}{\partial x}\left(\frac{\partial F}{\partial u_x} \right) - \frac{\partial}{\partial y}\left(\frac{\partial F}{\partial u_y} \right) = 0,$$

$$\frac{\partial F}{\partial v} - \frac{\partial}{\partial x}\left(\frac{\partial F}{\partial v_x} \right) - \frac{\partial}{\partial y}\left(\frac{\partial F}{\partial v_y} \right) = 0,$$

where the first equation is the same as equation (2.49), and the second equation simply replaces u with v. Thus, we have two coupled, second-order partial differential equations for $u(x,y)$ and $v(x,y)$.

The fixed or natural boundary conditions on the contour C are

$$u(x,y) = u_0(x,y) \quad \text{or} \quad \frac{\partial F}{\partial u_x}\frac{dy}{ds} - \frac{\partial F}{\partial u_y}\frac{dx}{ds} = 0,$$

$$v(x,y) = v_0(x,y) \quad \text{or} \quad \frac{\partial F}{\partial v_x}\frac{dy}{ds} - \frac{\partial F}{\partial v_y}\frac{dx}{ds} = 0.$$

EXAMPLE 2.10 Determine the stationary functions $u(x)$ and $v(x)$ for the functional

$$I[u,v] = \int_0^{\pi/4} \left(v'^2 + v'u' + u^2\right) dx,$$

with the end conditions

$$u(0) = 1, \quad v(0) = \frac{3}{2}, \quad u\left(\frac{\pi}{4}\right) = 2.$$

Solution: For an integrand of the form $F = F(u,v,u',v')$, the Euler equations are

$$\frac{\partial F}{\partial u} - \frac{d}{dx}\left(\frac{\partial F}{\partial u'}\right) = 0, \quad \frac{\partial F}{\partial v} - \frac{d}{dx}\left(\frac{\partial F}{\partial v'}\right) = 0.$$

Evaluating the first of these gives the ordinary differential equation

$$v'' - 2u = 0, \tag{2.63}$$

and the second yields

$$u'' + 2v'' = 0. \tag{2.64}$$

Solving equation (2.63) for v'' and substituting into equation (2.64) gives

$$u'' + 4u = 0,$$

which has the general solution

$$u(x) = c_1 \cos(2x) + c_2 \sin(2x). \tag{2.65}$$

Applying the two end conditions for $u(0)$ and $u(\pi/4)$ leads to the stationary function

$$u(x) = \cos(2x) + 2\sin(2x). \tag{2.66}$$

Substituting this result into equation (2.63) and integrating twice to obtain $v(x)$ results in

$$v(x) = -\frac{1}{2}\cos(2x) - \sin(2x) + c_3 x + c_4. \tag{2.67}$$

Applying the fixed condition for $v(0)$ requires that $c_4 = 2$. Because no fixed condition is provided for $v(\pi/4)$, we apply the natural boundary condition, which is

$$\left.\frac{\partial F}{\partial v'}\right|_{x=\pi/4} = 0,$$

or

$$\left[u' + 2v'\right]_{x=\pi/4} = 0.$$

Substituting the stationary functions $u(x)$ and $v(x)$ and evaluating at $x = \pi/4$ yields $c_3 = 0$. Therefore, the stationary function $v(x)$ is

$$v(x) = -\frac{1}{2}\cos(2x) - \sin(2x) + 2. \tag{2.68}$$

Hence, the two stationary functions are given by equations (2.66) and (2.68).

2.7 Constrained Functionals

Recall the example in Section 1.5 involving finding the minor and major axes of an ellipse. We find the minimum and maximum of the square of the distance from the origin

$$d = x_1^2 + x_2^2$$

constrained to be on the ellipse

$$3x_1^2 - 2x_1x_2 + 3x_2^2 = 8.$$

Because we seek the extrema of a function, differential calculus is used along with the Lagrange multiplier method.

Similarly, in many applications of calculus of variations, we seek to determine a stationary function of a functional subject to a constraint. The constraint may be integral, differential, or algebraic. An integral constraint provides a requirement on the final properties of a stationary function $u(x)$, while algebraic and differential constraints supply a relationship that applies throughout the domain between two (or more) dependent variables contained within a functional. Just as in the ellipse example, these constrained functional problems are solved using *Lagrange multipliers*.

2.7.1 Integral Constraints

Consider the functional

$$I[u] = \int_{x_0}^{x_1} F(x, u, u')dx \tag{2.69}$$

subject to a constraint expressed as a definite integral

$$\int_{x_0}^{x_1} G(x, u, u')dx = K, \tag{2.70}$$

where K is a specified constant. This is the case, for example, in Section 1.3.2, where the volume of the liquid drop is specified.

We follow the Lagrange multiplier method for functions outlined in Section 1.5 and apply it to functionals. From the constraint (2.70), we can write

$$\Lambda\left[\int_{x_0}^{x_1} G(x,u,u')dx - K\right] = 0,$$

where Λ is an arbitrary constant called the Lagrange multiplier. Taking the variation (with K being a specified constant) gives

$$\delta\left\{\Lambda\left[\int_{x_0}^{x_1} G(x,u,u')dx - K\right]\right\} = 0,$$

or

$$\Lambda\delta\int_{x_0}^{x_1} G(x,u,u')dx = 0.$$

Combining this result with the variation of equation (2.69), which is zero, we may write

$$\delta\left[\int_{x_0}^{x_1} F(x,u,u')dx + \Lambda\int_{x_0}^{x_1} G(x,u,u')dx\right] = 0,$$

or

$$\delta\int_{x_0}^{x_1} \tilde{F}(\Lambda,x,u,u')dx = 0, \quad \tilde{F} = F + \Lambda G.$$

Note that we are adding two quantities that are each equal to zero, in which case their sum is also zero. Therefore, finding the stationary function of the functional (2.69) subject to the integral constraint (2.70) is equivalent to finding the stationary function of a functional with the augmented integrand $\tilde{F} = F + \Lambda G$ subject to no constraint.

The procedure for obtaining the stationary function of a functional with an integral constraint is as follows:

1. Write the *augmented integrand*

$$\tilde{F}(\Lambda,x,u,u') = F(x,u,u') + \Lambda G(x,u,u'),$$

   where Λ is the constant Lagrange multiplier and G is the integrand of the constraint.
2. Determine the stationary function for the augmented functional

$$\tilde{I}[u] = \int_{x_0}^{x_1} \tilde{F}(\Lambda,x,u,u')dx,$$

   where the Lagrange multiplier Λ is an additional parameter to be determined because it varies. The Euler equation for $\tilde{F} = \tilde{F}(\Lambda,x,u,u')$ is of the form

$$\frac{\partial \tilde{F}}{\partial u} - \frac{d}{dx}\left(\frac{\partial \tilde{F}}{\partial u'}\right) = 0.$$

3. Solve the Euler equation for $\tilde{F}$ and use the constraint (2.70) and end conditions, for example, $u(x_0) = u_0$ and $u(x_1) = u_1$, to determine Λ and the constants of integration.

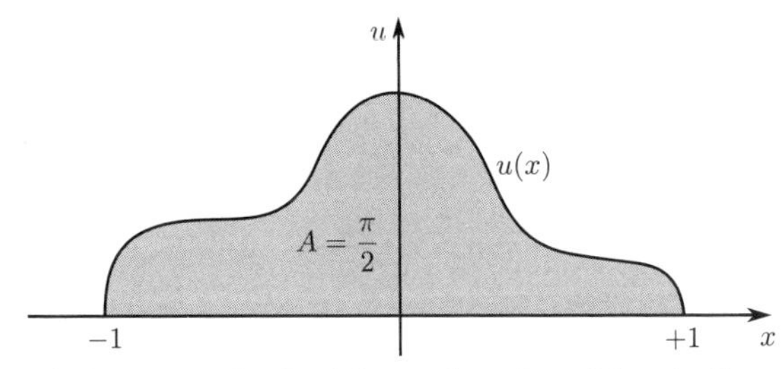

Figure 2.13. Schematic for constrained minimum length problem in Example 2.11.

This constrained functional procedure also applies if there are more than one independent variable, for example, x and y, and more than one constraint.

EXAMPLE 2.11 Determine the curve $u(x)$ from $(x,u) = (-1,0)$ to $(1,0)$ that has minimum length, but encloses an area $A = \frac{\pi}{2}$ with the x-axis, as illustrated in Figure 2.13.

Solution: As in previous examples, we must minimize the length of the curve $u(x)$ given by the functional

$$I[u] = \int_{x=-1}^{x=1} ds = \int_{-1}^{1} \sqrt{1+u'(x)^2}dx,$$

but in this case subject to the integral constraint

$$A = \int_{-1}^{1} udx = \frac{\pi}{2}.$$

The augmented integrand is

$$\tilde{F}(\Lambda,x,u,u') = F + \Lambda G = \sqrt{1+(u')^2} + \Lambda u.$$

Observe that the independent variable x does not appear explicitly, thereby corresponding to special case II in Section 2.1.4. However, in this case it is more straightforward to use the usual approach as follows. Evaluating the terms in the Euler equation yields

$$\frac{\partial \tilde{F}}{\partial u} = \Lambda, \quad \frac{\partial \tilde{F}}{\partial u'} = \frac{u'}{\sqrt{1+(u')^2}}.$$

Substituting into Euler's equation

$$\frac{\partial \tilde{F}}{\partial u} - \frac{d}{dx}\left(\frac{\partial \tilde{F}}{\partial u'}\right) = 0$$

results in

$$\Lambda - \frac{d}{dx}\left(\frac{u'}{\sqrt{1+(u')^2}}\right) = 0.$$

Integrating once with respect to x and solving for $u'(x)$ gives

$$\frac{u'}{\sqrt{1+(u')^2}} = \Lambda(x+c_1)$$

$$\frac{(u')^2}{1+(u')^2} = \Lambda^2(x+c_1)^2$$

$$(u')^2\left[1-\Lambda^2(x+c_1)^2\right] = \Lambda^2(x+c_1)^2$$

$$u'(x) = \frac{\Lambda(x+c_1)}{\sqrt{1-\Lambda^2(x+c_1)^2}}.$$

Integrating again with respect to x leads to

$$u(x) = -\frac{1}{\Lambda}\sqrt{1-\Lambda^2(x+c_1)^2} - c_2, \tag{2.71}$$

or

$$(x+c_1)^2 + (u+c_2)^2 = \frac{1}{\Lambda^2}, \tag{2.72}$$

which is the equation for a circular arc with center at $(-c_1, -c_2)$ and radius $1/|\Lambda|$. This provides the general result that the curve with the shortest length enclosing a given area is a circle.

The integration constants c_1 and c_2 and the constant Lagrange multiplier Λ are obtained from the boundary conditions

$$u(-1) = 0, \quad u(1) = 0,$$

and the constraint

$$\int_{-1}^{1} u dx = \frac{\pi}{2}.$$

Applying the end condition $u(-1) = 0$ leads to

$$(0+c_2)^2 + (-1+c_1)^2 = \frac{1}{\Lambda^2}$$

$$c_2^2 + c_1^2 - 2c_1 + 1 = \frac{1}{\Lambda^2}. \tag{2.73}$$

Then from $u(1) = 0$, we have

$$(0+c_2)^2 + (1+c_1)^2 = \frac{1}{\Lambda^2}$$

$$c_2^2 + c_1^2 + 2c_1 + 1 = \frac{1}{\Lambda^2}. \tag{2.74}$$

Subtracting (2.73) and (2.74) yields $c_1 = 0$. Adding (2.73) and (2.74), with $c_1 = 0$, gives

$$c_2 = \sqrt{\frac{1}{\Lambda^2} - 1}. \tag{2.75}$$

Observe that $|\Lambda| \leq 1$, so for c_2, which is the y-coordinate of the center of the circular arc, to be real, the radius of the circular arc must be $1/|\Lambda| \geq 1$. This is consistent with the fact that there is no circular arc that passes through $(-1,0)$ and $(1,0)$ with radius less than one.

Substituting equations (2.74) and (2.75) into equation (2.71) leads to

$$u(x) = -\frac{1}{\Lambda}\left[\sqrt{1-\Lambda^2x^2}+\sqrt{1-\Lambda^2}\right]. \tag{2.76}$$

From the constraint

$$\int_{-1}^{1} u(x)dx = A,$$

$$-\frac{1}{\Lambda}\int_{-1}^{1}\left[\sqrt{1-\Lambda^2x^2}+\sqrt{1-\Lambda^2}\right]dx = A$$

$$-\frac{1}{\Lambda}\left[\frac{1}{2}x\sqrt{1-\Lambda^2x^2}+\frac{1}{2\Lambda}\sin^{-1}(\Lambda x)+\sqrt{1-\Lambda^2}x\right]_{-1}^{1} = A$$

$$-\frac{1}{\Lambda}\left\{\left[\frac{1}{2}\sqrt{1-\Lambda^2}+\frac{1}{2\Lambda}\sin^{-1}(\Lambda)+\sqrt{1-\Lambda^2}\right]\right.$$
$$\left.-\left[-\frac{1}{2}\sqrt{1-\Lambda^2}+\frac{1}{2\Lambda}\sin^{-1}(-\Lambda)-\sqrt{1-\Lambda^2}\right]\right\} = A$$

$$-\frac{1}{\Lambda}\left[3\sqrt{1-\Lambda^2}+\frac{1}{\Lambda}\sin^{-1}(\Lambda)\right] = A.$$

Recall that $|\Lambda| \leq 1$. One can confirm that $\Lambda = -1$ satisfies this relationship for $A = \frac{\pi}{2}$. Then from equation (2.75), $c_2 = 0$, and we have

$$u(x) = \sqrt{1-x^2},$$

or

$$u^2 + x^2 = 1^2,$$

which is a circular arc centered at the origin with radius one. The curve $u(x)$ is a semi-circle, which does indeed have area $A = \frac{\pi}{2}$.

Note that for $A < \frac{\pi}{2}$, the center of the circular arc $(0, -c_2)$ becomes increasingly negative ($c_2 > 0$), and the radius is greater than one ($|\Lambda| < 1$). For $A > \frac{\pi}{2}$, the solution remains a circular arc; however, it is not a single-valued function limited to the range $-1 \leq x \leq 1$.

Observe that Dido's problem described in the preamble of Chapter 1 is essentially Example 2.11 with the functional and integral constraint being switched. That is, we maximize the area under a curve subject to the length of the curve being fixed. Such problems are called *isoperimetric* as the perimeter is constant.

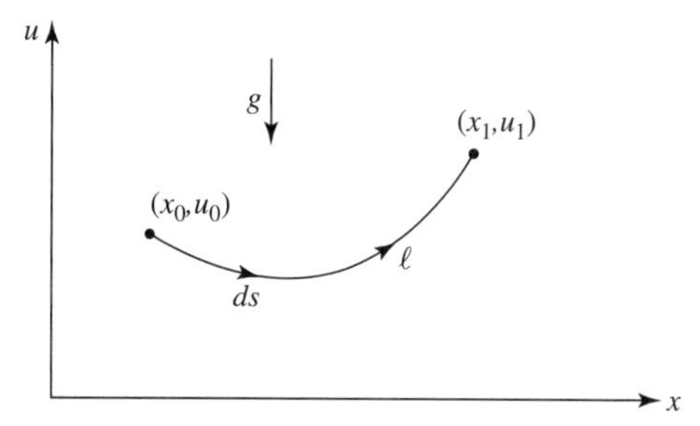

Figure 2.14. Hanging-cable problem in Example 2.12.

EXAMPLE 2.12 As illustrated in Figure 2.14, a cable with mass per unit length ρ is suspended at two points (x_0,u_0) and (x_1,u_1). Obtain the equilibrium position $u(x)$ for a cable of length ℓ.[6]

Solution: The differential arc length is

$$ds = \sqrt{1+(u')^2}dx.$$

The static equilibrium position is that which minimizes the *total potential energy* of the cable, which can be expressed as a functional of the form

$$I[u] = \rho g \int_{(x_0,u_0)}^{(x_1,u_1)} u ds = \rho g \int_{x_0}^{x_1} u\sqrt{1+(u')^2}dx.$$

The end conditions are

$$u(x_0) = u_0, \quad u(x_1) = u_1,$$

and the constraint on the length of the cable is

$$\int_{x_0}^{x_1} ds = \int_{x_0}^{x_1} \sqrt{1+(u')^2}dx = \ell,$$

where ℓ is fixed.

Forming the augmented integrand (with the Lagrange multiplier chosen as $\rho g\Lambda$ for convenience) gives

$$\tilde{F}(\Lambda,x,u,u') = \rho g u\sqrt{1+(u')^2} + \rho g\Lambda\sqrt{1+(u')^2} = \rho g(u+\Lambda)\sqrt{1+(u')^2},$$

where we observe that the independent variable x does not appear explicitly. Applying special case II in Section 2.1.4, we have

$$\bar{F}(u,x') = \rho g x'(u+\Lambda)\sqrt{1+(x')^{-2}} = \rho g(u+\Lambda)\sqrt{(x')^2+1},$$

and the Euler equation (2.17) is

$$\frac{\partial \bar{F}}{\partial x'} = c,$$

[6] As with the brachistochrone problem, this "hanging-cable problem" preoccupied the Bernoulli family. In this case, it was Jacob Bernoulli who posed the problem and his younger brother Johann who solved it a year later.

or

$$\frac{1}{2}\rho g(u+\Lambda)\left[(x')^2+1\right]^{-1/2}\left[2x'\right]=\rho g c_1 \quad (c=\rho g c_1).$$

Solving for x' yields

$$(u+\Lambda)^2(x')^2=c_1^2\left[(x')^2+1\right]$$

$$\left[(u+\Lambda)^2-c_1^2\right](x')^2=c_1^2$$

$$\frac{dx}{du}=\sqrt{\frac{c_1^2}{(u+\Lambda)^2-c_1^2}}.$$

Separating variables and integrating again gives

$$c_1\int\frac{du}{\sqrt{(u+\Lambda)^2-c_1^2}}=\int dx=x+c_2.$$

Making the substitution $u+\Lambda=c_1\cosh\theta$ ($du=c_1\sinh\theta d\theta$) leads to

$$c_1\int\frac{\sinh\theta}{\sqrt{\cosh^2\theta-1}}d\theta=x+c_2,$$

but recalling that $\cosh^2\theta-1=\sinh^2\theta$, we have

$$c_1\int d\theta=x+c_2.$$

Thus,

$$\theta=\frac{x+c_2}{c_1}.$$

Taking the hyperbolic cosine of both sides and applying the substitution, we obtain the solution as

$$u+\Lambda=c_1\cosh\frac{x+c_2}{c_1}. \tag{2.77}$$

This shape is known as a *catenary*.

The integration constants c_1 and c_2 and the constant Lagrange multiplier Λ are obtained from the boundary conditions

$$u(x_0)=u_0, \quad u(x_1)=u_1,$$

and the constraint

$$\int_{x_0}^{x_1}\sqrt{1+(u')^2}\,dx=\ell.$$

For example, let us consider the symmetric case with $(x_0,u_0)=(-1,1)$ and $(x_1,u_1)=(1,1)$, in which case $\ell>2$. Then

$$u(-1)=1: \quad 1+\Lambda=c_1\cosh\left(\frac{-1+c_2}{c_1}\right), \tag{2.78}$$

$$u(1)=1: \quad 1+\Lambda=c_1\cosh\left(\frac{1+c_2}{c_1}\right). \tag{2.79}$$

Equations (2.78) and (2.79) require that

$$\cosh\left(\frac{-1+c_2}{c_1}\right) = \cosh\left(\frac{1+c_2}{c_1}\right),$$

which can only be true if $c_2 = 0$. To obtain c_1, consider the constraint with

$$u' = \sinh\left(\frac{x}{c_1}\right).$$

Then the constraint requires

$$\begin{aligned}
\int_{x_0}^{x_1} \left[1+(u')^2\right]^{1/2} dx &= \ell \\
\int_{-1}^{1} \left[1+\sinh^2\left(\frac{x}{c_1}\right)\right]^{1/2} dx &= \ell \\
\int_{-1}^{1} \left[\cosh^2\left(\frac{x}{c_1}\right)\right]^{1/2} dx &= \ell \\
\int_{-1}^{1} \cosh\left(\frac{x}{c_1}\right) dx &= \ell \\
c_1 \sinh\left(\frac{x}{c_1}\right)\Big|_{-1}^{1} &= \ell \\
c_1\left[\sinh\left(\frac{1}{c_1}\right) - \sinh\left(\frac{-1}{c_1}\right)\right] &= \ell \\
2c_1 \sinh\left(\frac{1}{c_1}\right) &= \ell.
\end{aligned}$$

We cannot solve for c_1 exactly. Instead, we must determine c_1 numerically using a root-finding technique. There are two roots of the above equation, one positive and one negative. The negative root corresponds to that which maximizes the potential energy of the cable, in which case the cable is "hanging up." We are interested in the positive root corresponding to a minimum in potential energy. For example, for $\ell = 2.5$, we find the positive root for c_1 to be $c_1 = 0.845505$. It remains to determine the Lagrange multiplier Λ. To do so, recall equation (2.79)

$$\Lambda = c_1 \cosh\left(\frac{1}{c_1}\right) - 1,$$

which for $c_1 = 0.845505$ gives

$$\Lambda = 0.509098.$$

Thus, the solution (2.77) for $(x_0, u_0) = (-1, 1)$, $(x_1, u_1) = (1, 1)$, and $\ell = 2.5$ is

$$u(x) = c_1 \cosh\left(\frac{x}{c_1}\right) - \Lambda, \quad c_1 = 0.845505, \quad \Lambda = 0.509098.$$

This solution is plotted in Figure 2.15.

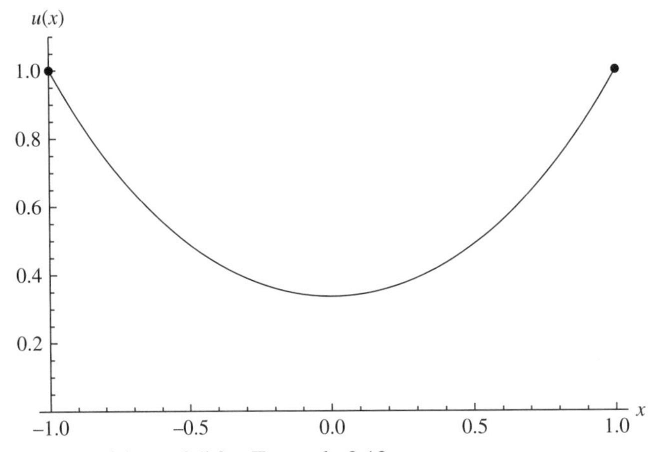

Figure 2.15. Catenary with $\ell = 2.5$ for Example 2.12.

REMARKS:

1. The catenary shape has the property that the only stresses in the hanging cable are in tension. This makes sense physically as a flexible cable cannot withstand shear stresses or bending moments.
2. The surface of revolution of the catenary around the x-axis is the *catenoid*, which is the solution of the "soap film problem" described in Section 2.1.4.
3. In Latin the word *catena* means "chain." Although Christiaan Huygens first used the term *catenaria* in a letter to Leibniz in 1690, Thomas Jefferson coined the English term *catenary*.
4. In 1671 Robert Hooke determined that the optimal shape of an arch of uniform density and thickness supporting its own weight is that of an inverted catenary. It is optimal in the sense that the arch is in compression only, and there are no shear stresses or bending moments. The Gateway Arch in St. Louis, Missouri, USA, is not a true catenary; instead it is a *flattened*, or *weighted, catenary*, which has the equation $u(x) = c_1 \cosh(c_2 x) + c_3$ (a true catenary is such that $c_2 = 1/c_1$).

2.7.2 Sturm-Liouville Problems

Differential equations of the Sturm-Liouville form are widespread in applications and have a well-developed theory. In particular, as shown in Section 1.8, for an ordinary differential equation to be self-adjoint, its operator must be of the Sturm-Liouville form. Furthermore, because such equations are self-adjoint, the eigenfunctions of their differential operators are all mutually orthogonal. The orthogonality property has numerous mathematical and practical consequences that are beyond the scope of this text. It is shown in this section that the Sturm-Liouville differential equation results from consideration of a functional of a particular form subject to an integral constraint, which has important consequences in the context of the calculus of variations.

We seek the stationary function $u(x)$ for the functional

$$I[u] = \int_{x_0}^{x_1} \left[q(x)u^2 - p(x)(u')^2 \right] dx \tag{2.80}$$

subject to the integral constraint

$$\int_{x_0}^{x_1} r(x)u^2 dx = 1, \tag{2.81}$$

where $p(x)$, $q(x)$, and $r(x)$ are known functions. The constraint requires that the norm of $u(x)$ with respect to the weight function $r(x)$ be equal to one.

Using the Lagrange multiplier method, the augmented integrand is

$$\tilde{F}(\Lambda, x, u, u') = q(x)u^2 - p(x)(u')^2 + \Lambda[r(x)u^2].$$

Evaluating the partial derivatives in the Euler equation yields

$$\frac{\partial \tilde{F}}{\partial u} = 2q(x)u + 2\Lambda r(x)u, \quad \frac{\partial \tilde{F}}{\partial u'} = -2p(x)u'.$$

Then Euler's equation is

$$\frac{\partial \tilde{F}}{\partial u} - \frac{d}{dx}\left(\frac{\partial \tilde{F}}{\partial u'}\right) = 0$$

$$q(x)u + \Lambda r(x)u - \frac{d}{dx}[-p(x)u'] = 0,$$

or

$$\frac{d}{dx}\left[p(x)\frac{du}{dx}\right] + [q(x) + \Lambda r(x)]u(x) = 0. \tag{2.82}$$

Therefore, the stationary function $u(x)$ for the functional (2.80) subject to the constraint (2.81) satisfies the Sturm-Liouville equation (2.82). In other words, Euler's equation is the Sturm-Liouville equation, and the Lagrange multiplier Λ is an eigenvalue of the Sturm-Liouville differential operator. One consequence of this result is that any equation having a self-adjoint, that is, Sturm-Liouville, differential operator can be expressed as a variational principle, a fact that is utilized in the Rayleigh-Ritz method in Section 3.1.2.

Recall that the natural boundary conditions for functionals of the form (2.80) are

$$\left.\frac{\partial \tilde{F}}{\partial u'}\right|_{x_0}^{x_1} = 0;$$

accordingly, for the Sturm-Liouville problem, the natural boundary conditions are

$$\left[p(x)\frac{du}{dx}\right]_{x_0}^{x_1} = 0,$$

in which case pu' must be zero at the end points.

EXAMPLE 2.13 Determine the Euler equation for the functional (2.80) subject to the integral constraint (2.81) with

$$p(x) = x, \quad q(x) = -\frac{\nu^2}{x}, \quad r(x) = x,$$

where ν is a constant parameter. The interval is $0 \leq x \leq 1$.
Solution: From the result (2.82), the Euler equation, which is a Sturm-Liouville equation, is

$$\frac{d}{dx}\left[x\frac{du}{dx}\right] + \left[-\frac{\nu^2}{x} + \Lambda x\right]u = 0.$$

This is the *Bessel equation*, for which the solutions are Bessel functions.

2.7.3 Algebraic and Differential Constraints

It is often the case that the constraint is algebraic or differential, rather than being a definite integral. For example, we may seek the stationary function of a functional

$$I[u,v] = \int_{x_0}^{x_1} F(x,u,v,u',v')dx \tag{2.83}$$

involving two functions $u(x)$ and $v(x)$ subject to an algebraic constraint on the relationship between the two unknown functions of the form

$$\phi(u,v) = 0,$$

or a constraint in the form of a differential equation, such as

$$\phi(u,v,u',v') = 0. \tag{2.84}$$

Note that an algebraic constraint is simply a special case of a differential constraint.

Whereas integral constraints are *global* constraints that apply over the entire domain as a whole, algebraic and differential constraints are *local* and apply at each point in the domain. Because the algebraic or differential constraint applies locally at every point in the domain, $x_0 \leq x \leq x_1$, effectively we must apply a constant Lagrange multiplier to the constraint at each point in the continuous domain. This is accomplished by having the Lagrange multiplier be a continuous function of the independent variable(s), in this case x, and summing through integration. Multiplying the constraint by $\Lambda(x)$, integrating over the interval $x_0 \leq x \leq x_1$, and combining with the functional (2.83) gives the augmented integrand

$$\tilde{F} = F + \Lambda(x)\phi.$$

We then proceed as before taking note that $\Lambda(x)$ is now a function of x and varies.

To verify this procedure, let us again use the Lagrange multiplier procedure to treat the case of a differential constraint, with the end conditions of the stationary function being specified. Begin by taking the variation of the functional (2.83) and setting it equal to zero

$$\delta \int_{x_0}^{x_1} F(x,u,v,u',v')dx = 0.$$

Following the procedure in Section 2.1.3, evaluating the variation produces

$$\int_{x_0}^{x_1} \left\{ \left[\frac{\partial F}{\partial u} - \frac{d}{dx}\left(\frac{\partial F}{\partial u'} \right) \right] \delta u + \left[\frac{\partial F}{\partial v} - \frac{d}{dx}\left(\frac{\partial F}{\partial v'} \right) \right] \delta v \right\} dx = 0. \tag{2.85}$$

Now we take the variation of the constraint (2.84) as follows

$$\delta \phi = \frac{\partial \phi}{\partial u}\delta u + \frac{\partial \phi}{\partial v}\delta v + \frac{\partial \phi}{\partial u'}\delta u' + \frac{\partial \psi}{\partial v'}\delta v' = 0.$$

Multiplying the Lagrange multiplier $\Lambda(x)$ by $\delta\phi$ and integrating over the domain leads to

$$\int_{x_0}^{x_1} \Lambda(x) \left(\frac{\partial \phi}{\partial u}\delta u + \frac{\partial \phi}{\partial v}\delta v + \frac{\partial \phi}{\partial u'}\delta u' + \frac{\partial \phi}{\partial v'}\delta v' \right) dx = 0. \tag{2.86}$$

Consider integration by parts for the last two terms as follows

$$\int_{x_0}^{x_1} \Lambda(x) \frac{\partial \phi}{\partial u'}\delta u' dx = \left[\Lambda(x) \frac{\partial \phi}{\partial u'} \cancelto{0}{\delta u} \right]_{x_0}^{x_1} - \int_{x_0}^{x_1} \frac{d}{dx}\left(\Lambda \frac{\partial \phi}{\partial u'} \right) \delta u dx. \tag{2.87}$$

Similarly,

$$\int_{x_0}^{x_1} \Lambda(x) \frac{\partial \phi}{\partial v'}\delta v' dx = \left[\Lambda(x) \frac{\partial \phi}{\partial v'} \cancelto{0}{\delta v} \right]_{x_0}^{x_1} - \int_{x_0}^{x_1} \frac{d}{dx}\left(\Lambda \frac{\partial \phi}{\partial v'} \right) \delta v dx. \tag{2.88}$$

Substituting (2.87) and (2.88) into equation (2.86) and adding to equation (2.85) yields

$$\int_{x_0}^{x_1} \left\{ \left[\frac{\partial F}{\partial u} - \frac{d}{dx}\left(\frac{\partial F}{\partial u'} \right) + \Lambda \frac{\partial \phi}{\partial u} - \frac{d}{dx}\left(\Lambda \frac{\partial \phi}{\partial u'} \right) \right] \delta u \right.$$
$$\left. + \left[\frac{\partial F}{\partial v} - \frac{d}{dx}\left(\frac{\partial F}{\partial v'} \right) + \Lambda \frac{\partial \phi}{\partial v} - \frac{d}{dx}\left(\Lambda \frac{\partial \phi}{\partial v'} \right) \right] \delta v \right\} dx = 0.$$

Because $u(x)$ and $v(x)$ are related through the constraint (2.84), they do not vary independently. Hence, we choose $\Lambda(x)$ such that the coefficient of δu vanishes leaving $v(x)$ to vary and requiring that the coefficient of δv also vanishes. This is why the Lagrange multiplier must be allowed to be a function of the independent variable for algebraic and differential constraints. Therefore, the Euler equations are

$$\frac{\partial F}{\partial u} - \frac{d}{dx}\left(\frac{\partial F}{\partial u'} \right) + \Lambda \frac{\partial \phi}{\partial u} - \frac{d}{dx}\left(\Lambda \frac{\partial \phi}{\partial u'} \right),$$

$$\frac{\partial F}{\partial v} - \frac{d}{dx}\left(\frac{\partial F}{\partial v'} \right) + \Lambda \frac{\partial \phi}{\partial v} - \frac{d}{dx}\left(\Lambda \frac{\partial \phi}{\partial v'} \right).$$

We may write the first equation as

$$\frac{\partial}{\partial u}[F + \Lambda\phi] - \frac{d}{dx}\left(\frac{\partial}{\partial u'}[F + \Lambda\phi] \right) = 0,$$

or

$$\frac{\partial \tilde{F}}{\partial u} - \frac{d}{dx}\left(\frac{\partial \tilde{F}}{\partial u'}\right) = 0,$$

where the augmented integrand is

$$\tilde{F} = F + \Lambda(x)\phi.$$

Likewise, the second equation is written in the form

$$\frac{\partial \tilde{F}}{\partial v} - \frac{d}{dx}\left(\frac{\partial \tilde{F}}{\partial v'}\right) = 0.$$

Therefore, we have the same form of Euler's equations as for the unconstrained functional (2.83), but for the augmented integrand $\tilde{F} = F + \Lambda(x)\phi$. Once again, we have converted a constrained problem into an equivalent unconstrained problem for which we follow the usual procedures.

EXAMPLE 2.14 Determine the stationary function for the functional

$$I[u,v] = \int_0^1 \left\{\frac{1}{2}\left[(u')^2 + (v')^2\right] + uv\right\} dx \tag{2.89}$$

subject to the differential constraint

$$\phi(u,v') = v' - u = 0. \tag{2.90}$$

The end conditions for $u(x)$ and $v(x)$ are

$$u(0) = 0, \quad u(1) = 0, \quad v(0) = 0, \quad v(1) = 1. \tag{2.91}$$

Solution: First, we form the augmented integrand

$$\tilde{F}(\Lambda,x,u,v,u',v') = F + \Lambda(x)\phi = \frac{1}{2}\left(u'^2 + v'^2\right) + uv + \Lambda(x)\left(v' - u\right).$$

The Euler equations are

$$\frac{\partial \tilde{F}}{\partial u} - \frac{d}{dx}\left(\frac{\partial \tilde{F}}{\partial u'}\right) = 0, \quad \frac{\partial \tilde{F}}{\partial v} - \frac{d}{dx}\left(\frac{\partial \tilde{F}}{\partial v'}\right) = 0.$$

Observe that because $\Lambda(x)$ also varies, there is a third Euler equation

$$\frac{\partial \tilde{F}}{\partial \Lambda} - \frac{d}{dx}\left(\frac{\partial \tilde{F}}{\partial \Lambda'}\right) = 0.$$

However, because $\partial \tilde{F}/\partial \Lambda' = 0$, this simply produces $v' - u = 0$, which is the differential constraint (2.90). Accordingly, it is not necessary to evaluate the Euler equation corresponding to the Lagrange multiplier. Similarly, it may seem necessary to include the boundary conditions (2.91) as constraints on the functional (2.89) along with the differential equation (2.90). However, this would simply reconfirm that the boundary conditions must be satisfied, which we already know.

Considering the first Euler equation, we have

$$v - \Lambda - \frac{d}{dx}(u') = 0$$

or

$$u'' - v = -\Lambda. \tag{2.92}$$

Similarly, for the second Euler equation, we have

$$u - \frac{d}{dx}(v' + \Lambda) = 0$$

or

$$v'' - u = -\Lambda'. \tag{2.93}$$

Thus, we have the three ordinary differential equations (2.90), (2.92), and (2.93) for the three unknowns $u(x)$, $v(x)$, and $\Lambda(x)$. We use these equations to reduce the system to a single equation for one of the three dependent variables. To do so, differentiate equation (2.92) and substitute into equation (2.93) to eliminate the Lagrange multiplier, giving

$$u''' - v' = v'' - u.$$

However, from the constraint (2.90), $v' = u$ and $v'' = u'$; therefore, we are left with a third-order equation for $u(x)$, which is

$$u''' - u' = 0. \tag{2.94}$$

Given a solution of (2.94) for $u(x)$, we integrate the constraint (2.90) to obtain $v(x)$. The solution to the third-order, constant coefficient ordinary differential equation (2.94) is

$$u(x) = c_1 e^{-x} + c_2 e^x + c_3. \tag{2.95}$$

Integrating to obtain $v(x)$ leads to

$$v(x) = -c_1 e^{-x} + c_2 e^x + c_3 x + c_4. \tag{2.96}$$

To satisfy the end conditions

$$\begin{aligned}
u(0) = 0: &\quad c_1 + c_2 + c_3 = 0, \\
u(1) = 0: &\quad c_1 e^{-1} + c_2 e + c_3 = 0, \\
v(0) = 0: &\quad -c_1 + c_2 + c_4 = 0, \\
v(1) = 1: &\quad -c_1 e^{-1} + c_2 e + c_3 + c_4 = 1,
\end{aligned}$$

or in matrix form

$$\begin{bmatrix} 1 & 1 & 1 & 0 \\ e^{-1} & e & 1 & 0 \\ -1 & 1 & 0 & 1 \\ -e^{-1} & e & 1 & 1 \end{bmatrix} \begin{bmatrix} c_1 \\ c_2 \\ c_3 \\ c_4 \end{bmatrix} = \begin{bmatrix} 0 \\ 0 \\ 0 \\ 1 \end{bmatrix}.$$

The solution of this system for the integration constants is

$$\begin{bmatrix} c_1 \\ c_2 \\ c_3 \\ c_4 \end{bmatrix} = \frac{1}{e-3} \begin{bmatrix} e \\ 1 \\ -e-1 \\ e-1 \end{bmatrix}.$$

As a result, we have the solution for the stationary function $u(x)$ and $v(x)$ given by (2.95) and (2.96), respectively, where the constants are provided above.

REMARKS:

1. Adjoining the integrand with the algebraic or differential constraint does not change the value of the functional for the stationary solution because the expression multiplied by the arbitrary Lagrange multiplier vanishes when $u(x)$ is the stationary function. Thus, I and $\tilde{I}$ have the same value for the stationary function $u(x)$.
2. If there is more than one constraint, multiply each by its own Lagrange multiplier and add to the augmented integrand.
3. In the case of algebraic and differential constraints that apply throughout the domain, the Lagrange multipliers are a function of *all* of the independent variables.
4. In some cases, algebraic and/or differential constraints are only applied at the boundaries of the domain, in which case the Lagrange multiplier is a function only of the locations in space or time at which the constraint is applied.
5. See Luenberger (1969) for a helpful geometric interpretation of Lagrange multipliers in the context of the theory of linear vector spaces.
6. Constrained functionals play a central role in optimization and control (see Chapter 10), where the Euler equations governing the system dynamics provide differential constraints on the optimization or control objective, which is typically expressed as a functional. In this context, the Lagrange multipliers used to augment the algebraic and/or differential constraints to the functional are sometimes referred to as *influence*, or *sensitivity*, *functions* that indicate how the functional I is “influenced by” or “sensitive to” changes in the corresponding variables.
7. See Section 5.1.2 for additional examples involving algebraic constraints in the context of particle dynamics.
8. See Sections 6.5 and 9.5.4 on transient growth stability analysis and Chapter 10 on optimization and control for additional examples with differential constraints.
9. In dynamics (see Section 5.1), algebraic (geometric) constraints are referred to as *holonomic* and differential (kinematic) constraints as *nonholonomic*.
10. In optimization and control, we often have occasion to impose inequality constraints. This eventuality is addressed in Sections 10.4.4 and 10.4.5.

2.8 Summary of Euler Equations

Throughout this chapter, we have derived several versions of the Euler equation(s) for various functionals. In order to summarize the developments of Chapter 2, we repeat these various versions of Euler's equation(s) along with the corresponding natural boundary conditions. This will aid the reader in seeing how these various forms are related, in particular, how each form of the Euler equation is simply a special case of more general situations.

1. $F = F(x,u,u')$:

$$\frac{\partial F}{\partial u} - \frac{d}{dx}\left(\frac{\partial F}{\partial u'}\right) = 0,$$

$$\frac{\partial F}{\partial u'} = 0 \quad \text{at} \quad x = x_0, x_1.$$

2. $F = F(x,u,u',u'')$:

$$\frac{\partial F}{\partial u} - \frac{d}{dx}\left(\frac{\partial F}{\partial u'}\right) + \frac{d^2}{dx^2}\left(\frac{\partial F}{\partial u''}\right) = 0,$$

$$\frac{d}{dx}\left(\frac{\partial F}{\partial u''}\right) - \frac{\partial F}{\partial u'} = 0 \quad \text{and} \quad \frac{\partial F}{\partial u''} = 0 \quad \text{at} \quad x = x_0, x_1.$$

3. $F = F(x,y,u,u_x,u_y)$:

$$\frac{\partial F}{\partial u} - \frac{\partial}{\partial x}\left(\frac{\partial F}{\partial u_x}\right) - \frac{\partial}{\partial y}\left(\frac{\partial F}{\partial u_y}\right) = 0,$$

$$\frac{\partial F}{\partial u_x}\frac{dy}{ds} - \frac{\partial F}{\partial u_y}\frac{dx}{ds} = 0 \quad \text{on} \quad C.$$

4. $F = F(x,y,u,v,u_x,v_x,u_y,v_y)$:

$$\frac{\partial F}{\partial u} - \frac{\partial}{\partial x}\left(\frac{\partial F}{\partial u_x}\right) - \frac{\partial}{\partial y}\left(\frac{\partial F}{\partial u_y}\right) = 0, \quad \frac{\partial F}{\partial v} - \frac{\partial}{\partial x}\left(\frac{\partial F}{\partial v_x}\right) - \frac{\partial}{\partial y}\left(\frac{\partial F}{\partial v_y}\right) = 0,$$

$$\frac{\partial F}{\partial u_x}\frac{dy}{ds} - \frac{\partial F}{\partial u_y}\frac{dx}{ds} = 0 \quad \text{and} \quad \frac{\partial F}{\partial v_x}\frac{dy}{ds} - \frac{\partial F}{\partial v_y}\frac{dx}{ds} = 0 \quad \text{on} \quad C.$$

EXERCISES

2.1 Obtain the Euler equation that governs the stationary function $u(x)$ of the functional

$$I[u] = \int_0^1 \left[(u')^2 + uu' + u^2\right] dx.$$

Assume that the values of $u(x)$ at the boundaries are specified.

2.2 Obtain the Euler equation that governs the stationary function $u(x)$ of the functional

$$I[u] = \int_0^1 \left[(u')^2 + \cos u\right] dx.$$

Assume that the values of $u(x)$ at the boundaries are specified.

2.3 Obtain the Euler equation that governs the stationary function $u(x)$ of the functional

$$I[u] = \int_0^1 \left[x(u')^2 - uu' + u\right] dx.$$

Assume that the values of $u(x)$ at the boundaries are specified.

2.4 In Example 2.2, we make use of special case I to obtain the shortest curve between two points.
   a) Obtain the shortest curve using special case II.
   b) Obtain the shortest curve using the usual method, that is, without appealing to either special case.

2.5 Determine the stationary function $u(x)$ for the functional

$$I[u] = \int_1^2 u'\left(1 + x^2 u'\right) dx,$$

with the boundary conditions

$$u(1) = 0, \quad u(2) = 1.$$

2.6 Determine the stationary function $u(x)$ for the functional

$$I[u] = \int_0^1 \left[(u')^2 + xu'\right] dx,$$

with the boundary conditions

$$u(0) = 0, \quad u(1) = 0.$$

2.7 Determine the stationary function $u(x)$ for the functional

$$I[u] = \frac{1}{2}\int_0^{\pi/2} \left[u^2 - (u')^2\right] dx,$$

with the boundary conditions

$$u(0) = 0, \quad u(\pi/2) = 1.$$

2.8 Determine the stationary function $u(x)$ for the functional

$$I[u] = \frac{1}{2}\int_0^1 \left[x^2 (u')^2 + 2u^2\right] dx,$$

with the boundary conditions

$$u(0) = 0, \quad u(1) = 2.$$

2.9 Determine the natural boundary conditions at $x = 0$ and $x = 1$ for the functional

$$I[u] = \int_0^1 \left[(u')^2 - 2u' - 2xu\right] dx.$$

2.10 Determine the stationary function $u(x)$ for the functional

$$I[u] = \int_0^1 \left[(u')^2 - 2uu' - 2u'\right] dx,$$

with the boundary condition $u(1) = 1$ and a natural boundary condition at $x = 0$.

2.11 Determine the stationary function $u(x)$ for the functional

$$I[u]=\int_0^1\left[\frac{1}{2}(u')^2-x^2u'\right]dx,$$

with the boundary condition $u(0)=0$ and a natural boundary condition at $x=1$.

2.12 Determine the stationary function $u(x)$ for the functional

$$I[u]=\int_0^{\pi/2}\left[(u')^2+u^2-2u\right]dx,$$

with the boundary condition $u(0)=0$ and a natural boundary condition at $x=\pi/2$.

2.13 Determine the stationary function $u(x)$ for the functional

$$I[u]=\frac{1}{2}\int_0^1\left[(u')^2+u^2-2xuu'\right]dx,$$

with the boundary condition $u(0)=1$ and a natural boundary condition at $x=1$.

2.14 Determine the stationary function $u(x)$ for the functional

$$I[u]=\frac{1}{2}\int_0^1\left[(u')^2+2xu+2\right]dx,$$

with the boundary condition $u(0)=1$ and a natural boundary condition at $x=1$.

2.15 Show that no continuous stationary function $u(x)$ exists for the functional

$$I[u]=\int_0^1\left[u^2+xu-2u^2u'\right]dx,$$

with the boundary conditions

$$u(0)=1, \quad u(1)=2.$$

2.16 Determine the stationary function $u(x)$ for the dual functional

$$I[u]=u(1)+u(0)u'(0)+\frac{1}{2}\int_0^1(u')^2dx.$$

2.17 Determine the stationary function $u(x)$ for the dual functional

$$I[u]=-u(1)+\frac{1}{2}\int_0^1\left[(u')^2-u^2-2u\right]dx.$$

2.18 Determine the stationary function $u(x)$ for the dual functional

$$I[u]=-u(\pi)u'(\pi)+\frac{1}{2}\int_0^{\pi}\left[(u')^2-u^2+2u\right]dx.$$

2.19 Determine the stationary function $u(x)$ for the dual functional

$$I[u]=\frac{1}{2}[u(1)]^2-u(1)+\frac{1}{2}\int_0^1\left[(u')^2+u^2\right]dx.$$

2.20 Considering functionals of the form

$$I[u] = \int_{x_0}^{x_1} F(x,u,u',u'')dx,$$

obtain the associated Euler equation (2.46) and natural boundary conditions (2.47).

2.21 Determine the general form of the stationary function $u(x)$ for the functional

$$I[u] = \int_0^1 \left[16u^2 - (u'')^2 + x^2\right] dx,$$

where the boundary conditions are such that the values of $u(x)$ and derivatives of $u(x)$ are all specified.

2.22 Convert the differential equation

$$u'' - u = 0, \quad u(0) = 0, \quad u(1) = 0,$$

into its equivalent variational form.

2.23 In Section 2.5.3 we derive the variational form of the Laplace equation by evaluating the inverse problem. Starting with the resulting variational form (2.62), show that the Euler equation is Laplace's equation (2.60).

2.24 Obtain the Euler equation for the functional

$$I[u] = \iint_A \left[a(x,y)\left(\frac{\partial u}{\partial x}\right)^2 + b(x,y)\left(\frac{\partial u}{\partial y}\right)^2 - c(x,y)u^2 \right] dxdy = 0.$$

2.25 Consider the functional

$$I[u] = \iiint_{\bar{V}} \left[\left(\frac{\partial u}{\partial x}\right)^2 + \left(\frac{\partial u}{\partial y}\right)^2 + \left(\frac{\partial u}{\partial z}\right)^2 + 2fu \right] dxdydz,$$

where $f = f(x,y,z)$ is a specified function. Show that the stationary function $u(x)$ must satisfy the Poisson equation.

2.26 Consider the functional

$$I[u] = \frac{1}{2}\int_0^1 \int_0^{2\pi} \left[\left(\frac{\partial u}{\partial r}\right)^2 + \frac{1}{r^2}\left(\frac{\partial u}{\partial \theta}\right)^2 \right] r d\theta dr,$$

with the boundary condition

$$u(1,\theta) = u_0(\theta),$$

where $u_0(\theta)$ is specified. Show that the stationary function $u(x)$ must satisfy the Laplace equation in polar coordinates.

2.27 The total kinetic and potential energy of a rotating elastic bar is given by the functional

$$I[u] = \int_{t_1}^{t_2} \int_0^1 \left[\left(\frac{\partial u}{\partial t}\right)^2 + \Omega^2 (r+u)^2 - K^2 \left(\frac{\partial u}{\partial r}\right)^2 \right] drdt,$$

where $u(r,t)$ is the radial displacement, Ω is the constant angular velocity, and K is a constant. The radial location of the end of the bar at $r=0$ is fixed, in which case $u(0,t)=0$. Obtain Euler's equations, that is, the equations of motion, along with the boundary conditions at $r=0$ and $r=1$.

2.28 Consider the transverse deflection of a string of length ℓ with mass per unit length ρ subject to a transverse load $f(x,t)$ and constant tension force P. The governing differential equation is

$$\rho\frac{\partial^2 u}{\partial t^2}-P\frac{\partial^2 u}{\partial x^2}-f(x,t)=0,$$

where $u(x,t)$ is the transverse deflection of the string. Show that the correspond ing variational form is

$$I[u]=\int_{t_1}^{t_2}\int_0^{\ell}\left[\frac{1}{2}\rho\left(\frac{\partial u}{\partial t}\right)^2-\frac{1}{2}P\left(\frac{\partial u}{\partial x}\right)^2+fu\right]dxdt.$$

2.29 Consider the functional

$$I[u,v]=\int_0^1\left[(u')^2+(v')^2-2uv+2xu\right]dx.$$

Determine the Euler equations for $u(x)$ and $v(x)$. Assume that the values of $u(x)$ and $v(x)$ at the boundaries are specified.

2.30 Consider the functional

$$I[u,v]=\frac{1}{2}\int_0^1\left[(u')^2+(v')^2+2u'v'\right]dx,$$

with the boundary conditions

$$u(0)=0,\quad v(1)=1,$$

and natural boundary conditions for all unprescribed conditions. Obtain the Euler equations and all necessary boundary conditions for $u(x)$ and $v(x)$.

2.31 Determine the stationary functions $u(x)$ and $v(x)$ for the functional

$$I[u,v]=\int_0^1\left[(u')^2+(v')^2+u'v'\right]dx,$$

with the boundary conditions

$$u(0)=0,\quad u(1)=1,\quad v(0)=2,\quad v(1)=0.$$

2.32 In Section 12.4.2, we will consider two-dimensional grid generation using the *weighted-length functional*

$$I[u,v]=\frac{1}{2}\iint_A\left\{\frac{1}{\phi}\left[\left(\frac{\partial u}{\partial x}\right)^2+\left(\frac{\partial v}{\partial x}\right)^2\right]+\frac{1}{\psi}\left[\left(\frac{\partial u}{\partial y}\right)^2+\left(\frac{\partial v}{\partial y}\right)^2\right]\right\}dxdy,$$

where the two weight functions $\phi(x,y)>0$ and $\psi(x,y)>0$ are given. Show that the resulting Euler equations are

$$\frac{\partial}{\partial x}\left(\frac{1}{\phi}\frac{\partial u}{\partial x}\right)+\frac{\partial}{\partial y}\left(\frac{1}{\psi}\frac{\partial u}{\partial y}\right)=0,\quad \frac{\partial}{\partial x}\left(\frac{1}{\phi}\frac{\partial v}{\partial x}\right)+\frac{\partial}{\partial y}\left(\frac{1}{\psi}\frac{\partial v}{\partial y}\right)=0.$$

2.33 Determine the natural boundary condition at $x = 1$ for the functional

$$I[u] = \int_0^1 (u')^2 dx,$$

subject to the boundary condition

$$u(0) = 0,$$

and the constraint

$$\int_0^1 u^2 dx = 1.$$

2.34 Determine the natural boundary conditions for the functional

$$I[u] = \int_0^1 \left[(u')^2 - 4u^2\right] dx,$$

subject to the constraint

$$\int_0^1 u^2 dx = 1.$$

2.35 Determine the natural boundary conditions for the functional

$$I[u] = \int_1^2 \left[x^2(u')^2 - 4u^2\right] dx,$$

subject to the constraint

$$\int_1^2 u^2 dx = 1.$$

2.36 Determine the Euler equation satisfied by the stationary function $u(x)$ for the functional

$$I[u] = \int_0^1 \left[\left(\sqrt{x}u'\right)^2 - (xu)^2\right] dx,$$

subject to the boundary conditions

$$u(0) = 0, \quad u(1) = 0,$$

and the constraint

$$\int_0^1 (xu)^2 dx = 1.$$

2.37 Determine the Euler equation satisfied by the stationary function $u(x)$ for the functional

$$I[u] = \int_0^1 \left[xu^2 - x^2(u')^2\right] dx,$$

subject to the boundary conditions

$$u(0) = 0, \quad u(1) = 0,$$

and the constraint

$$\int_0^1 xu^2 dx = 1.$$

2.38 Determine the stationary function $u(x)$ for the functional

$$I[u] = \int_0^1 (u')^2 dx,$$

subject to the boundary conditions

$$u(0) = 0, \quad u(1) = 0,$$

and the constraint

$$\int_0^1 u dx = 1.$$

2.39 Determine the stationary function $u(x)$ for the functional

$$I[u] = \int_0^\pi (u')^2 dx,$$

subject to the boundary conditions

$$u(0) = 0, \quad u(\pi) = 0,$$

and the constraint

$$\int_0^\pi u^2 dx = 1.$$

2.40 Consider the functional

$$I[u] = \int_{x_0}^{x_1} \left[s(x)(u'')^2 - p(x)(u')^2 + q(x)u^2 \right] dx,$$

subject to the constraint

$$\int_{x_0}^{x_1} r(x)u^2 dx = 1.$$

If $u(x)$ and its derivative are specified at the boundaries, show that the corresponding Euler equation is the Sturm-Liouville equation

$$\frac{d^2}{dx^2}\left[s(x)\frac{d^2u}{dx^2} \right] + \frac{d}{dx}\left[p(x)\frac{du}{dx} \right] + [q(x) + \Lambda r(x)]u(x) = 0,$$

where Λ is a constant.

2.41 Determine the stationary function $u(x)$ for the functional

$$I[u] = \int_0^\pi (u'')^2 dx,$$

subject to the boundary conditions

$$u(0) = u''(0) = 0, \quad u(\pi) = u''(\pi) = 0,$$

and the constraint

$$\int_0^\pi u^2 dx = 1.$$

2.42 Determine the stationary function $u(x)$ for the functional

$$I[u] = \frac{1}{2}\int_0^\infty \left[(u')^2 - 2xuu'\right] dx,$$

subject to the boundary conditions

$$u(0) = 1, \quad u \to 0 \quad \text{as} \quad x \to \infty,$$

and the constraint

$$\frac{1}{2}\int_0^\infty u^2 dx = \frac{1}{4}.$$

2.43 Determine the Euler equations for the functional

$$I[u,v] = \int_0^1 \left[(u')^2 + (v')^2\right] dx,$$

subject to the constraint

$$u + v^2 = 2.$$

2.44 Determine the Euler equations for the functional

$$I[u,v] = \int_0^1 \left[(u')^2 + u + v'\right] dx,$$

subject to the constraint

$$u' + (v')^2 = 1.$$

2.45 Determine the stationary functions $u(x)$ and $v(x)$ for the functional

$$I[u,v] = \int_0^1 \left[(u')^2 + (v')^2 - 4xv' - 4v\right] dx,$$

subject to the boundary conditions

$$u(0) = 0, \quad u(1) = 1, \quad v(0) = 0, \quad v(1) = 1,$$

and the constraint

$$\int_0^1 \left[(u')^2 - xu' - (v')^2\right] dx = 2.$$

2.46 Determine the stationary functions $u(x)$ and $v(x)$ for the functional

$$I[u,v] = \int_0^1 \left(v^2 - 2u^2\right) dx,$$

subject to the boundary condition

$$u(0) = 1,$$

and a natural boundary condition at $x = 1$, with the constraint

$$u' - u - v = 0.$$

2.47 Consider the functional

$$I[u,v] = \int_0^1 \frac{1}{2}\left[(u')^2 - (v')^2\right] dx,$$

subject to the constraint

$$v'' + u = 0.$$

Given that the end conditions for $u(x)$ and $v(x)$ are specified at both $x = 0$ and $x = 1$, determine the general form of the stationary functions $u(x)$ and $v(x)$.

2.48 Determine the stationary functions $u(x)$ and $v(x)$ for the dual functional

$$I[u,v] = \frac{1}{2}[u(\pi)]^2 + \int_0^\pi \left\{\frac{1}{2}\left[(u')^2 + (v')^2\right] + 1\right\} dx,$$

subject to the boundary conditions

$$u(0) = 0, \quad v(0) = 0, \quad u(\pi) + v(\pi) = 1,$$

with the constraint

$$u' - v = 0.$$

Because this is a dual functional, impose the boundary conditions as constraints.

2.49 Recall Fermat's principle of optics considered in Section 1.3.1. Determine the general path of light in a medium in which the index of refraction only depends upon the vertical coordinate such that $n(u) = \sqrt{u}$. The starting and ending points of the light path are (x_0, u_0) and (x_1, u_1), respectively.

2.50 Recall the liquid drop shape considered in Section 1.3.2. Show that the Euler equation to be solved in order to minimize the total energy of the liquid drop is

$$\frac{u''}{\left[1 + (u')^2\right]^{3/2}} + \frac{u'}{r\left[1 + (u')^2\right]^{1/2}} + \frac{\rho g}{\sigma} u = -\frac{\Lambda}{\sigma},$$

subject to the boundary conditions

$$u'(0) = 0, \quad u(R) = 0, \quad u'(R) = -\tan\alpha,$$

where α is the given contact angle. The constraint on the bubble volume and the extra boundary condition allow us to find the Lagrange multiplier Λ and the bubble radius R, respectively.

3 Rayleigh-Ritz, Galerkin, and Finite-Element Methods

> At a purchase cost of the order of, say, $300,000 or $400,000, the university can operate a computer for the education of its students quite economically ... In a few years it is not unlikely that the computer may have settled immutably into our thought as an absolutely essential part of any university research program in the physical, psychological, and economic sciences.
>
> (Alan J. Perlis, *Computers and the World of the Future*, 1962)

Before proceeding to the applications of variational methods that will occupy our attention throughout the remainder of the text, it is worthwhile to pause and consider how the solution of variational problems is accomplished when a closed form solution is not possible. We have already encountered several such scenarios, particularly when dealing with functionals with multiple independent variables that produce partial differential equations, and we will encounter numerous more situations in the balance of the text.

If an Euler equation cannot be solved in closed form, we may obtain approximate solutions using the following techniques:

1. Solve the differential Euler equation using numerical methods, such as finite-difference methods, spectral methods, or finite-element methods, where the latter two are based on the Galerkin (or other method of weighted residual) approach.
2. Solve the integral variational form approximately using the Rayleigh-Ritz method or finite-element methods based on the Rayleigh-Ritz method.

In this chapter, a brief introduction is supplied for the Rayleigh-Ritz and Galerkin approximate methods, as they provide the basis for finite-element methods. The Rayleigh-Ritz method is applied directly to the variational form of the equations, which is related to the so called *weak form*, while the Galerkin method begins with the differential Euler equation, called the *strong form*. Finite-difference and spectral methods based on solving the differential Euler equation are further removed from the variational approach and are beyond the scope of this text. However, it should be noted that these approaches may be more generally applicable and computationally efficient in certain settings where such solutions are sought.

The emphasis throughout this chapter is on providing a clear account of the variational underpinnings of these methods, primarily through consideration of one-dimensional problems; the details of applying such methods in more general contexts is left to the numerous dedicated texts on finite-element methods applied to initial- and boundary-value problems governed by ordinary and partial differential equations.

3.1 Rayleigh-Ritz Method

3.1.1 Basic Procedure

The Rayleigh-Ritz method, which was introduced by Lord Rayleigh in 1877 and extended by Walther Ritz in 1909, is a procedure used to obtain approximate solutions to variational problems. It can be used as a standalone approximate method or as the basis for a more general numerical method, such as the finite-element method. Because it is applied directly to the variational form rather than the differential form, the Rayleigh-Ritz method is often classified as a *direct method*.

Consider the variational problem

$$\delta \int_{x_0}^{x_1} F(x,u,u',\dots)dx = 0, \tag{3.1}$$

where the stationary function $u(x)$ is to be determined. We approximate the unknown exact stationary function $u(x)$ by $\bar{u}(x)$ using a linear combination of n preselected *basis functions* $\phi_0(x),\phi_1(x),\dots,\phi_n(x)$ as follows:

$$\bar{u}(x) = \sum_{i=0}^{n} c_i\phi_i(x) = \phi_0(x) + c_1\phi_1(x) + \cdots + c_i\phi_i(x) + \dots + c_n\phi_n(x), \tag{3.2}$$

where $c_0 = 1$. This is often called a *trial function*. The solution method consists of determining the c_i coefficients that produce the closest approximation to the exact stationary function $u(x)$ in terms of the preselected basis functions. If the basis functions are linearly independent, then there is a unique solution for the constants $c_1,c_2,\dots,c_n$.

If the end points are specified as

$$u(x_0) = u_0, \quad u(x_1) = u_1,$$

we choose the first basis function $\phi_0(x)$ such that it satisfies both boundary conditions, and the remaining basis functions are homogeneous at the end points according to

$$\phi_0(x_0) = u_0, \quad \phi_1(x_0) = \phi_2(x_0) = \cdots = \phi_n(x_0) = 0,$$
$$\phi_0(x_1) = u_1, \quad \phi_1(x_1) = \phi_2(x_1) = \cdots = \phi_n(x_1) = 0.$$

For example, if $0 \le x \le \ell$ and $u(0) = 0, u(\ell) = h$, one possibility is to let

$$\phi_0(x) = \frac{h}{\ell}x,$$

which is a straight line between the end conditions. Specified boundary conditions are called *essential boundary conditions* because it is essential that they be satisfied by the trial function.

Substituting the trial function (3.2) into the functional (3.1) produces the variational problem

$$\delta \int_{x_0}^{x_1} \bar{F}(x, c_1, c_2, \ldots, c_n) dx = 0,$$

where the c_i constants vary. Therefore, we seek $c_i, i = 1, \ldots, n$ in the trial function (3.2) that corresponds to the stationary function of this functional, which is an approximation of the exact stationary function of (3.1).

Whereas all smooth functions are considered in order to find the exact stationary function $u(x)$, an approximate representation is obtained by prescribing a general form of a trial function and finding the coefficients that most closely approximate the exact stationary function for the given trial function. Increasing n results in a more accurate approximation to the exact stationary function $u(x)$, but it requires more work to find the c_i's in the trial function $\bar{u}(x)$. Specifically, the Rayleigh-Ritz method requires the solution of an $n \times n$ system of algebraic equations, where n is the number of constants sought in the trial function.

EXAMPLE 3.1 Apply the Rayleigh-Ritz method to the differential equation

$$\frac{d^2u}{dx^2} - u = -x, \tag{3.3}$$

with the boundary conditions

$$u(0) = 0, \quad u(\ell) = 0.$$

Solution: First, we must convert the differential Euler equation into its equivalent variational form so that it can be solved using the Rayleigh-Ritz method. To do this so-called *inverse problem*, multiply the equation by the variation of the dependent variable, $\delta u(x)$, and integrate over the domain as follows:

$$\int_0^\ell \left(u'' - u + x\right) \delta u dx = 0.$$

Performing integration by parts on the first term yields

$$\left[u'\delta u\right]_0^\ell + \int_0^\ell \left(-u'\delta u' - u\delta u + x\delta u\right) dx = 0.$$

Because the boundary conditions are specified, the first term evaluated at the end points vanishes. The variation can then be taken out of the integral to give

$$\delta \int_0^\ell \left[-\frac{1}{2}(u')^2 - \frac{1}{2}u^2 + xu\right] dx = 0,$$

which can be written in the variational form

$$\delta \int_0^\ell \left[(u')^2 + u^2 - 2xu\right] dx = 0. \tag{3.4}$$

In the Rayleigh-Ritz method, we select the first basis function $\phi_0(x)$ to satisfy the end conditions; therefore, in this case the simplest choice is

$$\phi_0(x) = 0.$$

Let us use polynomials for the remaining basis functions of the form

$$\phi_1(x) = x(\ell - x), \quad \phi_2(x) = x^2(\ell - x), \quad \ldots, \quad \phi_n(x) = x^n(\ell - x),$$

which is equivalent to the trial function

$$\bar{u}(x) = \phi_0(x) + x(\ell - x)(c_1 + c_2 x + \cdots + c_n x^{n-1}).$$

Note that the basis functions $\phi_1(x), \phi_2(x), \ldots$ are chosen in this way so that they are zero at the end points $x = 0$ and $x = \ell$.

In order to illustrate the Rayleigh-Ritz method, take $\ell = 1$ and $n = 1$, which gives the trial function

$$\bar{u}(x) = c_1 x(1 - x).$$

Differentiating yields

$$\bar{u}'(x) = c_1(1 - x) - c_1 x = c_1(1 - 2x).$$

Substituting into the variational form (3.4) and integrating leads to

$$\delta \int_0^1 \left[c_1^2(1-2x)^2 + c_1^2 x^2(1-x)^2 - 2c_1 x^2(1-x)\right] dx = 0$$

$$\delta \int_0^1 \left[c_1^2\left(x^4 - 2x^3 + 5x^2 - 4x + 1\right) + 2c_1\left(x^3 - x^2\right)\right] dx = 0$$

$$\delta \left[c_1^2\left(\frac{1}{5}x^5 - \frac{1}{2}x^4 + \frac{5}{3}x^3 - 2x^2 + x\right) + 2c_1\left(\frac{1}{4}x^4 - \frac{1}{3}x^3\right)\right]_{x=0}^{x=1} = 0$$

$$\delta \left[\frac{11}{30}c_1^2 - \frac{1}{6}c_1\right] = 0.$$

Only c_1 is varied; thus, evaluating the variation gives

$$\frac{11}{30}(2c_1 \delta c_1) - \frac{1}{6}\delta c_1 = 0,$$

or

$$\left(\frac{11}{15}c_1 - \frac{1}{6}\right)\delta c_1 = 0.$$

The variation of c_1, that is, δc_1, is arbitrary; as a result, its coefficient must vanish according to

$$\frac{11}{15}c_1 - \frac{1}{6} = 0.$$

Solving for c_1 gives

$$c_1 = \frac{5}{22}.$$

The approximate solution with $n = 1$ is then

$$\bar{u}(x) = \frac{5}{22}x(1 - x).$$

Observe that the approximate solution satisfies the end conditions exactly.

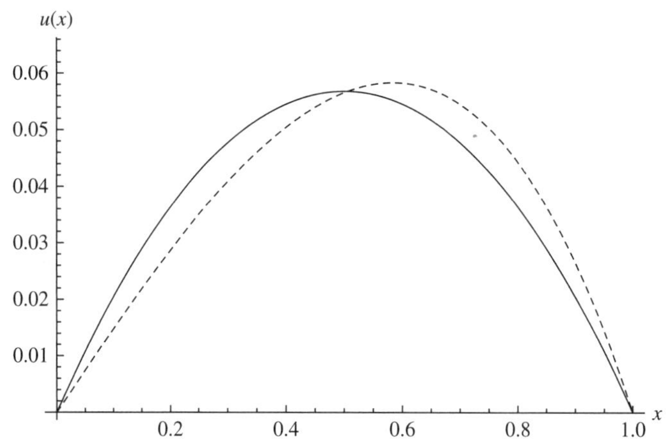

Figure 3.1. Comparison of approximate solution $\bar{u}(x)$ with $n = 1$ (solid line) and exact solution $u(x)$ (dashed line) for Example 3.1.

For comparison, the exact solution of the Euler equation (3.3) is

$$u(x) = \frac{e}{e^2 - 1}\left(e^{-x} - e^x\right) + x.$$

The approximate and exact solutions are compared in Figure 3.1. Note that the comparison is not very good because we only retained one term, which is a parabola, in the trial function; however, it is the "best" parabola!

Observe that the Rayleigh-Ritz method reduces the variational problem to solving a system of algebraic equations for the coefficients in the trial function, rather than solving the differential Euler equation. By using polynomials for the basis functions in the trial function, the integration step becomes particularly straightforward. We only address application of specified, that is, *essential* or *Dirichlet*, boundary conditions here. For the treatment of more general boundary conditions, such as *nonessential* or *Neumann*, see Arpaci (1966).

3.1.2 Self-Adjoint Differential Operators

As discussed in Section 1.8, a differential operator $\mathcal{L}$ is said to be *self-adjoint* if

$$\langle u, \mathcal{L}v \rangle = \langle v, \mathcal{L}u \rangle,$$

where $u(x)$ and $v(x)$ are arbitrary functions. In order for a linear differential operator to be self-adjoint, it must be of the Sturm-Liouville form

$$\mathcal{L}u = \frac{d}{dx}\left[p(x)\frac{du}{dx}\right] + q(x)u = f(x), \tag{3.5}$$

or

$$\mathcal{L}u = \frac{d^2}{dx^2}\left[s(x)\frac{d^2u}{dx^2}\right] + \frac{d}{dx}\left[p(x)\frac{du}{dx}\right] + q(x)u = f(x). \tag{3.6}$$

As in the previous example, to convert the differential equation (3.5) or (3.6) into its equivalent variational form, multiply by δu and integrate over the domain

$$\int_{x_0}^{x_1} (\mathcal{L}u - f)\,\delta u dx = 0. \tag{3.7}$$

Essentially, we are taking the inner product of δu with the differential equation, that is, $\langle \mathcal{L}u - f, \delta u\rangle = 0$. In doing so, we are taking the orthogonal projection of the equation onto the space of admissible functions.

As before, we could perform the required integration by parts in order to convert this to the proper variational form

$$\delta \int_{x_0}^{x_1} F dx = 0. \tag{3.8}$$

For self-adjoint differential operators, however, it generally requires less effort to apply the Rayleigh-Ritz method directly to the *reduced variational form* given by equation (3.7), rather than first obtaining the proper variational form (3.8) and then applying the Rayleigh-Ritz method. This is possible in the case of self-adjoint differential operators because we know that the proper variational form (3.8) does exist. In other words, $\mathcal{L}u = f$ is the Euler equation for a proper variational problem (3.8) for which the reduced form (3.7) is equivalent. This is not always the case for differential equations for which the differential operator is not self-adjoint.

EXAMPLE 3.2 Let us reconsider the differential equation from Example 3.1

$$\frac{d^2u}{dx^2} - u = -x, \quad u(0) = 0, \quad u(1) = 0. \tag{3.9}$$

Determine a two-constant Rayleigh-Ritz approximation to the stationary function using the reduced variational form.

Solution: This is in the self-adjoint form given by (3.5) with $p(x) = 1, q(x) = -1$ and $f(x) = -x$; therefore, the reduced variational form (3.7) is

$$\int_0^1 (u'' - u + x)\delta u dx = 0. \tag{3.10}$$

We set $\phi_0(x) = 0$ to satisfy the end conditions and use a trial function with $n = 2$ based on polynomials according to

$$\bar{u}(x) = x(1-x)(c_1 + c_2 x),$$

or equivalently

$$\bar{u}(x) = (x - x^2)c_1 + (x^2 - x^3)c_2.$$

Then

$$\bar{u}''(x) = -2c_1 + 2(1-3x)c_2.$$

Substituting into the reduced variational form (3.10) with (both c_1 and c_2 vary)

$$\delta u = \frac{\partial u}{\partial c_1}\delta c_1 + \frac{\partial u}{\partial c_2}\delta c_2,$$

gives

$$\int_0^1 \left[-2c_1+2(1-3x)c_2-(x-x^2)c_1-(x^2-x^3)c_2+x\right] \cdot\left[(x-x^2)\delta c_1+(x^2-x^3)\delta c_2\right]dx=0.$$

Multiplying and collecting terms yields

$$\begin{aligned}\int_0^1 \Big\{&\Big[2(c_2-c_1)x+(c_1-8c_2+1)x^2+(2c_1+5c_2-1)x^3\\ &+(2c_2-c_1)x^4-c_2x^5\Big]\delta c_1\\ &+\Big[2(c_2-c_1)x^2+(c_1-8c_2+1)x^3+(2c_1+5c_2-1)x^4\\ &+(2c_2-c_1)x^5-c_2x^6\Big]\delta c_2\Big\}\,dx=0.\end{aligned}$$

Integrating with respect to x and applying the limits, we have

$$\left(-\frac{11}{30}c_1-\frac{11}{60}c_2+\frac{1}{12}\right)\delta c_1+\left(-\frac{11}{60}c_1-\frac{1}{7}c_2+\frac{1}{20}\right)\delta c_2=0.$$

Both δc_1 and δc_2 are arbitrary; therefore, we have two coupled algebraic equations for the constants c_1 and c_2

$$\begin{aligned}22c_1+11c_2&=5,\\ 77c_1+60c_2&=21.\end{aligned}\tag{3.11}$$

Solving this 2×2 system of equations for the unknown coefficients, we obtain

$$c_1=\frac{69}{473},\quad c_2=\frac{7}{43}.$$

Thus, the approximate solution with $n=2$ is

$$\bar{u}(x)=\frac{69}{473}x(1-x)+\frac{7}{43}x^2(1-x).\tag{3.12}$$

Observe from comparison with the previous example that c_1 is different for the one- and two-constant trial functions. That is, changing the trial function requires repeating the Rayleigh-Ritz calculation to obtain all of the coefficients. Comparison of the $n=2$ Rayleigh-Ritz approximation with the exact solution is provided in Figure 3.2. Observe that the two-constant trial function provides a significantly improved approximation to the exact stationary function as compared to the one-constant trial function shown in Figure 3.1. However this improved accuracy is at the expense of additional work to obtain the constants.

3.1.3 Estimating Eigenvalues of Differential Operators

In addition to approximating the solutions to differential equations, the Rayleigh-Ritz method can be used to estimate an *upper bound* for the *smallest* eigenvalue of

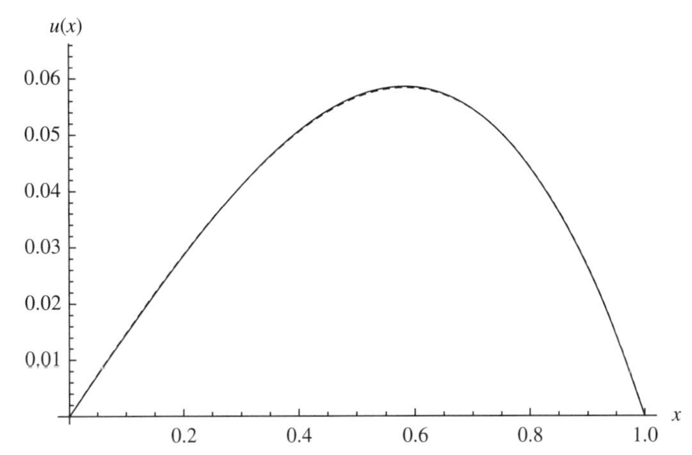

Figure 3.2. Comparison of approximate solution $\bar{u}(x)$ with $n = 2$ (solid line) and exact solution $u(x)$ (dashed line) for Example 3.2.

a differential operator. Recall that for a second-order linear differential operator to be self-adjoint, it must be of the Sturm-Liouville form

$$\mathcal{L} = \frac{1}{r(x)}\left\{\frac{d}{dx}\left[p(x)\frac{d}{dx}\right] + q(x)\right\}.$$

Then the eigenproblem $\mathcal{L}u(x) = -\lambda u(x)$ for the differential operator is the Sturm-Liouville equation

$$\frac{d}{dx}\left[p(x)\frac{du}{dx}\right] + q(x)u(x) + \lambda r(x)u(x) = 0,$$

where λ are the associated eigenvalues of the differential operator.

To obtain the reduced variational form, multiply the Sturm-Liouville equation by δu and integrate over the domain $x_0 \leq x \leq x_1$. Solving for the eigenvalue yields

$$\lambda = -\frac{\displaystyle\int_{x_0}^{x_1}\left\{\left[p(x)u'\right]' + q(x)u\right\}\delta u dx}{\displaystyle\int_{x_0}^{x_1} r(x)u\delta u dx}, \tag{3.13}$$

which is referred to as *Rayleigh's quotient*. In this form we can see that an eigenvalue λ may itself be regarded as a functional; therefore, minimizing this functional will result in the smallest eigenvalue. It is often the smallest eigenvalue that is of physical significance, as in the beam buckling example considered in Section 6.6. In the context of vibrations, for example, where the eigenvalues correspond to natural frequencies, the eigenfunction corresponding to the smallest eigenvalue, which is the fundamental frequency, is the fundamental vibrational mode (see Example 3.4 for the vibrating membrane).

Formation of the functional (3.13) suggests that one approach to estimating the smallest eigenvalue of a Sturm-Liouville differential operator is to utilize a Rayleigh-Ritz approximation of the above reduced functional for λ. The result would be an upper bound on the smallest eigenvalue because any Rayleigh-Ritz estimate will result in an estimate for the eigenvalue that is greater than the exact eigenvalue corresponding to the exact stationary function. In fact, Rayleigh's quotient can be used to obtain the first m eigenvalues in succession by imposing constraints that account for the first $m-1$ eigenvalues and corresponding eigenfunctions (Smith 1974).

EXAMPLE 3.3 Consider the Sturm-Liouville case with $p(x)=1, q(x)=0$ and $r(x)=1$ and boundary conditions $u(0)=0, u(1)=0$. Use the Rayleigh-Ritz method to obtain an upper bound on the smallest eigenvalue and compare with the exact value $\lambda_1=\pi^2$.

Solution: With $p(x)=1, q(x)=0$ and $r(x)=1$, the Rayleigh quotient for the smallest eigenvalue (3.13) is

$$\lambda = -\frac{\int_0^1 u''\delta u dx}{\int_0^1 u\delta u dx}. \tag{3.14}$$

Let us consider the trial function

$$\bar{u}(x) = c_1 x(1-x),$$

in which case

$$\bar{u}''(x) = -2c_1, \quad \delta\bar{u} = x(1-x)\delta c_1.$$

Substituting into equation (3.14) and integrating the numerator and denominator leaves

$$\lambda \approx -\frac{-\int_0^1 2c_1 x(1-x)\delta c_1 dx}{\int_0^1 c_1 x(1-x)x(1-x)\delta c_1 dx} = -\frac{-\frac{1}{3}c_1\delta c_1}{\frac{1}{30}c_1\delta c_1} = 10,$$

which provides an upper bound on the smallest eigenvalue and is very close to the exact value of $\lambda_1=\pi^2=9.8696$.

EXAMPLE 3.4 Calculate an upper bound on the fundamental frequency of vibration of a square membrane clamped along its perimeter as illustrated in Figure 3.3.

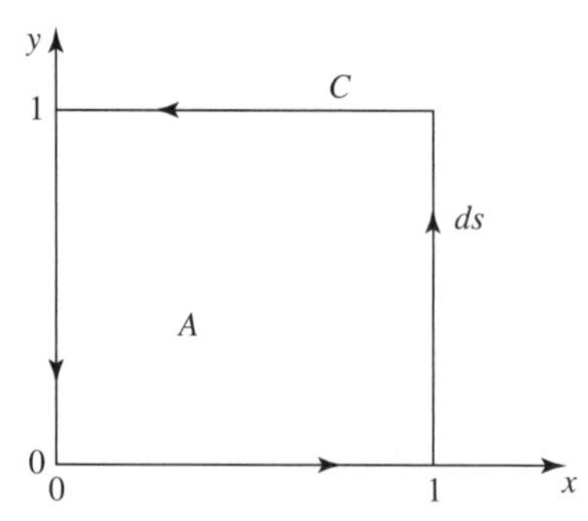

Figure 3.3. Domain for vibration of a square membrane in Example 3.4.

A membrane is a two-dimensional analog of a string in that neither can withstand moments.

Solution: The vibrating membrane is governed by the two-dimensional wave equation

$$\frac{\partial^2 u}{\partial x^2}+\frac{\partial^2 u}{\partial y^2}=\frac{1}{\gamma^2}\frac{\partial^2 u}{\partial t^2}, \tag{3.15}$$

where $u(x,y,t)$ is the lateral displacement of the membrane, and γ is the wave speed in the membrane (see Section 5.4).

As we are seeking a natural frequency, consider harmonic motion of the form

$$u(x,y,t)=e^{i\omega t}\phi(x,y),$$

where ω is the frequency of vibration, and $\phi(x,y)$ is the corresponding vibrational mode shape. Harmonic motion results by taking $\text{Re}(u)=\cos(\omega t)\phi(x,y)$ or $\text{Im}(u)=\sin(\omega t)\phi(x,y)$. Substituting into (3.15) gives

$$\frac{\partial^2 \phi}{\partial x^2}+\frac{\partial^2 \phi}{\partial y^2}=-\frac{\omega^2}{\gamma^2}\phi, \tag{3.16}$$

which is an eigenproblem of the form $\mathcal{L}\phi=-\lambda\phi$, with $\lambda=\omega^2/\gamma^2$. Equation (3.16) is a Helmholtz equation.

To obtain the reduced variational form, multiply (3.16) by $\delta\phi$ and integrate over the domain A as follows:

$$\int_A \nabla^2\phi\delta\phi dA=-\frac{\omega^2}{\gamma^2}\int_A \phi\delta\phi dA.$$

Solving for the eigenvalues, we have

$$\lambda=\frac{\omega^2}{\gamma^2}=-\frac{\displaystyle\int_A \nabla^2\phi\delta\phi dA}{\displaystyle\int_A \phi\delta\phi dA}. \tag{3.17}$$

Applying the Rayleigh-Ritz method, consider the following two-dimensional trial function for the mode shapes

$$\bar{\phi}(x,y)=c_1x(x-1)y(y-1), \tag{3.18}$$

which satisfies zero displacement at the boundaries as required by the boundary conditions. Noting that

$$\begin{aligned}\nabla^2\bar{\phi} &= 2c_1[y(y-1)+x(x-1)],\\ \delta\bar{\phi} &= x(x-1)y(y-1)\delta c_1,\end{aligned}$$

substitution of the trial function into (3.17) gives

$$\lambda=\frac{\omega^2}{\gamma^2}=-\frac{-\frac{1}{45}c_1\delta c_1}{\frac{1}{900}c_1\delta c_1}.$$

Thus,

$$\omega^2 = 20\gamma^2,$$

and the upper bound for the fundamental frequency from the Rayleigh-Ritz approximation is

$$\omega = 2\sqrt{5}\gamma = 4.4721\gamma. \tag{3.19}$$

Let us consider the exact solution and determine the fundamental frequency for comparison with our Rayleigh-Ritz approximation. The general solution of (3.16) for the vibrational mode shapes is

$$\phi(x,y) = c_1 \sin\left(\frac{\omega x}{\sqrt{2}\gamma}\right) \sin\left(\frac{\omega y}{\sqrt{2}\gamma}\right) + c_2 \cos\left(\frac{\omega x}{\sqrt{2}\gamma}\right) \cos\left(\frac{\omega y}{\sqrt{2}\gamma}\right).$$

Enforcing the boundary conditions $\phi(0,y) = \phi(x,0) = 0$ requires that the constant $c_2 = 0$. Then from the other boundary conditions

$$\phi(1,y) = 0 \quad \Rightarrow \quad \sin\left(\frac{\omega}{\sqrt{2}\gamma}\right) \sin\left(\frac{\omega y}{\sqrt{2}\gamma}\right) = 0,$$

$$\phi(x,1) = 0 \quad \Rightarrow \quad \sin\left(\frac{\omega x}{\sqrt{2}\gamma}\right) \sin\left(\frac{\omega}{\sqrt{2}\gamma}\right) = 0.$$

Both expressions lead to the eigenvalues

$$\frac{\omega}{\sqrt{2}\gamma} = n\pi, \quad n = 0,1,2,\ldots.$$

We disregard $n = 0$ as it produces the trivial solution. Therefore, the natural frequencies are

$$\omega = \sqrt{2}\pi\gamma n, \quad n = 1,2,3,\ldots.$$

The exact fundamental frequency, corresponding to $n = 1$, is then

$$\omega_1 = \sqrt{2}\pi\gamma = 4.4429\gamma,$$

which is only slightly smaller than our Rayleigh-Ritz estimate (3.19). Including only one term in the trial function (3.18) has produced a surprisingly accurate upper bound for the smallest frequency.

3.2 Galerkin Method

The Galerkin method was introduced in 1915 by Boris Galerkin and is based on the *method of weighted residuals*. The primary advantage over the Rayleigh-Ritz method is that it is not necessary to write, or even be able to write, the equation in variational form prior to applying the Galerkin method as is the case when using the Rayleigh-Ritz method. This allows the Galerkin method to be applied to a much wider class of problems.

We are again seeking an approximation to the solution of the differential equation, that is, the strong form,

$$\mathcal{L}u = f.$$

As in the Rayleigh-Ritz method, this approximation is in the form of a *trial function* comprised of a linear combination of basis functions as follows

$$\bar{u}(x) = \sum_{i=0}^{n} c_i\phi_i(x) = \phi_0(x) + c_1\phi_1(x) + \cdots + c_i\phi_i(x) + \ldots + c_n\phi_n(x). \tag{3.20}$$

Given this approximate solution $\bar{u}(x)$, we can define the *residual*

$$R(x) = \mathcal{L}\bar{u} - f,$$

which is a measure of the error of the approximation $\bar{u}(x)$. If $\bar{u}(x) = u(x)$, then the residual vanishes. In the method of weighted residuals, we multiply the residual by a set of weight functions $w_i(x), i = 1, \ldots, N$ and integrate over the domain as follows:

$$\int_{x_0}^{x_1} (\mathcal{L}\bar{u} - f)\, w_i\, dx = 0.$$

This is equivalent to setting the inner product of the residual with each weight function to zero, that is, $\langle R(x), w_i(x)\rangle = 0$. Thus, we are taking an orthogonal projection of the residual onto a subspace spanned by the weight functions, or one can think of it as averaging the error (residual) in the weighted differential equation over the domain (for the exact solution, the integrand is zero). After carrying out the necessary integration by parts to shift derivatives from the trial to the weight functions, we arrive at the weak form of the problem.

In the general method of weighted residuals, we can choose different functions for the weight functions $w_i(x)$ and the basis functions $\phi_i(x)$ within the trial function $\bar{u}(x)$. For example, in the *collocation method*, Dirac delta functions are used for the weight functions, and in the *method of least squares*, the residual itself is used as the weight function. In the Galerkin method, however, we use the same functions for the weight functions as we use for the basis functions in the trial function such that $w_i(x) = \phi_i(x)$.

EXAMPLE 3.5 Let us again consider the differential equation in Examples 3.1 and 3.2

$$\frac{d^2u}{dx^2} - u = -x, \quad u(0) = 0, \quad u(1) = 0. \tag{3.21}$$

Using a polynomial trial function with $n = 2$, obtain an approximate solution for the stationary function $u(x)$ using the Galerkin method.

Solution: As in Example 3.2, we will use the basis functions

$$\phi_0(x) = 0, \quad \phi_1(x) = x(1-x), \quad \phi_2(x) = x^2(1-x),$$

which give the trial function

$$\bar{u}(x) = (x - x^2)c_1 + (x^2 - x^3)c_2.$$

Differentiating once gives

$$\bar{u}'(x) = (1 - 2x)c_1 + (2x - 3x^2)c_2.$$

In the Galerkin method, the weight functions are set equal to the basis functions in the trial function such that $w_i(x) = \phi_i(x)$. In general, we then have that

$$\langle R(x), \phi_i(x) \rangle = 0,$$

or

$$\int_0^1 (\mathcal{L}\bar{u} - f)\,\phi_i dx = 0.$$

In our case, this is

$$\int_0^1 (\bar{u}'' - \bar{u} + x)\,\phi_i dx = 0.$$

Upon integration by parts of the first term (and accounting for the specified, or essential, boundary conditions), the weak form is

$$\int_0^1 (-\bar{u}'\phi_i' - \bar{u}\phi_i + x\phi_i)\, dx = 0.$$

We use this weak form because it only includes first-order derivatives and is less restrictive with respect to the continuity of derivatives as compared to the second-order derivatives that appear in the reduced variational form.

For $i = 1$, with $\phi_1(x) = x - x^2$, we have

$$\int_0^1 \Big\{ -\Big[(1-2x)c_1 + (2x - 3x^2)c_2\Big][1-2x]$$
$$-\Big[(x-x^2)c_1 + (x^2 - x^3)c_2\Big]\Big[x - x^2\Big] + x\Big[x - x^2\Big]\Big\}\, dx = 0.$$

Multiplying out the polynomials and collecting terms yields

$$\int_0^1 \Big[\Big(-1 + 4x - 5x^2 + 2x^3 - x^4\Big)c_1 + \Big(-2x + 7x^2 - 7x^3 + 2x^4 - x^5\Big)c_2$$
$$+x^2 - x^3\Big]\, dx = 0.$$

Integrating with respect to x and applying the limits leads to the algebraic equation

$$22c_1 + 11c_2 = 5,$$

which is the same as the first equation in (3.11) using the Rayleigh-Ritz method. A similar procedure with $i = 2$ leads to the second of these two algebraic equations for c_1 and c_2. As in Example 3.2, we solve the two algebraic equations for the two constants in the trial function to provide the approximate solution (3.12).

It turns out that it is no coincidence that the Galerkin and Rayleigh-Ritz methods have produced precisely the same approximate solutions. This is always the case for equations having self-adjoint operators when the same trial function is used. In addition, note the similarities, but subtle differences, between the Galerkin method illustrated here and the Rayleigh-Ritz method applied to the reduced variational form in Example 3.2. For example, taking the inner product of the variation of the dependent variable with the differential equation is essentially the same as taking the inner product with the weight function. In addition, we perform integration by parts in both cases. Recall, however, that the Rayleigh-Ritz method requires that we be able to write the equation in a proper variational form. The Galerkin method, and other methods based on weighted residuals, on the other hand, can be applied to a much broader set of equations as we simply need to obtain the less restrictive weak form.

3.3 Finite-Element Methods

One of the most popular numerical methods in widespread use today is the *finite-element method*. In its original form, it was based on variational methods and originated in the 1950s to facilitate analysis of aerospace structures. It expanded to other fields beginning in the 1960s. Finite-element methods are now employed extensively in both research and practical settings, with a number of commercial software codes based on it. For example, it is now routinely applied to structural mechanics and dynamics, heat transfer, fluid mechanics, and electrodynamics, among numerous other fields.

When implementing finite-element methods, the continuous domain is discretized into a set of interconnected *finite elements* defined by a network of *nodes*. Depending on the geometry of the domain, whether it is one-, two-, or three-dimensional, and the physics being modeled, there could be tens or hundreds of elements for simple problems or thousands, or even millions, for more complex problems. Whereas the trial function is applied over the entire domain for Rayleigh-Ritz and weighted residual methods, it is applied over each individual element in finite-element methods. Essentially, the Rayleigh-Ritz or Galerkin approximation described previously is applied over each element to produce a very large system of linear algebraic equations involving the constants in the trial functions applied to each element. Because the domain is *discretized* using a large number of elements, it is typically sufficient to use low-order polynomials as trial functions for each element. Linear or quadratic polynomials are most common. The larger the number of elements, the greater the resolution of the physics throughout the domain; however, a larger system of algebraic equations is produced that takes longer to solve numerically.

One of the primary advantages of finite-element methods is the ability to accurately approximate complex geometries owing to the flexibility to arrange a sufficient number of finite elements into almost any domain shape. There is a great deal of flexibility in terms of their shape, size, and arrangement. The primary drawback of the Rayleigh–Ritz-based finite-element method is that its application is limited to settings for which the governing equations can be expressed in a proper variational form. For partial differential equations, this can only be proven to be the case for self-adjoint differential operators (see Section 1.8). As we have seen in the previous section, this limitation can be overcome through the use of methods based on weighted residuals, for example, the Galerkin method. Weighted residual methods have been applied within the context of finite-element methods beginning in the late 1960s, which dramatically increased their scope. For example, the governing equations of viscous fluid mechanics (see Chapter 9) do not have an equivalent proper variational form. Therefore, they cannot be solved using Rayleigh-Ritz–based finite-element methods; however, Galerkin and other weighted residual methods can be applied to fluid flows.

In finite-element methods, the differential Euler equation is again referred to as the *strong form*, and the integral equation is the *weak form*. The terminology arises from the fact that the smoothness, that is, continuity, requirements are weaker for the integral form than the differential form owing to, for example, second-order derivatives in the differential Euler equation corresponding to first-order derivatives in the weak form (after integration by parts).

Although the real versatility of finite-element methods is revealed in multidimensional problems, we begin by illustrating the basic approach in a one-dimensional setting. We use the Rayleigh-Ritz method as the underlying approach to approximate the stationary function as it adheres most closely to the variational methods on which this text is focused. It is more common in practice, however, to base finite-element methods on a variant of the method of weighted residuals, such as the Galerkin method.

3.3.1 Rayleigh-Ritz–Based Finite-Element Method

We seek the stationary function of the functional

$$I[u] = \int_a^b F(x,u,u')dx$$

by taking its variation and setting it equal to zero according to

$$\delta \int_a^b F(x,u,u')dx = 0.$$

Dividing the domain $a \leq x \leq b$ into N intervals, we have $N+1$ *nodal points* located at $x_i, i = 0,1,\ldots,N$ as illustrated in Figure 3.4. Consider the ith element as shown in Figure 3.5. Whereas x is the *global variable*, let ξ be a *local variable* across each element, in which case $\xi = 0$ corresponds to the left node x_{i-1}, where $u(x) = u_{i-1}$, and $\xi = 1$ corresponds to the right node x_i, where $u(x) = u_i$. If $h_i = x_i - x_{i-1}$, then

$$\xi = \frac{x - x_{i-1}}{h_i}, \quad x_{i-1} \leq x \leq x_i;$$

consequently, we transform the problem from one in x to one in ξ according to

$$x = x_{i-1} + h_i\xi, \quad dx = h_i d\xi, \quad \frac{d}{dx} = \frac{d\xi}{dx}\frac{d}{d\xi} = \frac{1}{h_i}\frac{d}{d\xi}. \tag{3.22}$$

Over the interval $0 \leq \xi \leq 1$, $u(\xi)$ may be approximated using the Rayleigh-Ritz method in terms of a trial function, called a *shape*, or *interpolation*, *function* when

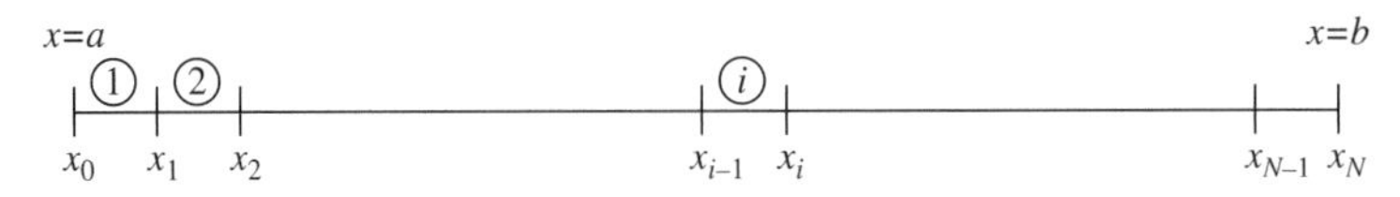

Figure 3.4. One-dimensional discretized domain.

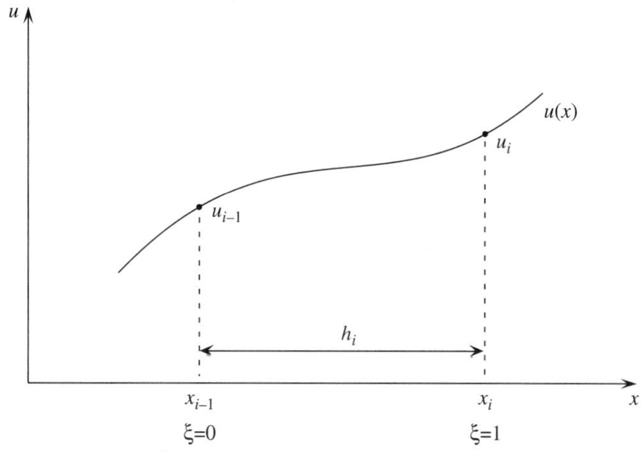

Figure 3.5. Schematic of the ith element.

applied to individual elements. A shape function is given by

$$u(\xi) = u_{i-1}\phi_L(\xi) + u_i\phi_R(\xi).$$

Note that u_{i-1} and u_i replace the c_i constants in the Rayleigh-Ritz method of Section 3.1. We specify the shape functions $\phi_L(\xi)$ and $\phi_R(\xi)$ and seek the nodal values of the constants u_{i-1} and u_i between elements.

In principle, any function can be used for the shape functions. However, for ease of carrying out the integration, polynomials are typically used. In order to improve accuracy, the choice is between a low-order polynomial with more finite elements or a higher-order polynomial with fewer elements. For simplicity, let us use linear shape functions

$$\phi_L(\xi) = 1 - \xi, \quad \phi_R(\xi) = \xi,$$

in which case the function $u(\xi)$ over the ith element is approximated by

$$u(\xi) = u_{i-1}(1-\xi) + u_i\xi = u_{i-1} + \xi(u_i - u_{i-1}). \tag{3.23}$$

Note that $u(0) = u_{i-1}$ and $u(1) = u_i$. Then the variational problem for the entire domain is the sum of the Rayleigh-Ritz approximations over each element, that is,

$$\delta I \approx \sum_{i=1}^{N} \delta \int_0^1 F_i(\xi, u_{i-1}, u_i)d\xi = 0,$$

where $I = I(u_0, u_1, \ldots, u_N)$. Whereas the Rayleigh-Ritz method is a *global* approximation method, using a single trial function for the entire domain, the finite-element method is a *local* approximation method with each element being approximated by a trial function.

EXAMPLE 3.6 Apply the finite-element method to obtain an approximate solution to the differential equation

$$\frac{d}{dx}\left(x\frac{du}{dx}\right) = -x, \tag{3.24}$$

with the boundary conditions

$$u(1)=0, \quad u'(2)=0.$$

The specified boundary condition at $x=1$ is called *essential* because it must be satisfied by the trial function. The derivative boundary condition at $x=2$ is *nonessential* because it arises naturally and is not required to be satisfied by the trial function.
Solution: In order to obtain the variational form, multiply the equation by δu and integrate over the domain

$$\int_1^2 \left[(xu')'+x\right]\delta u dx=0.$$

Integrating by parts, the first term becomes

$$\int_1^2 (xu')'\delta u dx=\left[\cancelto{0}{xu'\delta u}\right]_1^2-\int_1^2 xu'\delta u' dx=-\int_1^2 \frac{1}{2}x\delta(u')^2 dx.$$

The resulting variational form of (3.24) is

$$\delta\int_1^2\left[\frac{1}{2}x(u')^2-xu\right]dx=0.$$

We subdivide this integral into the sum of integrals over each element as follows ($x_0=1, x_N=2$):

$$\delta\int_1^{x_1} F dx+\delta\int_{x_1}^{x_2} F dx+\cdots+\delta\int_{x_{i-1}}^{x_i} F dx+\cdots+\delta\int_{x_{N-1}}^{2} F dx=0, \tag{3.25}$$

where $F=F(x,u,u')$. Consider the ith element

$$\delta\int_{x_{i-1}}^{x_i} F dx=\delta\int_{x_{i-1}}^{x_i}\left[\frac{1}{2}x\left(\frac{du}{dx}\right)^2-xu\right]dx. \tag{3.26}$$

Now we transform this expression from one in terms of the global variable x to one in terms of the local variable ξ according to (3.22) and (3.23). Consider the first term of the integrand in (3.26):

$$\begin{aligned}\frac{1}{2}x\left(\frac{du}{dx}\right)^2 &= \frac{1}{2}(x_{i-1}+h_i\xi)\left\{\frac{1}{h_i}\frac{d}{d\xi}\left[u_{i-1}+\xi(u_i-u_{i-1})\right]\right\}^2\\ &= \frac{1}{2h_i^2}(x_{i-1}+h_i\xi)(u_i-u_{i-1})^2.\end{aligned}$$

Substituting this along with (3.22) and (3.23) into equation (3.26) gives the expression for the ith element in terms of ξ as

$$\delta\int_0^1 F_i d\xi=\delta\int_0^1\left[\frac{1}{2h_i^2}(x_{i-1}+h_i\xi)\Delta u_i^2-(x_{i-1}+h_i\xi)(u_{i-1}+\xi\Delta u_i)\right]h_i d\xi,$$

where $\Delta u_i = u_i - u_{i-1}$ and $F_i = F_i(\xi, u_{i-1}, u_i)$. Multiplying out the integrand and collecting terms in powers of ξ, we have

$$\delta \int_0^1 F_i d\xi = \delta \int_0^1 \left[\frac{1}{2h_i} x_{i-1} \Delta u_i^2 - h_i x_{i-1} u_{i-1} \right.$$
$$\left. + \left(\frac{1}{2} \Delta u_i^2 - h_i x_{i-1} \Delta u_i - h_i^2 u_{i-1} \right) \xi - h_i^2 \Delta u_i \xi^2 \right] d\xi.$$

Carrying out the integration and applying the limits of integration gives

$$\delta \int_0^1 F_i d\xi = \delta \left[\frac{1}{2h_i} x_{i-1} \Delta u_i^2 - h_i x_{i-1} u_{i-1} \right.$$
$$\left. + \frac{1}{4} \Delta u_i^2 - \frac{1}{2} h_i x_{i-1} \Delta u_i - \frac{1}{2} h_i^2 u_{i-1} - \frac{1}{3} h_i^2 \Delta u_i \right].$$

The term in brackets $[\cdot]$ varies with respect to u_{i-1} and u_i; therefore,

$$\delta[\cdot] = \frac{\partial[\cdot]}{\partial u_{i-1}} \delta u_{i-1} + \frac{\partial[\cdot]}{\partial u_i} \delta u_i.$$

Noting that $\partial \Delta u_i / \partial u_{i-1} = -1$ and $\partial \Delta u_i / \partial u_i = 1$, we obtain the following after some simplification:

$$\delta \int_0^1 F_i d\xi = \left[-\left(\frac{1}{2} + \frac{x_{i-1}}{h_i} \right) \Delta u_i - \frac{1}{2} h_i x_{i-1} - \frac{1}{6} h_i^2 \right] \delta u_{i-1}$$
$$+ \left[\left(\frac{1}{2} + \frac{x_{i-1}}{h_i} \right) \Delta u_i - \frac{1}{2} h_i x_{i-1} - \frac{1}{3} h_i^2 \right] \delta u_i.$$

For convenience, we write this in the form

$$\delta \int_0^1 F_i d\xi = (a_i u_{i-1} - a_i u_i - b_i) \delta u_{i-1} + (-a_i u_{i-1} + a_i u_i - c_i) \delta u_i, \tag{3.27}$$

where

$$a_i = \frac{1}{2} + \frac{x_{i-1}}{h_i}, \quad b_i = \frac{1}{2} h_i x_{i-1} + \frac{1}{6} h_i^2, \quad c_i = \frac{1}{2} h_i x_{i-1} + \frac{1}{3} h_i^2.$$

Therefore, the *local matrix* for the ith element is

$$a_i \begin{bmatrix} 1 & -1 \\ -1 & 1 \end{bmatrix} \begin{bmatrix} u_{i-1} \\ u_i \end{bmatrix} = \begin{bmatrix} b_i \\ c_i \end{bmatrix}.$$

The result (3.27) is for the ith element as shown in Figure 3.6.

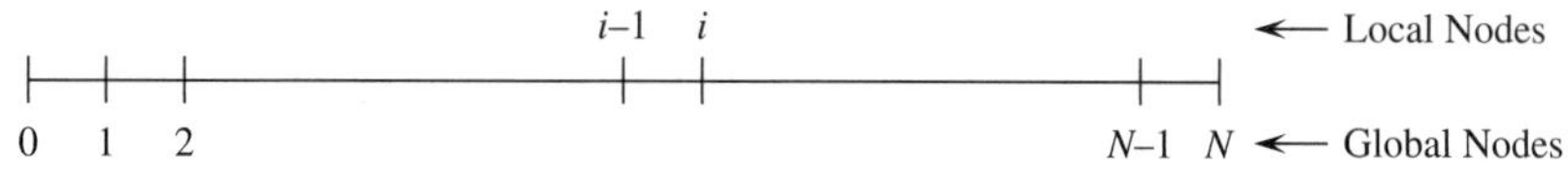

Figure 3.6. Local and global nodes for the one-dimensional finite-element method.

As in equation (3.25), we sum the integrals (3.27) for each of $i = 1,2,\ldots,N$ and set it equal to zero in order to obtain the *global matrix* as follows:

$$
\begin{aligned}
&(a_1u_0 - a_1u_1 - b_1)\delta u_0 + (-a_1u_0 + a_1u_1 - c_1)\delta u_1 \\
&+(a_2u_1 - a_2u_2 - b_2)\delta u_1 + (-a_2u_1 + a_2u_2 - c_2)\delta u_2 \\
&+(a_3u_2 - a_3u_3 - b_3)\delta u_2 + (-a_3u_2 + a_3u_3 - c_3)\delta u_3 \\
&\quad\vdots \\
&+(a_{N-1}u_{N-2} - a_{N-1}u_{N-1} - b_{N-1})\delta u_{N-2} \\
&\qquad +(-a_{N-1}u_{N-2} + a_{N-1}u_{N-1} - c_{N-1})\delta u_{N-1} \\
&+(a_Nu_{N-1} - a_Nu_N - b_N)\delta u_{N-1} + (-a_Nu_{N-1} + a_Nu_N - c_N)\delta u_N = 0.
\end{aligned}
$$

However, the nodal values, $u_0, u_1, \ldots, u_N$, are arbitrary; therefore, setting the coefficients of the variation of each nodal value equal to zero gives the following tridiagonal system of equations:

$$
\begin{array}{cccccc}
a_1u_0 & -a_1u_1 & & & = & b_1 \\
-a_1u_0 & +(a_1+a_2)u_1 & -a_2u_2 & & = & b_2 + c_1 \\
 & -a_2u_1 & +(a_2+a_3)u_2 & -a_3u_3 & = & b_3 + c_2 \\
 & \vdots & & & & \\
 & -a_{N-1}u_{N-2} & +(a_{N-1}+a_N)u_{N-1} & -a_Nu_N & = & b_N + c_{N-1} \\
 & & -a_Nu_{N-1} & +a_Nu_N & = & c_N.
\end{array}
$$

In general, this *global matrix* for the entire domain is a tridiagonal system of $N+1$ equations for $N+1$ unknown nodal values $u_0, u_1, u_2, \ldots u_N$. In this case, however, $u_0 = u(1) = 0$; therefore, we solve the final N equations for the remaining unknown node points.

For comparison with the finite-element solution, the exact solution to the differential equation (3.24) is

$$
u(x) = -\frac{1}{4}x^2 + 2\ln x + \frac{1}{4}, \quad 1 \le x \le 2.
$$

The results are summarized in Figure 3.7. The finite-element results are indicated by dots at each node and the exact solution is the solid line. As would be expected, two elements (three nodes) is insufficient and does not compare well with the exact solution. For ten elements (eleven nodes), however, the finite-element solution compares very well with the exact solution for this smoothly varying one-dimensional example.

There is a trade-off for accuracy between the number of elements and the order of the shape functions used. If n is the number of terms in the Rayleigh-Ritz approximation, that is, the form of the shape functions, and N is the number of elements, then the choice is between:

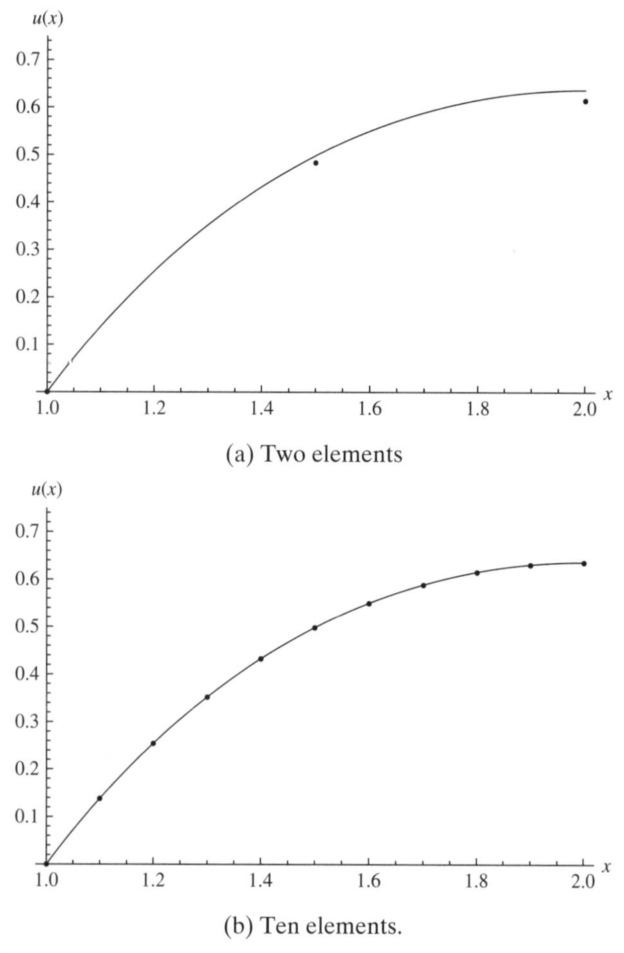

(a) Two elements

(b) Ten elements.

Figure 3.7. Finite-element (dots) and analytical (solid line) solutions for Example 3.6.

$$\text{Large } n, \text{ Small } N \quad \text{vs.} \quad \text{Small } n, \text{ Large } N.$$

Recall that the Rayleigh-Ritz and Galerkin methods produce the same approximate trial function when applied to self-adjoint equations using the same trial functions. Likewise, finite-element methods based on these approaches also produce the same approximate trial function when applied to self-adjoint equations using the same trial functions. Once again, however, the finite-element method based on the Galerkin method (or other weighted residual methods) can be applied to a much broader set of differential equations because it is not necessary to have a proper variational form as is the case when using Rayleigh-Ritz–based finite-element methods. For this reason, most finite-element methods in use today are based on the method of weighted residuals.

3.3.2 Finite-Element Methods in Multidimensions

The real power and flexibility of finite-element methods is realized when applied to multidimensional problems. For two-dimensional domains, the elements are typically quadrilateral, defined by four nodes, or triangular, defined by three nodes. For example, the simplest two-dimensional shape functions associated with quadrilateral elements are linear and of the form

$$u(x,y) = c_1 + c_2 x + c_3 y + c_4 xy.$$

Linear shape functions for triangular elements would be represented by

$$u(x,y) = c_1 + c_2 x + c_3 y.$$

Higher-order shape functions, such as quadratic elements, can also be used to provide additional accuracy with the same number of elements (but more computation per element). In three-dimensional domains, the elements would be generalized to become hexahedral or tetrahedral. In some cases, particularly in structural mechanics, a domain may be thin compared to its major dimension, such as its length. In such cases, *beam* or *shell elements* may be used to simplify the finite-element analysis.

In addition to being able to accommodate very complex domain shapes in two- or three-dimensions, finite-element methods allow one to adaptively improve the numerical resolution in areas where the solution experiences large gradients by simply adding additional elements in such regions. Smaller elements and/or higher-order shape functions are able to produce more accurate approximations to the solutions of partial differential equations (strong forms) or their equivalent variational (weak) forms.

There are numerous practical issues that arise in the application of finite-element methods in multidimensions. For example, the local element information must be mapped to the global domain as they are not necessarily arranged in a predefined orderly fashion as in the one-dimensional case illustrated here; this is done using an *element connectivity matrix*, which defines the equivalence between local element node numbers and global node numbers.

There are many books on finite-element methods. However, it is common for authors, particularly of modern texts, to ignore the variational underpinnings of finite-element methods. Instead, they focus on the use of weighted residual approaches, such as the Galerkin method. See Chung (1978) and Fish and Belytschko (2007) for more details regarding the use of finite-element methods, including its variational roots. A particularly promising approach to solving partial differential equations numerically is the *spectral-element method*. It combines the flexibility of finite-element methods in terms of being able to treat complex domain shapes and adaptively refine the elements as necessary with the high-accuracy, fast-convergence properties of spectral methods.

EXERCISES

3.1 Consider the functional

$$I[u] = \int_0^1 \left[(xu')^2 + 2u^2 \right] dx,$$

with the boundary conditions $u(0) = 0$ and $u(1) = 2$.

a) Use the Rayleigh-Ritz method, with the trial function

$$\bar{u}(x) = 2x + x(1-x)(c_1 + c_2 x),$$

to determine an approximation to the stationary function $u(x)$. Note that $\phi_0(x) = 2x$ is chosen in order to satisfy the boundary conditions.

b) Compare the approximate solution from part (a) with the exact solution from Example 2.1.

3.2 Consider the functional

$$I[u] = \int_0^{\pi/2} \left[(u')^2 + u^2 - 2u \right] dx,$$

with the end condition $u(0) = 0$. A natural boundary condition is applied at $x = \pi/2$.

a) Use the Rayleigh-Ritz method, with the trial function

$$\bar{u}(x) = c_1 \sin x,$$

to determine an approximation to the stationary function $u(x)$. Note that the trial function satisfies both boundary conditions.

b) Compare the approximate solution from part (a) evaluated at the mid-point $x = \pi/4$ with the exact solution from Exercise 2.12 evaluated at the same point.

3.3 Consider the differential equation

$$\frac{d}{dx}\left(x \frac{du}{dx} \right) + u = x,$$

along with the boundary conditions

$$u(0) = 0, \quad u(1) = 1.$$

a) Convert the differential equation into its equivalent variational form.

b) Use the Rayleigh-Ritz method to find an approximate solution with the trial function

$$\bar{u}(x) = x + c_1 x(1 - x).$$

Note that $\phi_0(x) = x$ is chosen in order to satisfy the boundary conditions.

3.4 Consider the differential equation

$$\frac{d^2u}{dx^2} + u = x,$$

along with the boundary conditions

$$u(0) = 0, \quad u(1) = 0.$$

a) Convert the differential equation into its equivalent variational form.

b) Use the Rayleigh-Ritz method to find an approximate solution with the trial function

$$\bar{u}(x) = c_1 x(1 - x).$$

3.5 Consider the differential equation

$$\frac{d^4u}{dx^4} + \omega^2 u = 0,$$

along with the boundary conditions

$$u(0) = 0, \quad u'(0) = 0, \quad u''(1) = 0, \quad u'''(1) = 0.$$

a) Convert the differential equation into its equivalent variational form.
b) Employ the Rayleigh-Ritz method to approximate ω^2 using the trial function

$$\bar{u}(x) = c_1 x^2.$$

3.6 Consider the differential equation

$$\frac{d^2u}{dx^2} - \frac{2}{x^2}u = 0,$$

along with the boundary conditions

$$u(0) = 0, \quad u(1) = 1.$$

a) Convert the differential equation into its equivalent variational form.
b) Using the Rayleigh-Ritz method, determine which of the following trial functions gives the *best* solution:

$$\bar{u}_1 = c_1 x, \quad \bar{u}_2 = c_1 x^2, \quad \bar{u}_3 = c_1 x^3.$$

Note that in all three cases, in order to satisfy the boundary conditions, we must have $c_1 = 1$.
c) Check to see if the best trial function is also an exact solution of the differential equation.

3.7 Minimization of potential energy for a simply supported beam of length ℓ with a uniformly distributed load P requires that the functional

$$V[u] = \int_0^\ell \left[\frac{1}{2}EI(u'')^2 - Pu\right] dx$$

be minimized, where EI is the bending stiffness, which is constant, and $u(x)$ is the static deflection of the beam. The boundary conditions for a simply supported beam are

$$u(0) = 0, \quad u''(0) = 0, \quad u(\ell) = 0, \quad u''(\ell) = 0.$$

a) Obtain the exact deflection distribution $u(x)$.
b) Use the Rayleigh-Ritz method, with the trial function

$$\bar{u}(x) = c_1 \sin(\pi x/\ell),$$

to determine an approximate deflection for the beam.

c) Compare the approximate solution for the deflection at the center of the beam $u(\ell/2)$ from part (b) with the exact value from part (a).

3.8 Consider the differential equation

$$\frac{d}{dx}\left[(1+x)\frac{du}{dx}\right]+\lambda u=0,$$

along with the boundary conditions

$$u(0)=0, \quad u(1)=0.$$

Obtain an approximation to the smallest eigenvalue λ using the trial function

$$\bar{u}(x)=c_1 x(1-x).$$

3.9 Consider the differential equation

$$\mathcal{L}u=x^2\frac{d^2u}{dx^2}-2x\frac{du}{dx}+2u=0.$$

a) Obtain the adjoint operator of $\mathcal{L}$ assuming homogeneous boundary conditions. Is $\mathcal{L}$ self-adjoint?

b) Now consider the above differential equation subject to the following nonhomogeneous boundary conditions

$$u(-1)=-2, \quad u(1)=0.$$

The boundary conditions may be made homogeneous, that is, $\hat{u}(-1)=0, \hat{u}(1)=0$, using the transformation

$$u(x)=\hat{u}(x)+x-1.$$

In addition, through multiplication by a suitable factor, the differential operator $\mathcal{L}$ may be converted to the Sturm-Liouville form

$$\mathcal{L}_{SL}=\frac{d}{dx}\left[p(x)\frac{d}{dx}\right]+q(x),$$

where $p(x)=1/x^2$ and $q(x)=2/x^4$. After converting the differential equation

$$\mathcal{L}_{SL}\hat{u}=\frac{d}{dx}\left[\frac{1}{x^2}\frac{d\hat{u}}{dx}\right]+\frac{2}{x^4}\hat{u}=2$$

to its equivalent (reduced or proper) variational form, determine an approximate solution for $u(x)$ using the Rayleigh-Ritz method with the trial function

$$\bar{u}(x)=c_1(-1-x)(1-x).$$

c) Obtain the exact solution of the original differential equation $\mathcal{L}u=0$, with the nonhomogeneous boundary conditions, and compare it to the Rayleigh-Ritz approximate solution obtained in part (b).

PART II

PHYSICAL APPLICATIONS

4 Hamilton's Principle

Nature loves simplicity and unity.

(Johannes Kepler)

Calculus of variations, more than any other branch of mathematics, is intimately connected with the physical world in which we live. Nature favors extremum principles, and calculus of variations provides the mathematical framework in which to express such principles. As a result, many of the laws of physics find their most natural mathematical expression in variational form. In Chapter 1, we considered the cases of Fermat's principle of optics and minimization of total energy to determine the shape of a liquid drop on a solid surface. The objective of Part II is to provide a brief introduction to a variety of physical phenomena from a unified variational point of view. The emphasis is on illustrating the wide range of applications of the calculus of variations, and the reader is referred to dedicated texts for more complete treatments of each topic. The centerpiece of these seemingly disparate subjects is Hamilton's principle, which provides a compact form of the dynamical equations of motion – its traditional area of application – and the governing equations for many other physical phenomena as illustrated throughout this and subsequent chapters. Much of the historical development of the calculus of variations is centered around its application to dynamical systems; therefore, a number of the important principles and historical figures intimately connected with the calculus of variations will be highlighted in this chapter.

We begin by stating Hamilton's principle without proof for discrete and continuous systems and showing that the corresponding Euler equation is Newton's second law. Several examples are included in order to assist in developing some intuition about Hamilton's principle before providing a formal derivation from the first law of thermodynamics. Along the way, we will encounter the Euler-Lagrange equations of motion, the Hamiltonian formulation, and show that the Euler-Lagrange equations are independent of the coordinate system. The chapter closes with a discussion of the powerful theorem of Noether, which draws a connection between mathematical symmetries in Hamilton's principle and physical conservation laws, and a brief discussion of the philosophy of science.

4.1 Hamilton's Principle for Discrete Systems

Let us begin by considering a single $O(1)$-sized rigid particle, a point mass m, acted upon by a force $\mathbf{f}(t)$, where $\mathbf{r}(t)$ is the position vector of the system and is the dependent variable. In order to determine the actual trajectory of the system $\mathbf{r}(t)$ over the finite time interval $t_1 \leq t \leq t_2$, Hamilton's principle asserts that the trajectory must be such that $\mathbf{r}(t)$ is a stationary function of the functional

$$\int_{t_1}^{t_2} (\delta T + \mathbf{f} \cdot \delta \mathbf{r}) dt = 0. \tag{4.1}$$

The *kinetic energy* of the system as a whole is

$$T = \frac{1}{2} m \mathbf{v}^2 = \frac{1}{2} m \dot{\mathbf{r}}^2,$$

where dots denote differentiation with respect to time, and $\mathbf{v} = \dot{\mathbf{r}}$ is the velocity of the system. Note that $\mathbf{v}^2 = \mathbf{v} \cdot \mathbf{v}$, which is a scalar. In the second term, $\mathbf{f} \cdot \delta \mathbf{r}$ is known as the *virtual work* owing to application of the actual resultant external force $\mathbf{f}(t)$ acting on the system through the infinitesimal virtual (hypothetical) displacement $\delta \mathbf{r}$ leading to, for example:

- *Gravitational potential energy* owing to the action of gravity on the system.
- Potential energy of deformation, or *strain energy*, owing to deformation of the system by external forces.
- *Electrostatic potential* owing to a distribution of charges.

Equation (4.1) is the general form of *Hamilton's variational principle for a single particle*. We will formally derive Hamilton's principle in Section 4.5.

We speak in terms of a *virtual work* owing to a *virtual displacement* instead of an actual work, because we must consider all of the forces acting on the system whether they actually produce work, which is a force acting through a distance, or not. For example, a system has gravitational potential energy whether the force of gravity mg acts through a distance, thereby producing work, or not. More specifically, virtual displacement is a hypothetical (possible) displacement that is consistent with the geometric constraints on a system; it is in the direction that the system could move, whether it actually does so or not.

In order to determine the Euler equation corresponding to Hamilton's principle (4.1), consider the first term for the kinetic energy

$$\begin{aligned}
\int_{t_1}^{t_2} \delta T dt &= \int_{t_1}^{t_2} \delta \left(\frac{1}{2} m \dot{\mathbf{r}}^2 \right) dt \\
&= m \int_{t_1}^{t_2} \dot{\mathbf{r}} \cdot \delta \dot{\mathbf{r}} \, dt \\
\int_{t_1}^{t_2} \delta T dt &= m \left(\left[\dot{\mathbf{r}} \cdot \delta \mathbf{r} \right]_{t_1}^{t_2} - \int_{t_1}^{t_2} \ddot{\mathbf{r}} \cdot \delta \mathbf{r} \, dt \right),
\end{aligned}$$

where integration by parts is employed in the final step. The first term vanishes if we insist that the virtual displacements $\delta \mathbf{r}$ are zero at t_1 and t_2, in which case the varied and actual paths coincide at the end points. Substituting into (4.1) leads to

$$\int_{t_1}^{t_2} (-m\ddot{\mathbf{r}} + \mathbf{f}) \cdot \delta\mathbf{r}\, dt = 0.$$

Because $\delta\mathbf{r}$ is arbitrary, by the fundamental lemma of the calculus of variations, we have the Euler equation

$$m\ddot{\mathbf{r}} - \mathbf{f} = 0,$$

which is the familiar *Newton's Second Law*

$$\mathbf{f} = m\mathbf{a}, \tag{4.2}$$

where $\mathbf{a} = \ddot{\mathbf{r}}$ is the linear acceleration of the system. Accordingly, Newton's second law, which in the form (4.2) is typically regarded as conservation of linear momentum, is the Euler equation corresponding to Hamilton's variational principle applied to discrete systems.

Let us return to Hamilton's principle (4.1). In general, the forces $\mathbf{f}$ acting on the system can be either *conservative*, $\mathbf{f}_c$, or *nonconservative*, $\mathbf{f}_{nc}$. The virtual work done by conservative forces, W_c, is independent of the path and only depends on the initial and final positions. Such work is recoverable, that is, the process is *reversible*. The work done is all converted to kinetic energy or stored as potential energy in the system. For example, the work done in compressing or stretching an elastic spring is stored as *strain energy* within the spring. In this manner, mechanical energy is conserved. Conservative forces include those due to gravity, stretching of an elastic spring, deformation of an elastic body, and electric forces, for example. In contrast, the virtual work done by nonconservative forces, W_{nc}, is dependent on the path, and it is not recoverable. In the case of nonconservative forces, although total energy is conserved according to the first law of thermodynamics, mechanical energy is not conserved; typically, some energy is converted to heat. Examples of nonconservative forces include friction, drag, damping, deformation of an inelastic body, and magnetic forces. Note that nonconservative forces, which are often associated with some form of dissipation, can act on a conservative system within the context of Hamilton's principle as long as any associated heat transfer between the system and its surroundings or changes in internal energy are neglected.

For nonconservative forces, we must leave the virtual work term in Hamilton's principle in the form $\mathbf{f}_{nc} \cdot \delta\mathbf{r}$. For conservative forces, however, the work may be expressed as an exact differential of a scalar function $\Gamma(\mathbf{r})$, such that $\mathbf{f}_c \cdot \delta\mathbf{r} = \delta\Gamma$. Then the work done by a conservative force is

$$W_c = \int_{\mathbf{r}_1}^{\mathbf{r}_2} \mathbf{f}_c \cdot \delta\mathbf{r} = \int_{\mathbf{r}_1}^{\mathbf{r}_2} \delta\Gamma = \Gamma(\mathbf{r}_2) - \Gamma(\mathbf{r}_1),$$

in which case the work only depends on the initial and final positions $\mathbf{r}_1$ and $\mathbf{r}_2$, respectively, and not the path taken between them. In order to regard the scalar function Γ as the *potential energy* V, we let $\Gamma = -V$. The negative sign is introduced because, for example, the gravitational potential energy, which equals the work required to raise a mass against gravity to its current location, is equal to the negative of the virtual work $\mathbf{f}_c \cdot \delta\mathbf{r}$ acting on the body while being raised. As a result, the force on the object acts in the direction of decreasing potential. Therefore, for conservative forces we have

$$\mathbf{f}_c \cdot \delta\mathbf{r} = -\delta V. \tag{4.3}$$

Note that in Cartesian coordinates, for example,

$$\delta V = \frac{\partial V}{\partial x}\delta x + \frac{\partial V}{\partial y}\delta y + \frac{\partial V}{\partial z}\delta z = \nabla V \cdot \delta\mathbf{r}, \tag{4.4}$$

where ∇V is the gradient of V given by

$$\nabla V = \frac{\partial V}{\partial x}\mathbf{i} + \frac{\partial V}{\partial y}\mathbf{j} + \frac{\partial V}{\partial z}\mathbf{k},$$

and the virtual displacement is

$$\delta\mathbf{r} = \delta x\mathbf{i} + \delta y\mathbf{j} + \delta z\mathbf{k}.$$

Substituting equation (4.4) into (4.3), we see that

$$\mathbf{f}_c \cdot \delta\mathbf{r} = -\nabla V \cdot \delta\mathbf{r};$$

therefore,

$$\mathbf{f}_c = -\nabla V,$$

showing that a conservative force may be expressed as the gradient of a scalar function. Consequently, the total virtual work done by all of the forces $\mathbf{f}(t)$ acting on the system may be expressed as

$$W_f = W_c + W_{nc} = -V + \int_{\mathbf{r}_1}^{\mathbf{r}_2} \mathbf{f}_{nc} \cdot \delta\mathbf{r},$$

where the first term represents the potential energy (virtual work) owing to conservative forces, and the second term is virtual work owing to nonconservative (and/or conservative) forces. In other words, the virtual work for both conservative and nonconservative forces may be expressed in the form $W_f = \int \mathbf{f} \cdot \delta\mathbf{r}$, whereas conservative forces alternatively may be expressed as $W_c = -V$.

Substituting equation (4.3) for conservative forces into the general form of *Hamilton's principle for conservative and nonconservative forces* (4.1) leads to

$$\delta \int_{t_1}^{t_2} (T - V)dt + \int_{t_1}^{t_2} \mathbf{f} \cdot \delta\mathbf{r}dt = 0, \tag{4.5}$$

where V is the potential energy owing to the conservative forces, and the $\mathbf{f} \cdot \delta\mathbf{r}$ contribution can be used for either conservative or nonconservative forces $\mathbf{f}$. If only conservative forces act on the system, we may write

$$\delta I = \delta \int_{t_1}^{t_2} Ldt = 0, \tag{4.6}$$

where $L = T - V$ is called the *Lagrangian*. This is *Hamilton's principle for conservative force fields*. Of all the possible paths from t_1 to t_2, the actual path of a particle is that for which the integral $\int_{t_1}^{t_2} Ldt$ is stationary, but not necessarily a minimum.

Hence, the system moves in such a way as to make the time average of the difference between the kinetic and potential energy stationary. Lanczos (1970) describes the Lagrangian $L = T - V$ as "the excess of kinetic energy over potential energy."

Because it is expressed solely in terms of scalar quantities, namely energies, the variational form of Hamilton's principle for conservative systems (4.6) is independent of the coordinate system. This is one of its primary advantages over simply using Newton's second law. Recall that Newton's second law (4.2) is a vector equation; therefore, when using Newton's second law directly, the choice of coordinate system is of primary consideration. In particular, it determines the form of Newton's second law to be used. In contrast, the choice of coordinate system is a secondary consideration in the variational approach. See Example 4.3 for an illustration. Why is this so important? Because coordinate systems (reference frames) are arbitrary and chosen by us for convenience, physical laws should not be dependent on our choice of coordinate system. In other words, the physics of the system remains the same no matter what our vantage point.

We regard Hamilton's principle and Newton's second law as equivalent variational and differential forms, respectively, of the same physical principle. However, there is one important aspect in which they appear to differ, where it is not so clear that they are indeed equivalent. A differential equation is thought of as governing the behavior at each point in space and time, and in the case of initial-value problems as we have here, initial conditions are specified at the initial time t_1 only. In particular, Newton's second law requires two initial conditions at the initial time t_1, typically specifying the initial position and velocity. Given the governing differential equation and initial conditions, the solution may be found for all time. In contrast, Hamilton's variational principle is formulated over a specified time range $t_1 \leq t \leq t_2$ and requires that the position vector be taken as known at both the initial time t_1, for which it truly is known, but also at the final time t_2, at which it is not known in general. Thus, Hamilton's principle presumes that we know the system's final time and position. Whereas the differential form applies the two initial conditions on position and velocity at t_1, the variational form assumes that the position is specified at both t_1 and t_2, and no velocity condition appears to be necessary. In other words, the differential form is framed as an initial-value problem, whereas the variational form is framed as a boundary-value problem. This distinction is largely philosophical as it is the differential form that is typically solved to obtain the stationary function of Hamilton's principle; however, quantum theory, in the form of Feynman's path-integral, offers a possible resolution (see Section 8.2.3).

Implied by the above description of Hamilton's principle and Newton's second law is an important fundamental distinction between variational and differential forms of the governing equations that deserves highlighting. The differential form, in the form of the Euler equation, is posed as a *local law* that enforces the physical principle for each rigid particle in the system at each point within the domain. On the other hand, the variational form of Hamilton's principle is posed as a *global law* that enforces the physical principle globally over the entire system or domain. Although the two forms are certainly equivalent mathematically, the global variational form leads to a greater intuitive understanding of the physical principle involved and provides a platform for determining certain properties of the physical laws (for example, see Noether's theorem in Section 4.7), while the local differential form is typically the one used to actually obtain solutions.

Hamilton's principle in the form (4.1), (4.5), or (4.6) may be extended to systems having multiple discrete particles through summation. For a system of N particles (point masses), the kinetic energy of the moving system is

$$T = \sum_{k=1}^{N} T_k = \sum_{k=1}^{N} \frac{1}{2} m_k \dot{\mathbf{r}}_k^2 = \sum_{k=1}^{N} \frac{1}{2} m_k \mathbf{v}_k^2.$$

For conservative forces, the potential energy, which is the negative of the virtual work required to move the particles, is

$$V = \sum_{k=1}^{N} V_k = \sum_{k=1}^{N} \int \delta V_k = -\sum_{k=1}^{N} \int_{\mathbf{r}_{1k}}^{\mathbf{r}_k} \mathbf{f}_{ck} \cdot \delta \mathbf{r}_k,$$

where $\mathbf{r}_{1k}$ are the position vectors for the system at the initial time t_1, and $\mathbf{r}_k$ are the position vectors at the current time t.

REMARKS:

1. A system is said to have n degrees of freedom, where n is the minimum number of scalar dependent variables required to fully describe the motion or state of the system. For example, a single particle system ($N = 1$) free to move in three-dimensional space has three degrees of freedom ($n = 3$).
2. $S = \int_{t_1}^{t_2} L dt$ is sometimes called the *action integral*, in which case equation (4.6) is called the *principle of least action* developed by Pierre Louis Moreau de Maupertuis, a student of Johann Bernoulli. However, this is somewhat misleading as it implies a minimum, which does not necessarily occur. It would be better referred to as the *principle of stationary action*. In fact, the calculus of variations has shown that there has been an overemphasis on minimum principles during the period in which such methods were being developed in the seventeenth and eighteenth centuries.
3. Hamilton's principle, which was formulated in 1834 by William Rowan Hamilton, was a formalization of the imprecisely stated principle of least action of Maupertuis.
4. Observe that Hamilton's principle for conservative force fields (4.6) is in the proper variational form $\delta \int_{t_1}^{t_2} F dt = 0$. However, it is not possible to write the general form of Hamilton's principle (4.1) or (4.5) for nonconservative forces in such a proper variational form.
5. The mechanical energy of a system represents its capacity to do work. The *kinetic energy* of the system is the energy due to its motion, whereas the *potential energy* of the system is due to its position in a force field, such as gravitational potential energy or strain energy contained within the system owing to deformation of a spring.
6. We refer to systems acted upon by nonconservative forces as nonconservative systems. Whereas all systems must obey conservation of energy according to the first law of thermodynamics, a nonconservative system is such that the *mechanical energy* in the form of kinetic and potential energy is not conserved.

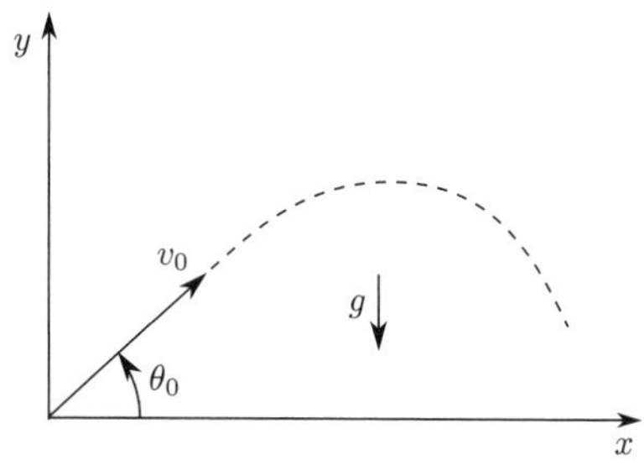

Figure 4.1. Schematic of two-dimensional projectile motion.

The lost mechanical energy is converted to some other form, such as heat, which is not accounted for in Hamilton's principle. Likewise, we say that a system is conservative when its mechanical energy is conserved.

As illustrated in the following three examples, application of Hamilton's principle to systems of discrete particles leads to the *equations of motion* for dynamical systems. We introduce a more general approach in Section 4.3.

EXAMPLE 4.1 Consider the two-dimensional projectile motion of an object of mass m leaving the origin at angle θ_0 from the x-axis with velocity v_0 as illustrated in Figure 4.1. Determine the positions $x(t)$ and $y(t)$.

Solution: We will neglect drag, which is caused by friction owing to the motion of the object through the air. The corresponding initial conditions for position and velocity are

$$x(0) = 0, \quad y(0) = 0, \quad \dot{x}(0) = v_0 \cos\theta_0, \quad \dot{y}(0) = v_0 \sin\theta_0.$$

The system has one particle ($N = 1$), the projectile, and two degrees of freedom ($n = 2$) as it requires $x(t)$ and $y(t)$ in order to fully specify the motion of the projectile.

With the velocity being given by $\mathbf{v} = \dot{\mathbf{r}} = v_x\mathbf{i} + v_y\mathbf{j} = \dot{x}\mathbf{i} + \dot{y}\mathbf{j}$, the kinetic energy of the projectile is

$$T = \frac{1}{2}m\mathbf{v}^2 = \frac{1}{2}m\dot{\mathbf{r}}\cdot\dot{\mathbf{r}} = \frac{1}{2}m(\dot{x}^2 + \dot{y}^2).$$

In the absence of drag, the only force acting on the projectile is that due to gravity, which is conservative. Consequently, $\mathbf{f}_c = 0\mathbf{i} - mg\mathbf{j}$. The projectile is free to move in both the x and y directions; therefore, the virtual displacement is $\delta\mathbf{r} = \delta x\mathbf{i} + \delta y\mathbf{j}$. The potential energy is then

$$V = -\int_{\mathbf{r}_1}^{\mathbf{r}} \mathbf{f}_c \cdot \delta\mathbf{r} = -\int_{\bar{y}=0}^{\bar{y}=y} (-mg)\delta\bar{y} = mgy.$$

Forming the Lagrangian, we have

$$L = T - V = m\left[\frac{1}{2}\left(\dot{x}^2 + \dot{y}^2\right) - gy\right].$$

Applying Hamilton's principle for conservative systems (4.6) gives

$$\delta \int_{t_1}^{t_2} L(t,y,\dot{x},\dot{y})dt = 0$$

$$\int_{t_1}^{t_2} \left(\frac{\partial L}{\partial y}\delta y + \frac{\partial L}{\partial \dot{x}}\delta \dot{x} + \frac{\partial L}{\partial \dot{y}}\delta \dot{y} \right) dt = 0$$

$$m \int_{t_1}^{t_2} (-g\delta y + \dot{x}\delta \dot{x} + \dot{y}\delta \dot{y})dt = 0$$

$$\int_{t_1}^{t_2} (-g\delta y - \ddot{x}\delta x - \ddot{y}\delta y)dt = 0$$

$$\int_{t_1}^{t_2} [\ddot{x}\delta x + (\ddot{y} + g)\delta y]\, dt = 0,$$

where integration by parts has been employed, with x and y at t_1 and t_2 taken as known. Because δx and δy are arbitrary, the Euler equations are then

$$\ddot{x} = 0, \quad \ddot{y} = -g.$$

Integrating the equations of motion and applying the initial conditions leads to the equations for projectile motion without drag

$$x(t) = v_0 \cos\theta_0 t, \quad y(t) = v_0 \sin\theta_0 t - \frac{1}{2}gt^2.$$

Note that this is equivalent to applying Newton's second law, $\mathbf{f} = m\mathbf{a}$, to the projectile.

One must be careful in determining the signs of the potential energy. In this example, the gravitational potential energy equals the work required to raise the projectile to its current y-location against gravity, which is $V = mgy$. The virtual work acting on the projectile is the negative of this, or $\int \mathbf{f} \cdot \delta\mathbf{r} = -V = -mgy$. The difference in sign results from the fact that the force due to gravity acts in the negative y-direction, whereas the force required to raise the mass is in the positive y-direction.

EXAMPLE 4.2 Determine the position $x(t)$ for the undamped harmonic oscillator shown in Figure 4.2. The location of the mass $x(t)$ is taken as zero when the spring is in its unstretched position, that is, when no force is exerted by the spring on the mass.

Solution: We label the mass as particle ① and the spring as particle ② ($N = 2, n = 1$). Note that the mass does not have any gravitational potential energy if we define the vertical reference level as that of the mass. In other words, $V_1 = 0$, and the kinetic

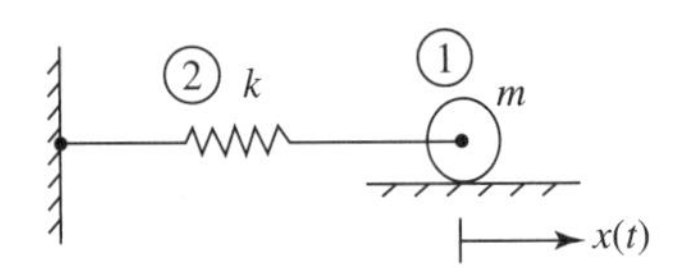

Figure 4.2. Schematic of an undamped harmonic oscillator.

energy of the spring is zero as it does not have any mass, that is, $T_2 = 0$. Consequently, the kinetic energy of the system is

$$T = T_1 = \frac{1}{2}m\mathbf{v}^2 = \frac{1}{2}m\dot{x}^2.$$

The mass can only move in the x-direction, so the virtual displacement is $\delta\mathbf{r} = \delta x\mathbf{i}$. The potential energy due to the strain energy U required to deform the spring is then

$$V = V_2 = U = -\int_{\mathbf{r}_1}^{\mathbf{r}} \mathbf{f}\cdot\delta\mathbf{r} = -\int_{\bar{x}=0}^{\bar{x}=x} (-k\bar{x})\delta\bar{x} = \frac{1}{2}kx^2,$$

where the strain energy is the negative of the virtual work done by the spring. In addition, note that we use the restoring force, which is the force exerted by the spring on the mass, $\mathbf{f} = -kx\mathbf{i}$, whether in tension ($x > 0$) or compression ($x < 0$).

The Lagrangian of the system is then

$$L = T_1 - V_2 = \frac{1}{2}m\dot{x}^2 - \frac{1}{2}kx^2,$$

and Hamilton's principle leads to

$$\begin{aligned}
\delta\int_{t_1}^{t_2} L(t,x,\dot{x})dt &= 0\\
\int_{t_1}^{t_2}\left(\frac{\partial L}{\partial x}\delta x + \frac{\partial L}{\partial \dot{x}}\delta\dot{x}\right)dt &= 0\\
\int_{t_1}^{t_2}(-kx\delta x + m\dot{x}\delta\dot{x})dt &= 0\\
\int_{t_1}^{t_2}(-kx - m\ddot{x})\delta x dt &= 0.
\end{aligned}$$

Therefore, the Euler equation (equation of motion) is

$$\ddot{x} + \frac{k}{m}x = 0,$$

which has the solution

$$x(t) = c_1\cos\left(\sqrt{\frac{k}{m}}t\right) + c_2\sin\left(\sqrt{\frac{k}{m}}t\right).$$

The constants c_1 and c_2 would be determined from initial conditions on position and velocity if provided. Regardless of the initial conditions, however, the solution for this simple harmonic oscillator exhibits sinusoidal oscillation at a single frequency, which in this case is $\sqrt{k/m}$. This is known as the *natural frequency* of oscillation. When an entire system oscillates at a single frequency, it is known as a *normal mode*.

As in the previous example, let us carefully consider the signs of the potential energy. In this example, the potential (strain) energy owing to deformation of the spring equals the work required to stretch (or compress) the spring, which is $V_2 = U = \frac{1}{2}kx^2$. The virtual work acting on the spring is the negative of this, which is

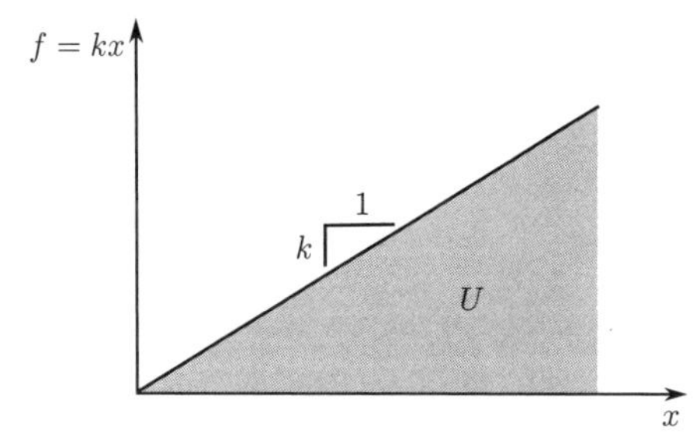

Figure 4.3. Force versus elongation plot for the linear elastic spring.

$\mathbf{f} \cdot \delta \mathbf{r} = -U = -\frac{1}{2}kx^2$. The difference in sign results from the fact that the restoring force, $\mathbf{f} = -kx\mathbf{i}$, must be used when calculating the virtual work, whereas the force acting on the spring is $\mathbf{f} = kx\mathbf{i}$.

Observe also that the strain energy in the spring, $U = \frac{1}{2}kx^2$, is the area under a force versus elongation plot, which for a linear elastic spring is equivalent to the product of the average force $\frac{1}{2}kx$ and the elongation x. This is illustrated in Figure 4.3.

The next example illustrates the independence of Hamilton's principle with respect to the coordinate system, which is established formally in Section 4.4.

EXAMPLE 4.3 Consider a mass m hanging from a massless rod of length ℓ as shown in Figure 4.4. The rotation rate around the vertical axis is constant and given by $\dot{\theta} = \Omega =$ constant, thereby causing the rod to rotate at a constant angle α to the vertical. Determine this angle.

Solution: Observe that the location of the mass with respect to θ does not affect the dynamics of the system because the angular velocity is a constant, that is, the kinetic and potential energies are independent of θ; therefore, the rotation is *axisymmetric*. In order to fix the state of the system, we need either the radial distance r of the mass from the vertical axis, the elevation y, or the angle α. As a result, our one particle system ($N = 1$) has one degree of freedom ($n = 1$), and we can choose either of the three possible variables as our dependent variable.

We begin by writing the kinetic and potential energies in terms of their most convenient coordinate system in the form

$$T = \frac{1}{2}m\mathbf{v}^2 = \frac{1}{2}m\left(\cancelto{0}{v_r^2} + v_\theta^2 + \cancelto{0}{v_y^2}\right) = \frac{1}{2}m(r\dot{\theta})^2 = \frac{1}{2}m(\Omega r)^2, \quad V = mgy,$$

where we note that $v_r = v_y = 0$ for $\dot{\theta} = \Omega =$ const. We then choose our dependent variable(s) to be consistent with the number of degrees of freedom and all other variables using geometric considerations. Hence, the Lagrangian is

$$L = T - V = \frac{1}{2}m(\Omega r)^2 - mgy = \frac{1}{2}m\Omega^2(\ell^2 - y^2) - mgy,$$

where we have recognized from the geometry that $r^2 = \ell^2 - y^2$ for the one degree-of-freedom system. Thus, the Lagrangian is written solely in terms of y. Although the form of the Lagrangian changes if written in terms of y, r, or α, its scalar value

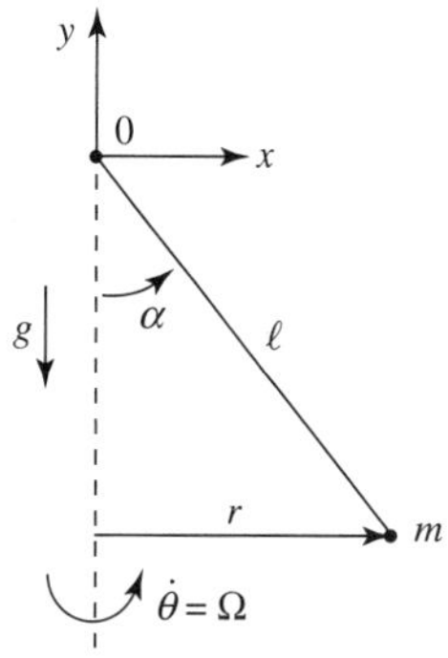

Figure 4.4. Schematic of the rotating pendulum.

remains the same at each state of the system. This results from its coordinate system independence. Hamilton's principle requires that

$$\delta \int_{t_1}^{t_2} L(t,y)dt = 0$$

$$\int_{t_1}^{t_2} \frac{\partial L}{\partial y}\delta y dt = 0$$

$$m\int_{t_1}^{t_2} (-\Omega^2 y - g)\delta y dt = 0,$$

and the equation of motion is

$$\Omega^2 y + g = 0,$$

which is an algebraic, rather than a differential, equation. This is because there is no $\dot{y}$ in the Lagrangian for the axisymmetric case considered here and indicates that y (along with α and r) does not depend upon time; that is, the solution is time invariant, or steady. To determine the angle α, note from the geometry that $-y = l\cos\alpha$; therefore, solving for α and substituting for y leads to

$$\alpha = \cos^{-1}\left(\frac{g}{\ell\Omega^2}\right).$$

Interestingly, there is no influence of the mass on the angle α.

Observe that no special consideration has been necessary owing to the use of cylindrical coordinates. Moreover, it has not been necessary to determine the force in the massless rod. These are two of the primary advantages of using Hamilton's principle in place of Newton's second law. This example is unusual in that the equation of motion is algebraic. Typically, equations of motion for dynamics of discrete systems are ordinary differential equations in time as found in the previous examples.

We can envision systems of any size, including $O(1)$-sized systems comprised of *discrete* rigid (nondeformable) particles and bodies, or infinitesimally small systems, such as *continuous* (deformable) differential elements in a solid or fluid. The former case has been considered in this section, and we consider the continuous case in the next section.

4.2 Hamilton's Principle for Continuous Systems

In the previous section, we considered *discrete systems* comprised of a finite number N of particles (point masses). Such systems are characterized by a finite number n of degrees of freedom, and we apply Hamilton's principle (4.6)

$$\delta \int_{t_1}^{t_2} L dt = 0 \tag{4.7}$$

using a discrete sum for the Lagrangian as follows:

$$L = T - V = \sum_{k=1}^{N} (T_k - V_k). \tag{4.8}$$

Because the particles are nondeformable, the virtual work is only that from the external forces acting on the system. The corresponding Euler-Lagrange equations for such discrete systems typically produce ordinary differential equations of motion for the positions of the discrete objects as continuous functions of time, as in Examples 4.1 and 4.2.

We now turn our attention to *continuous systems* comprised of deformable bodies in which the mass is distributed throughout the body. In such cases we apply the discrete form of Hamilton's principle to differential elements of the continuous system. The system may be thought of as having an infinite number of particles ($N \to \infty$) and an infinite number of degrees of freedom ($n \to \infty$), and we sum over the entire volume $\bar{V}$ through integration. Accordingly, we replace the discrete sum (4.8) in the Lagrangian for discrete systems with the volume integral

$$L = T - V = \iiint_{\bar{V}} \mathcal{L} d\bar{V}, \tag{4.9}$$

where $\mathcal{L} = \mathcal{T} - \mathcal{V}$ is the *Lagrangian density*, with $\mathcal{T}$ and $\mathcal{V}$ being the kinetic and potential energies per unit volume, respectively. Alternatively, the Lagrangian density may be defined per unit length or area as appropriate. In addition to the external forces acting on the body, we must also account for the virtual work from the internal forces that deform the continuous system. Observe that substitution of (4.9) into Hamilton's principle (4.7) produces an integral over space and time. Therefore, the corresponding Euler equations for such continuous systems typically produce partial differential equations for the field variables as continuous functions of space and time.

The application of Hamilton's principle to continuous systems is intimately connected with the *continuum* approach to mechanics. The continuum approach is described in more detail for solid mechanics and fluid mechanics in Sections 5.3 and 9.1, respectively. Also see the books by Berdichevsky (2009), Lanczos (1970), Morse and Feshback (1953), Nair (2009), Oden and Reddy (1983), and Reddy (2002), for example.

EXAMPLE 4.4 Consider the static deflection of an elastic string shown in Figure 4.5, where P is the tension in the string, $f(x)$ is the downward distributed

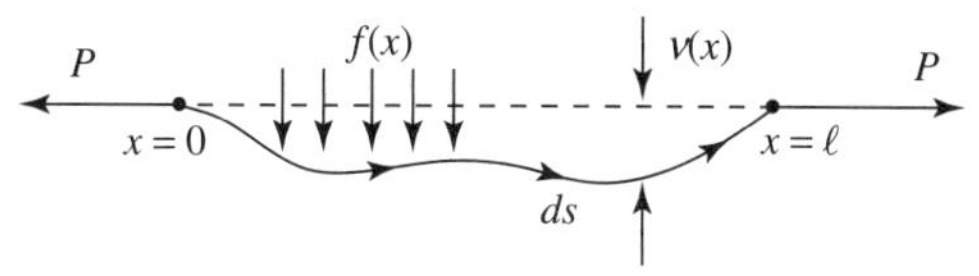

Figure 4.5. Static deflection of an elastic string.

transverse load per unit length, and $v(x)$ is the downward transverse deflection, which is assumed small. Determine the governing equation for the lateral deflection $v(x)$.
Solution: Note that because the transverse deflection is assumed small, differences in the gravitational potential energy may be neglected, and we may consider the tension P to be constant because the additional tension owing to deflection of the string is small compared to P.

For the one-dimensional static ($T = 0$) case considered here, the Lagrangian is

$$L = -V = -\int_0^\ell \mathcal{V} dx.$$

Consider the potential energy per unit length of a differential length of string ds required to stretch and deflect the string

$$\mathcal{V} = \frac{1}{dx}\left[dU - f(x)v dx\right], \tag{4.10}$$

where dU is the *strain energy*, which is the virtual work required to stretch (deform) the differential length of string, and $f(x)v(x)dx$ is the virtual work done by the transverse load in deflecting the differential length of string. The strain energy (with a negative restoring force as for springs) is

$$\begin{aligned} dU &= -(-P)\cdot(\text{elongation}) \\ &= P(ds - dx) \\ dU &= P\left[\sqrt{1+(v')^2} - 1\right]dx. \end{aligned}$$

Recall that we are assuming that the deflection $v(x)$ is small; therefore, the slope of the deflection curve is small enough such that $(v')^2 \ll 1$. It follows from the binomial theorem, which for $|w| < 1$ and α real is

$$(1+w)^\alpha = 1 + \alpha w + \frac{\alpha(\alpha-1)}{2!}w^2 + \frac{\alpha(\alpha-1)(\alpha-2)}{3!}w^3 + \cdots,$$

that $\left[w = (v')^2, \alpha = \frac{1}{2}\right]$

$$\left[1+(v')^2\right]^{1/2} = 1 + \frac{1}{2}(v')^2 + \cdots.$$

Because $(v')^2$ is small, we can truncate the expansion after the first two terms, and the strain energy for small deflections may be approximated as

$$dU \approx \frac{1}{2}P(v')^2 dx.$$

Substituting into (4.10) and integrating over the length of the string gives

$$V[v]=\int_0^\ell \mathcal{V} dx=\int_0^\ell \left[\frac{1}{2}P(v')^2-f(x)v\right]dx. \tag{4.11}$$

As a result, Hamilton's principle for a static body requires that we minimize the potential energy by taking $\delta V=0$. Doing so leads to

$$\delta\int_0^\ell \left[\frac{1}{2}P(v')^2-f(x)v\right]dx=0$$

$$\int_0^\ell \left[Pv'\delta v'-f(x)\delta v\right]dx=0$$

$$P\left[v'\delta v\right]_0^\ell+\int_0^\ell \left[-Pv''-f(x)\right]\delta v dx=0,$$

where the term evaluated at the end points resulting from the integration by parts vanishes because $v=0$ at $x=0$ and $x=\ell$. Therefore, the Euler equation governing the static deflection of an elastic string owing to a transverse load $f(x)$ is

$$Pv''+f(x)=0. \tag{4.12}$$

Because this is a one-dimensional statics problem, there is no time dependence, and the Euler equation is an ordinary differential equation, whereas typically continuous systems produce partial differential equations of motion for position and time.

In some cases, we may wish to seek the equivalent variational form for a known governing differential equation. In the following example, the procedure outlined in Section 2.5.3 for the *inverse problem* is followed to obtain the variational form for the rotating string example, which is a generalization of the previous example.

EXAMPLE 4.5 Consider the deflection of a rotating string, as illustrated in Figure 4.6. Here, $f(x)$ is the outward radial force per unit length acting on the string as it rotates, ρ is the density (mass per unit length) of the string, and Ω is the angular velocity of the rotating string. Given the differential equation governing the displacement $v(x)$ from the axis of rotation, determine the equivalent variational form.

Solution: The governing differential equation for small deflections $v(x)$ is [note the equivalence with equation (4.12) when $\Omega=0$]

$$Pv''+\rho\Omega^2 v+f(x)=0, \tag{4.13}$$

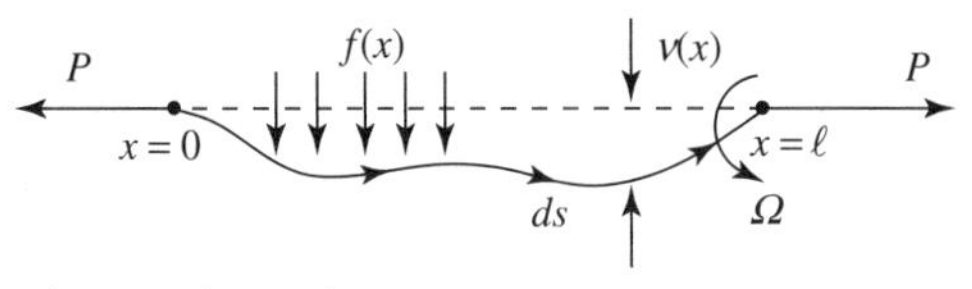

Figure 4.6. Deflection of a rotating string.

which is Euler's equation. To obtain a proper variational problem of the form $\delta \int_0^\ell F(x,v,v')dx = 0$, multiply the differential equation (4.13) by the variation of the dependent variable δv and integrate over the domain according to

$$P\int_0^\ell v''\delta v dx + \rho\Omega^2 \int_0^\ell v\delta v dx + \int_0^\ell f(x)\delta v dx = 0. \tag{4.14}$$

Integrating the first term in (4.14) by parts ($p = \delta v, dq = v''dx$) gives

$$P\int_0^\ell v''\delta v dx = P\Big[v'\delta v\Big]_0^\ell - P\int_0^\ell v'\delta v' dx.$$

Suppose we impose the end conditions

$$\begin{aligned} v = v_0 \quad &\text{or} \quad v' = 0 \quad \text{at} \quad x = 0, \\ v = v_\ell \quad &\text{or} \quad v' = 0 \quad \text{at} \quad x = \ell, \end{aligned} \tag{4.15}$$

where $v' = 0$ corresponds to a natural boundary condition. Recall that for any ϕ

$$\phi\delta\phi = \frac{1}{2}\delta\phi^2.$$

With $\phi = v'$ and the boundary conditions (4.15), we have

$$P\int_0^\ell v''\delta v dx = -P\int_0^\ell \frac{1}{2}\delta\left[(v')^2\right]dx. \tag{4.16}$$

Similarly, for the second term in equation (4.14), with $\phi = v$, observe that

$$v\delta v = \frac{1}{2}\delta v^2. \tag{4.17}$$

Rewriting (4.14) with (4.16) and (4.17) yields

$$\delta\int_0^\ell \left[-\frac{1}{2}P(v')^2 + \frac{1}{2}\rho\Omega^2 v^2 + f(x)v\right]dx = 0. \tag{4.18}$$

This is the variational form of the rotating string problem for which the stationary function $v(x)$ satisfies the differential equation (4.13). The terms in equation (4.18) represent (per unit length):

$\frac{1}{2}P(v')^2$ = potential (strain) energy owing to tension in the string,

$\frac{1}{2}\rho\Omega^2 v^2$ = kinetic energy of the rotating string,

$-f(x)v$ = potential energy due to the radial force.

4.3 Euler-Lagrange Equations

Now that we have developed some experience with the use of Hamilton's principle for discrete and continuous systems, it is advantageous to introduce *generalized coordinates* and obtain the *Euler-Lagrange equations*, which provide a more convenient approach for addressing applications of Hamilton's principle throughout the remainder of Part II. We focus our attention on the discrete case.

The dependent variables for the three examples considered in Section 4.1 are geometric coordinates. In order to accommodate different coordinate systems and/or facilitate dependent variables that are not coordinate directions, it is convenient to work in terms of generalized coordinates. For a dynamical system with n degrees of freedom, we define the following generalized coordinates

$$q_1(t), q_2(t), \dots, q_n(t),$$

which are the dependent variables. The *generalized velocities* are then

$$\dot{q}_1(t), \dot{q}_2(t), \dots, \dot{q}_n(t).$$

Infinitesimally small variations, called *virtual displacements*, in each of the generalized coordinate directions are given by

$$\delta q_1, \delta q_2, \dots, \delta q_n.$$

The n-dimensional space defined by the n generalized coordinates for the n degree-of-freedom system defines the *configuration manifold*, which may or may not correspond to the coordinate directions.

As an example, consider a single particle that may move freely in three-dimensional Cartesian coordinates having the position vector

$$\mathbf{r} = x\mathbf{i} + y\mathbf{j} + z\mathbf{k}.$$

Alternatively, the position vector could be written in terms of generalized coordinates $q_1 = x, q_2 = y$, and $q_3 = z$ as

$$\mathbf{r} = q_1\mathbf{i} + q_2\mathbf{j} + q_3\mathbf{k}.$$

The velocity vector, which is the time rate of change of the position vector, is given by

$$\mathbf{v} = \dot{\mathbf{r}} = \dot{x}\mathbf{i} + \dot{y}\mathbf{j} + \dot{z}\mathbf{k} = \dot{q}_1\mathbf{i} + \dot{q}_2\mathbf{j} + \dot{q}_3\mathbf{k},$$

where the actual velocity components and generalized velocities are the same for Cartesian coordinates, that is, $\dot{q}_1 = \dot{x}, \dot{q}_2 = \dot{y}$, and $\dot{q}_3 = \dot{z}$ are the velocities in each of the coordinate directions. The kinetic energy is then

$$T = \frac{1}{2}m\mathbf{v}^2 = \frac{1}{2}m\dot{\mathbf{r}} \cdot \dot{\mathbf{r}} = \frac{1}{2}m\left(\dot{x}^2 + \dot{y}^2 + \dot{z}^2\right) = \frac{1}{2}m\left(\dot{q}_1^2 + \dot{q}_2^2 + \dot{q}_3^2\right). \tag{4.19}$$

Alternatively, in cylindrical coordinates, the position vector is

$$\mathbf{r} = r\mathbf{e}_r + z\mathbf{e}_z,$$

where the generalized coordinates are now $q_1 = r, q_2 = \theta, q_3 = z$. The actual velocity components and generalized velocities are not the same in the case of cylindrical coordinates. The vector of generalized velocities is

$$\dot{\mathbf{q}} = \dot{q}_1\mathbf{e}_r + \dot{q}_2\mathbf{e}_\theta + \dot{q}_3\mathbf{e}_z = \dot{r}\mathbf{e}_r + \dot{\theta}\mathbf{e}_\theta + \dot{z}\mathbf{e}_z,$$

whereas the actual velocity[1] is

$$\mathbf{v} = \dot{\mathbf{r}} = v_r \mathbf{e}_r + v_\theta \mathbf{e}_\theta + v_z \mathbf{e}_z = \dot{r}\mathbf{e}_r + r\dot{\theta}\mathbf{e}_\theta + \dot{z}\mathbf{e}_z = \dot{q}_1 \mathbf{e}_r + q_1\dot{q}_2 \mathbf{e}_\theta + \dot{q}_3 \mathbf{e}_z.$$

The kinetic energy is then

$$T = \frac{1}{2}m\mathbf{v}^2 = \frac{1}{2}m\left[v_r^2 + v_\theta^2 + v_z^2\right] = \frac{1}{2}m\left[\dot{r}^2 + (r\dot{\theta})^2 + \dot{z}^2\right] = \frac{1}{2}m\left[\dot{q}_1^2 + (q_1\dot{q}_2)^2 + \dot{q}_3^2\right]. \tag{4.20}$$

Observe that generalized coordinates may be distances, such as r and z, or angles, such as θ. Therefore, the *generalized velocities* are simply the time derivatives of the *generalized coordinates*, but they are not necessarily velocities. For example, for the generalized coordinate $q_2(t) = \theta(t)$, which is an angle, the generalized velocity is $\dot{q}_2(t) = \dot{\theta}(t)$; however, this is not the same as the angular velocity, which is $v_\theta = r\dot{\theta}$.

Returning to the general case, we note that the kinetic energy depends on the position and velocity[2] as follows

$$T = T(q_1, \ldots, q_n, \dot{q}_1, \ldots, \dot{q}_n),$$

whereas the potential energy for conservative forces depends only on position according to

$$V = V(q_1, q_2, \ldots, q_n).$$

We apply Hamilton's principle for discrete conservative systems

$$\delta I = \delta \int_{t_1}^{t_2} (T - V)dt = 0,$$

where the integrand is

$$F = L(t, q_i, \dot{q}_i) = T(t, q_i, \dot{q}_i) - V(t, q_i), \quad i = 1, \ldots, n.$$

Then Euler's equations for conservative systems, which from Chapter 2 are

$$\frac{\partial F}{\partial q_i} - \frac{d}{dt}\left(\frac{\partial F}{\partial \dot{q}_i}\right) = 0, \quad i = 1, \ldots, n,$$

become the following

$$\frac{\partial L}{\partial q_i} - \frac{d}{dt}\left(\frac{\partial L}{\partial \dot{q}_i}\right) = 0, \quad i = 1, \ldots, n. \tag{4.21}$$

These are the *Euler-Lagrange equations of motion* for an n degree-of-freedom system, where $L = T - V$ is the Lagrangian. Alternatively, recognizing that

$$\frac{\partial F}{\partial \dot{q}_i} = \frac{\partial L}{\partial \dot{q}_i} = \frac{\partial T}{\partial \dot{q}_i} - \cancelto{0}{\frac{\partial V}{\partial \dot{q}_i}} = \frac{\partial T}{\partial \dot{q}_i},$$

[1] Because the unit vector $\mathbf{e}_r$ changes with time as the particle moves, such that $\dot{\mathbf{e}}_r = \dot{\theta}\mathbf{e}_\theta$, we note that

$$\dot{\mathbf{r}} = \frac{d}{dt}(r\mathbf{e}_r + z\mathbf{e}_z) = \dot{r}\mathbf{e}_r + r\dot{\mathbf{e}}_r + \dot{z}\mathbf{e}_z = \dot{r}\mathbf{e}_r + r\dot{\theta}\mathbf{e}_\theta + \dot{z}\mathbf{e}_z.$$

[2] In Cartesian coordinates, the kinetic energy (4.19) only depends on velocity.

Table 4.1. *Comparison of Hamilton's principle and the general form in Chapter 2.*

	General Form	Hamilton's Principle
Independent Variable	x	t
Dependent Variable(s)	$u(x)$	$q_i(t)$
Integrand	$F(x,u,u')$	$L(t,q_i,\dot{q}_i)$
Euler Equation	$\frac{\partial F}{\partial u} - \frac{d}{dx}\left(\frac{\partial F}{\partial u'}\right) = 0$	$\frac{\partial L}{\partial q_i} - \frac{d}{dt}\left(\frac{\partial L}{\partial \dot{q}_i}\right) = 0$

these may be written

$$\frac{\partial L}{\partial q_i} - \frac{d}{dt}\left(\frac{\partial T}{\partial \dot{q}_i}\right) = 0, \quad i = 1,\ldots,n. \tag{4.22}$$

Finally, observe that

$$\frac{\partial F}{\partial q_i} = \frac{\partial L}{\partial q_i} = \frac{\partial T}{\partial q_i} - \frac{\partial V}{\partial q_i},$$

in which case we may write equations (4.22) in the form

$$\frac{\partial T}{\partial q_i} - \frac{d}{dt}\left(\frac{\partial T}{\partial \dot{q}_i}\right) = \frac{\partial V}{\partial q_i}, \quad i = 1,\ldots,n, \tag{4.23}$$

where for conservative forces, *generalized forces* Q_i may be defined such that $\partial V/\partial q_i = -Q_i$.

Equations (4.21), (4.22), and (4.23) are three equivalent forms of the *Euler-Lagrange equations of motion* for conservative systems. They result in n simultaneous (coupled) ordinary differential equations corresponding to the n degrees of freedom. Observe the comparison of notations in Table 4.1 between the general form of Euler's equation in Chapter 2 and Hamilton's principle for conservative systems in the form (4.21).

For both conservative and nonconservative forces, the virtual work is

$$\mathbf{f} \cdot \delta \mathbf{r} = \sum_{i=1}^{n} \mathbf{f} \cdot \frac{\partial \mathbf{r}}{\partial q_i} \delta q_i = \sum_{i=1}^{n} Q_i \delta q_i, \tag{4.24}$$

where the *generalized forces* are

$$Q_i = \mathbf{f} \cdot \frac{\partial \mathbf{r}}{\partial q_i}, \tag{4.25}$$

and $Q_i \delta q_i$ is the virtual work done by the generalized force Q_i owing to the virtual displacement δq_i. For conservative forces, it may also be written in terms of the scalar potential in the form

$$Q_i = -\frac{\partial V}{\partial q_i}, \tag{4.26}$$

as is the case in equation (4.23). If q_i is a linear displacement, then Q_i is a force. If q_i is an angular displacement, then Q_i is a torque or moment. Although the q_i's and Q_i's do not always have the dimensions of length and force, respectively, the virtual work $Q_i \delta q_i$ always has the dimensions of work.

Corresponding to Hamilton's principle in the form (4.5), if nonconservative forces act on the system, the Euler-Lagrange equations of motion (4.21) are modified as follows:

$$\frac{\partial L}{\partial q_i} - \frac{d}{dt}\left(\frac{\partial L}{\partial \dot{q}_i}\right) = -Q_i, \quad i = 1,\ldots,n, \tag{4.27}$$

where Q_i are the generalized forces (4.25) owing to nonconservative forces acting on the system, and the potentials owing to conservative forces are included in the Lagrangian $L = T - V$, specifically in the potential energy V. We will formally establish the form (4.27) for nonconservative systems in Section 4.5.

It is sometimes convenient to work in terms of *generalized momenta*, in particular when using the Hamiltonian form of the equations of motion to be introduced in Section 4.6. The generalized momenta are defined by

$$p_i = \frac{\partial T}{\partial \dot{q}_i} = \frac{\partial L}{\partial \dot{q}_i}, \quad i = 1,\ldots,n,$$

where the $\dot{q}_i$ are the generalized velocities. Then the Euler-Lagrange equations of motion (4.21) become

$$\frac{\partial L}{\partial q_i} - \dot{p}_i = 0, \quad i = 1,\ldots,n.$$

For a single particle of mass m in Cartesian coordinates $(q_1 = x, q_2 = y, q_3 = z)$, the generalized momenta are

$$p_i = \frac{\partial L}{\partial \dot{q}_i} = \frac{\partial T}{\partial \dot{q}_i} = \frac{\partial}{\partial \dot{q}_i}\left[\frac{1}{2}m\left(\dot{q}_1^2 + \dot{q}_2^2 + \dot{q}_3^2\right)\right].$$

Then for $i = 1,2$, and 3:

$$\begin{aligned} i = 1: &\quad p_1 = m\dot{q}_1 = m\dot{x}, \\ i = 2: &\quad p_2 = m\dot{q}_2 = m\dot{y}, \\ i = 3: &\quad p_3 = m\dot{q}_3 = m\dot{z}, \end{aligned}$$

which are the components of the linear momentum in the x-, y-, and z-directions, respectively.

For a single particle of mass m in cylindrical coordinates $(q_1 = r, q_2 = \theta, q_3 = z)$, the generalized momenta are

$$p_i = \frac{\partial L}{\partial \dot{q}_i} = \frac{\partial T}{\partial \dot{q}_i} = \frac{\partial}{\partial \dot{q}_i}\left[\frac{1}{2}m\left(\dot{q}_1^2 + q_1^2\dot{q}_2^2 + \dot{q}_3^2\right)\right].$$

Then for $i = 1,2$, and 3:

$$\begin{aligned} i = 1: &\quad p_1 = m\dot{q}_1 = m\dot{r}, \\ i = 2: &\quad p_2 = mq_1^2\dot{q}_2 = mr^2\dot{\theta}, \\ i = 3: &\quad p_3 = m\dot{q}_3 = m\dot{z}, \end{aligned}$$

where p_2 is the angular momentum.

4.4 Invariance of the Euler-Lagrange Equations

As stated previously, one of the primary advantages of using the Euler-Lagrange equations as compared to Newton's second law is that they are invariant, that is, they are independent of the coordinate system. Thus far, this has been justified on the basis of Hamilton's principle and the associated Euler-Lagrange equations being formulated in terms of energies, which are scalars, as compared to the vector form of Newton's second law. Here, we explicitly show that this is indeed the case by transforming from one set of generalized coordinates to another.

Recall that in terms of the original generalized coordinates (dependent variables) $q_1(t), q_2(t), \ldots, q_i(t), \ldots, q_n(t)$, the Euler-Lagrange equations of motion for a conservative system are

$$\frac{\partial L}{\partial q_i} - \frac{d}{dt}\left(\frac{\partial L}{\partial \dot{q}_i}\right) = 0, \quad i = 1, \ldots, n, \tag{4.28}$$

where the Lagrangian is $L(t, q_i, \dot{q}_i)$. We seek the corresponding equations of motion in terms of an alternative set of generalized coordinates $q_1^*(t), q_2^*(t), \ldots, q_i^*(t), \ldots, q_n^*(t)$, for which the Lagrangian is $L^*(t, q_i^*, \dot{q}_i^*)$. Transforming the derivatives in (4.28) with $L = L^*(t, q_i^*, \dot{q}_i^*)$ gives

$$\frac{\partial L}{\partial q_i} = \sum_{j=1}^{n}\left[\frac{\partial L^*}{\partial q_j^*}\frac{\partial q_j^*}{\partial q_i} + \frac{\partial L^*}{\partial \dot{q}_j^*}\frac{\partial \dot{q}_j^*}{\partial q_i}\right] = \sum_{j=1}^{n}\left[\frac{\partial L^*}{\partial q_j^*}\frac{\partial q_j^*}{\partial q_i} + \frac{\partial L^*}{\partial \dot{q}_j^*}\frac{d}{dt}\left(\frac{\partial q_j^*}{\partial q_i}\right)\right], \quad i = 1, \ldots, n,$$

and

$$\frac{\partial L}{\partial \dot{q}_i} = \sum_{j=1}^{n}\left[\frac{\partial L^*}{\partial q_j^*}\overbrace{\frac{\partial q_j^*}{\partial \dot{q}_i}}^{0} + \frac{\partial L^*}{\partial \dot{q}_j^*}\frac{\partial \dot{q}_j^*}{\partial \dot{q}_i}\right] = \sum_{j=1}^{n}\frac{\partial L^*}{\partial \dot{q}_j^*}\frac{\partial q_j^*}{\partial q_i}, \quad i = 1, \ldots, n.$$

The final step results from the fact that because $q_j^* = q_j^*(q_1, q_2, \ldots, q_i, \ldots, q_n)$, we can write

$$\frac{dq_j^*}{dt} = \frac{\partial q_j^*}{\partial q_1}\frac{dq_1}{dt} + \frac{\partial q_j^*}{\partial q_2}\frac{dq_2}{dt} + \cdots + \frac{\partial q_j^*}{\partial q_i}\frac{dq_i}{dt} + \cdots + \frac{\partial q_j^*}{\partial q_n}\frac{dq_n}{dt} = \sum_{i=1}^{n}\frac{\partial q_j^*}{\partial q_i}\frac{dq_i}{dt},$$

or

$$\dot{q}_j^* = \sum_{i=1}^{n}\frac{\partial q_j^*}{\partial q_i}\dot{q}_i.$$

Therefore, taking $\partial/\partial \dot{q}_i$ of both sides yields the result

$$\frac{\partial \dot{q}_j^*}{\partial \dot{q}_i} = \sum_{i=1}^{n}\frac{\partial q_j^*}{\partial q_i}.$$

Substituting the transformations into the Euler-Lagrange equation (4.28) yields

$$\sum_{j=1}^{n}\left[\frac{\partial L^*}{\partial q_j^*}\frac{\partial q_j^*}{\partial q_i} + \frac{\partial L^*}{\partial \dot{q}_j^*}\frac{d}{dt}\left(\frac{\partial q_j^*}{\partial q_i}\right) - \frac{d}{dt}\left(\frac{\partial L^*}{\partial \dot{q}_j^*}\frac{\partial q_j^*}{\partial q_i}\right)\right] = 0, \quad i = 1, \ldots, n.$$

After evaluating the time derivative in the last term and simplifying, we have

$$\sum_{j=1}^{n}\left[\frac{\partial L^*}{\partial q_j^*}\frac{\partial q_j^*}{\partial q_i}-\frac{\partial q_j^*}{\partial q_i}\frac{d}{dt}\left(\frac{\partial L^*}{\partial \dot{q}_j^*}\right)\right]=0, \quad i=1,\ldots,n,$$

which simplifies to

$$\sum_{j=1}^{n}\left[\frac{\partial L^*}{\partial q_j^*}-\frac{d}{dt}\left(\frac{\partial L^*}{\partial \dot{q}_j^*}\right)\right]\frac{\partial q_j^*}{\partial q_i}=0, \quad i=1,\ldots,n.$$

If the Jacobian of the transformation is nonzero, then the matrix $\partial q_j^*/\partial q_i$ is nonsingular, and the Euler-Lagrange equations in terms of the new generalized coordinates are

$$\frac{\partial L^*}{\partial q_j^*}-\frac{d}{dt}\left(\frac{\partial L^*}{\partial \dot{q}_j^*}\right)=0, \quad j=1,\ldots,n, \tag{4.29}$$

which is the same as equation (4.28). Therefore, the equations of motion are the same when written in terms of any generalized coordinates as long as the Jacobian of the transformation from $q_i(t)$ to $q_i^*(t)$ is nonzero.

4.5 Derivation of Hamilton's Principle from the First Law of Thermodynamics

Throughout this chapter we have been hinting that Hamilton's principle (4.1) is a consequence of the first law of thermodynamics. By providing a proof of this, we will clearly see the assumptions that are made in formulating Hamilton's principle, thereby placing it into its broader context and indicating its scope of applicability. Our derivation includes the possibility of including both conservative and nonconservative forces.

A number of important laws of physics can be derived from the *first law of thermodynamics*, which simply, but powerfully, states that the total energy of a system must be conserved. The first law of thermodynamics is so fundamental that it cannot be proven mathematically; in other words, it cannot be shown to be a consequence of other proven physical principles. We regard it as a *law*, however, because no known physical processes violate the first law.

Let us consider a *closed system*, in which no matter is moving across the system boundary C. The system is at a location given by the position vector $\mathbf{r}(t)$ with respect to some coordinate system as illustrated in Figure 4.7. Conservation of mass requires that the mass m of the closed system be a constant, or in rate form

$$\dot{m}=0. \tag{4.30}$$

The first law of thermodynamics stated in words is

$$\left\{\begin{array}{c}\text{Time rate of change of}\\ \text{total energy of the system}\end{array}\right\}=\left\{\begin{array}{c}\text{Rate of energy transfer}\\ \text{across the system boundary}\end{array}\right\}.$$

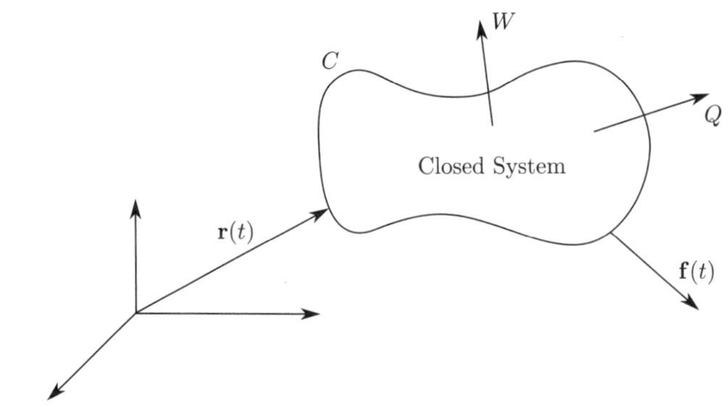

Figure 4.7. A closed system.

The energy transfer across the system boundary can be in the form of heat[3] Q or work W, both of which will be assumed to be zero in the present context. There is no heat transfer across C if the system is *adiabatic*, which is the case if the system is at the same temperature as its surroundings or is well insulated. By assuming $W = 0$, we are not excluding work done *on* the system by external forces $\mathbf{f}(t)$, only work entering or exiting the system across the system boundary, for example, via an electric current. With no energy transfer across the system boundary, the first law then requires that

$$\dot{E} = 0, \tag{4.31}$$

where E is the total energy of the system. Therefore, the total energy of the system remains constant according to

$$E = \text{constant}. \tag{4.32}$$

There are various forms of energy that must be accounted for in the first law of thermodynamics (4.31) or (4.32). This includes the kinetic energy T and potential energy V of the system as a whole and the virtual work W_{nc} done by nonconservative forces. Recall that the potential energy V is due to any conservative forces acting on the system. In addition, there is the thermodynamic *internal energy* owing to the kinetic (random motion of molecules) and potential (intermolecular bonds) energies of the molecules *within* the system. It will be assumed that there is no change with respect to time of the thermodynamic internal energy, which is a good approximation for *isothermal* (constant temperature) solids, liquids, and ideal gases.

Neglecting heat transfer, work crossing the system boundary, and changes in thermodynamic internal energy, the first law of thermodynamics for a closed system (4.32) becomes

$$E = T + V - W_{nc}, \tag{4.33}$$

where the total energy of the system E is a constant. Because the Lagrangian is defined by $L = T - V$, we can write

$$E = 2T - L - W_{nc}. \tag{4.34}$$

[3] It is customary to use Q for heat in thermodynamics. This is not to be confused with the generalized forces Q_i.

Recall that the kinetic energy for an n degree-of-freedom system is

$$T = \sum_{i=1}^{n} \frac{1}{2} m \dot{q}_i^2,$$

where we consider only one particle ($N = 1$) for illustrative purposes. Also note that because potential energy does not depend on $\dot{q}_i$, that is, $L = T(q_i, \dot{q}_i) - V(q_i)$, we have

$$\sum_{i=1}^{n} \dot{q}_i \frac{\partial L}{\partial \dot{q}_i} = \sum_{i=1}^{n} \dot{q}_i \frac{\partial T}{\partial \dot{q}_i}. \tag{4.35}$$

Consequently, we can write

$$2T = \sum_{i=1}^{n} 2\left(\frac{1}{2} m \dot{q}_i^2\right) = \sum_{i=1}^{n} \dot{q}_i \frac{\partial}{\partial \dot{q}_i}\left(\frac{1}{2} m \dot{q}_i^2\right) = \sum_{i=1}^{n} \dot{q}_i \frac{\partial T}{\partial \dot{q}_i} = \sum_{i=1}^{n} \dot{q}_i \frac{\partial L}{\partial \dot{q}_i}. \tag{4.36}$$

Thus, the first law (4.34) becomes

$$E = \sum_{i=1}^{n} \dot{q}_i \frac{\partial L}{\partial \dot{q}_i} - L - W_{nc}. \tag{4.37}$$

Because the total energy of the system is a constant, differentiating with respect to time yields

$$\frac{d}{dt}\left(\sum_{i=1}^{n} \dot{q}_i \frac{\partial L}{\partial \dot{q}_i} - L - W_{nc}\right) = 0, \tag{4.38}$$

or evaluating the time derivatives and solving for dL/dt yields

$$\frac{dL}{dt} = \frac{d}{dt}\left(\sum_{i=1}^{n} \dot{q}_i \frac{\partial L}{\partial \dot{q}_i}\right) - \frac{dW_{nc}}{dt}.$$

Then

$$\frac{dL}{dt} = \sum_{i=1}^{n}\left[\ddot{q}_i \frac{\partial L}{\partial \dot{q}_i} + \dot{q}_i \frac{d}{dt}\left(\frac{\partial L}{\partial \dot{q}_i}\right)\right] - \frac{dW_{nc}}{dt}. \tag{4.39}$$

Furthermore, recall that $L = L(t, q_i, \dot{q}_i)$, that is, the Lagrangian depends upon the position of the particles and their velocities. Let us consider the common case of autonomous systems in which the Lagrangian does not depend explicitly on time. Therefore, differentiating the Lagrangian with respect to time gives

$$\frac{dL}{dt} = \sum_{i=1}^{n}\left(\frac{\partial L}{\partial q_i}\frac{dq_i}{dt} + \frac{\partial L}{\partial \dot{q}_i}\frac{d\dot{q}_i}{dt}\right) = \sum_{i=1}^{n}\left(\frac{\partial L}{\partial q_i}\dot{q}_i + \frac{\partial L}{\partial \dot{q}_i}\ddot{q}_i\right). \tag{4.40}$$

Given these two expressions for the time derivative of the Lagrangian, we can subtract equation (4.40) from equation (4.39) to obtain

$$\sum_{i=1}^{n}\left[\frac{d}{dt}\left(\frac{\partial L}{\partial \dot{q}_i}\right) - \frac{\partial L}{\partial q_i}\right]\dot{q}_i - \frac{dW_{nc}}{dt} = 0. \tag{4.41}$$

Now let us consider more carefully the work done by any nonconservative forces. We know from Section 4.1 that

$$dW_{nc} = \mathbf{f}_{nc} \cdot d\mathbf{r},$$

where $\mathbf{f}_{nc}$ are the nonconservative forces. Then

$$\frac{dW_{nc}}{dt} = \frac{\mathbf{f}_{nc} \cdot d\mathbf{r}}{dt} = \frac{\mathbf{f}_{nc} \cdot \left(\sum_{i=1}^{n} \frac{\partial \mathbf{r}}{\partial q_i} dq_i \right)}{dt} = \sum_{i=1}^{n} \left(\mathbf{f}_{nc} \cdot \frac{\partial \mathbf{r}}{\partial q_i} \right) \frac{dq_i}{dt} = \sum_{i=1}^{n} Q_i \dot{q}_i,$$

where Q_i are the generalized forces defined by equation (4.25). Substituting into equation (4.41) yields the first law of thermodynamics in the form

$$\sum_{i=1}^{n} \left[\frac{d}{dt} \left(\frac{\partial L}{\partial \dot{q}_i} \right) - \frac{\partial L}{\partial q_i} - Q_i \right] \dot{q}_i = 0.$$

Because the generalized coordinates q_i are defined to be independent of one another, the generalized velocities $\dot{q}_i, i = 1, \ldots, n$ are linearly independent. Therefore, the coefficient $[\cdot]$ of each of the generalized velocities in the summation above must each vanish independently. This leads to the Euler-Lagrange equations (4.27)

$$\frac{\partial L}{\partial q_i} - \frac{d}{dt} \left(\frac{\partial L}{\partial \dot{q}_i} \right) = -Q_i, \quad i = 1, \ldots, n. \tag{4.42}$$

Having derived the Euler-Lagrange equations from the first law of thermodynamics, we now show that the general form of Hamilton's principle (4.1) arises from performing the *inverse problem* as in Sections 2.5.3 and 3.1.2. To do so, we multiply the Euler-Lagrange equations (4.42) by δq_i, sum, and integrate over the time interval $t_0 \leq t \leq t_1$ as follows:

$$\int_{t_0}^{t_1} \sum_{i=1}^{n} \left[\frac{\partial L}{\partial q_i} - \frac{d}{dt} \left(\frac{\partial L}{\partial \dot{q}_i} \right) + Q_i \right] \delta q_i dt = 0.$$

Integrating the second term by parts yields

$$-\left[\sum_{i=1}^{n} \frac{\partial L}{\partial \dot{q}_i} \delta q_i \right]_{t_0}^{t_1} \stackrel{\nearrow 0}{} + \int_{t_0}^{t_1} \sum_{i=1}^{n} \left[\frac{\partial L}{\partial q_i} \delta q_i + \frac{\partial L}{\partial \dot{q}_i} \delta \dot{q}_i + Q_i \delta q_i \right] dt = 0.$$

Because the Lagrangian is such that

$$\delta L(q_i, \dot{q}_i) = \sum_{i=1}^{n} \left(\frac{\partial L}{\partial q_i} \delta q_i + \frac{\partial L}{\partial \dot{q}_i} \delta \dot{q}_i \right),$$

and from equation (4.24), this can be written

$$\int_{t_0}^{t_1} (\delta L + \mathbf{f}_{nc} \cdot \delta \mathbf{r})\, dt = 0.$$

With $L = T - V$ and allowing for the virtual work term to account for both conservative and nonconservative forces, we recover the general form of Hamilton's principle (4.1)

$$\int_{t_0}^{t_1} (\delta T + \mathbf{f} \cdot \delta \mathbf{r})\, dt = 0.$$

Hence, we have confirmed that Hamilton's principle for conservative and nonconservative forces follows directly from conservation of energy in the form of the first law of thermodynamics. The assumptions that we have made are that there is no energy crossing the system boundary in the form of heat or work, the system is isothermal, and it is autonomous.

Enforcement of the adiabatic and isothermal assumptions would seem to preclude employing Hamilton's principle in whole classes of problems in thermodynamics and heat transfer. While this is true at the macroscopic, or continuum, level, they are unified with Hamilton's principle at the molecular level, where we simply have the mechanics of many moving and colliding molecules. Temperature, nonconservative forces, and irreversibility of dissipative phenomena, for example, are macroscopic consequences of the reversible and conservative mechanics that take place at the molecular level. See Chapter 2 of Berdichevsky (2009) for a discussion of the relationship between thermodynamics and mechanics at the macroscopic and microscopic levels.

Even with these restrictions at the macroscopic level, Hamilton's principle casts a wide net across a broad spectrum of application areas as we will see throughout the remainder of Part II. The fact that such a wide variety of physical phenomena can be captured in a single variational principle highlights the remarkable generality of the variational approach and allows for extension of approaches and results from one field to another by analogy.

4.6 Conservation of Mechanical Energy and the Hamiltonian

Having derived the general form of Hamilton's principle and the Euler-Lagrange equations for the case of conservative and nonconservative forces, we now consider the special case of systems subject to conservative forces only and show in what sense we call such systems *conservative*. In doing so, we will obtain an alternative formulation of Hamilton's principle involving the *Hamiltonian*, instead of the *Lagrangian*.

If only conservative forces act on a system, in which case $Q_i = 0$ and $W_{nc} = 0$, then the Euler-Lagrange equations for an n degree-of-freedom, conservative system are

$$\frac{\partial L}{\partial q_i} - \frac{d}{dt}\left(\frac{\partial L}{\partial \dot{q}_i}\right) = 0, \quad i = 1, \dots, n. \tag{4.43}$$

In addition, the first law of thermodynamics (4.33) becomes

$$E = H = T + V,$$

where for conservative systems, it is customary to designate the energy E as the *Hamiltonian* H. The constant H is the total *mechanical energy* of a conservative system.[4] Thus, the sum of the kinetic and potential energies does not change with time, that is, $\dot{H} = 0$, and the mechanical energy is conserved for a conservative dynamical system. This is the case if the Lagrangian, and thus the Hamiltonian, do not explicitly depend on time, in which case the system is autonomous. Viewed slightly differently, if the Lagrangian or Hamiltonian is invariant to changes in time (because it does not appear explicitly), then mechanical energy is conserved. This connection between symmetries in the form of an invariance of the system to a coordinate transformation is the essence of *Noether's theorem* to be considered in Section 4.7.

The definition of the Hamiltonian can be used to write the Euler-Lagrange equations in an alternative form in terms of the Hamiltonian instead of the Lagrangian. Recall from Section 4.3 that the Euler-Lagrange equations can be written in the form

$$\frac{\partial L}{\partial q_i} - \dot{p}_i = 0, \quad i = 1, \ldots, n, \tag{4.44}$$

where the generalized momenta are defined by

$$p_i = \frac{\partial L}{\partial \dot{q}_i}, \quad i = 1, \ldots, n. \tag{4.45}$$

Using the definition (4.45), the Hamiltonian from equation (4.37) with $W_{nc} = 0$ (and $E = H$) can be related to the Lagrangian as follows:

$$H = \sum_{i=1}^{n} \dot{q}_i \frac{\partial L}{\partial \dot{q}_i} - L = \sum_{i=1}^{n} \dot{q}_i p_i - L(q_i, \dot{q}_i). \tag{4.46}$$

Alternatively, solving for the Lagrangian yields

$$L = \sum_{i=1}^{n} \dot{q}_i p_i - H(q_i, p_i).$$

Substituting into the Euler-Lagrange equations (4.44) leads to

$$\frac{\partial H}{\partial q_i} = -\dot{p}_i, \quad i = 1, \ldots, n. \tag{4.47}$$

From equation (4.46), we also have

$$\frac{\partial H}{\partial p_i} = \dot{q}_i, \quad i = 1, \ldots, n. \tag{4.48}$$

Equations (4.47) and (4.48) are the Euler-Lagrange equations of motion written in terms of the Hamiltonian, rather than the Lagrangian. They are often referred to as *Hamilton's canonical equations*. The transformation (4.46) from the Lagrangian to the Hamiltonian is known as a *Legendre transformation* and applies for all autonomous systems.

[4] When one appeals to *conservation of energy* in mechanics, we mean conservation of mechanical energy. It is important to distinguish between conservation of (all) energy in the context of the first law of thermodynamics and conservation of mechanical energy in dynamical systems. The former always holds, while the latter only holds for conservative, autonomous systems.

Note that whereas the Euler-Lagrange equations of motion are in terms of generalized positions and velocities when formulated in terms of the Lagrangian $L(q_i, \dot{q}_i)$, they are in terms of generalized positions and momenta in the Hamiltonian form $H(q_i, p_i)$. Moreover, whereas the Euler-Lagrange equations of motion formulated in terms of the Lagrangian are a system of n second-order ordinary differential equations for $q_i(t)$, the equations of motion in terms of the Hamiltonian produce a system of $2n$ first-order ordinary differential equations for $q_i(t)$ and $p_i(t)$, sometimes called the *state-space* representation. Of course, the two systems are mathematically equivalent. Note that Hamilton's canonical equations are formulated for conservative forces only, whereas the Euler-Lagrange equations in the form (4.27) can accommodate nonconservative forces as well.

Because of its more direct connection with the general theory of the calculus of variations, we predominantly formulate our problems using the Lagrangian. However, in some fields, such as quantum mechanics (see Section 8.2), it is more straightforward to formulate the physics in terms of the Hamiltonian. A third formulation of the dynamical equations of motion is given by the *Hamilton-Jacobi equation*, which is in the form of a first-order, nonlinear partial differential equation for the action in place of the Lagrangian or Hamiltonian.

4.7 Noether's Theorem – Connection Between Conservation Laws and Symmetries in Hamilton's Principle

The centerpiece of our discussion is Hamilton's principle, which will be used throughout Part II to obtain the equations governing various physical phenomena involving discrete and continuous systems. Hamilton's principle contains an even more fundamental utility in that it can be used directly to identify various conservation laws. This is facilitated by *Noether's theorem* (Noether 1918) that makes it possible to discern conservation laws directly from Hamilton's principle rather than needing to consider the equations of motion. As summarized by Ian Stewart:[5]

> Symmetry is deeply involved in every area of mathematics, and it underlies most of the basic ideas of mathematical physics. Symmetries express underlying regularities of the world, and those are what drive physics. Continuous symmetries such as rotations are closely related to the nature of space, time, and matter; they imply various conservation laws, such as the law of conservation of energy, which states that a closed system can neither gain nor lose energy. This connection was worked out by Emmy Noether.

One of the remarkable aspects of Noether's theorem is that it allows us to learn a great deal about the physics of a system, namely what quantity is actually conserved, simply by seeking invariances and associated symmetries in the Lagrangian of a system. This allows us to discern important fundamental properties of the system without actually obtaining any solutions of the equations of motion.

Before articulating Noether's theorem, it is important to understand the difference between a quantity being *invariant* and being *conserved*. A quantity is invariant if it does not change when a coordinate transformation (change in reference frame) is applied. A quantity is conserved if it does not change with time during a process

[5] *Why Beauty Is Truth: A History of Symmetry* by Ian Stewart.

considered within a given reference frame. In the present context, *symmetry* is defined as invariance with respect to a transformation in one of the independent or dependent variables.

We have already seen how the dynamics of a system may be expressed in the system's Lagrangian $L(t,q_i,\dot{q}_i), i=1,\ldots,n$, where $L=T-V$. Noether's theorem asserts that if the Lagrangian exhibits some mathematical symmetry, for which it is invariant (unchanged) under a transformation in one of the variables on which it depends, denoted by s, then an associated property C exists that is conserved such that $dC/dt=0$. That is, C is independent of time throughout its motion. If the Lagrangian is invariant to s, then $\partial L/\partial s=0$. Such a circumstance is clearly the case if s does not appear explicitly in the Lagrangian of the system. The variable s could either be an independent variable, such as t, or a dependent variable q_k.

First, let us show that Hamilton's principle for conservative systems is invariant to time with $s=t$. Recall equation (4.38), which is repeated here (with $W_{nc}=0$ for a conservative system):

$$\frac{d}{dt}\left(\sum_{i=1}^{n}\dot{q}_i\frac{\partial L}{\partial \dot{q}_i}-L\right)=0.$$

Because the derivative with respect to time of $(\cdot)$ is zero, this expression in $(\cdot)$, call it C, must be conserved. What is C? It is none other than the total mechanical energy of the system. Recall from equation (4.36) that

$$\sum_{i=1}^{n}\dot{q}_i\frac{\partial L}{\partial \dot{q}_i}=2T,$$

in which case we have

$$C=2T-L,$$

but because $L=T-V$,

$$C=H=T+V.$$

Once again we see that conservation of mechanical energy is a consequence of Hamilton's principle for *autonomous systems*, in which the Lagrangian does not depend explicitly on time, and only conservative forces act on the system. This has been the case for all of the scenarios considered thus far. In Section 6.4 we consider the case of the forced pendulum, for which the Lagrangian depends explicitly on time; therefore, mechanical energy is not conserved. Such systems are called *nonautonomous*.

The above case is a consequence for systems that are invariant to (symmetric over) time, that is, $s=t$, which is true for all conservative systems. What about if a system is symmetric over one of its generalized coordinates $s=q_k$? From the Euler-Lagrange equations of motion, we can write

$$\frac{d}{dt}\left(\frac{\partial L}{\partial \dot{q}_k}\right)=\frac{\partial L}{\partial q_k}.$$

Hence, if the system is symmetric about $s=q_k$, in which case the right-hand side of the above expression is zero, then

$$C=\frac{\partial L}{\partial \dot{q}_k}=p_k, \tag{4.49}$$

where p_k is a generalized momentum. It is clear that this is the case if $s = q_k$ does not appear explicitly in the Lagrangian, in which case the Lagrangian is invariant to changes in $s = q_k$, and the corresponding linear momentum is conserved. We say that the Lagrangian is symmetric about $s = q_k$.

For example, consider a particle of mass m undergoing linear, horizontal motion in the x-direction. The Lagrangian is

$$L = T - \cancelto{0}{V} = \frac{1}{2}m\dot{x}^2 = \frac{1}{2}m\dot{q}^2,$$

in which case

$$C = \frac{\partial L}{\partial \dot{q}} = m\dot{q} = m\dot{x}.$$

Wherein $C = m\dot{x} = p$ is the linear momentum of the mass, we see that linear momentum must be conserved.

Similarly, for a particle of mass m undergoing rotational horizontal motion in the $(q_1,q_2) = (r,\theta)$ plane, the Lagrangian is

$$L = T - \cancelto{0}{V} = \frac{1}{2}m\left[\dot{r}^2 + (r\dot{\theta})^2\right] = \frac{1}{2}m\left[\dot{q}_1^2 + (q_1\dot{q}_2)^2\right].$$

Because $s = q_2 = \theta$ does not appear in the Lagrangian, it is rotationally symmetric, and (4.49) gives

$$C = \frac{\partial L}{\partial \dot{q}_2} = mq_1^2\dot{q}_2 = mr^2\dot{\theta},$$

which is the angular momentum of the mass. Therefore, angular momentum must be conserved.

The above procedure straightforwardly reveals conservation of energy and momentum principles. However, it is not so clear how these conservation laws correspond to symmetries in the Lagrangian, where symmetries correspond to the manner in which the Lagrangian is invariant to a transformation. In order to expose additional symmetries and corresponding conservation laws, and to more explicitly elucidate mathematical symmetries in the Lagrangian, we require a more general procedure.

To illustrate this procedure, let us reconsider conservation of angular momentum. As before, the Lagrangian is

$$L(r,\theta) = \frac{1}{2}m\left[\dot{r}^2 + (r\dot{\theta})^2\right]. \tag{4.50}$$

In order to show that this Lagrangian is invariant to angle changes, transform $\theta(t)$ by a constant angle

$$\theta(t) = \bar{\theta}(t) + \epsilon\alpha,$$

where $\epsilon \ll 1$. Because only the derivative of the angle $\theta(t)$ appears in the Lagrangian, it is invariant to the constant angle change. In order to determine the conservation law that this symmetry exposes, Noether put forth the following procedure. We modify the above transformation to allow for α to depend on time according to

$$\theta(t) = \bar{\theta}(t) + \epsilon\alpha(t), \tag{4.51}$$

but where $\alpha(t_1) = \alpha(t_2) = 0$. Substituting into the Lagrangian (4.50) yields

$$\begin{aligned}\bar{L}(r,\bar{\theta},\alpha) &= \frac{1}{2}m\left[\dot{r}^2 + r^2\left(\dot{\bar{\theta}} + \epsilon\dot{\alpha}\right)^2\right] \\ &= \frac{1}{2}m\left[\dot{r}^2 + r^2\left(\dot{\bar{\theta}}^2 + 2\epsilon\dot{\bar{\theta}}\dot{\alpha}\right)\right] + O(\epsilon^2) \\ &= \frac{1}{2}m\left[\dot{r}^2 + r^2\dot{\bar{\theta}}^2 + 2\epsilon r^2\dot{\bar{\theta}}\dot{\alpha}\right] \\ \bar{L}(r,\bar{\theta},\alpha) &= L(r,\bar{\theta}) + m\epsilon r^2\dot{\bar{\theta}}\dot{\alpha},\end{aligned}$$

where we neglect terms quadratic in the small parameter ϵ. Observe that the first term, which is $O(1)$, is the same as the original Lagrangian (4.50), but in terms of $\bar{\theta}(t)$. Therefore, it produces the original equations of motion. Let us then determine the consequences of the $O(\epsilon)$ term in Hamilton's principle. Specifically, we are interested in the variation of Hamilton's principle with respect to the transformed variable (4.51). Substituting into Hamilton's principle and integrating by parts gives

$$\epsilon\int_{t_1}^{t_2} mr^2\dot{\bar{\theta}}\dot{\alpha}dt = -\epsilon\int_{t_1}^{t_2}\left[\frac{d}{dt}\left(mr^2\dot{\bar{\theta}}\right)\right]\alpha dt,$$

where it has been taken into account that $\alpha(t_1) = \alpha(t_2) = 0$. Because $\alpha(t)$ varies, the term in $[\cdot]$ must vanish for $t_1 \leq t \leq t_2$. Hence,

$$\frac{d}{dt}\left(mr^2\dot{\bar{\theta}}\right) = 0,$$

in which case

$$C = mr^2\dot{\bar{\theta}}$$

must be a constant during the system's motion as it does not change with time. Because C is the angular momentum, invariance with respect to the angle change (4.51) corresponds to conservation of angular momentum. Observe that this more general procedure explicitly links the symmetry expressed in (4.51) to the corresponding conservation law directly through the Lagrangian and Hamilton's principle without appealing to the Euler-Lagrange equations of motion.

Here, we have applied Noether's theorem to discrete systems; however, it applies equally well to continuous systems. The fact that we can use a purely mathematical technique involving checking for mathematical symmetries to expose physical conservation laws in both discrete and continuous systems is a powerful endorsement for the use of Hamilton's principle over using Newton's second law directly (see Table 4.2). In addition, it has played an integral role in solidifying special and general relativity theory during the early part of the twentieth century. For more on Noether's theorem, see Lanczos (1970) and Neuenschwander (2011).

4.8 Summary

In the remaining chapters of Part II we will derive several important equations in mathematical physics using the appropriate form of Hamilton's variational principle applied to various *discrete* and *continuous* systems. The topics covered are

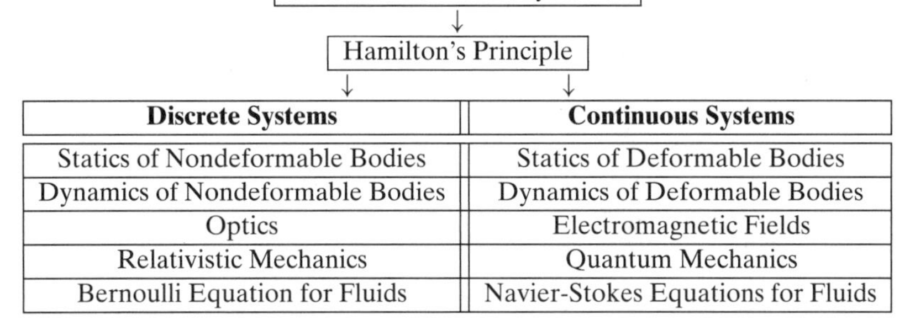

Discrete Systems	Continuous Systems
Statics of Nondeformable Bodies	Statics of Deformable Bodies
Dynamics of Nondeformable Bodies	Dynamics of Deformable Bodies
Optics	Electromagnetic Fields
Relativistic Mechanics	Quantum Mechanics
Bernoulli Equation for Fluids	Navier-Stokes Equations for Fluids

Figure 4.8 Topics to be treated in subsequent chapters.

illustrated in Figure 4.8. In so far as Hamilton's principle follows from the first law of thermodynamics, each of these physical phenomena can be traced back to conservation of energy. The power and simplicity of Hamilton's principle cannot be overstated with regard to its ability to encapsulate such a wide variety of physical phenomena. The number and variety of physical phenomena derivable from Hamilton's principle are truly remarkable and reflect the fundamental nature of the law of conservation of energy on which it is based. As we explore these various physical settings, observe that situations involving *discrete systems* typically lead to ordinary differential equations, and *continuous systems* typically lead to partial differential equations (unlike Examples 4.3, 4.4, and 4.5).

Throughout this chapter, we have used a number of interrelated terms for which it is important to clearly distinguish their applicability:

- *First law of thermodynamics – conservation of energy*: Applies to all systems. Insufficient to directly determine the unique solution trajectory.
- *Conservation of mechanical energy*: Applies to autonomous, adiabatic systems with no work crossing the system boundary. Insufficient to directly determine the unique solution trajectory.
- *Newton's second law*: Applies to autonomous and nonautonomous systems with conservative and nonconservative forces. Sufficient to directly determine the unique solution trajectory.
- *Hamilton's principle*: Applies to autonomous and nonautonomous systems with conservative and nonconservative forces. Sufficient to directly determine the unique solution trajectory. Conservation of mechanical energy is a consequence of Hamilton's principle for autonomous systems (see Noether's theorem).

Although Newton's second law is more familiar and came first historically, Hamilton's principle is more fundamental. In the case of classical (nonrelativistic) Newtonian mechanics, they are equivalent with Newton's second law being the Euler equation for Hamilton's variational principle. Note, however, that it is only through the calculus of variations that this equivalence can be proven. Although Newton's second law is limited in its scope, Hamilton's principle extends to relativistic and quantum mechanics as well. Therefore, it is best to think of Newton's second law as a consequence of Hamilton's principle in the case of classical mechanics. In addition to the remarkable generality of Hamilton's principle, there are several

Table 4.2. *A comparison of Newton's second law and Hamilton's principle.*

Newton's Second Law	Hamilton's Principle
Deals with forces and moments in the form of vectors (vector mechanics)	Deals with energy and work in the form of scalars (analytical dynamics)
Form is dependent on coordinate system	Form is independent of coordinate system
Considers each object in system separately	Considers system as a whole
Forces of constraint must be incorporated	Virtual displacements take constraints into account implicitly
No distinction between how conservative and nonconservative forces are treated	Conservative and nonconservative forces treated differently
Primarily applied to discrete systems	Natural application to discrete and continuous systems
Applies to classical mechanics	Intuitive unification of classical and relativistic mechanics
	Explicit connection between mathematical symmetry and conservation laws

practical advantages of solving problems using Hamilton's principle over doing so using Newton's second law. Table 4.2 summarizes these differences.

4.9 Brief Remarks on the Philosophy of Science

Owing to its fundamental nature and wide applicability, which will become even more apparent throughout the remainder of the text, Hamilton's principle has invited an almost metaphysical[6] quality. Being developed during the Age of Enlightenment (see Section 1.1), this was initially due to the perspective of scientists of the day, many of whom viewed their scientific endeavors as directly discerning the mind of God. For example, Leonhard Euler is reported to have said, "Since the fabric of the universe is most perfect and the work of a most wise Creator, nothing at all takes place in the universe in which some rule of maximum or minimum does not appear." Similarly, the mathematician Pierre Louis Moreau de Maupertuis, who authored the *principle of least action*, wrote, "These simple and beautiful laws are perhaps the only ones that the Creator and Organizer of all things has put in place to carry out the workings of the visible world" (Essai de cosmologie 1759) (see more of his quotes at the beginning of Chapter 5). It was common for scientists and mathematicians to regard their discoveries as directly revealing the underlying workings of nature. Indeed, Maupertuis sought to unify a mathematical and theological justification for his principle of least action. This intimate connection between science and an overall view of the universe, including its origins, is reflected in the fact that what we now call *science* was referred to as *natural philosophy* during the time of Newton. A more typical twentieth century perspective is reflected by Yourgrau and Mandelstam (1968), who remark that, "It is not too much to say that, if certain scientists are constrained to indulge in metaphysical daydreams about the action principle, their meditations, though perhaps interesting, are nevertheless scientifically unintelligible."

[6] *Metaphysical* means transcending physical matter or the laws of nature.

Because of the strong association between Hamilton's principle and many of the fundamental laws of physics, it is worthwhile to provide the author's perspective on the relationship between scientific principles, along with their mathematical expression, and the reality of how the universe functions. Although mathematics certainly can be pursued devoid of physical considerations, and science can be pursued devoid of mathematical considerations, it is the intersection of mathematics and physics that the author, and many others, have found remarkable and suggestive of a deep fundamental connection between the two.

Science proceeds from *observation* to *explanation* to *prediction*. The mark of a lasting physical law is not just that it effectively explains observed phenomena, but that it accurately predicts unobserved, but observable, events. We develop theories that explain our observations and ultimately can be used as predictive tools in scientific and engineering pursuits. As a result, scientific fields tend to become increasingly fundamental and mathematical as they mature. In particular, whereas we generally derive less fundamental physical principles from more fundamental principles, more often than not, discovery of the more fundamental principles follows that of the less fundamental ones. Sometimes a theory is developed that succeeds in explaining a wide range of observed behavior and unifies multiple scientific fields. Hamilton's principle was recognized to be just such a unifying framework very early in its development.

Throughout history, science has sought to explain an ever-expanding collection of observed and postulated[7] phenomena within an ever-consolidating set of scientific laws and principles. It is remarkable that these seemingly dichotomous aims have actually facilitated this trend; as we discover new physics, we see how more and more physics can be unified within a reduced set of increasingly universal laws. Likewise, it is equally remarkable that the diversity of approaches to science – including experimental and observational, theoretical and postulative, and computational and approximate – have all agreed to submit to a common scientific method and standard of acceptance. As much as we embrace this common standard, we must also embrace the diversity of thought and approach to science. By its very nature science is ever changing both in terms of its techniques and its contemporary principles; therefore, it is imperative that it rests on a set of unchanging (or at least slowly changing) standards of acceptance in order to maintain its steady march toward the convergence of scientific principle and reality.

While those who *develop* scientific theories may be well served by regarding them as the actual workings of the universe, those who *use* those theories are best served by regarding them as mathematical models of physical behavior. Seeking the way nature and the universe actually operate drives the scientist to develop increasingly fundamental explanations for observed phenomena, never settling for imperfect or incomplete theories. New scientific discoveries are constantly pushing, and exposing, the limitations of established theories. In fact, breakthrough theories, such as Einstein's theory of relativity, often arise to resolve the tensions that build over time as an existing theory proves inadequate to explain newly observed phenomena. Newton's laws served extremely well for over one hundred years in quantifying celestial mechanics, dynamics of order-one-sized bodies that we

[7] A *postulate* is a claim made without proof as the basis for reasoning.

experience everyday, and many other physical phenomena. However, scientific observations in the nineteenth century concerning the very large, the very small, and the very fast exposed the limitations of Newtonian mechanics, thereby leading Einstein to seek deeper explanations that encompassed all of the observed behavior and, indeed, much that had not yet been observed up to that time.

Mathematical models, on the other hand, are like an analogy; they are a useful, but imperfect, representation of reality, not reality itself. Does the development of Einstein's theories render those of Newton useless or superseded? In terms of their ability to explain the widest range of observed behavior, yes, indeed, Einstein's theories supersede those of Newton. However, as a mathematical model that can be used to predict certain physical behavior in a straightforward and concise way, Newtonian mechanics is more useful in practice than ever before. For example, the satellites, spacecraft, and space probes launched since the middle of the twentieth century have been designed to operate based on the principles of Newtonian mechanics. Similarly, while quantum mechanics is more fundamental than Newtonian mechanics, the latter still has great utility as a model to be applied to many everyday systems. Likewise, physicists now theorize that all matter is comprised of an alphabet soup of discrete subatomic particles; however, it remains extremely useful and productive to model most solids, fluids, and gases as continuous substances in space and time. Fermat's principle succeeded in illuminating the field of optics long before photons were fully understood, and the discovery of photons does not invalidate the use of Fermat's principle as a model that allows us to determine the "path" of light. In other words, light does not "ask" Fermat's principle where to go; it is a model that explains the observed behavior. Likewise, a particle does not ask Hamilton's principle where to go, yet Hamilton's principle produces sufficiently accurate results for predicting such paths under many useful conditions.

In summary, it is often instructive for the scientist to view fundamental theories as "true," that is, actually describing natural behavior, in order to expose their weaknesses and lead to even more fundamental and generally applicable theories. Over time, such a pursuit has led, and will no doubt continue to lead, to greater unification of the increasing inventory of observed behavior in terms of a reduced, but more widely applicable, set of theories. While it is instructive for scientists to view their efforts as deciphering how nature actually works, however, it is more accurate for those using such theories in practice to view them as models, selecting the one that is most appropriate for the situation. In other words, a healthy dose of scientific arrogance is useful when developing physical theories, but a healthy measure of scientific humility is called for when applying them.

Not only are we interested in the resulting theories produced by science, we are also interested in the process by which such theories are developed. Of course, there are whole books written on the subject; we merely seek to clarify the role of basic principles that have guided science. In particular, we collectively address the value of Ockham's razor; the Plato and Pythagoras convictions of an ordered universe; notions of simplicity, repeatability, and predictability; the expectation that a grand unification theory of physics exists; and the scientific method itself. Rather than being absolutes, these are working principles that have served the advancement of science exceptionally well over the centuries. These guiding principles provide the bounds on our everyday science as we march one experiment or postulate at a time toward a better understanding of how the universe works and the physical laws that govern it. Although they have proven remarkably robust given the increasing scope of science and the growth of available scientific techniques, they are open to reevaluation, reinterpretation, and refinement as science continues to advance. There is no way to prove scientifically that such notions must be true in an absolute sense, that is, they are not required properties of the universe, and there is room for disagreement as to their roles or priority in the advancement of science. However, they serve as useful working principles that guide the efforts of the scientific community, providing appropriate checks and balances on the advancement of scientific understanding.

When we first encounter the scientific method, for example, it is suggested – or we are explicitly told – that this rigorous procedure leads to all scientific knowledge. Although all scientific principles must submit to the scientific method for confirmation, this naive view discounts the importance of scientific intuition. The source of scientific intuition is the intersection of scientific observations, known physical laws, and formal mathematics. In the hands – more accurately the mind – of an intelligent scientist, this intuition sometimes leads to profound postulates that change the way we think about the universe in which we live. Such postulates are then subject to the unforgiving scrutiny of the scientific method and formal analysis to test their consistency against new scientific observations and the library of known physical laws, thereby laying the foundation for the next cycle of scientific discovery.

The ultimate test of a new theory is its correspondence with observable reality along with its consistency with the lexicon of established theories and its ability to consolidate such prevailing theories. The most useful scientific theories combine an ability to encompass a wide set of scientific observations with the somewhat intangible quality of mathematical simplicity, sometimes expressed tangibly through symmetries, and the less precise quality of mathematical beauty. Why do we prefer a heliocentric (Sun-centered) model of the solar system over a geocentric (Earth-centered) viewpoint, for example? All of the same physics can be modeled and explained; it is simply a matter of reference frame. It is because of the inherent mathematical and geometric simplicity of the planet's elliptical orbits as compared to the far more complex shapes and mathematics required in the geocentric model. Just as we are best served by regarding physical laws as models and not as reality itself, it is important not to elevate the above-mentioned working principles of science to being inherent properties of the universe. On the other hand, however, they should be taken very seriously as they have stood the test of time and serve the scientific community extremely well in a broad set of disciplines.

Let us conclude this section where we started it. Despite the fact that modern science has largely purged us of the need to reconcile the scientific process and its results with religious belief (and sometimes scolded previous generations for having sought to do so), one cannot help but wonder whether the concerted efforts by the likes of Fermat, Newton, Leibniz, Maupertuis, Euler, and others to do so have not played an essential role in laying the foundation for modern science, including its methods and physical principles. For example, we continue to uphold simplicity and beauty in our mathematical and scientific pursuits having largely disparaged a significant reason for such subjective criteria being introduced into our vernacular in the first place. Why do we still emphasize these qualities today but that they have served us so well for many years in many fields? Why did many of the patriarchs of the scientific age pursue such intangible qualities but that they were largely motivated by a conception of a perfect Creator and their quest to better understand His creation? Did such an effort to integrate science and belief hinder or help their scientific pursuits?

While science and religion certainly encroach on each other's turf occasionally, in particular when addressing the origins of the universe and life, they are addressing different questions related to such matters. In so far as science is addressing the *what* and the *how*, religion addresses the *who* and the *why*. Science is based on observed fact, while religion is rooted in faith. Having affirmed this, however, we must be honest and acknowledge that as scientists we place some measure of faith in our approach to science as discussed above. Likewise, our religious beliefs must be informed by, and consistent with, observed facts.

Although the scientific community as a whole necessarily pursues science devoid of religious belief, each of us as individuals must harmonize our science with our beliefs (or lack thereof), reconciling any apparent inconsistencies. If we seek ultimate truths, which should be the goal of both science and religion, the truths of each – properly understood – cannot contradict one another (the story of Galileo and the Church serving as an important warning in this regard). By "properly understood," I mean that one must be careful not to impose requirements and standards on texts and ideas that are outside of their purview. For example, ancient religious texts, such as the Bible, should not be read as if they are modern science textbooks. Likewise, the science text does not address the questions of why the universe is the way it is or what our purpose for existing in it is.

Much more could be, and has been, said about such matters, and I have intentionally avoided speculating on questions for which I do not presume to even know if there is a knowable answer, let alone having a strong opinion on the matter. For example, is it ever possible to know the ultimate truths of how the universe operates, or is it even possible to know whether we know such truths? The book by Yourgrau and Mandelstam (1968) is highly recommended for its accurate historical accounts of the development of variational principles (including the math!) and the philosophy of science.[8] However, we turn our attention to how Hamilton's principle and the calculus of variations unify a wide variety of physical phenomena and optimization topics and the practical issues that arise in such fields.

EXERCISES

4.1 Repeat Example 4.1 involving projectile motion using the Euler-Lagrange equations.

4.2 A particle of mass m is constrained to move along the horizontal x-axis. Its potential energy is $\frac{1}{2}m\omega^2x^2$, where ω is a constant. Obtain the equation of motion for the particle, and determine the position as a function of time if

[8] The interested reader may also want to inform their thinking about such matters by reading books from the popular literature. Some recommended titles include *Why Beauty Is Truth: A History of Symmetry* by Ian Stewart, *Is God a Mathematician?* by Mario Livio, and *The Mind of God: The Scientific Basis for a Rational World* by Paul Davies. All three of these have been written by mathematicians/scientists for a general audience.

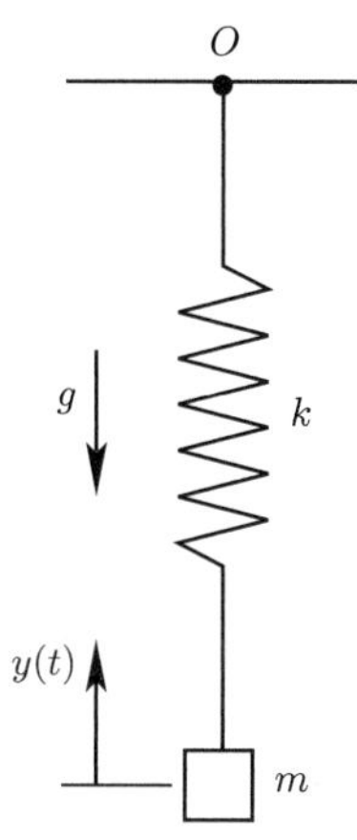

Figure 4.9. Schematic for Example 4.4.

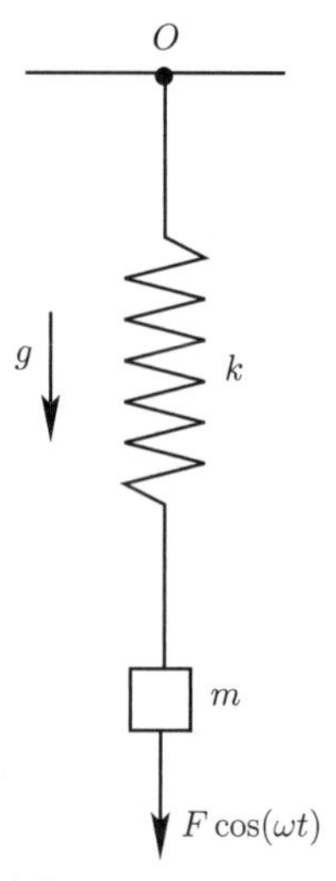

Figure 4.10. Schematic for Example 4.5.

at $t = 0$, the particle is at $x = 0$ and moving in the positive x-direction with velocity U.

4.3 Repeat Example 4.2 involving the undamped harmonic oscillator using the Euler-Lagrange equation.

4.4 Consider the spring-mass system shown in Figure 4.9. The location of the mass $y(t)$ is taken as zero when the spring is in its unstretched position.
   a) Determine the Lagrangian for the spring-mass system.
   b) Obtain the equation of motion for the mass.

4.5 As shown in Figure 4.10, consider a mass m hanging vertically from a linear spring, with spring constant k, which is fixed at its upper end O. A prescribed periodic force $F\cos(\omega t)$ acts downward on the mass along with the force due to gravity.
   a) Obtain the equation(s) of motion for this system.
   b) Solve the equation(s) of motion for the case when the forcing frequency ω is *not* the same as the natural frequency $\sqrt{k/m}$ of the spring-mass.

c) Consider the case when the forcing frequency *is* the same as the natural frequency, that is, $\omega = \sqrt{k/m}$. Discuss how the solution is different from that in part (b) both mathematically and physically. It is *not* necessary to obtain the general solution for this case; simply describe the general behavior and explain why it occurs mathematically.

4.6 Obtain the equation of motion for the simple pendulum of mass m and length ℓ in terms of the angle $\theta(t)$ measured from the pendulum's static equilibrium position hanging straight down.

4.7 Consider the simple pendulum of mass m and length ℓ by employing two coordinates of the mass in terms of the radial distance $r(t)$ from the pivot point to the mass and the angle $\theta(t)$ from the vertical.

a) Show that the equations of motion for $r(t)$ and $\theta(t)$ are

$$m\left[\ddot{r} - r\left(\dot{\theta}\right)^2 - g\cos\theta\right] = \Lambda,$$

$$m\left[r^2\ddot{\theta} + 2r\dot{r}\dot{\theta} + gr\sin\theta\right] = 0.$$

b) By introducing the constraint $r = 1$, obtain the usual form of the pendulum equation in terms of $\theta(t)$.

c) As discussed further in Section 5.1.2, the term containing the Lagrange multiplier in the augmented Lagrangian may be interpreted as the potential energy owing to the force required to maintain the constraint, that is, the tension in the pendulum. Show that the tension S in the string is given by

$$S(t) = -\Lambda(t) = m\left[1\dot{\theta}^2 + g\cos\theta\right].$$

4.8 Consider the simple pendulum of mass m and length ℓ in which the pivot point is restrained from moving in the vertical direction as usual, but where the pivot point can move in the horizontal direction with resistance owing to a horizontal spring with spring constant k. The horizontal displacement of the pivot point from the spring's neutral position is given by $x(t)$.

a) Obtain the equations of motion of the suspended mass m in terms of $x(t)$ and $\theta(t)$.

b) Show that if the system is limited to small amplitude oscillations, that is, $\theta(t) \ll 1$, the equations of motion may be approximated as follows

$$x = \frac{mg}{k}\theta, \quad \left(1 + \frac{mg}{k}\right)\ddot{\theta} + g\theta = 0,$$

such that the linearized system is equivalent to a simple pendulum of length $\left(1 + \frac{mg}{k}\right)$ with a fixed support.

4.9 Repeat Example 5.6 for the double pendulum using constraints as in Exercise 4.7.

4.10 Two identical pendulums of mass m and length ℓ hang from pivot points that are a horizontal distance ℓ apart. A spring having spring constant k connects the two masses with its neutral position occurring for a spring length of ℓ. Determine the equations of motion of the pendulums in terms of the angles $\theta_1(t)$ and $\theta_2(t)$ measured from each pendulum's static equilibrium position hanging straight down.

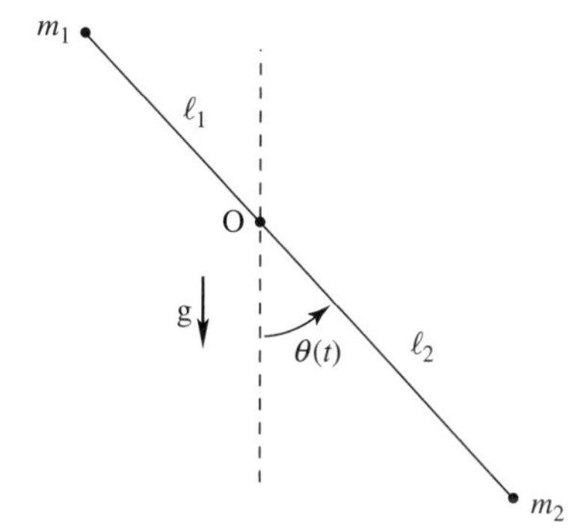

Figure 4.11. Schematic for Example 4.11.

4.11 Consider the dumbbell pendulum shown in Figure 4.11. Determine the equation of motion in terms of the angle $\theta(t)$.

4.12 Consider a spherical pendulum of mass m and length ℓ that is free to move in three-dimensional space. Show that the equations of motion in terms of the two angles $\theta(t)$ and $\phi(t)$ in spherical coordinates are

$$\ddot{\phi} - (\dot{\theta})^2 \sin\phi \cos\phi + \frac{g}{\ell} \sin\phi = 0 \quad \dot{\theta} \sin^2\phi = C,$$

where C is a constant.

4.13 Consider the orbit of a satellite of mass m around the Earth of mass M. The force of gravity between the two bodies is given by

$$\mathbf{f} = -\frac{GmM}{r^2}\mathbf{e}_r,$$

where $r = \sqrt{x^2 + y^2}$ is measured from the center of the Earth, and G is Newton's gravitational constant. Show that the equations of motion of the satellite are given by

$$\ddot{x} + GM\frac{x}{r^3} = 0, \quad \ddot{y} + GM\frac{y}{r^3} = 0.$$

4.14 An Atwood machine, which is a simple device that can be used to measure the acceleration due to gravity, consists of two masses of mass m_1 and m_2 connected by a massless string of length ℓ that hangs over a massless and frictionless pulley of radius R. The masses hang vertically downward in the gravitational field g, and the vertical position of mass 1 is the generalized coordinate q_1 measured from the rotational axis of the pulley. Determine the equation of motion for q_1.

4.15 Consider the pulley-weight system shown in Figure 4.12. A string of length ℓ_1 passes over the upper pulley of radius R_1 and has a mass $4m$ attached on the left side. On the right a pulley of mass m and radius R_2 is attached. A string of length ℓ_2 passes over this pulley and has a mass m attached on the left and a mass $2m$ on the right. The pulleys are frictionless, and the strings are massless. Using the generalized coordinates $q_1(t)$ and $q_2(t)$ shown in the figure, perform the following:

a) Determine the equations of motion for the system in terms of $q_1(t)$ and $q_2(t)$.

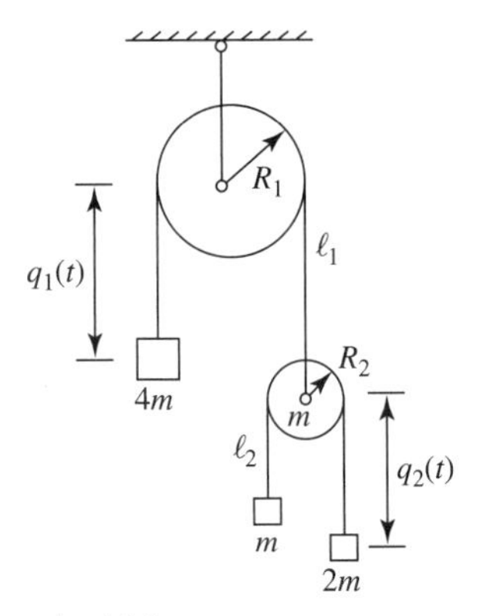

Figure 4.12. Schematic for Example 4.15.

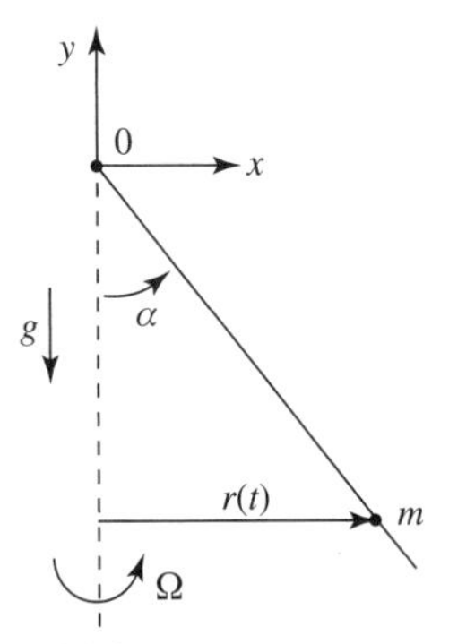

Figure 4.13. Schematic for Example 4.16.

b) If the system starts from rest and $q_1 = 1$ at $t = 0$, determine the motion $q_1(t)$ of the mass $4m$.

4.16 Consider the motion of a bead with mass m sliding without friction on a rigid massless wire that is rotating with constant angular velocity Ω at a fixed angle α about a vertical axis as shown in Figure 4.13.

a) Obtain the Lagrangian of the mass in terms of its radial distance $r(t)$ from the vertical axis (note that this is a one degree-of-freedom system).

b) Determine the equation of motion for the mass in terms of $r(t)$.

c) Solve the equation of motion subject to the initial conditions

$$r = 0, \dot{r} = 0 \quad \text{at} \quad t = 0.$$

That is, the bead is initially at rest at the end of the wire coinciding with the vertical axis of rotation.

4.17 Consider the motion of the two-particle system shown in Figure 4.14. One end of a light string of length ℓ is fastened to a particle of mass m_1, and the other end is attached to a particle of mass m_2. The mass m_2 is constrained to slide on a smooth, horizontal surface, and the string passes through a hole in the surface such that the mass m_1 hangs directly below the hole without swinging.

a) Construct the Lagrangian for this two-particle system, and obtain the equations of motion for the two dependent variables $r(t)$, which is the distance of the mass m_2 from the hole, and $\theta(t)$.

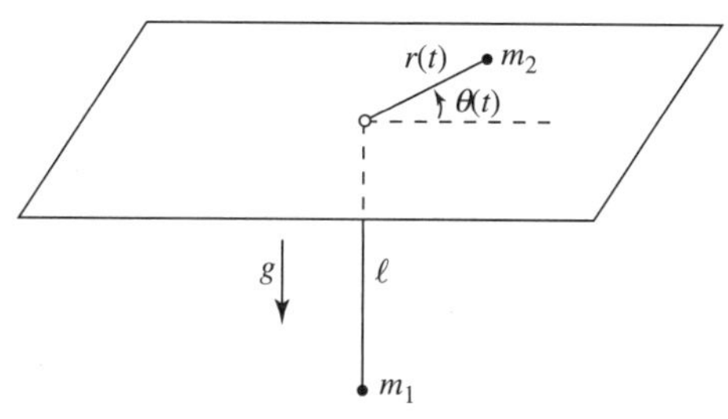

Figure 4.14. Schematic for Example 4.17.

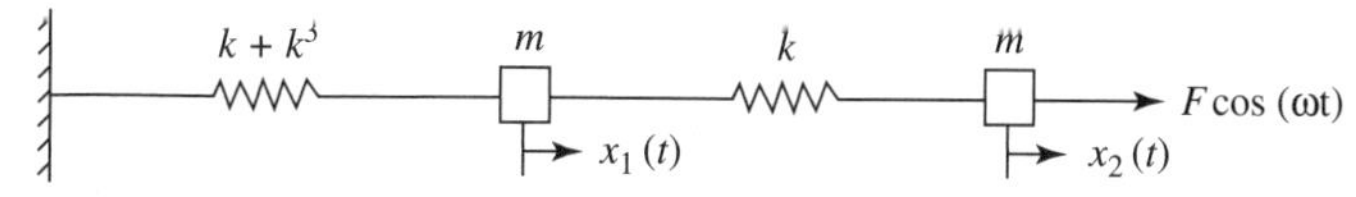

Figure 4.15. Schematic for Example 4.20.

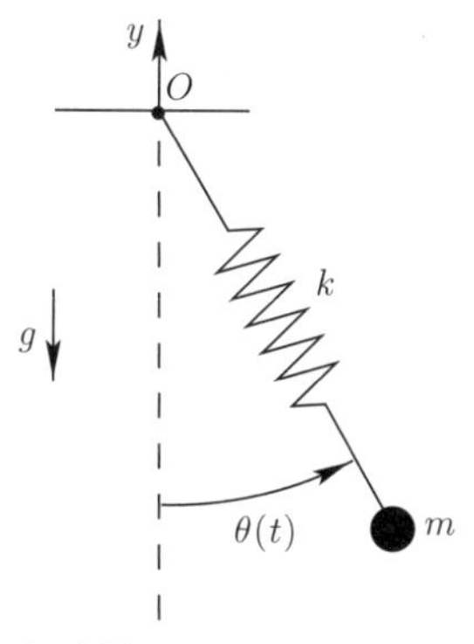

Figure 4.16. Schematic for Example 4.21.

b) For the special case in which the angular velocity of mass m_2 is $\Omega = \text{constant}$, show that the path of mass m_2 must be circular, and find the radius of this circle.

4.18 Repeat Example 5.1 involving a spring-mass system using the Euler-Lagrange equation.

4.19 Repeat Example 5.1 involving a spring-mass system if the mass hangs freely as a pendulum of length ℓ and mass m.

4.20 Consider the spring-mass system shown in Figure 4.15. The two masses have mass m, and a prescribed periodic force $F\cos(\omega t)$ acts to the right on the right mass. The left spring has spring constant $k + k^3$, and the right spring has spring constant k. Obtain the Euler-Lagrange equation(s) of motion for this system.

4.21 Consider the planar motion of an elastic pendulum consisting of a mass m suspended under gravity g by a weightless elastic rod of unstretched length ℓ and having spring constant k as shown in Figure 4.16. Let the position of the mass be given by the coordinates $r(t)$, the radial distance from the pivot point O, and $\theta(t)$, the angle from the vertical.

a) Obtain the general equation(s) of motion for this system.

b) Solve the equation(s) of motion for the r-mode, for which $\theta(t) = 0$.

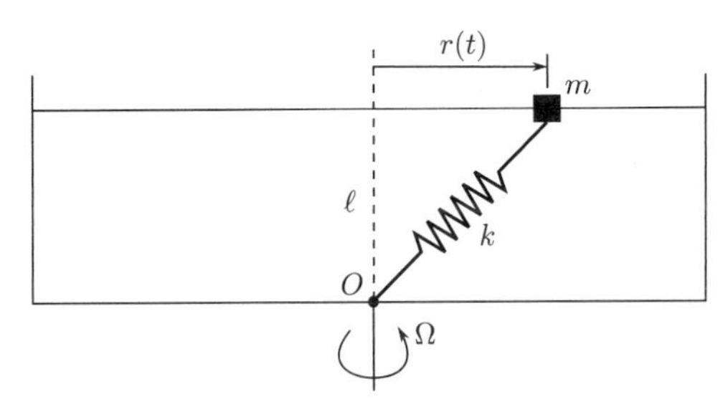

Figure 4.17. Schematic for Example 4.23.

4.22 Repeat Example 5.2 involving a linear damped oscillator using the appropriate form of the Euler-Lagrange equation.

4.23 A bead of mass m is free to slide without friction along a straight, horizontal wire as shown in Figure 4.17. The wire is rotating about a vertical axis with constant angular velocity Ω, and the mass is connected to the point O by a linear spring with spring constant k. The wire is located a distance ℓ vertically above the point O. The position of the mass is given by the radial distance $r(t)$ from the vertical axis of rotation. The spring has unstretched length a, which may be greater than, less than, or equal to ℓ.

a) Determine the Euler-Lagrange equation of motion for the mass in terms of the radial distance $r(t)$ from the rotational axis.

b) Determine the equilibrium position(s) of the mass by finding the radial positions $r_0(t)$ for which the radial velocity and acceleration vanish.

c) Consider stability of the equilibrium position(s) r_0 by introducing the perturbation

$$r(t) = r_0 + \epsilon u(t), \quad \epsilon \ll 1,$$

and obtaining the equation for $u(t)$. Determine whether the system is stable for the equilibrium position(s).

4.24 A bead of mass m is free to slide on a circular wire of radius R. The circular loop lies in a vertical plane and rotates with constant angular velocity Ω about a vertical axis passing through the center of the loop. The vertical coordinate y is measured from the center of the circular loop, r is measured radially out from the axis of rotation, and the angle θ gives the angular position of the mass in the loop measured from the base of the circular loop.

a) Determine the equation of motion of the mass in terms of the angle $\theta(t)$.

b) Show that the equilibrium positions of the mass, for which $\dot{\theta} = \ddot{\theta} = 0$, are given by $\theta_0 = 0, \pi$.

c) Consider stability of the $\theta_0 = 0$ equilibrium position by introducing the perturbation

$$\theta(t) = \theta_0 + \epsilon u(t), \quad \epsilon \ll 1,$$

and obtaining the equation for $u(t)$. Determine whether the system is stable for the following two cases: *i*) slow rotation $\Omega < \sqrt{g/R}$, and *ii*) rapid rotation $\Omega > \sqrt{g/R}$.

4.25 Repeat Example 4.4 involving the static deflection of an elastic string using the Euler-Lagrange equation.

4.26 Consider the deflection of a rotating string as in Example 4.5. Starting with the functional form (4.18), obtain the equation of motion (4.13) using the Euler-Lagrange equation.

4.27 A simply supported beam of length ℓ with a concentrated load P at the center has deflection $u(x)$ that is symmetric about the center of the beam at $x = \ell/2$. The deflection corresponding to equilibrium conditions minimizes the potential energy of the beam

$$V[u] = -Pu(\ell/2) + \int_0^{\ell/2} \frac{1}{2} EI(u'')^2 dx,$$

where EI is the constant bending stiffness. Note that we only consider half of the beam $0 \leq x \leq \ell/2$ owing to symmetry. The boundary conditions at the simply supported end are $u(0) = 0$ and $u''(0) = 0$, and natural boundary conditions apply at the midpoint $x = \ell/2$. Determine the deflection curve $u(x)$ for the beam.

5 Classical Mechanics

> Nature is thrifty in all its actions.
>
> Now, here is this principle, so wise, so worthy of the Supreme Being: when some change occurs in Nature, the amount of Action used for this change is always the smallest possible.
>
> The laws of movement thus deduced [from the principle of least action], being found to be precisely the same as those observed in nature, we can admire the application of it to all phenomena, in the movement of animals, in the vegetation of plants, in the revolution of the heavenly bodies: and the spectacle of the universe becomes so much the grander, so much the more beautiful, so much more worthy of its Author, when one knows that a small number of laws, most wisely established, suffice for all movements.
>
> (Pierre Louis Moreau de Maupertuis)

Classical[1] mechanics encompasses those areas of physics that originated prior to the development of relativistic mechanics at the beginning of the twentieth century. Its primary focus is on the application of Newtonian mechanics to macroscopic systems. Classical mechanics provides the basis for many of the important fields in engineering, including solid mechanics, fluid mechanics, transport phenomena, and dynamics. These fields are employed broadly in the design of almost all devices that make modern life possible, including being sure that your mobile phone can withstand a fall on a hard surface, placing and maintaining communication satellites in their orbits, and both terrestrial and extraterrestrial transportation systems.

This chapter considers the classical applications of Hamilton's principle and the Euler-Lagrange equations to statics and dynamics of nondeformable and deformable bodies. Statics and dynamics of nondeformable bodies follow from Hamilton's principle for discrete systems and lead to the *principle of virtual work* and *Newton's second law*, respectively. Likewise, Hamilton's principle for continuous systems applied to statics and dynamics of deformable bodies leads to the *theorem of minimum potential energy* and canonical partial differential equations, such as the

[1] In general, the term *classical* refers to things that are "old, but still worthwhile" (for example, classical music); that is, they have stood the test of time. It also carries the connotation of having originated during the seminal period of a field's development.

wave equation, respectively. While it is customary to consider the simpler statics problem before dynamics, here we view statics as a special case of the more general dynamics problem to be considered first.

5.1 Dynamics of Nondeformable Bodies

The traditional application of Hamilton's principle for discrete systems (4.6) is in obtaining the equations of motion for dynamical systems comprised of particles and rigid bodies. A particle is such that its mass is concentrated at a point, that is, a *point mass*. It is small relative to the $O(1)$ scale of the overall system, but it is still large relative to the molecular scale. In practice, the mass is always distributed throughout some region in space, in which case we are approximating the object as a nondeformable body with mass concentrated at the object's center of mass. This is a good approximation for small bodies or those with small mass relative to other objects in the system. For $O(1)$-sized bodies with $O(1)$ distributed mass, it becomes necessary to consider moments of inertia of the rigid body (see, for example, Goldstein 1959 and José and Saletan 1998). The dynamics of the case in which an $O(1)$-sized body with distributed mass is regarded as deformable is considered in Section 5.4.

Consideration of moving particles and rigid bodies is encompassed by the subjects of *dynamics* and *dynamical systems*. The variational approach taken herein is classically referred to as *analytical dynamics*. Entire courses and textbooks are devoted to these subjects.

In dynamics problems, the potential energy is the total virtual work done by conservative forces required to move the system from its unloaded condition to its current, fully loaded state at time t. The framework for addressing such problems is established in Chapter 4; therefore, we primarily focus on examples to illustrate the basic approach. Recall from equation (4.6) that Hamilton's principle for discrete systems with conservative force fields is

$$\delta I = \delta \int_{t_1}^{t_2} L dt = 0, \tag{5.1}$$

where $L = T - V$ is called the *Lagrangian*, T is the *kinetic energy*, and V is the *potential energy*.

5.1.1 Applications of Hamilton's Principle

We first consider two examples by applying Hamilton's principle and determining the equations of motion directly from the variational form as in the derivation in Section 2.1.3 and the examples in Section 4.1.

EXAMPLE 5.1 Obtain the equation of motion for the system shown in Figure 5.1, where the rod is massless. The spring is unloaded when $\theta = 0$, corresponding to $x_1 = x = 0, y_2 = y = 0$.

Solution: This is a single degree-of-freedom system ($n = 1$) requiring only one coordinate to describe the motion of the entire system. We could use $x(t)$, $y(t)$,

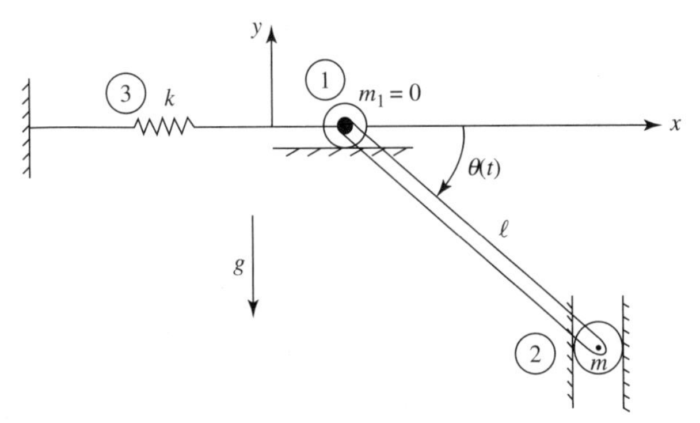

Figure 5.1. Schematic of spring-mass system in Example 5.1.

or $\theta(t)$ as our dependent variable; it is most convenient to use $\theta(t)$. The coordinates of each particle can be expressed in terms of θ as follows:

$$x_1 = x = \ell(1 - \cos\theta),$$

$$y_2 = y = -\ell\sin\theta.$$

The forces on both particle two due to gravity and particle one due to the spring are conservative (the potential energy of the spring is the *strain energy* due to its extension).

Recalling our treatment of the spring in Example 4.2, the Lagrangian ($N = 3$), where each term is written using its most natural variable, is

$$\begin{aligned} L &= T_1 + T_2 + T_3 - V_1 - V_2 - V_3 \\ &= 0 + \frac{1}{2}m\dot{y}^2 + 0 - 0 + \int_{\bar{y}=0}^{\bar{y}=y} (-mg)\,\delta\bar{y} + \int_{\bar{x}=0}^{\bar{x}=x} (-k\bar{x})\,\delta\bar{x} \\ L &= \frac{1}{2}m\dot{y}^2 - mgy - \frac{1}{2}kx^2. \end{aligned}$$

Note that the gravitational potential energy V_2 and strain energy V_3 are determined by obtaining the work required to move the system from its unloaded state, $\bar{x} = 0$ and $\bar{y} = 0$, to its current loaded state, $\bar{x} = x$ and $\bar{y} = y$. Writing the Lagrangian in terms of θ ($\dot{y} = -\ell\dot{\theta}\cos\theta$) yields

$$\begin{aligned} L &= \frac{1}{2}m\left[-\ell\dot{\theta}\cos\theta\right]^2 - mg[-\ell\sin\theta] - \frac{1}{2}k\left[\ell(1-\cos\theta)\right]^2 \\ L &= \frac{1}{2}m\ell^2\left[\dot{\theta}^2\cos^2\theta + \frac{2g}{\ell}\sin\theta - \frac{k}{m}(1-\cos\theta)^2\right]. \end{aligned}$$

Applying Hamilton's principle leads to

$$\delta I = \int_{t_1}^{t_2} \delta L(t,\theta,\dot{\theta})dt = 0$$

$$\int_{t_1}^{t_2} \left(\frac{\partial L}{\partial \theta}\delta\theta + \frac{\partial L}{\partial \dot{\theta}}\delta\dot{\theta}\right) dt = 0$$

$$m\ell^2 \int_{t_1}^{t_2} \left\{\left[\dot{\theta}^2\cos\theta\sin\theta - \frac{g}{\ell}\cos\theta + \frac{k}{m}(1-\cos\theta)\sin\theta\right]\delta\theta - \dot{\theta}\cos^2\theta\,\delta\dot{\theta}\right\} dt = 0.$$

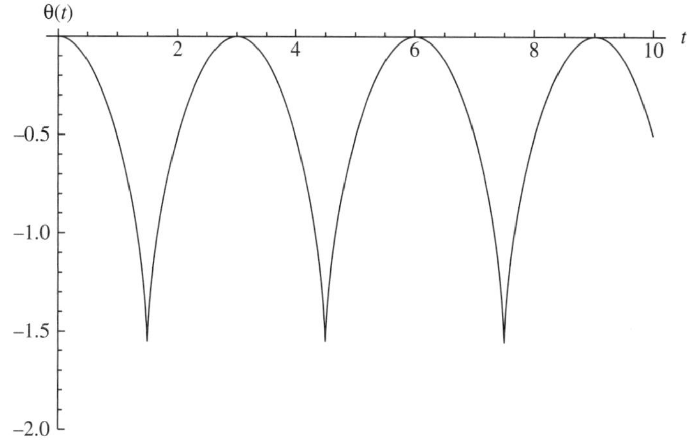

Figure 5.2. Solution $\theta(t)$ of the spring-mass system for the case with $k = 2, m = 1, \ell = 10, g = 9.81$, and initial conditions $\theta(0) = 0, \dot{\theta}(0) = 0$.

Noting that

$$\delta\dot{\theta} = \frac{d}{dt}(\delta\theta),$$

and integrating the last term by parts ($p = \dot{\theta}\cos^2\theta, dq = \frac{d}{dt}(\delta\theta)dt$) gives

$$\int_{t_1}^{t_2} \dot{\theta}\cos^2\theta\delta\dot{\theta}dt = -\int_{t_1}^{t_2} \frac{d}{dt}\left(\dot{\theta}\cos^2\theta\right)\delta\theta dt.$$

Substituting yields

$$\int_{t_1}^{t_2}\left[\frac{d}{dt}\left(\dot{\theta}\cos^2\theta\right) + \dot{\theta}^2\cos\theta\sin\theta - \frac{g}{\ell}\cos\theta + \frac{k}{m}(1-\cos\theta)\sin\theta\right]\delta\theta dt = 0,$$

but $\delta\theta$ is arbitrary; therefore, the term in brackets must be zero for $t_1 \leq t \leq t_2$. With some rearrangement, this gives the equation of motion as

$$\ddot{\theta}\cos^2\theta - \dot{\theta}^2\cos\theta\sin\theta + \frac{k}{m}(1-\cos\theta)\sin\theta - \frac{g}{\ell}\cos\theta = 0.$$

Note that this is a second-order, nonlinear ordinary differential equation. A sample numerical solution for $\theta(t)$ is shown in Figure 5.2. As in Example 4.3, it is not necessary to determine the force in the rod.

In the previous example, the forces are all conservative. In the next example, we illustrate how to account for nonconservative forces for the case of a damped oscillator.

EXAMPLE 5.2 Consider the linear damped oscillator shown in Figure 5.3. The oscillatory motion of the mass is damped such that the damping force is proportional to the velocity of the mass according to $\mathbf{f} = -c\dot{\mathbf{y}}$, where c is the damping coefficient. Note that the y-position is defined such that it is zero when there is no force in the spring. We seek the equation of motion for the position of the mass $y(t)$.

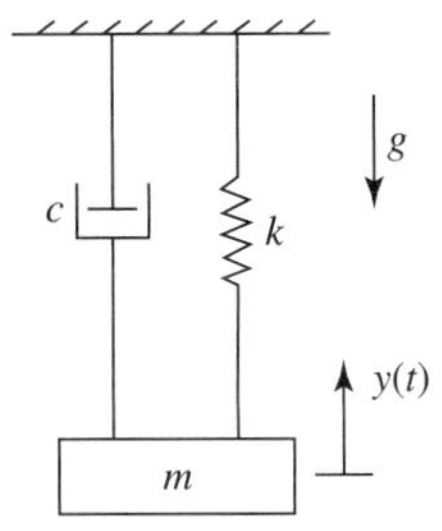

Figure 5.3. Schematic of the linear damped oscillator in Example 5.2.

Solution: The kinetic energy is

$$T = \frac{1}{2}m\dot{y}^2,$$

and the potential energy due to the conservative forces of gravity and the linear elastic spring is

$$V = mgy + \frac{1}{2}ky^2.$$

The nonconservative force due to damping is proportional to the velocity according to

$$\mathbf{f}_{nc} = -c\dot{\mathbf{y}} = -c\dot{y}\mathbf{j},$$

and the virtual displacement is $\delta\mathbf{r} = \delta y\mathbf{j}$. Applying the general form of Hamilton's principle (4.5), we obtain

$$\delta\int_{t_1}^{t_2}(T-V)dt + \int_{t_1}^{t_2}\mathbf{f}\cdot\delta\mathbf{r}dt = 0$$

$$\delta\int_{t_1}^{t_2}\left(\frac{1}{2}m\dot{y}^2 - mgy - \frac{1}{2}ky^2\right)dt + \int_{t_1}^{t_2}(-c\dot{y}\delta y)dt = 0$$

$$\int_{t_1}^{t_2}(m\dot{y}\delta\dot{y} - mg\delta y - ky\delta y)\,dt - \int_{t_1}^{t_2}c\dot{y}\delta y dt = 0$$

$$\int_{t_1}^{t_2}(-m\ddot{y} - mg - ky - c\dot{y})\,\delta y dt = 0,$$

where integration by parts has been used for the first term as usual. Therefore, the equation of motion is

$$\ddot{y} + \frac{c}{m}\dot{y} + \frac{k}{m}y = -g.$$

Note that the first-derivative term results from the inclusion of damping, and if there is no damping ($c = 0$), then this reduces to the governing equation for the undamped harmonic oscillator considered in Example 4.2 (without the effect of gravity owing to its horizontal orientation).

Let us solve the equation of motion with $c = 1, m = \frac{1}{2}, k = 5$ and the initial conditions $y(0) = 0, \dot{y}(0) = 0$. The equation of motion is then

$$\ddot{y} + 2\dot{y} + 10y = -g,$$

which has the solution

$$y(t) = -\frac{1}{10}g\left\{1 - e^{-t}\left[\frac{1}{3}\sin(3t) + \cos(3t)\right]\right\}.$$

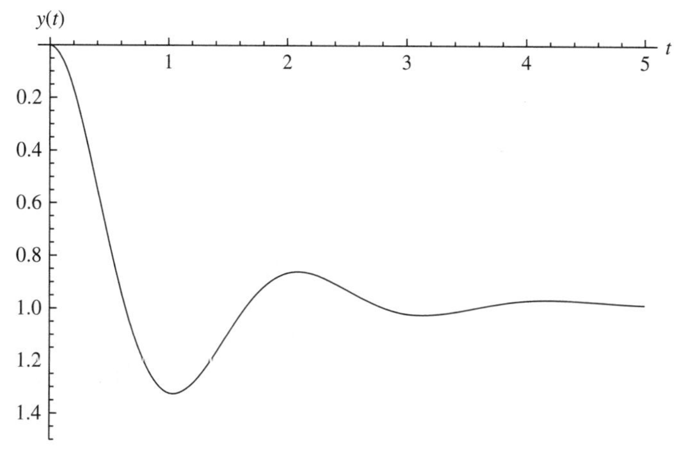

Figure 5.4. Solution $y(t)$ for the damped oscillator.

This solution with $g = 9.81$ is plotted in Figure 5.4, exhibiting a damped oscillation.

The static equilibrium state can be found by taking the solution as $t \to \infty$, from the equation of motion with $\dot{y} = \ddot{y} = 0$, or from the Euler-Lagrange equation of motion with $T = 0$, that is, $\partial V / \partial y = 0$. In either case, the static equilibrium state for the given parameters is

$$y = -\frac{1}{10} g,$$

which is consistent with the solution shown in Figure 5.4 as $t \to \infty$.

Rather than solving dynamics problems directly from Hamilton's principle as illustrated in the previous two examples, we can simplify obtaining the equations of motion by using the general form of the Euler-Lagrange equations of motion. For conservative systems, the Euler-Lagrange equations (4.21) are

$$\frac{\partial L}{\partial q_i} - \frac{d}{dt}\left(\frac{\partial L}{\partial \dot{q}_i}\right) = 0, \quad i = 1, \ldots, n. \tag{5.2}$$

We use the more general form (4.27) if nonconservative forces act on the system.

EXAMPLE 5.3 Let us consider the unconstrained motion of a particle of mass m in three-dimensional Cartesian coordinates as shown in Figure 5.5 and determine the equations of motion.

Solution: Recall from equation (4.19) that with the generalized coordinates $q_1 = x, q_2 = y$, and $q_3 = z$, the kinetic energy is given by

$$T = \frac{1}{2} m \left(\dot{x}^2 + \dot{y}^2 + \dot{z}^2\right) = \frac{1}{2} m \left(\dot{q}_1^2 + \dot{q}_2^2 + \dot{q}_3^2\right).$$

Therefore, the Euler-Lagrange equations (5.2) are

$$\frac{\partial L}{\partial q_1} - \frac{d}{dt}\left(\frac{\partial L}{\partial \dot{q}_1}\right) = 0, \tag{5.3}$$

$$\frac{\partial L}{\partial q_2} - \frac{d}{dt}\left(\frac{\partial L}{\partial \dot{q}_2}\right) = 0, \tag{5.4}$$

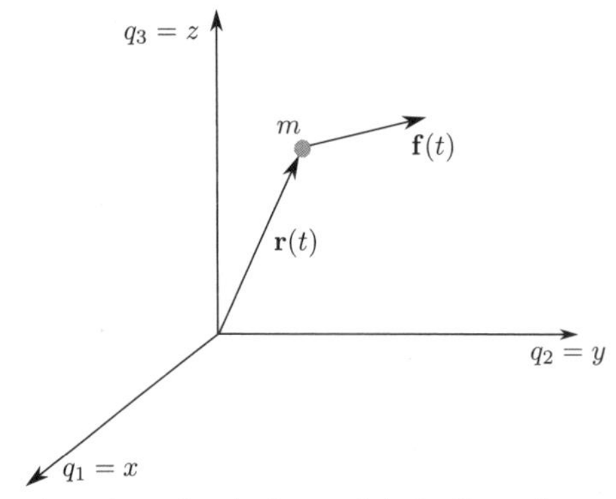

Figure 5.5. Unconstrained motion of a single particle in Cartesian coordinates.

$$\frac{\partial L}{\partial q_3} - \frac{d}{dt}\left(\frac{\partial L}{\partial \dot{q}_3}\right) = 0. \tag{5.5}$$

Consider equation (5.3) with $L = T - V$, which yields

$$-\frac{\partial V}{\partial q_1} - \frac{d}{dt}\left[\frac{1}{2}m(2\dot{q}_1)\right] = 0,$$

$$m\ddot{q}_1 = Q_1,$$

where the generalized force is $Q_1 = -\partial V/\partial q_1$. Similarly, equations (5.4) and (5.5) produce

$$m\ddot{q}_2 = Q_2, \quad m\ddot{q}_3 = Q_3.$$

These three equations are the component forms of Newton's second law in Cartesian coordinates, which can also be written in the form

$$m\ddot{x} = f_x, \quad m\ddot{y} = f_y, \quad m\ddot{z} = f_z,$$

where $f_x = Q_1, f_y = Q_2$, and $f_z = Q_3$ are the components of the resultant force $\mathbf{f}$ in the x-, y-, and z-directions, respectively.

EXAMPLE 5.4 Now consider the unconstrained motion of a particle of mass m in three-dimensional cylindrical coordinates as shown in Figure 5.6, for which we seek the equations of motion.

Solution: Recall from equation (4.20) that the kinetic energy is given by

$$T = \frac{1}{2}m\left[\dot{r}^2 + (r\dot{\theta})^2 + \dot{z}^2\right] = \frac{1}{2}m\left[\dot{q}_1^2 + (q_1\dot{q}_2)^2 + \dot{q}_3^2\right].$$

Therefore, with the generalized coordinates $q_1 = r$, $q_2 = \theta$, and $q_3 = z$, the Euler-Lagrange equations (5.2) are

$$\frac{\partial L}{\partial q_1} - \frac{d}{dt}\left(\frac{\partial L}{\partial \dot{q}_1}\right) = 0, \tag{5.6}$$

$$\frac{\partial L}{\partial q_2} - \frac{d}{dt}\left(\frac{\partial L}{\partial \dot{q}_2}\right) = 0, \tag{5.7}$$

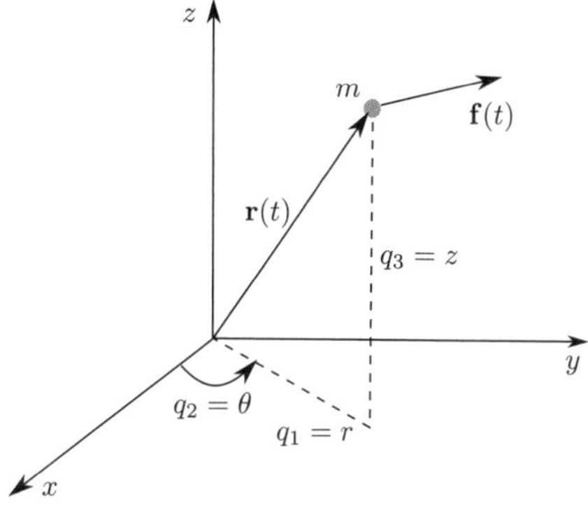

Figure 5.6. Unconstrained motion of a single particle in cylindrical coordinates.

$$\frac{\partial L}{\partial q_3} - \frac{d}{dt}\left(\frac{\partial L}{\partial \dot{q}_3}\right) = 0. \tag{5.8}$$

Observe once again that the Euler-Lagrange equations of motion are independent of the coordinate system. Consider equation (5.6) with $L = T - V$, which yields

$$\frac{1}{2}m(2q_1\dot{q}_2^2) - \frac{\partial V}{\partial q_1} - \frac{d}{dt}\left[\frac{1}{2}m(2\dot{q}_1)\right] = 0$$

$$m\ddot{q}_1 - mq_1\dot{q}_2^2 = Q_1, \tag{5.9}$$

where the generalized force is $Q_1 = -\partial V/\partial q_1$. The second term on the left-hand side is referred to as the *centrifugal acceleration*. From equation (5.7), we obtain

$$-\frac{\partial V}{\partial q_2} - \frac{d}{dt}\left[\frac{1}{2}m(2q_1^2\dot{q}_2)\right] = 0$$

$$mq_1^2\ddot{q}_2 + 2mq_1\dot{q}_1\dot{q}_2 = Q_2, \tag{5.10}$$

where the generalized force is $Q_2 = -\partial V/\partial q_2$. The second term on the left-hand side is the *Coriolis acceleration*. Finally, equation (5.8) produces

$$-\frac{\partial V}{\partial q_3} - \frac{d}{dt}\left[\frac{1}{2}m(2\dot{q}_3)\right] = 0$$

$$m\ddot{q}_3 = Q_3, \tag{5.11}$$

where the generalized force is $Q_3 = -\partial V/\partial q_3$.

These three equations are the component forms of Newton's second law in cylindrical coordinates. Observe that Q_1 and Q_2 have units of force because $q_1 = r$ and $q_3 = z$ are linear displacements, whereas Q_2 is a torque or moment because $q_2 = \theta$ is an angular displacement. Note that some authors refer to the additional centrifugal and Coriolis terms as "forces"; however, they arise from the kinetic energy, not the potential energy. Consequently, they are more accurately referred to as "accelerations" rather than "forces." It is important to realize, however, that these "accelerations" are a byproduct of the coordinate system used, and no new physics has been introduced. However, centrifugal and Coriolis forces (in Q_1 and/or Q_2) are required in order to maintain the motion that gives rise to these accelerations.

Table 5.1. *The analogy between coupled mechanical and electrical systems*

Mechanical Systems		Electrical Systems
displacement (x)	$\longleftrightarrow$	charge
mass (m)	$\longleftrightarrow$	inductance
spring constant (k)	$\longleftrightarrow$	1/capacitance
damping (c)	$\longleftrightarrow$	resistance
force ($\mathbf{f}$)	$\longleftrightarrow$	voltage
potential energy (V)	$\longleftrightarrow$	electromagnetic energy
kinetic energy (T)	$\longleftrightarrow$	magnetic energy

For example, the tension in a rotating string constrains the motion of a mass to a circular path.

REMARKS:

1. For single-particle systems, the generalized coordinates would typically be the components of the position vector $\mathbf{r}$, as illustrated in the previous two examples for Cartesian and cylindrical coordinates. More generally, however, the generalized coordinates are the dependent variables, for example, positions or angles, that fix the state of the system and may or may not be the components of position vectors.
2. The equations of motion may be written in the general matrix form

$$\mathbf{M}(\mathbf{q})\ddot{\mathbf{q}} + \mathbf{C}(\mathbf{q},\dot{\mathbf{q}})\dot{\mathbf{q}} + \mathbf{K}(\mathbf{q}) = \mathbf{f}(t), \tag{5.12}$$

   where $\mathbf{q} = [q_1, q_2, \ldots, q_n]$ is the generalized coordinate vector, $\dot{\mathbf{q}}$ is the generalized velocity vector, $\ddot{\mathbf{q}}$ is the generalized acceleration vector, and $\mathbf{f}(t)$ is the applied force vector. The matrices are the *mass matrix* $\mathbf{M}$, the *damping matrix* $\mathbf{C}$, and the *stiffness matrix* $\mathbf{K}$.
3. Observe that in most of the examples performed thus far in Chapters 4 and 5, the Lagrangians do not explicitly depend upon the independent variable time. This would suggest application of special case II in Section 2.1.4. However, there is typically no practical advantage to invoking this special case for obtaining the Euler-Lagrange equations of motion. In fact, we normally seek the differential equations of motion, which would be bypassed if applying the special cases. Therefore, we simply treat such problems in the usual way.
4. There is an analogy between coupled mechanical and electrical systems in which all of the preceding developments hold for coupled electrical systems if the substitutions given in Table 5.1 are made.
5. In the present text, we refer to the differential equation that results from taking the extremum of a functional in a general context as an *Euler equation*; we reserve the name *Euler-Lagrange equation* for when it is the equation of motion that results from Hamilton's variational principle. Likewise, we only use the term *Lagrangian* for the integrand in the context of Hamilton's principle and its applications.

6. See Yourgrau and Mandelstram (1968) and Lanczos (1970) for a treatment of calculus of variations put into its historical context as applied to analytical dynamics. Lanczos's text also includes insightful comparisons of the variational approach to dynamics as compared to using Newton's second law directly.

EXAMPLE 5.5 Consider the spring-mass system shown in Figure 5.7. The system is such that there is no force in any of the springs when the masses are located at $x_1 = 0, x_2 = 0$. Determine the equations of motion for the system.

Solution: This spring-mass system has two degrees of freedom ($n = 2$), in which case the generalized coordinates are $q_1 = x_1(t)$ and $q_2 = x_2(t)$. Regarding the masses as particles $k = 1$ and $k = 2$ and the springs, from left to right, as particles $k = 3$, $k = 4$, and $k = 5$, the kinetic energy is ($T_3 = T_4 = T_5 = 0$):

$$T = T_1 + T_2 = \frac{1}{2}m\dot{x}_1^2 + \frac{1}{2}(2m)\dot{x}_2^2 = m\left(\frac{1}{2}\dot{x}_1^2 + \dot{x}_2^2\right).$$

The potential (strain) energy in the springs is ($V_1 = V_2 = 0$):

$$\begin{aligned} V &= V_3 + V_4 + V_5 \\ &= -\int \mathbf{f}_3 \cdot \delta\mathbf{r}_3 - \int \mathbf{f}_4 \cdot \delta\mathbf{r}_4 - \int \mathbf{f}_5 \cdot \delta\mathbf{r}_5 \\ &= -\int_{\bar{x}_1=0}^{\bar{x}_1=x_1} (-2k\bar{x}_1)\delta\bar{x}_1 - \int_{\bar{u}=0}^{\bar{u}=u} (-k\bar{u})\delta\bar{u} - \int_{\bar{x}_2=0}^{\bar{x}_2=x_2} (-k\bar{x}_2)\delta\bar{x}_2 \\ V &= kx_1^2 + \frac{1}{2}k(x_2 - x_1)^2 + \frac{1}{2}kx_2^2, \end{aligned}$$

where we recall that the restoring force in a linear spring, whether in tension or compression, is $f = -ks$, where s is the elongation of the spring. For example, the elongation of the middle spring $k = 4$ is $s = x_2 - x_1$. The Lagrangian is then

$$L = T - V = m\left(\frac{1}{2}\dot{x}_1^2 + \dot{x}_2^2\right) - kx_1^2 - \frac{1}{2}k(x_2 - x_1)^2 - \frac{1}{2}kx_2^2.$$

For our two degree-of-freedom system, the Euler-Lagrange equations of motion are

$$\frac{\partial L}{\partial x_1} - \frac{d}{dt}\left(\frac{\partial L}{\partial \dot{x}_1}\right) = 0, \quad \frac{\partial L}{\partial x_2} - \frac{d}{dt}\left(\frac{\partial L}{\partial \dot{x}_2}\right) = 0.$$

Substituting the Lagrangian into the first equation gives

$$-2kx_1 + k(x_2 - x_1) - \frac{d}{dt}(m\dot{x}_1) = 0$$

$$m\frac{d^2x_1}{dt^2} = -3kx_1 + kx_2,$$

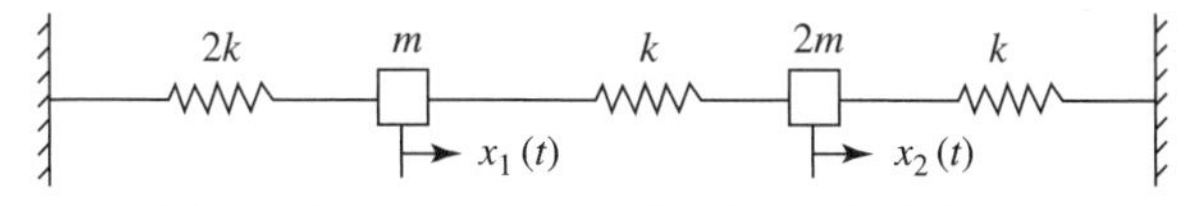

Figure 5.7. Schematic of the spring-mass system in Example 5.5.

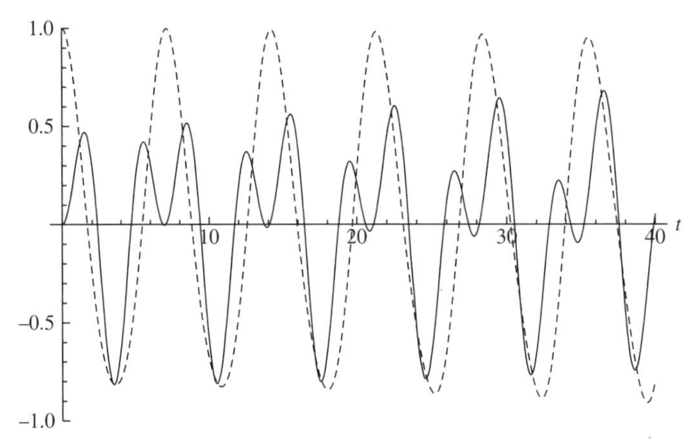

Figure 5.8. Solution of the spring-mass system for the case with $k = m$ and initial conditions $x_1(0) = 0, x_2(0) = 1, \dot{x}_1(0) = 0, \dot{x}_2(0) = 0$ [solid line is $x_1(t)$ and dashed line is $x_2(t)$].

and substituting into the second equation leads to

$$-k(x_2 - x_1) - kx_2 - \frac{d}{dt}(2m\dot{x}_2) = 0$$

$$2m\frac{d^2x_2}{dt^2} = kx_1 - 2kx_2.$$

In the matrix form (5.12), the equations of motion are

$$\begin{bmatrix} m & 0 \\ 0 & 2m \end{bmatrix}\begin{bmatrix} \ddot{x}_1 \\ \ddot{x}_2 \end{bmatrix} + \begin{bmatrix} 3k & -k \\ -k & 2k \end{bmatrix}\begin{bmatrix} x_1 \\ x_2 \end{bmatrix} = \begin{bmatrix} 0 \\ 0 \end{bmatrix}, \tag{5.13}$$

where the matrix in the first term is the mass matrix, and that in the second term is the stiffness matrix. These are the same equations of motion that would be obtained by evaluation of Newton's second law for free-body diagrams of the two masses. A sample solution is shown in Figure 5.8.

Any general solution of the coupled system of equations, such as that in Figure 5.8, is a linear combination of the two *natural*, or *resonant, frequencies* of the system. These occur when the entire system oscillates sinusoidally with the same frequency. In the present system, this corresponds to the two masses oscillating with the same frequency. Note that the masses may be oscillating in or out of phase; however, the frequency must be the same. Let us assume periodic motion of the form:[2]

$$x_1(t) = A_1 e^{i\omega t}, \quad x_2(t) = A_2 e^{i\omega t},$$

where ω is the natural frequency, and A_1 and A_2 are the amplitudes of masses 1 and 2, respectively. It is understood that we are taking the real part of the final expressions for $x_1(t)$ and $x_2(t)$. For example,

$$x_1(t) = \text{Re}\left[A_1 e^{i\omega t}\right] = \text{Re}\left[A_1\cos(\omega t) + A_1 i \sin(\omega t)\right] = A_1\cos(\omega t).$$

[2] This form is better than setting $x_1(t) = A_1\cos(\omega t)$ directly, for example, which does not work if we have odd-order derivatives.

Evaluating the derivatives gives

$$\frac{d^2x_1}{dt^2} = -A_1\omega^2 e^{i\omega t}, \quad \frac{d^2x_2}{dt^2} = -A_2\omega^2 e^{i\omega t}.$$

Substituting into the governing equations (5.13) and canceling $e^{i\omega t}$ from each term gives

$$-m\omega^2 A_1 = -3kA_1 + kA_2,$$
$$-2m\omega^2 A_2 = kA_1 - 2kA_2,$$

or upon rearranging

$$3A_1 - A_2 = \lambda A_1,$$
$$-\tfrac{1}{2}A_1 + A_2 = \lambda A_2,$$

where $\lambda = m\omega^2/k$. Thus, we have two equations for the two unknown amplitudes A_1 and A_2. In matrix form, this is

$$\begin{bmatrix} 3 & -1 \\ -\frac{1}{2} & 1 \end{bmatrix}\begin{bmatrix} A_1 \\ A_2 \end{bmatrix} = \lambda \begin{bmatrix} A_1 \\ A_2 \end{bmatrix}.$$

Observe that this is an eigenvalue problem in which the eigenvalues λ are related to the natural frequencies ω of the system, which is typical of dynamical systems. Solving, we find that the eigenvalues are

$$\lambda_1 = \frac{m\omega_1^2}{k} = 2 - \frac{\sqrt{6}}{2}, \quad \lambda_2 = \frac{m\omega_2^2}{k} = 2 + \frac{\sqrt{6}}{2}.$$

Therefore, the two *natural frequencies* of the system, corresponding to the two degrees of freedom, are

$$\omega_1 = \sqrt{2 - \frac{\sqrt{6}}{2}}\sqrt{\frac{k}{m}}, \quad \omega_2 = \sqrt{2 + \frac{\sqrt{6}}{2}}\sqrt{\frac{k}{m}}.$$

These are called *normal modes*, for which the solution is a sinusoidal oscillation; that is, each normal mode behaves like a simple harmonic oscillator (see Section 4.1). The number of normal modes (natural frequencies) is equal to the number of degrees of freedom of the system. The smallest of the natural frequencies, ω_1, is known as the *fundamental mode*. The important physical characteristic of normal modes is that they do not interact. The modes are normal in the sense that they can move independently. In other words, each normal mode can be excited independently without exciting the others. Once again, the general motion of the system that results from imposition of any set of initial conditions is a linear superposition of the natural frequencies.

Recall that the kinetic energy $T = \sum \frac{1}{2}\dot{\mathbf{q}}^T\mathbf{A}\dot{\mathbf{q}}$ is always a quadratic. The potential energy for harmonic oscillators is also a quadratic. Specifically, the potential (strain) energy for a system of N linear springs is a quadratic of the form

$$V = \sum_{i=1}^{N}\sum_{j=1}^{N}\frac{1}{2}s_j\bar{K}_{ij}s_j = \frac{1}{2}\mathbf{s}^T\bar{\mathbf{K}}\mathbf{s},$$

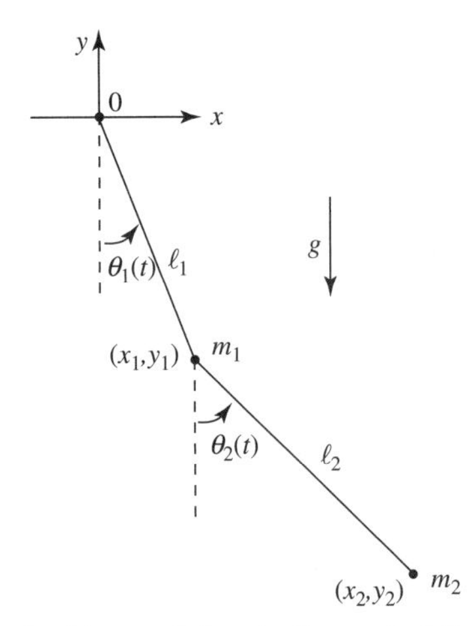

Figure 5.9. Schematic of the double pendulum in Example 5.6.

where $\bar{\mathbf{K}} = \bar{K}_{ij}$ is a symmetric matrix that contains the spring constants, and $\mathbf{s}$ is the elongation vector. Note that $\bar{\mathbf{K}}$ is not the same as the stiffness matrix $\mathbf{K}$ in the general equation of motion (5.12). In the previous example,

$$\mathbf{s} = \begin{bmatrix} x_1 \\ x_2 - x_1 \\ -x_2 \end{bmatrix}, \quad \bar{\mathbf{K}} = \begin{bmatrix} 2k & 0 & 0 \\ 0 & k & 0 \\ 0 & 0 & k \end{bmatrix}.$$

EXAMPLE 5.6 Determine the equations of motion for the double pendulum shown in Figure 5.9.

Solution: In terms of Cartesian coordinates (x,y), the double pendulum would require four values to specify the system at a given time, that is, the location of mass m_1, say (x_1,y_1), and the location of mass m_2, say (x_2,y_2). Alternatively, we can fully specify the system using $q_1 = \theta_1$ and $q_2 = \theta_2$; therefore, it is a two degree-of-freedom system $(n = 2)$. The geometric relationships between (x_1,y_1), (x_2,y_2), and (θ_1,θ_2) are

$$x_1 = \ell_1 \sin\theta_1, \qquad x_2 = \ell_1 \sin\theta_1 + \ell_2 \sin\theta_2,$$

$$y_1 = -\ell_1 \cos\theta_1, \quad y_2 = -\ell_1 \cos\theta_1 - \ell_2 \cos\theta_2.$$

The kinetic energy is $(N = 2)$

$$T = T_1 + T_2 = \frac{1}{2}m_1\left(\dot{x}_1^2 + \dot{y}_1^2\right) + \frac{1}{2}m_2\left(\dot{x}_2^2 + \dot{y}_2^2\right).$$

Substituting the relationships between x, y, and θ and simplifying yields

$$T = \frac{1}{2}(m_1 + m_2)\ell_1^2\dot{\theta}_1^2 + m_2\ell_1\ell_2\dot{\theta}_1\dot{\theta}_2\cos(\theta_1 - \theta_2) + \frac{1}{2}m_2\ell_2^2\dot{\theta}_2^2.$$

The potential energy is

$$V = V_1 + V_2 = m_1 g y_1 + m_2 g y_2,$$

which in terms of θ_1 and θ_2 becomes

$$V = -(m_1+m_2)g\ell_1\cos\theta_1 - m_2 g\ell_2\cos\theta_2.$$

Observe once again that we begin by writing T and V in terms of the most convenient coordinates, here x_1, y_1, x_2, and y_2, and then transforming to the chosen generalized coordinates, here $q_1 = \theta_1$ and $q_2 = \theta_2$.

The Euler-Lagrange equations of motion are then

$$\frac{\partial L}{\partial \theta_1} - \frac{d}{dt}\left(\frac{\partial L}{\partial \dot{\theta}_1}\right) = 0, \quad \frac{\partial L}{\partial \theta_2} - \frac{d}{dt}\left(\frac{\partial L}{\partial \dot{\theta}_2}\right) = 0.$$

Evaluating the partial derivatives in the first equation and substituting yields

$$\begin{aligned} &-m_2\ell_1\ell_2\dot{\theta}_1\dot{\theta}_2\sin(\theta_1-\theta_2) - (m_1+m_2)g\ell_1\sin\theta_1 \\ &-\frac{d}{dt}\left[(m_1+m_2)\ell_1^2\dot{\theta}_1 + m_2\ell_1\ell_2\dot{\theta}_2\cos(\theta_1-\theta_2)\right] = 0. \end{aligned}$$

After differentiating and simplifying, we have the equation of motion

$$\begin{aligned} &(m_1+m_2)\ell_1\ddot{\theta}_1 + m_2\ell_2\left[\dot{\theta}_2^2\sin(\theta_1-\theta_2) + \ddot{\theta}_2\cos(\theta_1-\theta_2)\right] \\ &+ (m_1+m_2)g\sin\theta_1 = 0. \end{aligned} \tag{5.14}$$

Similarly, the second equation produces

$$\begin{aligned} &m_2\ell_1\ell_2\dot{\theta}_1\dot{\theta}_2\sin(\theta_1-\theta_2) - m_2 g\ell_2\sin\theta_2 \\ &-\frac{d}{dt}\left[m_2\ell_1\ell_2\dot{\theta}_1\cos(\theta_1-\theta_2) + m_2\ell_2^2\dot{\theta}_2\right] = 0. \end{aligned}$$

Again evaluating the derivative and simplifying yields the other equation of motion

$$\ell_2\ddot{\theta}_2 - \ell_1\dot{\theta}_1^2\sin(\theta_1-\theta_2) + \ell_1\ddot{\theta}_1\cos(\theta_1-\theta_2) + g\sin\theta_2 = 0. \tag{5.15}$$

Equations (5.14) and (5.15) are the equations of motion for the double pendulum, which are two coupled, second-order, nonlinear ordinary differential equations for $\theta_1(t)$ and $\theta_2(t)$ corresponding to the two degrees of freedom. Written in the matrix form (5.12), they are

$$\begin{bmatrix} M_{11} & M_{12} \\ M_{21} & M_{22} \end{bmatrix}\begin{bmatrix} \ddot{\theta}_1 \\ \ddot{\theta}_2 \end{bmatrix} + \begin{bmatrix} C_{11} & C_{12} \\ C_{21} & C_{22} \end{bmatrix}\begin{bmatrix} \dot{\theta}_1 \\ \dot{\theta}_2 \end{bmatrix} + \begin{bmatrix} K_{11} & K_{12} \\ K_{21} & K_{22} \end{bmatrix} = \begin{bmatrix} 0 \\ 0 \end{bmatrix},$$

where

$$M_{11} = (m_1+m_2)\ell_1,\ M_{12} = m_2\ell_2\cos(\theta_1-\theta_2),\ M_{21} = \ell_1\cos(\theta_1-\theta_2),\ M_{22} = \ell_2,$$

$$C_{11} = 0, \quad C_{12} = m_2\ell_2\dot{\theta}_2\sin(\theta_1-\theta_2), \quad C_{21} = -\ell_1\dot{\theta}_1\sin(\theta_1-\theta_2), \quad C_{22} = 0,$$

$$K_{11} = (m_1+m_2)g\sin\theta_1, \quad K_{12} = 0, \quad K_{21} = 0, \quad K_{22} = g\sin\theta_2.$$

Note that although there is no physical damping in the system, there are damping-like terms that result from the coupled motions of the two pendulums. Again, it is

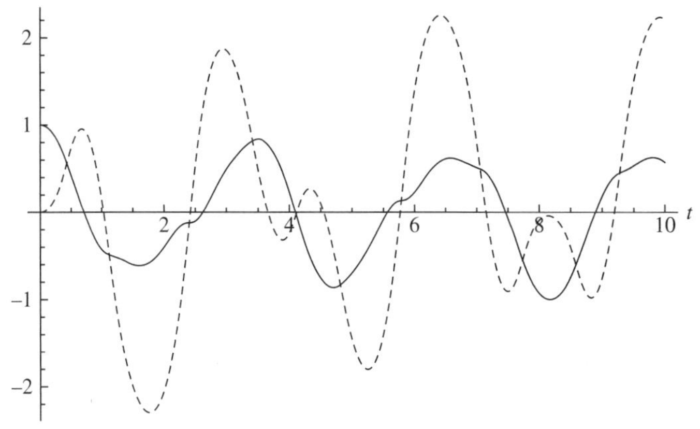

Figure 5.10. Solution of the double pendulum for the case with $m_1 = 2$, $m_2 = 1$, $\ell_1 = 2$, $\ell_2 = 1$, $g = 9.81$ and initial conditions $\theta_1(0) = 1$, $\theta_2(0) = 0$, $\dot{\theta}_1(0) = 0$, $\dot{\theta}_2(0) = 0$ [solid line is $\theta_1(t)$ and dashed line is $\theta_2(t)$].

not necessary to determine the forces in the pendulums as would be the case if using Newton's second law. An example numerical solution is shown in Figure 5.10.

Hildebrand (1965), Section 2.15, considers the solution of the *linearized* form of the equations of motion for the double pendulum problem. This form results from restricting the motion of the pendulums to small amplitude oscillations close to the equilibrium position $\theta_1 = 0, \theta_2 = 0$.

5.1.2 Dynamics Problems with Constraints

In the previous examples, we explicitly write the Lagrangian $L = T - V$ in terms of the n generalized coordinates $q_1, q_2, \ldots, q_n$, where n is the number of degrees of freedom. In other words, we eliminate any other variables prior to evaluating the Euler-Lagrange equations. One consequence of this approach is that it is not necessary to determine the forces of constraint, such as the force in a pendulum, because they do no net work.

Alternatively, an n degree-of-freedom system may be expressed in terms of $n+k$ generalized coordinates, $q_1, q_2, \ldots, q_n, \ldots, q_{n+k}$, with k geometric constraints

$$\begin{aligned}
\phi_1(q_1, \ldots, q_{n+k}) &= 0, \\
\phi_2(q_1, \ldots, q_{n+k}) &= 0, \\
&\vdots \\
\phi_k(q_1, \ldots, q_{n+k}) &= 0.
\end{aligned}$$

Algebraic constraints of this form, which constrain the coordinates of the system, are sometimes called *holonomic constraints*. We illustrate this approach through two examples, the second of which will be completed using both approaches for comparison.

EXAMPLE 5.7 Reconsider the rotating pendulum problem solved in Example 4.3 using constraints.

Solution: Recall that the Lagrangian in terms of r and y is

$$L = T - V = \frac{1}{2}m\Omega^2 r^2 - mgy. \tag{5.16}$$

Rather than using the geometry to eliminate the variable r and writing the Lagrangian solely in terms of y, here we will impose the algebraic (geometric) constraint

$$r^2 + y^2 = \ell^2. \tag{5.17}$$

As in Section 2.7.3, we form the augmented Lagrangian as follows:

$$\tilde{L} = \frac{1}{2}m\Omega^2 r^2 - mgy + \Lambda(t)\left(r^2 + y^2 - \ell^2\right),$$

where we recall that the Lagrange multiplier $\Lambda(t)$ must be allowed to be a function of the independent variable for an algebraic constraint. In this case, $n = 1$, $k = 1$, and $n + k = 2$; therefore, we require two Euler-Lagrange equations of motion, which are

$$\frac{\partial \tilde{L}}{\partial r} - \frac{d}{dt}\left(\frac{\partial \tilde{L}}{\partial \dot{r}}\right) = 0, \quad \frac{\partial \tilde{L}}{\partial y} - \frac{d}{dt}\left(\frac{\partial \tilde{L}}{\partial \dot{y}}\right) = 0.$$

Substituting the augmented Lagrangian into each of these equations in turn gives

$$m\Omega^2 r + 2\Lambda r = 0$$

$$\Lambda = -\frac{1}{2}m\Omega^2,$$

and

$$-mg + 2\Lambda y = 0$$

$$y = \frac{mg}{2\Lambda}.$$

Thus, the solution for the vertical location of the mass is

$$y = \frac{1}{2}mg\left(-\frac{2}{m\Omega^2}\right) = -\frac{g}{\Omega^2},$$

which is the same as that obtained in Example 4.3.

REMARKS:

1. Observe that we use the two Euler-Lagrange equations to eliminate the Lagrange multiplier and obtain the equation for the dependent variable y.
2. Physically, we may interpret the term with the Lagrange multiplier in the augmented Lagrangian as the potential energy owing to the force required to maintain the constraint, which is the tension in the massless rod.
3. It is important to realize that because the additional term in the augmented Lagrangian owing to the constraint vanishes when the constraint is satisfied (as

it must be), the underlying physical properties of the system remain unchanged except that the trajectory itself is forced to be consistent with the imposed constraint.

EXAMPLE 5.8 Consider a mass m sliding without friction in a parabola as shown in Figure 5.11. Determine the equation of motion for the position $x(t)$.

Solution: This is a single degree-of-freedom system with $q_1 = x(t)$. First, let us determine the equation of motion using the usual approach without constraints. Because $y = x^2$, we have

$$\dot{y} = 2x\dot{x}.$$

The Lagrangian is then

$$\begin{aligned} L &= T - V \\ &= \frac{1}{2}m(\dot{x}^2 + \dot{y}^2) - mgy \\ &= \frac{1}{2}m(\dot{x}^2 + 4x^2\dot{x}^2) - mgx^2 \\ L &= \frac{1}{2}m(1 + 4x^2)\dot{x}^2 - mgx^2. \end{aligned}$$

Note that we begin by writing the Lagrangian in terms of $x(t)$ and $y(t)$, but that the final version is only in terms of $q_1 = x(t)$. In order to evaluate the Euler-Lagrange equation for $L = L(t, x, \dot{x})$, we have that

$$\frac{\partial L}{\partial x} = 4mx\dot{x}^2 - 2mgx, \quad \frac{\partial L}{\partial \dot{x}} = m(1 + 4x^2)\dot{x}.$$

Therefore, the Euler-Lagrange equation is

$$\begin{gathered} \frac{\partial L}{\partial x} - \frac{d}{dt}\left(\frac{\partial L}{\partial \dot{x}}\right) = 0 \\ 4mx\dot{x}^2 - 2mgx - \frac{d}{dt}[m(1 + 4x^2)\dot{x}] = 0 \\ 4x\dot{x}^2 - 2gx - 8x\dot{x}^2 - (1 + 4x^2)\ddot{x} = 0 \\ (1 + 4x^2)\ddot{x} + 4x\dot{x}^2 + 2gx = 0. \end{gathered}$$

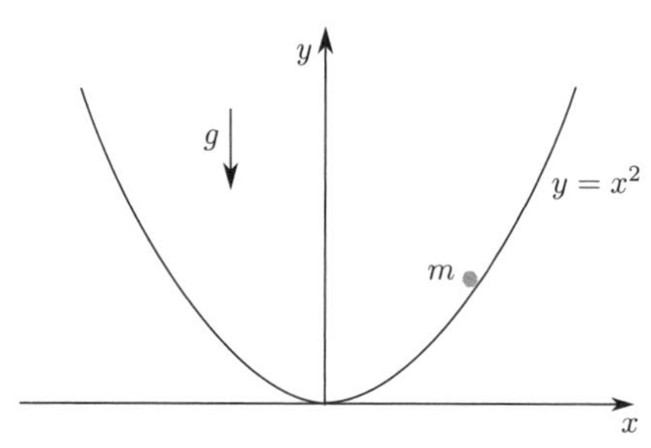

Figure 5.11. Schematic for the mass sliding in a parabola in Example 5.8.

Hence, the equation of motion is a nonlinear, second-order ordinary differential equation.

Now we revisit the system by considering the problem in terms of two generalized coordinates $q_1 = x(t)$ and $q_2 = y(t)$ and a geometric constraint. Recall that the Lagrangian for this system is

$$L = \frac{1}{2}m(\dot{x}^2 + \dot{y}^2) - mgy.$$

Because we have two generalized coordinates and a one degree-of-freedom system ($n=1$), we need one constraint ($k=1$) to close the problem ($n+k=2$). The constraint in this case is that the mass must remain on the parabola; therefore, we have the algebraic geometric constraint $y = x^2$. The augmented Lagrangian is then

$$\tilde{L} = \frac{1}{2}m(\dot{x}^2 + \dot{y}^2) - mgy + \Lambda(t)(y - x^2). \tag{5.18}$$

In this case, with $\tilde{L} = \tilde{L}(t,x,y,\dot{x},\dot{y})$, we have two Euler-Lagrange equations

$$\frac{\partial \tilde{L}}{\partial x} - \frac{d}{dt}\left(\frac{\partial \tilde{L}}{\partial \dot{x}}\right) = 0, \quad \frac{\partial \tilde{L}}{\partial y} - \frac{d}{dt}\left(\frac{\partial \tilde{L}}{\partial \dot{y}}\right) = 0.$$

The derivatives are

$$\frac{\partial \tilde{L}}{\partial x} = -2\Lambda x, \quad \frac{\partial \tilde{L}}{\partial \dot{x}} = m\dot{x}, \quad \frac{\partial \tilde{L}}{\partial y} = -mg + \Lambda, \quad \frac{\partial \tilde{L}}{\partial \dot{y}} = m\dot{y},$$

which produce the equations of motion

$$\ddot{x} + \frac{2\Lambda}{m}x = 0, \quad \ddot{y} - \frac{\Lambda}{m} + g = 0.$$

To eliminate the Lagrange multiplier $\Lambda(t)$, multiply the second equation by $2x$ and add to the first equation to give

$$\ddot{x} + 2x\ddot{y} + 2gx = 0. \tag{5.19}$$

From the constraint $y = x^2$, however, we have

$$\ddot{y} = 2\dot{x}^2 + 2x\ddot{x}.$$

Substituting into equation (5.19) leads to

$$\ddot{x} + 2x(2\dot{x}^2 + 2x\ddot{x}) + 2gx = 0,$$

or

$$(1 + 4x^2)\ddot{x} + 4x\dot{x}^2 + 2gx = 0,$$

which is the same equation of motion obtained using the previous approach. A sample solution for $x(t)$ is shown in Figure 5.12.

Again, we may interpret the term with the Lagrange multiplier in the augmented Lagrangian (5.18) as being the potential energy owing to the force required to maintain the system such that the constraint is satisfied. In the present example,

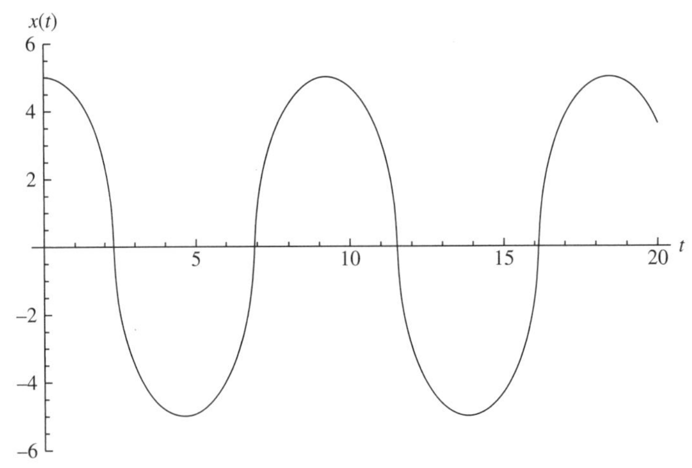

Figure 5.12. Solution of the mass in a parabola for the case with $g = 9.81$ and the initial conditions $x(0) = 5, \dot{x}(0) = 0$.

it is the potential energy owing to the force exerted on the mass required to maintain its parabolic path. There is no need to directly determine this force as would be the case if applying Newton's second law directly. That is, because of the constraint, there is no virtual displacement in the direction of the force; consequently, there is no virtual work. In other words, the mass' movement is orthogonal to the force exerted by the surface, so no virtual work is done. This is the case for algebraic (holonomic) constraints.

As illustrated in Examples 5.7 and 5.8, algebraic geometric (holonomic) constraints can be incorporated using Lagrange multipliers or by eliminating any variables beyond the required n generalized coordinates. It is advantageous to use Lagrange multipliers for algebraic constraints when: 1) it is difficult to eliminate an extraneous coordinate, and/or 2) we seek the forces of constraint.

Differential kinematic (nonholonomic) constraints are far more subtle and remain an active area of research. First, it is often the case that apparently nonholonomic constraints are actually holonomic. This is the case if the differential constraint is *integrable*, in which case it can be integrated to produce an algebraic constraint. We might expect that truly nonholonomic constraints could be enforced using Lagrange multipliers. Unfortunately, this is not the case in general, despite many claims to the contrary in the literature. An example of a situation that produces a nonholonomic constraint is a wheel rolling without slipping. Nonholonomic constraints are beyond the scope of the present text; the interested reader is referred to the book by Bloch (2003) and the paper by Bloch, Fernandez, and Mestdag (2009) for their treatment.

5.2 Statics of Nondeformable Bodies

We consider statics of nondeformable bodies as a special case of dynamics of nondeformable bodies treated in the previous section. Because the bodies are in static equilibrium, the kinetic energy is now zero. Consideration of systems of structural members in static equilibrium is encompassed by the subject of *statics*, in which it is applied to beams, trusses, and machine elements, for example. Note that statics is primarily concerned with determination of the forces acting *on* the members of the system assumed to be nondeformable. The resulting stress, strain, and displacement distributions *within* each member is the scope of *mechanics of solids* (elasticity), which is considered in Section 5.3, wherein Hamilton's principle is applied to continuous systems of elastic bodies.

First, we briefly review statics of nondeformable bodies using the vectorial approach taught in typical undergraduate courses on statics. Although this approach is very straightforward, it can become rather tedious for systems comprised of a large number of individual members. As a complement – not a replacement – to this approach, we then introduce the virtual work approach, which follows directly from Hamilton's variational principle.

5.2.1 Vectorial Approach

From Newton's second law applied to a body in static equilibrium, for which there is no linear or angular acceleration, the forces and moments must sum to zero according to

$$\mathbf{f} = \mathbf{0}, \quad \mathbf{M} = \mathbf{0}.$$

This applies to the entire system as a whole along with each individual component of the system. In practice, these vector equations are typically applied in component form in each coordinate direction.

EXAMPLE 5.9 For a given force P applied to the system shown in Figure 5.13, determine the angle θ for which the system is in static equilibrium. The bars of length ℓ are considered rigid and with negligible mass compared to m.

Solution: Using the vectorial approach, we first apply $\mathbf{f} = \mathbf{0}$ to a free-body diagram of the system as a whole to determine the reaction forces. This is illustrated in

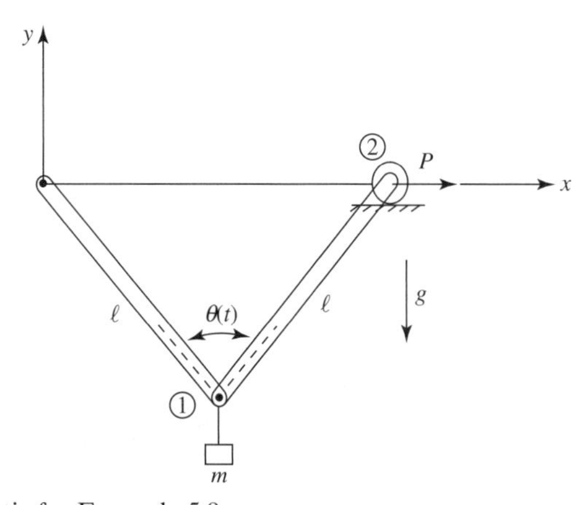

Figure 5.13. Schematic for Example 5.9.

Figure 5.14. Balancing forces in the x-direction requires that

$$R_x + P = 0;$$

thus,

$$R_x = -P.$$

Owing to symmetry (or from $\mathbf{M} = \mathbf{0}$ for the entire system), the vertical reaction forces R_y must be the same. Balancing the forces in the y-direction gives

$$2R_y - mg = 0,$$

or

$$R_y = \frac{1}{2}mg.$$

Now let us consider the free-body diagram for the left bar as shown in Figure 5.15. In order to balance forces in the x-direction, $f_x = P$. Summing moments about the origin requires

$$P\ell\cos\left(\frac{\theta}{2}\right) - \frac{1}{2}mg\ell\sin\left(\frac{\theta}{2}\right) = 0,$$

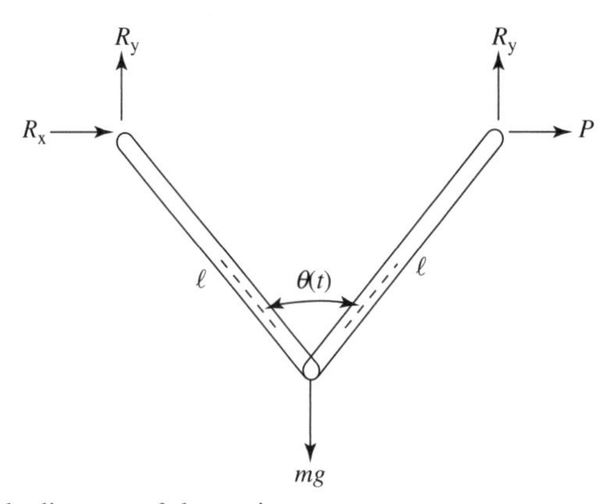

Figure 5.14. Free-body diagram of the entire system.

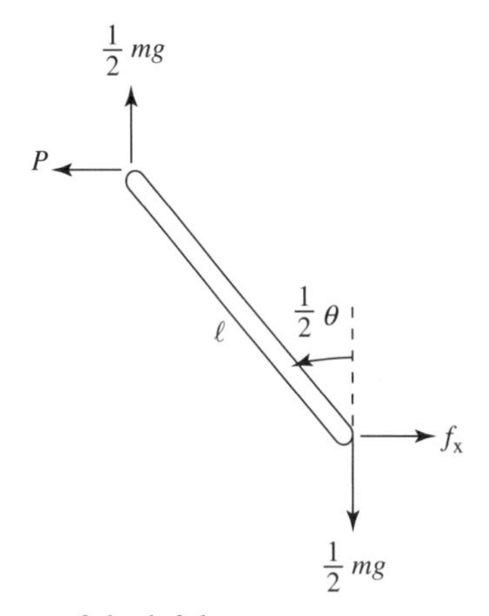

Figure 5.15. Free-body diagram of the left bar.

or

$$\tan\left(\frac{\theta}{2}\right) = \frac{2P}{mg}.$$

Therefore, the angle θ is given by

$$\theta = 2\tan^{-1}\left(\frac{2P}{mg}\right).$$

5.2.2 Virtual Work Approach

Bodies in static equilibrium do not move and as such do not have any kinetic energy ($T = 0$). Therefore, Hamilton's principle (4.6) for static conservative systems reduces to

$$\delta \int_{t_1}^{t_2} V dt = 0.$$

From equation (4.23), the associated Euler-Lagrange equations for an n degree-of-freedom system are

$$\frac{\partial V}{\partial q_i} = 0, \quad i = 1, \ldots, n, \tag{5.20}$$

where the potential energy $V(q_i)$ for conservative systems is an algebraic equation that depends only on the positions q_i. Therefore, the static equilibrium positions of conservative systems correspond to stationary points of the algebraic potential energy $V(q_i)$. Recall from Section 4.1 that the potential energy is the negative of the virtual work done in loading a conservative system as follows:

$$V = -\sum_{k=1}^{N} \int_{\mathbf{r}_{ik}}^{\mathbf{r}_k} \mathbf{f}_k \cdot \delta\mathbf{r}_k, \tag{5.21}$$

where $\mathbf{r}_{ik}$ are the position vectors for the initially unloaded system, and $\mathbf{r}_k$ are the position vectors for the system in static equilibrium after the forces are applied.

Alternatively, the equations governing static equilibrium can be expressed in the form of the principle of virtual work. From Hamilton's principle (4.1) with $T = 0$, we have

$$\int_{t_1}^{t_2} \mathbf{f} \cdot \delta\mathbf{r} dt = 0.$$

Because the forces, and the system itself, do not move with time for a static system, the virtual work must be zero

$$\mathbf{f} \cdot \delta\mathbf{r} = 0. \tag{5.22}$$

This is the *principle of virtual work*, which applies to the system as a whole as well as each individual component of the system. Because the sum of the forces $\mathbf{f}$ is zero for a system in static equilibrium, the principle of virtual work indicates that if the system is given an arbitrary virtual displacement, the net virtual work done by the forces acting on the system will be zero.

This is a good context in which to further clarify the notions of *virtual displacements* and *virtual work*. Observe that for nondeformable bodies in static equilibrium, there is no actual work done by the forces acting on the body as the forces do not act through a distance. However, we still would like to be able to apply conservation of energy to such systems, thereby making use of Hamilton's principle and the *theorem of minimum potential energy* (see Section 5.3). Therefore, we must appeal to a *virtual work* $\mathbf{f} \cdot \delta\mathbf{r}$ as the actual force is imagined to act through a small *virtual displacement* $\delta\mathbf{r}$ from its equilibrium position. We appeal to such virtual (hypothetical) displacements in the spirit of the variational approach in which we consider variations away from the stationary solution – which in this case is the solution for which the body is in static equilibrium – in order to obtain the actual equilibrium solution.

The virtual displacements must be consistent with any external constraints placed on the possible equilibrium positions of the system. That is, the virtual displacements are only considered in directions in which the system *could* move, even though it does not owing to the fact that it is in static equilibrium. In statics problems, therefore, the potential energy is the total virtual work that would be required to move the system through a virtual displacement from its unloaded condition to its fully loaded condition, where it is in static equilibrium. This is given by equation (5.21).

Strictly speaking, the continuously varying force required to move a system from its unloaded to loaded state should be used when determining the virtual work in statics problems. For an axial rod undergoing a tensile force, for example, this is equivalent to using the average force $\frac{1}{2}P$ if the material is linearly elastic, where P is the final static load. This is not necessary, however, for two reasons:

1. The one-half factor generally cancels from all terms in statics problems.
2. In dynamics problems, where actual forces are applied throughout the dynamical motion, the one-half factor would not appear. Because we may have occasion to consider both static and dynamic cases for the same system, neglecting the one-half factor will allow us to have the same virtual work (potential energy) terms in both static and dynamic cases.

The virtual work approach is somewhat different from the vectorial approach typically taken in statics, whereby free-body diagrams are drawn for each member of the system and the forces and moments are summed to zero as in Example 5.9. The free-body diagram approach, however, is necessary when considering nondeformable bodies in static equilibrium for which the unloaded and loaded states are the same, that is, where no virtual work is possible given the constraints. The approach using virtual work is illustrated for the same example as in the previous section.

EXAMPLE 5.10 Repeat Example 5.9 using the virtual work approach.

Solution: The unloaded state of the system, with $m = 0$ and $P = 0$, is taken as corresponding to $y_{i1} = -\ell$, $x_{i2} = 0$. This corresponds to the case where the two bars of length ℓ have small mass compared to m and P such that they hang vertically downward when unloaded. The loaded state is $y - y_{i1} = y - (-\ell)$, $x - x_{i2} = x$. Therefore, the potential energy of the system, which is the virtual work done on the system in taking it through the virtual displacement from its unloaded to loaded condition, is

$$\delta \int_{t_1}^{t_2} V dt = 0,$$

where we consider the potential energy for each of the two points in Figure 5.13.

Observe that particle 1 may move in the x-direction; therefore, we may imagine a virtual displacement δx. However, there is no force on particle 1 in that direction, and the virtual work is zero and need not be included. That is, $\mathbf{f}_1 = 0\mathbf{i} - mg\mathbf{j}$, and $\delta\mathbf{r}_1 = \delta x\mathbf{i} + \delta y\mathbf{j}$; therefore, $\mathbf{f}_1 \cdot \delta\mathbf{r}_1 = -mg\delta y$. Hence, we have

$$\begin{aligned}
V &= V_1 + V_2 \\
&= -\int_{\mathbf{r}_{i1}}^{r_1} \mathbf{f}_1 \cdot \delta\mathbf{r}_1 - \int_{\mathbf{r}_{i2}}^{r_2} \mathbf{f}_2 \cdot \delta\mathbf{r}_2 \\
&= -\int_{\bar{y}=-\ell}^{\bar{y}=y+\ell} (-mg)\delta\bar{y} - \int_{\bar{x}=0}^{\bar{x}=x} P\delta\bar{x} \\
&= mg\bar{y}\Big|_{\bar{y}=-\ell}^{\bar{y}=y+\ell} - P\bar{x}\Big|_{\bar{x}=0}^{\bar{x}=x} \\
V &= mg(y+2\ell) - Px.
\end{aligned}$$

From the geometry, we may express x and y in terms of the angle θ as follows:

$$x = 2\ell\sin\left(\frac{\theta}{2}\right), \quad y = -\ell\cos\left(\frac{\theta}{2}\right),$$

from which we may express the potential energy solely in terms of θ as

$$V(\theta) = mg\left[-\ell\cos\left(\frac{\theta}{2}\right) + 2\ell\right] - 2P\ell\sin\left(\frac{\theta}{2}\right).$$

From equation (5.20), the Euler-Lagrange equation is

$$\frac{\partial V}{\partial\theta} = 0.$$

Evaluating the Euler-Lagrange equation we obtain

$$mg\ell\sin\left(\frac{\theta}{2}\right) - 2P\ell\cos\left(\frac{\theta}{2}\right) = 0,$$

or

$$\tan\left(\frac{\theta}{2}\right) = \frac{2P}{mg}.$$

For the system to be in static equilibrium, the angle for a given applied force P is then

$$\theta = 2\tan^{-1}\left(\frac{2P}{mg}\right).$$

As illustrated in Example 5.9, solving this example by summing forces and moments to zero would require one to: 1) determine all reaction forces, and 2) analyze each member individually based on its free-body diagram. Therefore, the virtual work approach has allowed us to obtain the required information in a far more direct manner without necessitating a complete analysis of all forces at each reaction and within each member. However, the vectorial approach is conceptually far easier.

REMARKS:

1. The time dependence inherent in Hamilton's principle does not show up in the corresponding Euler-Lagrange equations as the state of the system does not depend upon time for statics problems. As a result, the Euler-Lagrange equations are algebraic.
2. The requirement that the potential energy satisfy

$$\frac{\partial V}{\partial q_i} = 0, \quad i = 1, \ldots, n$$

for static equilibrium simply indicates that the potential energy is stationary. We speak in terms of a *minimum* in potential energy because it is only when the potential energy is a minimum that the static equilibrium state is *stable*. This will be established in Section 6.1.

5.3 Statics of Deformable Bodies

We entitle this section *statics of deformable bodies* in order to maintain consistent terminology with respect to the other physical phenomena that are being addressed in this chapter. More typical terms for this subject matter are *elasticity, strength of materials*, or *mechanics of solids* when applied to single-member systems, such as beams, plates, and membranes, and *structural mechanics* or *structural analysis* when applied to multi-member systems. More specifically, the variational approach considered here falls under the heading of *energy methods* in these subjects. Along with fluid mechanics (see Chapter 9), the underlying fundamental theory of such deformable media is the purview of *continuum mechanics*.

As in Section 5.2, we are concerned with systems in static equilibrium; therefore, the kinetic energy is zero ($T = 0$), and we seek to establish the equations governing deformation of elastic bodies based on the *theorem of minimum potential energy*. Because such bodies are continuous and deformable, we switch attention to Hamilton's principle as applied to continuous systems in Section 4.2. In addition to considering the virtual work owing to the forces that could cause the system to move, as for nondeformable bodies, we now must also include the virtual work done in deforming the body in the form of strain energy.

5.3.1 Static Deflection of an Elastic Membrane – Poisson Equation

The two-dimensional analog of the static deflection of a string subject to a transverse load $f(x)$ considered in Example 4.4 is the deflection of an elastic membrane subject to a lateral load per unit area $f(x,y)$. Consider a membrane fixed at the boundary C of an area A with a uniform tension P applied at the edges. We assume that the lateral deflection of the membrane is small such that the additional tension owing to deflection of the membrane is small in comparison with P. We seek the lateral deflection $v(x,y)$ owing to the lateral force per unit area $f(x,y)$.

Similar to that for a string, the potential energy per unit area of the membrane is

$$\mathcal{V} = \frac{1}{dxdy}\left[dU - f(x,y)v\,dxdy\right]. \tag{5.23}$$

The area of the differential element $dA = dxdy$ becomes $d\sigma$ upon loading; therefore, the strain energy is

$$\begin{aligned} dU &= -(-P)\cdot(\text{elongation}) \\ &= P(d\sigma - dxdy) \\ dU &= P\left(\sqrt{1+v_x^2+v_y^2}-1\right)dxdy. \end{aligned}$$

Note the similarity between this expression for the strain energy in the membrane and the minimum area functional (2.57) in Section 2.5.2. Assuming that the deflection and its slope are small, we may approximate the strain energy in the form (see static string deflection)

$$dU \approx \frac{1}{2}P(v_x^2+v_y^2)dxdy.$$

Substituting into equation (5.23) and integrating over the area of the membrane gives

$$V[v] = \iint_A \mathcal{V}dxdy = \iint_A \left[\frac{1}{2}P(v_x^2+v_y^2) - f(x,y)v\right]dxdy.$$

This is equation (4.9) for continuous systems, but with $T=0$. Minimizing the potential energy by taking $\delta V = 0$ leads to the Euler-Lagrange equation

$$P\left(\frac{\partial^2 v}{\partial x^2}+\frac{\partial^2 v}{\partial y^2}\right)+f(x,y)=0,$$

or $P\nabla^2 v + f(x,y) = 0$, which is a Poisson equation and is the two-dimensional analog of equation (4.12) for static string deflection.

Strings and membranes cannot withstand bending moments or shear forces; therefore, the only contributions to the potential energy are the work done in deflection and the strain energy owing to tangential elongation. As a result, the Euler-Lagrange equations for such systems are of second order. In contrast, systems that can withstand moments and shear, such as beams, produce fourth-order Euler-Lagrange equations as shown in the next section.

5.3.2 Static Deflection of a Beam

Consider bending of a static beam owing to a downward transverse load per unit length $f(x)$ as shown in Figure 5.16. The downward deflection of the beam is $v(x)$, which is assumed small, and there is no axial load on the beam.

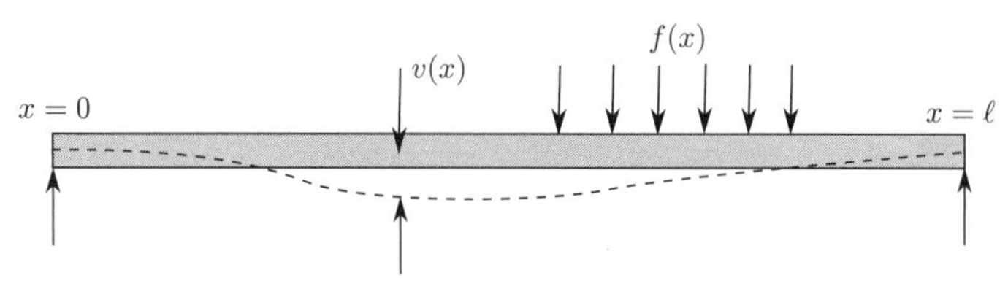

Figure 5.16. Static deflection of a beam under transverse load.

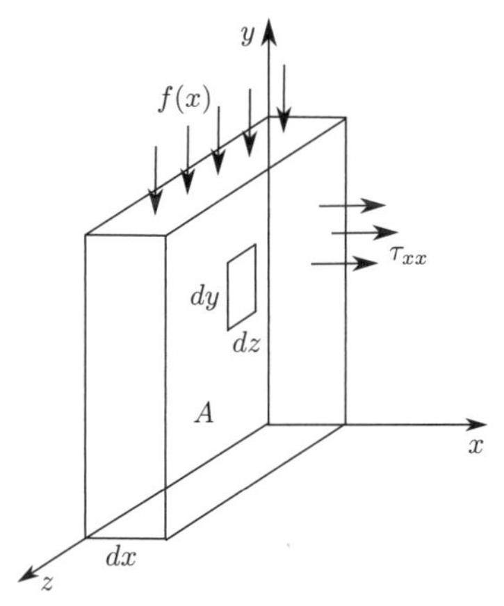

Figure 5.17. Stresses on the cross-section of a beam.

Similar to the static string deflection problem in Example 4.4 and the membrane example in the previous section, the potential energy per unit length of the beam is

$$\mathcal{V} = \frac{1}{dx}\left[dU - f(x)v\,dx\right], \tag{5.24}$$

where dU is the strain energy of a differential length of the beam owing to bending of the beam as a result of the lateral load $f(x)$. For bending, the strain energy of a differential length dx with cross-sectional area A is

$$dU = \left(\iint_A \frac{1}{2E}\tau_{xx}^2 dy dz\right) dx,$$

where integration is carried out over the cross-sectional area A of the beam shown schematically in Figure 5.17. The general form of this relationship will be established in the next section [see equation (5.33) for the strain energy in terms of stresses, with $\tau_{yy} = \tau_{xy} = 0$]. Here, E is Young's modulus, a property of the material, and $\tau_{xx}(y,z)$ is the stress (force per unit area) acting normal to the beam cross-section at x.

We would like to write the potential (strain) energy due to bending in terms of the beam deflection $v(x)$. For bending, elementary mechanics of solids indicates that the axial stress due to a bending moment M at an axial location x is given by the *flexure formula* for beams

$$\tau_{xx} = -\frac{My}{I},$$

where I is the moment of inertia of the cross-section of the beam, and y is measured vertically from the centroid of the cross-sectional area. The flexure formula indicates that the stress varies linearly with vertical distance from the centroid, being negative (in compression) above the centroid and positive (in tension) below the centroid for positive M. We also know from elementary mechanics of solids that the moment M is proportional to the curvature of the beam as follows

$$M = EIv''.$$

Substituting into the flexure formula yields

$$\tau_{xx} = -Ev''y,$$

and the strain energy may be expressed in terms of deflection $v(x)$ as

$$dU = \left[\iint_A \frac{1}{2}E(v'')^2 y^2 dydz\right]dx.$$

Because $v = v(x)$, and if we restrict changes in material properties such that $E = E(x)$, then

$$dU = \left[\frac{1}{2}E(v'')^2 \iint_A y^2 dydz\right]dx,$$

but the moment of inertia of the cross section is defined by

$$I = \iint_A y^2 dydz.$$

Therefore, the strain energy owing to bending of a differential length of the beam is

$$dU = \frac{1}{2}EI(v'')^2 dx.$$

Substituting into equation (5.24) and integrating along the length of the beam gives the total potential energy as

$$V[v] = \int_0^\ell \mathcal{V}dx = \int_0^\ell \left[\frac{1}{2}EI(v'')^2 - f(x)v\right]dx. \tag{5.25}$$

The individual terms represent:

$\frac{1}{2}EI(v'')^2$ = potential (strain) energy due to bending,

$-f(x)v$ = potential energy due to the lateral deflection.

The Euler equation for $F = F(x, v, v'')$ is of the form

$$\frac{\partial F}{\partial v} + \frac{d^2}{dx^2}\left(\frac{\partial F}{\partial v''}\right) = 0,$$

which leads to the beam equation

$$\frac{d^2}{dx^2}\left(EI\frac{d^2v}{dx^2}\right) - f(x) = 0, \tag{5.26}$$

where EI is the bending stiffness of the beam, which may vary with x. Observe that the beam equation is a fourth-order ordinary differential equation for the deflection $v(x)$ as compared to the second-order equations for static string and membrane deflection.

Often it is desirable to transform a known governing differential equation into its associated variational form. This inverse problem may be done, for example, in order to obtain an approximate solution to a differential equation using the *Rayleigh-Ritz method* as discussed in Section 3.1. To illustrate the process, let us begin with the beam equation (5.26) and convert it to its equivalent variational form (5.25) as accomplished for the Dirichlet problem in Section 2.5.3 and the rotating string in

Section 4.2. Here, we assume that EI = constant, that is, the beam has constant material properties and a constant cross-sectional area.

Multiplying equation (5.26) by the variation of the dependent variable δv and integrating gives

$$EI\int_0^\ell v''''\delta v dx - \int_0^\ell f(x)\delta v dx = 0. \tag{5.27}$$

We seek to extract the variation δ out of the integrand in order to obtain the variation of an integral. Integrating the first term by parts two times ($p_1 = \delta v, dq_1 = v''''dx$ and $p_2 = \delta v', dq_2 = v'''dx$), we obtain

$$\begin{aligned}\int_0^\ell v''''\delta v dx &= \left[v'''\delta v\right]_0^\ell - \int_0^\ell v'''\delta v' dx \\ &= \left[v'''\delta v - v''\delta v'\right]_0^\ell + \int_0^\ell v''\delta v'' dx.\end{aligned}$$

Hence, (5.27) becomes

$$EI\left[v'''\delta v - v''\delta v'\right]_0^\ell + EI\int_0^\ell v''\delta v'' dx - \int_0^\ell f(x)\delta v dx = 0.$$

Because $v''\delta v'' = \frac{1}{2}\delta\left[(v'')^2\right]$, and $f(x)\delta v = \delta[f(x)v]$ owing to $f(x)$ being specified and not varying, we have

$$EI\left[v'''\delta v - v''\delta v'\right]_0^\ell + EI\int_0^\ell \frac{1}{2}\delta\left[(v'')^2\right]dx - \int_0^\ell \delta[f(x)v]dx = 0.$$

In variational form

$$EI\left[v'''\delta v - v''\delta v'\right]_0^\ell + \delta\int_0^\ell \left[\frac{1}{2}EI(v'')^2 - f(x)v\right]dx = 0,$$

which is the same as equation (5.25) if the boundary conditions are such that the terms evaluated at the ends $x = 0$ and $x = \ell$ vanish.

In order to have

$$\left[v'''\delta v - v''\delta v'\right]_0^\ell = 0,$$

the boundary conditions must be such that

$$\left.\begin{array}{rcl} v = a & \text{or} & v''' = 0 \\ \text{and} \quad v' = b & \text{or} & v'' = 0 \end{array}\right\} \text{at } x = 0, \ell,$$

where a and b are specified constants. The four possible combinations of boundary conditions physically correspond to the following types of end conditions (with $a = b = 0$):

1. Clamped: $v = 0$ (zero deflection), $v' = 0$ (zero slope),
2. Simple (hinged): $v = 0$ (zero deflection), $v'' = 0$ (zero moment),
3. Free: $v'' = 0$ (zero moment), $v''' = 0$ (zero shear stress),
4. Guided: $v' = 0$ (zero slope), $v''' = 0$ (zero shear stress).

A guided support is a clamped support that is free to move vertically. Note that in each of these cases, no work is done at the end supports because either there is no force or because the force does not act through a distance.

EXAMPLE 5.11 Determine the deflection of a cantilever beam under its own weight. The beam has uniform properties and cross-sectional shape and is of length ℓ. It is clamped at $x = 0$ and free at $x = \ell$. The weight per unit length is $f(x) = w$.
Solution: Because the beam has uniform stiffness, $EI =$ constant, the beam equation (5.26) for the deflection $v(x)$ becomes

$$\frac{d^4 v}{dx^4} = \frac{w}{EI}.$$

Integrating four times produces

$$v(x) = \frac{1}{24}\frac{w}{EI}x^4 + \frac{1}{6}c_1 x^3 + \frac{1}{2}c_2 x^2 + c_3 x + c_4.$$

The boundary conditions at the clamped end are

$$v(0) = 0, \quad v'(0) = 0.$$

Hence,

$$c_3 = 0, \quad c_4 = 0.$$

The boundary conditions at the free end are

$$v''(\ell) = 0, \quad v'''(\ell) = 0.$$

Thus,

$$c_1 = -\frac{w\ell}{EI}, \quad c_2 = \frac{w\ell^2}{2EI}.$$

Substituting the constants of integration gives the deflection curve

$$v(x) = \frac{1}{24}\frac{w}{EI}\left(x^4 - 4\ell x^3 + 6\ell^2 x^2\right),$$

and the deflection of the free end ($x = \ell$) is then

$$v(\ell) = \frac{1}{8}\frac{w\ell^4}{EI}.$$

5.3.3 Governing Equations of Elasticity

Next, we extend the approach illustrated in the previous sections to the more general situation involving deformation of a two-dimensional elastic body. The approach can be applied to fully three-dimensional bodies; however, we limit our considerations to the two-dimensional case for simplicity. In doing so, we will arrive at the general form of the equations governing elasticity.

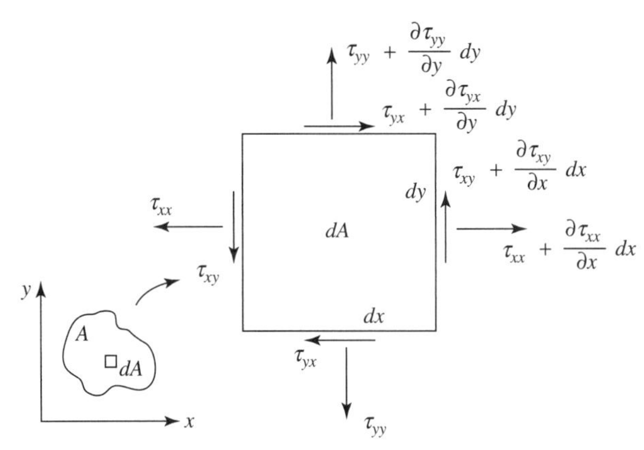

Figure 5.18. Two-dimensional differential element of an elastic body.

Consider a two-dimensional, thin elastic body in which the forces all act in the plane of the body of area A. This is referred to as the *plane stress* problem as all out of plane stresses are zero.[3] The stresses on a differential element dA within the body of area A are given in Figure 5.18. The stresses may be decomposed into normal stresses, τ_{xx} and τ_{yy}, and shear stresses, τ_{xy} and τ_{yx}, where we note that $\tau_{xy} = \tau_{yx}$ for the differential element to remain in static equilibrium. We use the usual convention for subscripts in which the first subscript denotes the face on which the stress or strain acts, designated by the normal to the surface, and the second subscript indicates the direction in which the stress or strain is acting. Truncated Taylor series are used to account for the change in each stress across the width and height of the differential element.

The body will deform from its unloaded state under the action of the applied forces with displacements $u(x,y)$ and $v(x,y)$ in the x and y directions, respectively. From kinematic[4] considerations, the strains ϵ, which are elongations per unit length, are related to the displacements according to

$$\epsilon_{xx} = \frac{\partial u}{\partial x}, \quad \epsilon_{yy} = \frac{\partial v}{\partial y}, \quad \epsilon_{xy} = \epsilon_{yx} = \frac{\partial v}{\partial x} + \frac{\partial u}{\partial y}. \tag{5.28}$$

For a *linearly elastic* material, the stresses τ and strains ϵ are proportional and obey Hooke's law according to

$$\epsilon_{xx} = \frac{1}{E}(\tau_{xx} - \nu\tau_{yy}), \quad \epsilon_{yy} = \frac{1}{E}(\tau_{yy} - \nu\tau_{xx}), \quad \epsilon_{xy} = \epsilon_{yx} = \frac{2(1+\nu)}{E}\tau_{xy}, \tag{5.29}$$

where E is Young's modulus, and ν is Poisson's ratio,[5] both of which are properties of the material. Hooke's law is a type of *constitutive relation*, which relates the response of a medium to applied forces.

[3] See Hartog (1952), p. 177, for a discussion of the *plane strain* problem in which a two-dimensional slice of an infinitely long body is considered.

[4] Kinematics is the branch of mechanics that describes the motion of objects without regard for the forces that lead to that motion.

[5] Poisson's ratio accounts for the fact that a stress in one direction not only results in a strain in that same direction but also in the normal direction(s) as well. That is, an elongation in one direction results in a contraction in the normal direction(s).

For the differential element to be in static equilibrium, the sum of the forces in the x and y directions must both be zero, or equivalently, the virtual work on the differential element is zero (see Section 5.3.4). Summing to zero the forces (stress times area) in the x-direction requires that

$$\left(\tau_{xx}+\frac{\partial \tau_{xx}}{\partial x}dx\right)dy-\tau_{xx}dy+\left(\tau_{yx}+\frac{\partial \tau_{yx}}{\partial y}dy\right)dx-\tau_{yx}dx=0,$$

and summing to zero the forces in the y-direction gives

$$\left(\tau_{yy}+\frac{\partial \tau_{yy}}{\partial y}dy\right)dx-\tau_{yy}dx+\left(\tau_{xy}+\frac{\partial \tau_{xy}}{\partial x}dx\right)dy-\tau_{xy}dy=0.$$

Canceling common terms and factors results in the two-dimensional *equilibrium equations* in the absence of body forces that must be satisfied by the stresses for static equilibrium,

$$\begin{aligned}\frac{\partial \tau_{xx}}{\partial x}+\frac{\partial \tau_{yx}}{\partial y}&=0,\\ \frac{\partial \tau_{xy}}{\partial x}+\frac{\partial \tau_{yy}}{\partial y}&=0.\end{aligned} \tag{5.30}$$

It is convenient to define the *Airy stress function* $\Phi(x,y)$ in the following manner such that the equilibrium equations are identically satisfied:

$$\tau_{xx}=\frac{\partial^2 \Phi}{\partial y^2}, \quad \tau_{yy}=\frac{\partial^2 \Phi}{\partial x^2}, \quad \tau_{xy}=\tau_{yx}=-\frac{\partial^2 \Phi}{\partial x \partial y}. \tag{5.31}$$

Observe, however, that the ability to define a general function in this manner indicates that there are an infinity of stress fields, each corresponding to a different Airy stress function, that satisfy the equilibrium equations. Therefore, the stress-strain field cannot be uniquely determined from the equilibrium equations alone, and we require an additional condition to uniquely determine the stress-strain field.

To the reader not familiar with mechanics of solids or elasticity, the following development may seem rather confusing owing to the regular switching back and forth between the various physical and mathematical quantities, namely, the Airy stress function Φ, stresses τ_{ij}, strains ϵ_{ij}, and displacements (u,v). Keep in mind the relationships between these quantities that allow us to switch between them when convenient as follows:

$$\Phi \leftarrow \text{Def. (5.31)} \rightarrow \tau_{ij} \leftarrow \text{Hooke's law (5.29)} \rightarrow \epsilon_{ij} \leftarrow \text{Kinematics (5.28)} \rightarrow (u,v).$$

From a stress field that satisfies the equilibrium equations (5.30), we can obtain a corresponding displacement field (u,v). Again, there is a different displacement field corresponding to each stress field (corresponding to each Airy stress function). However, we seek the displacement distribution that is *single-valued* (no overlapping material) and *continuous* (no cracks). If the displacements $u(x,y)$ and $v(x,y)$ are continuous (differentiable), then from equations (5.28) relating strains and displacements, we see that

$$\frac{\partial^2 \epsilon_{xx}}{\partial y^2}+\frac{\partial^2 \epsilon_{yy}}{\partial x^2}=\frac{\partial^3 u}{\partial x \partial y^2}+\frac{\partial^3 v}{\partial x^2 \partial y},$$

and similarly

$$\frac{\partial^2 \epsilon_{xy}}{\partial x \partial y} = \frac{\partial^3 u}{\partial x \partial y^2} + \frac{\partial^3 v}{\partial x^2 \partial y}.$$

Subtracting these two expressions gives the *compatibility equation in terms of strains* as

$$\frac{\partial^2 \epsilon_{xx}}{\partial y^2} + \frac{\partial^2 \epsilon_{yy}}{\partial x^2} = \frac{\partial^2 \epsilon_{xy}}{\partial x \partial y}. \tag{5.32}$$

Note that although there is only one compatibility equation in two-dimensions, there are six compatibility equations in the three-dimensional case. Of all the possible stress fields that satisfy the equilibrium equations (5.30), each of which corresponds to an Airy stress function, the actual stress field is that for which the resulting strains satisfy the compatibility equation (5.32), which ensures that the solution is single-valued and continuous.

Although the above formulation of the plane stress elasticity problem is closed in the sense that it ensures a unique solution for the stress–strain field, it is somewhat dissatisfying in that there is not a clear path to finding the stresses and strains given an elastic body with specified geometry and external loads. One could imagine through trial and error choosing various Airy stress functions and "backing out" the corresponding stresses, strains, and external loads seeking solutions that satisfy the compatibility equation.

To obtain a more direct approach, let us consider the work, which is the force acting through a distance, done in deforming the two-dimensional elastic body of area A from its unstressed and undeformed condition to its final state of stress and strain (see, for example, Hartog 1952, Section 33). We consider only perfectly elastic bodies in which no energy is dissipated during the loading process, and all of the work done in deforming the differential element is stored as strain energy, which may be recovered.

Referring to Figure 5.18, we construct the work required to deform each face of the differential element as a result of the normal and shear stresses. For example, the work on the right face due to the normal stress is

$$\begin{aligned}\frac{1}{2}\mathbf{f} \cdot (\text{displacement}) &= \frac{1}{2}\left(\tau_{xx} + \frac{\partial \tau_{xx}}{\partial x}dx\right)dy\left(u + \frac{\partial u}{\partial x}dx\right) \\ &= \frac{1}{2}\left(\tau_{xx}u + \tau_{xx}\frac{\partial u}{\partial x}dx + u\frac{\partial \tau_{xx}}{\partial x}dx\right)dy,\end{aligned}$$

where we have neglected higher-order-terms of $O(dx^2)$. The work in deforming $dA = dxdy$ from its unloaded to loaded condition is the area under the force-displacement curve, which is triangular for linear elastic materials giving rise to the one-half factor as with linear springs. The work due to the normal stress on the left face is

$$\frac{1}{2}\mathbf{f} \cdot (\text{displacement}) = -\frac{1}{2}\tau_{xx} dy\, u,$$

where the negative sign arises because the force and displacement are in opposite directions. Adding the previous two expressions gives the total work on the x-faces due to the normal stress as

$$dW_{xx} = \frac{1}{2}\left(\tau_{xx}\frac{\partial u}{\partial x} + u\frac{\partial \tau_{xx}}{\partial x}\right) dA.$$

Consideration of the total work on the x-faces due to the shear stresses leads to

$$dW_{xy} = \frac{1}{2}\left(\tau_{xy}\frac{\partial v}{\partial x} + v\frac{\partial \tau_{xy}}{\partial x}\right) dA.$$

Similarly, the work done on the y-faces by the normal and shear stresses are

$$dW_{yy} = \frac{1}{2}\left(\tau_{yy}\frac{\partial v}{\partial y} + v\frac{\partial \tau_{yy}}{\partial y}\right) dA,$$

and

$$dW_{yx} = \frac{1}{2}\left(\tau_{yx}\frac{\partial u}{\partial y} + u\frac{\partial \tau_{yx}}{\partial y}\right) dA,$$

respectively.

Combining each contribution to the work on the differential element, we obtain the total *strain energy*

$$\begin{aligned} dU &= dW_{xx} + dW_{yy} + dW_{xy} + dW_{yx} \\ &= \frac{1}{2}\left[\tau_{xx}\frac{\partial u}{\partial x} + \tau_{yy}\frac{\partial v}{\partial y} + \tau_{xy}\left(\frac{\partial v}{\partial x} + \frac{\partial u}{\partial y}\right)\right. \\ &\quad \left. + u\left(\frac{\partial \tau_{xx}}{\partial x} + \frac{\partial \tau_{yx}}{\partial y}\right) + v\left(\frac{\partial \tau_{xy}}{\partial x} + \frac{\partial \tau_{yy}}{\partial y}\right)\right] dA. \end{aligned}$$

However, the last two terms vanish upon substitution of the equilibrium equations (5.30), and substitution of the strains (5.28) gives the strain energy as

$$dU = \frac{1}{2}\left(\tau_{xx}\epsilon_{xx} + \tau_{yy}\epsilon_{yy} + \tau_{xy}\epsilon_{xy}\right) dA.$$

From Hooke's law (5.29), we may express the strain energy of dA in terms of the stresses only in the form

$$dU = \frac{1}{2E}\left[\tau_{xx}^2 + \tau_{yy}^2 - 2\nu\tau_{xx}\tau_{yy} + 2(1+\nu)\tau_{xy}^2\right] dA. \tag{5.33}$$

Integrating over the area A of the entire body gives the total strain energy for the plane stress case as

$$U[\tau_{xx}, \tau_{yy}, \tau_{xy}] = \frac{1}{2E}\iint_A \left[\tau_{xx}^2 + \tau_{yy}^2 - 2\nu\tau_{xx}\tau_{yy} + 2(1+\nu)\tau_{xy}^2\right] dA. \tag{5.34}$$

Alternatively, in terms of the Airy stress function (5.31), the total strain energy is

$$U[\Phi] = \frac{1}{2E}\iint_A \left[\left(\frac{\partial^2 \Phi}{\partial x^2}\right)^2 + \left(\frac{\partial^2 \Phi}{\partial y^2}\right)^2 - 2\nu\frac{\partial^2 \Phi}{\partial x^2}\frac{\partial^2 \Phi}{\partial y^2} + 2(1+\nu)\left(\frac{\partial^2 \Phi}{\partial x \partial y}\right)^2\right] dA, \tag{5.35}$$

or in short

$$U[\Phi] = \iint_A F\left(\Phi_{xx}, \Phi_{yy}, \Phi_{xy}\right) dA, \tag{5.36}$$

where subscripts on Φ denote partial differentiation with respect to the indicated variable, and the integrand is

$$F\left(\Phi_{xx}, \Phi_{yy}, \Phi_{xy}\right) = \Phi_{xx}^2 + \Phi_{yy}^2 - 2\nu\Phi_{xx}\Phi_{yy} + 2(1+\nu)\Phi_{xy}^2.$$

Consequently, we see that the strain energy U of the entire body is a functional involving the unknown function $\Phi(x,y)$, which is the Airy stress function [or stresses in equation (5.34)].

We are now in a position to prove the *theorem of minimum potential energy*. Let us minimize the potential (strain) energy of the body and determine the corresponding Euler-Lagrange equation. For functionals of the form (5.36), the corresponding Euler equation is

$$\frac{\partial^2}{\partial x^2}\left(\frac{\partial F}{\partial \Phi_{xx}}\right) + \frac{\partial^2}{\partial x \partial y}\left(\frac{\partial F}{\partial \Phi_{xy}}\right) + \frac{\partial^2}{\partial y^2}\left(\frac{\partial F}{\partial \Phi_{yy}}\right) = 0.$$

Therefore, the Euler-Lagrange equation takes the form

$$\frac{\partial^4 \Phi}{\partial x^4} + 2\frac{\partial^4 \Phi}{\partial x^2 \partial y^2} + \frac{\partial^4 \Phi}{\partial y^4} = 0, \tag{5.37}$$

or $\nabla^2\left(\nabla^2\Phi\right) = 0$, where $\nabla^2 = \nabla \cdot \nabla$ is the Laplacian operator. This is the *biharmonic equation*, which is the *compatibility equation in terms of the Airy stress function*; it must be satisfied in order for the potential energy to be a minimum.

To better understand the consequences of the fact that the Airy stress function must satisfy the biharmonic equation in order for the potential energy to be a minimum, let us recast it in terms of stresses. Rewriting the biharmonic equation in the form

$$\frac{\partial^2}{\partial x^2}\left(\frac{\partial^2 \Phi}{\partial x^2}\right) + \frac{\partial^2}{\partial x^2}\left(\frac{\partial^2 \Phi}{\partial y^2}\right) + \frac{\partial^2}{\partial y^2}\left(\frac{\partial^2 \Phi}{\partial x^2}\right) + \frac{\partial^2}{\partial y^2}\left(\frac{\partial^2 \Phi}{\partial y^2}\right) = 0,$$

and substituting the definitions of the Airy stress functions (5.31) and rearranging leads to

$$\left(\frac{\partial^2}{\partial x^2} + \frac{\partial^2}{\partial y^2}\right)\left(\tau_{xx} + \tau_{yy}\right) = 0. \tag{5.38}$$

Consequently, the sum of the normal stresses satisfies Laplace's equation, $\nabla^2(\tau_{xx} + \tau_{yy}) = 0$, which is the *compatibility equation in terms of stresses*.

Finally, let us rewrite equation (5.38) in terms of strains. In anticipation of using Hooke's law, we rewrite equation (5.38) in the form

$$\frac{\partial^2}{\partial y^2}(\tau_{xx} - \nu\tau_{yy}) + \frac{\partial^2}{\partial x^2}(\tau_{yy} - \nu\tau_{xx}) + (1+\nu)\left(\frac{\partial^2 \tau_{xx}}{\partial x^2} + \frac{\partial^2 \tau_{yy}}{\partial y^2}\right) = 0.$$

Observe from the equilibrium equations that

$$\frac{\partial \tau_{xx}}{\partial x} = -\frac{\partial \tau_{yx}}{\partial y}, \quad \frac{\partial \tau_{yy}}{\partial y} = -\frac{\partial \tau_{xy}}{\partial x};$$

therefore, recalling that $\tau_{xy} = \tau_{yx}$, we have

$$\frac{\partial^2}{\partial y^2}(\tau_{xx} - \nu\tau_{yy}) + \frac{\partial^2}{\partial x^2}(\tau_{yy} - \nu\tau_{xx}) - 2(1+\nu)\frac{\partial^2 \tau_{yx}}{\partial x \partial y} = 0.$$

Substituting Hooke's law (5.29) yields

$$\frac{\partial^2 \epsilon_{xx}}{\partial y^2} + \frac{\partial^2 \epsilon_{yy}}{\partial x^2} = \frac{\partial^2 \epsilon_{xy}}{\partial x \partial y},$$

which is the previously derived *compatibility equation in terms of strains* (5.32). Therefore, we have proven the *theorem of minimum potential energy* (or the *principle of least work*), which states that of all the possible stress distributions in an elastic body in static equilibrium, the one that minimizes the potential (strain) energy is the one that satisfies the compatibility equation, thereby ensuring a single-valued and continuous solution. In other words, although any Airy stress function $\Phi(x,y)$ (and corresponding stresses τ_{ij} and strains ϵ_{ij}) satisfies the equilibrium equations (5.30), the $\Phi(x,y)$ that minimizes U also satisfies the required compatibility equation (5.32).

In obtaining the governing equations for statics of deformable bodies, we have applied two variational principles. First, we have minimized the potential energy in a rigid-body sense to obtain the equilibrium equations. Second, we have minimized the potential (strain) energy of the body in a deformable-body sense to obtain the stress field that satisfies the compatibility equation.

The careful reader will recognize that, strictly speaking, we have only proven that the strain energy is stationary, not necessarily a minimum. However, it can be shown that $\delta V = 0$ does indeed produce a minimum for statics problems of nondeformable and deformable bodies (see Reddy 1984 for a proof or the discussion in Chapter 6 of the present text regarding stability of equilibrium states). This is also consistent with our physical intuition. For example, the hanging cable of Section 2.7.3 always hangs down (minimizing potential energy), not up (maximizing potential energy). The theorem of minimum potential energy leads to additional energy principles, such as Castigliano's theorems, that are of use in mechanics (see, for example, Reddy 1984). For comprehensive treatments of elasticity theory, see Love (1944), Sokolnikoff (1956), and Timoshenko and Goodier (1934).

There are several important extensions and special cases of elasticity theory as presented here. The most obvious extension is to the three-dimensional case, which is very straightforward except that there are six compatibility equations instead of the single one for the two-dimensional case. An important special case is that of thin plates and shells, in which the elastic material is thin in one direction in comparison with the other two directions. Thin, flat structures are referred to as *plates*, and *shells* are thin, curved structures. Plates and shells are capable of withstanding bending moments and shear forces perpendicular to their dominant plane. In contrast, a *membrane* is a plate or shell that cannot support bending moments or shear forces. Similarly, a *string* is a *beam* that cannot support bending moments or shear forces. Therefore, a membrane is a two-dimensional analog of a string, while plates and shells are two-dimensional analogs of beams, where a beam has two dimensions that are small compared to that in the other direction. See Den Hartog (1952) and Ventsel and Krauthammer (2001) for more on thin plates and shells.

5.3.4 Principle of Virtual Work

As with nondeformable bodies considered in Section 5.2.2, we can recast the static equilibrium equations for deformable bodies (5.30) in terms of the virtual work. In the case of deformable bodies, however, it is necessary to account for the internal forces (stresses) acting throughout the body of area A along with the external traction forces acting on the boundary C of the body. Note that the virtual work of the internal forces acting within the body are the negative of the strain energy. In order to show that the principle of virtual work is equivalent to the equations of static equilibrium (5.30), let us consider the inverse problem. Multiply the first of the equilibrium equations by the virtual displacement δu and the second equation by δv, sum, and integrate over the domain to give

$$\int_A \left[\left(\frac{\partial \tau_{xx}}{\partial x} + \frac{\partial \tau_{yx}}{\partial y} \right) \delta u + \left(\frac{\partial \tau_{xy}}{\partial x} + \frac{\partial \tau_{yy}}{\partial y} \right) \delta v \right] dxdy + \oint_C \mathbf{f} \cdot \delta \mathbf{u} ds = 0,$$

where s is along the boundary C. The last integral accounts for the virtual work of the traction forces $\mathbf{f}$ acting through the virtual displacements $\delta\mathbf{u}$ on the boundary owing to the boundary conditions acting on the body. The virtual work on the boundary can be written

$$\mathbf{f} \cdot \delta \mathbf{u} = f_x \delta u + f_y \delta v.$$

As in Section 2.5.1, we apply the divergence theorem for the two-dimensional case considered here in lieu of integration by parts. This leads to

$$-\int_A \left[\tau_{xx} \frac{\partial}{\partial x}(\delta u) + \tau_{yx} \frac{\partial}{\partial y}(\delta u) + \tau_{xy} \frac{\partial}{\partial x}(\delta v) + \tau_{yy} \frac{\partial}{\partial y}(\delta v) \right] dxdy$$
$$+ \oint_C \left(\tau_{xx}\delta u + \tau_{yx}\delta u + \tau_{xy}\delta v + \tau_{yy}\delta v \right) ds + \oint_C \left(f_x \delta u + f_y \delta v \right) ds = 0.$$

Because there are no stresses acting on the boundary other than those accounted for by the traction forces f_x and f_y, the second integral vanishes. Recall the kinematic relationships (5.28) between strain and displacement as well as the fact that derivatives and variations are commutative. With the requirement that $\tau_{xy} = \tau_{yx}$, we can then write

$$\frac{\partial}{\partial x}(\delta u) = \delta \left(\frac{\partial u}{\partial x} \right) = \delta \epsilon_{xx},$$
$$\frac{\partial}{\partial y}(\delta v) = \delta \left(\frac{\partial v}{\partial y} \right) = \delta \epsilon_{yy},$$
$$\frac{\partial}{\partial y}(\delta u) + \frac{\partial}{\partial x}(\delta v) = \delta \left(\frac{\partial u}{\partial y} + \frac{\partial v}{\partial x} \right) = \delta \epsilon_{xy} = \delta \epsilon_{yx}.$$

Substituting yields the *principle of virtual work* for deformable bodies

$$\int_A \left(\tau_{xx}\delta\epsilon_{xx} + \tau_{xy}\delta\epsilon_{xy} + \tau_{yy}\delta\epsilon_{yy} \right) dxdy = \oint_C \left(f_x \delta u + f_y \delta v \right) ds. \tag{5.39}$$

The term on the left-hand side arises from the virtual work accomplished *inside* the deformable body, and the term on the right-hand side accounts for the virtual work done *on* the body by the external forces. Thus, the principle of virtual work states that a deformable body is in static equilibrium if the total virtual work done by the internal stresses equals the total virtual work done by the external traction forces. Recall that this variational form is equivalent to application of the differential static equilibrium equations (5.30).

Alternatively, it is sometimes useful to formulate the *principle of complementary virtual work*. In this case, we start with the kinematic relationships (5.28) between strain and displacement. Similar to the above procedure, we multiply these by variations of the stresses, sum, integrate over the domain, and employ the divergence theorem. This leads to

$$\int_A \left(\epsilon_{xx}\delta\tau_{xx} + \epsilon_{xy}\delta\tau_{xy} + \epsilon_{yy}\delta\tau_{yy}\right) dxdy = \oint_C \left(u\delta f_x + v\delta f_y\right) ds. \tag{5.40}$$

Whereas the virtual work is the hypothetical work performed by an actual force (stress) acting through a virtual displacement (strain) and is the area under the force versus displacement curve, the complementary virtual work is the hypothetical work performed by a virtual force (stress) acting through an actual displacement (strain) and is the area under the displacement versus force curve. Whereas the principle of virtual work is equivalent to the differential static equilibrium equations, the principle of complementary virtual work is equivalent to the differential kinematic relationships between strain and displacement.

5.4 Dynamics of Deformable Bodies

Again, the title of this section is intended to draw a clear distinction from the other topics discussed in this chapter; it is material that is typically referred to as *mechanical vibrations*. Here, we only provide a brief introduction to this subject in order to highlight a few essential features. In particular, we focus on scenarios that lead to the wave equation.

5.4.1 Longitudinal Vibration of a Rod – Wave Equation

Consider an elastic rod of length ℓ as shown in Figure 5.19, vibrating in its longitudinal direction with small amplitude. Here, ρ is the mass per unit volume (density) of the rod, A is the cross-sectional area of the rod, P is the external tension applied to the rod, and $u(x,t)$ is the axial displacement within the rod.

To form the Lagrangian for this case, let us first consider the kinetic energy per unit volume of a differential element of the vibrating rod, which is

$$\mathcal{T} = \frac{1}{2}\rho\left(\frac{\partial u}{\partial t}\right)^2.$$

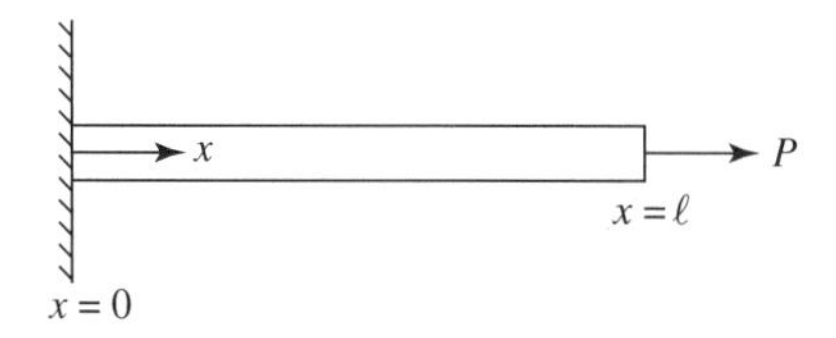

Figure 5.19. Vibration of a longitudinal rod.

The potential energy per unit volume of the differential element owing to longitudinal deformation becomes

$$\mathcal{V} = \frac{1}{Adx} dU,$$

which is similar to that for the spring considered earlier. As for the beam in Section 5.3.2, for axial stress and strain, the strain energy from equation (5.33) yields

$$dU = \frac{1}{2E} \tau_{xx}^2 A dx.$$

Alternatively, in terms of strain and displacement, from Hooke's law ($\tau_{xx} = E\epsilon_{xx}$) and the kinematic condition ($\epsilon_{xx} = \partial u/\partial x$), the strain energy is

$$dU = \frac{1}{2} E\epsilon_{xx}^2 A dx = \frac{1}{2} EA \left(\frac{\partial u}{\partial x} \right)^2 dx.$$

Then the potential energy per unit volume becomes

$$\mathcal{V} = \frac{1}{2} E \left(\frac{\partial u}{\partial x} \right)^2.$$

Combining the kinetic and potential energies, the Lagrangian density for a differential length of the rod is

$$\mathcal{L} = \mathcal{T} - \mathcal{V} = \frac{1}{2} \left[\rho \left(\frac{\partial u}{\partial t} \right)^2 - E \left(\frac{\partial u}{\partial x} \right)^2 \right],$$

and the Lagrangian for the entire rod is then

$$L(t,x,u_t,u_x) = \int_0^\ell \mathcal{L} dx = \frac{1}{2} \int_0^\ell \left[\rho \left(\frac{\partial u}{\partial t} \right)^2 - E \left(\frac{\partial u}{\partial x} \right)^2 \right] dx.$$

Therefore, Hamilton's variational form results in

$$\delta \int_{t_1}^{t_2} L dt = 0,$$

$$\delta \int_{t_1}^{t_2} \int_0^\ell \mathcal{L} dx dt = 0,$$

$$\frac{1}{2} \delta \int_{t_1}^{t_2} \int_0^\ell \left[\rho \left(\frac{\partial u}{\partial t} \right)^2 - E \left(\frac{\partial u}{\partial x} \right)^2 \right] dx dt = 0. \tag{5.41}$$

The corresponding Euler-Lagrange equation is

$$\frac{\partial^2 u}{\partial t^2} = \gamma^2 \frac{\partial^2 u}{\partial x^2}, \tag{5.42}$$

where $\gamma^2 = E/\rho$, and γ is the wave speed within the material. Note that the solution is independent of the applied tension P. Equation (5.42) is the familiar one-dimensional *wave equation*.

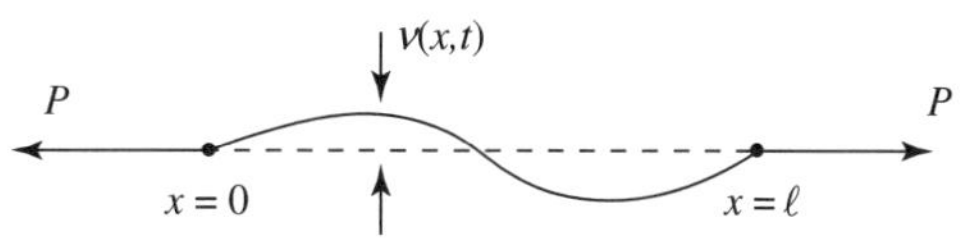

Figure 5.20. Vibration of an elastic string.

5.4.2 Lateral Vibration of a String – Wave Equation

In this section, we consider a derivation of the wave equation obtained in the previous section for an alternative setting. We evaluate the lateral vibration of the string of length ℓ shown in Figure 5.20. In this case, ρ is the mass per unit length of the string, and $v(x,t)$ is the lateral displacement of the string. The transverse deflection is assumed small such that we may take the tension P to be constant throughout the vibration of the string.

Let us consider the kinetic and potential energy of a differential length of string with initial length dx and stretched length ds. The kinetic energy per unit length of the differential string element is

$$\mathcal{T} = \frac{1}{2}\rho\left(\frac{\partial v}{\partial t}\right)^2.$$

The potential (strain) energy per unit length of the differential element due to stretching of the string is

$$\mathcal{V} = \frac{1}{dx}dU.$$

Similar to the static string deflection problem treated in Example 4.4, the strain energy becomes

$$\begin{aligned} dU &= -(-P)\cdot(\text{elongation}) \\ &= P(ds - dx) \\ &= P\left(\sqrt{1+v_x^2} - 1\right)dx \\ dU &\approx \frac{1}{2}Pv_x^2, \end{aligned}$$

where the final approximation applies for small deflections for which $v_x^2 \ll 1$. The Lagrangian density is then

$$\mathcal{L} = \mathcal{T} - \mathcal{V} = \frac{1}{2}\left[\rho\left(\frac{\partial v}{\partial t}\right)^2 - P\left(\frac{\partial v}{\partial x}\right)^2\right],$$

which is similar to that for the vibrating rod in the previous section. Therefore, the Euler-Lagrange equation is again the wave equation

$$\frac{\partial^2 v}{\partial t^2} = \gamma^2\frac{\partial^2 v}{\partial x^2}, \tag{5.43}$$

where $\gamma^2 = P/\rho$, and γ is the wave speed in the string. Unlike the vibrating horizontal rod, the tension P in the string does influence the solution to the wave equation through the wave speed.

Except for the fact that the independent variables are x and t, rather than x and y, the variational form of the wave equation (5.41) is nearly identical to that for Laplace's equation (2.62). Observe that the fundamental difference between the two equations is simply a minus sign. However, from comparison of the variational forms, we see that this minus sign leads to behavior that is qualitatively different, and is consistent with the different physical behavior governed by each equation. Recall that Laplace's equation governs diffusion processes and that the functional (2.62) indicates that the *sum* of the magnitudes of the gradients in both directions is minimized. Therefore, it represents a smoothing process. In contrast, the negative sign in the functional form of the wave equation (5.41) indicates that it is the *difference* between the magnitudes of the gradients, rather than the sum of the gradients, that is minimized. That is, if a spatial disturbance is introduced into a system governed by the wave equation, rather than smoothing out the disturbance as time advances, the system will generate a temporal disturbance, called a traveling wave, with its time rate of change set according to the proportionality constant γ^2, where γ is the wave speed. Furthermore, because it is the *magnitudes* of the spatial and temporal gradients that must match, both a positive and negative traveling wave are produced. The total energy contained within the left-moving and right-moving traveling waves (temporal disturbances) is equal to the energy contained in the initial spatial disturbance. Again, we see that the variational form provides additional physical insight that is not readily apparent from the differential form.

REMARKS:

1. Recall from Section 5.3.1 that the *static* deflection of an elastic membrane is governed by the Poisson equation

$$P\left(\frac{\partial^2 v}{\partial x^2}+\frac{\partial^2 v}{\partial y^2}\right)+f(x,y)=0.$$

By analogy to the vibrating string, if we allow the membrane to vibrate by including the kinetic energy $T=\frac{1}{2}\rho(\partial v/\partial t)^2$, the governing equation is the two-dimensional wave equation

$$\rho\frac{\partial^2 v}{\partial t^2}=P\left(\frac{\partial^2 v}{\partial x^2}+\frac{\partial^2 v}{\partial y^2}\right)+f(x,y). \tag{5.44}$$

See Section 3.1.3 for a method of approximating the fundamental frequency of vibration of a membrane (or other deformable body) using the Rayleigh-Ritz method.

2. Whereas the equations of motion in dynamics of nondeformable (discrete) bodies are a coupled system of ordinary differential equations of the form (5.12), dynamics of deformable (continuous) bodies produce equations of motion that are partial differential equations of the form

$$\mathcal{M}\ddot{v}+\mathcal{C}\dot{v}+\mathcal{K}v=f(x,y),$$

where $\mathcal{M}$, $\mathcal{C}$, and $\mathcal{K}$ are differential operators involving only spatial derivatives, and dots denote differentiation with respect to time. For the one-dimensional wave equation (5.43), for example,

$$\mathcal{M} = -\frac{1}{c^2}, \quad \mathcal{C} = 0, \quad \mathcal{K} = \frac{\partial^2}{\partial x^2}.$$

Similarly, for the two-dimensional wave equation (5.44),

$$\mathcal{M} = \rho, \quad \mathcal{C} = 0, \quad \mathcal{K} = -P\left(\frac{\partial^2}{\partial x^2} + \frac{\partial^2}{\partial y^2}\right).$$

3. For additional detailed discussion regarding the dynamics of nondeformable (discrete) and deformable (continuous) bodies, see Tse, Morse, and Hinkle (1978) and Goldstein (1959). For example, see Tse, Morse, and Hinkle (1978), Section 7–5, for the lateral vibration of a beam.

6 Stability of Dynamical Systems

> Philosophy is written in that great book which ever lies before our eyes (I mean the universe) but we cannot understand it if we do not first learn the language and grasp the characters in which it is written. It is written in the language of mathematics, and the characters are triangles, circles and other geometrical figures, without which it is humanly impossible to comprehend a single word of it, and without which one wanders in vain though a dark labyrinth.
>
> (Galileo Galilei)

In Sections 5.2 and 5.3, we considered techniques for determining the *static equilibrium* of nondeformable (discrete) and deformable (continuous) bodies, respectively. In principle, a system can remain in static equilibrium indefinitely. However, physical systems are generally subject to small perturbations, and it is of interest to determine whether a system is *stable* or *unstable* to such small disturbances. An equilibrium state is unstable if the response of the system is to cause the small amplitude of the perturbation to grow and become unbounded as $t \to \infty$; it is stable if the amplitude remains bounded, whereby the system returns to its equilibrium state or at least remains bounded about the equilibrium position.

6.1 Introduction

Recall that the Euler-Lagrange equations for an n degree-of-freedom system in static equilibrium ($T = 0$) reduce to

$$\frac{\partial V}{\partial q_i} = 0, \quad i = 1,2,\ldots,n,$$

in which case the potential energy V of the system in static equilibrium is stationary. A small perturbation applied to the system will push it away from its static equilibrium position causing a small change in potential energy. If the equilibrium state of the system corresponds to a minimum in potential energy, the perturbation will cause an increase in the potential energy. If the system is at a maximum in potential energy, the potential energy will decrease owing to the perturbation.

Recall that for a conservative system, the total mechanical energy remains constant and equals the Hamiltonian

$$H = T + V = \text{constant}.$$

Because $T = 0$ for static equilibrium, a small decrease in potential energy owing to the perturbation must be matched by a small increase in kinetic energy, thereby becoming positive. Likewise, a small increase in potential energy must be matched by a small decrease in kinetic energy, thereby becoming negative. The latter case is not possible, however, as kinetic energy is a quadratic and positive definite; accordingly, the system cannot allow for such a decrease in kinetic energy below its zero value while in static equilibrium. In other words, there is no excess potential energy that can be converted into kinetic energy, and the system returns (or remains close) to its static equilibrium position. The only additional energy is that introduced by the small perturbation itself. Consequently, a static equilibrium state corresponding to a minimum in potential energy is stable. This provides the missing piece referenced in Section 5.3.3 regarding the *theorem of minimum potential energy*. In contrast, when the potential energy is a maximum, there is a surplus of potential energy that can be converted into (positive) kinetic energy when the system is perturbed away from its equilibrium state, thereby becoming unstable.

Evaluating stability by considering extrema in potential energy is only applicable for systems having *static* equilibrium states, that is, having $T = 0$, in which case $\partial V/\partial q_i = 0, i = 1,\dots,n$. We will often encounter systems that do not have static equilibrium solutions because of the fact that there are no states for which $T = 0$. For such systems, we must generalize our notion of equilibrium states to include cases for which the system is moving ($T \neq 0$), but the generalized coordinates do not change with time. That is, they are *steady* or *nontransient*. In other words, equilibrium states (steady solutions) occur when the acceleration and velocity of the system's generalized coordinates both vanish, that is, $\ddot{q}_i = \dot{q}_i = 0, i = 1,\dots,n$. In this general context, we will formally perturb the system at its equilibrium state(s) $\mathbf{s}$ according to

$$\mathbf{q}(t) = \mathbf{s} + \epsilon \mathbf{u}(t),$$

where $\epsilon \ll 1$, and $\mathbf{u}(t)$ is the perturbation, and analyze the linearized behavior of the system near the equilibrium state(s). In the following section, we analyze stability of the simple pendulum, which does have static equilibrium states, using both techniques for comparison. The subsequent sections consider systems for which the more general analysis is required.

6.2 Simple Pendulum

To illustrate stability of discrete systems, let us consider stability of the simple pendulum shown in Figure 6.1. The kinetic and potential energies are

$$T = \frac{1}{2}m\left(\dot{x}^2 + \dot{y}^2\right) = \frac{1}{2}m(\ell\dot{\theta})^2,$$

and

$$V = mgy = -mg\ell\cos\theta,$$

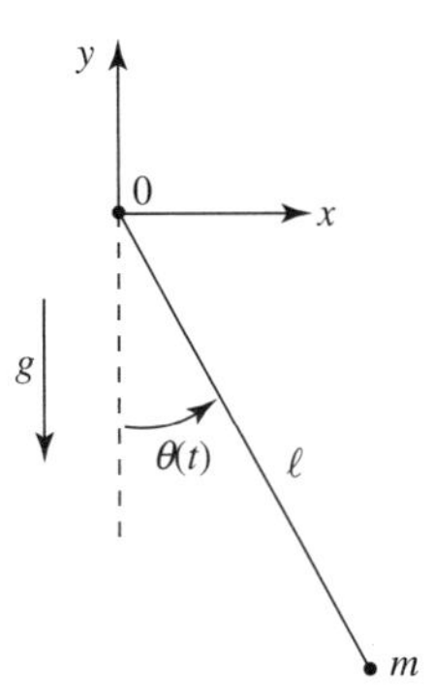

Figure 6.1. Schematic of the simple pendulum.

respectively. To determine the conditions for static equilibrium, recall that $(q = \theta)$

$$\frac{\partial V}{\partial \theta} = 0;$$

therefore,

$$mg\ell \sin\theta = 0.$$

The positions of static equilibrium (stationary points) $\theta = s$ require that $\sin s = 0$; thus, $s = n\pi, n = 0, 1, 2, \ldots$. The cases for which n is even correspond to $s = 0$, for which the pendulum is hanging vertically downward. Likewise, the cases for which n is odd correspond to $s = \pi$, for which the pendulum is located vertically above the pivot point.

Observe that the gravitational potential energy V is a minimum for n even $(s = 0)$, where $V = -mg\ell < 0$, and a maximum for n odd $(s = \pi)$, where $V = mg\ell > 0$. Therefore, the static equilibrium position corresponding to $s = 0$ is stable and that corresponding to $s = \pi$ is unstable. Although this conclusion is certainly consistent with our intuition, let us illustrate the more general form of stability analysis, which will show how the stable and unstable conditions appear in the solution.

With $L = T - V$, the equation of motion for the simple pendulum is

$$\ddot{\theta} + \frac{g}{\ell}\sin\theta = 0. \tag{6.1}$$

Let us introduce a small perturbation of magnitude ϵ about the static equilibrium positions $s = n\pi$ as follows

$$\theta(t) = s + \epsilon u(t) = n\pi + \epsilon u(t),$$

where $\epsilon \ll 1$. Substituting into (6.1) gives the equation

$$\epsilon\ddot{u} + \frac{g}{\ell}\sin(n\pi + \epsilon u) = 0 \tag{6.2}$$

governing the amplitude of the perturbations. Observe that

$$\sin(n\pi + \epsilon u_1) = \sin(n\pi)\cos(\epsilon u_1) + \cos(n\pi)\sin(\epsilon u_1) = \begin{cases} \sin(\epsilon u_1), & n \text{ even} \\ -\sin(\epsilon u_1), & n \text{ odd} \end{cases}.$$

Because ϵ is small, we may expand the sine function in terms of a Taylor series:

$$\sin(\epsilon u_1) = \epsilon u_1 - \frac{(\epsilon u_1)^3}{3!} + \cdots = \epsilon u_1 + O(\epsilon^3).$$

Essentially, we have linearized the system about the equilibrium points by considering an infinitesimally small perturbation to the equilibrium (stationary) solutions of the system.

For n even ($s = 0$), therefore, equation (6.2) produces the linearized equation

$$\ddot{u} + \frac{g}{\ell} u = 0,$$

which has the solution ($\lambda_{1,2} = \pm\sqrt{g/\ell}\, i$, that is, $\pm$ imaginary roots)

$$u(t) = c_1 \cos\left(\sqrt{\frac{g}{\ell}} t\right) + c_2 \sin\left(\sqrt{\frac{g}{\ell}} t\right).$$

This represents an oscillation with (small) constant amplitude about the equilibrium position $s = 0$. This motion is *stable* as it is bounded and does not grow with time. Observe that this equation (and its solution) is the same as that for the undamped harmonic oscillator discussed in Example 4.2.

Now for n odd ($s = \pi$), we have the linearized equation

$$\ddot{u} - \frac{g}{\ell} u = 0,$$

which has the solution ($\lambda_{1,2} = \pm\sqrt{g/\ell}$, that is, $\pm$ real roots)

$$u(t) = c_1 e^{\sqrt{\frac{g}{\ell}} t} + c_2 e^{-\sqrt{\frac{g}{\ell}} t}.$$

Although the $e^{-\sqrt{\frac{g}{\ell}} t}$ term decays with time, the $e^{\sqrt{\frac{g}{\ell}} t}$ term grows exponentially with time, and the equilibrium position $s = \pi$ is *unstable*.

Rather than using the second-order form of the governing equation (6.1), it is often convenient to transform the system to an equivalent set of first-order differential equations and evaluate stability using matrix methods. We repeat the analysis here for the simple pendulum to illustrate this approach.

In order to transform the nonlinear governing equation (6.1) to a system of first-order equations, known as its *state-space* representation, let us define a new set of dependent variables

$$x_1(t) = \theta(t),$$
$$x_2(t) = \dot{\theta}(t),$$

such that $x_1(t)$ and $x_2(t)$ give the angular position and velocity, respectively. Differentiating and substituting equation (6.1) produces

$$\begin{aligned} \dot{x}_1 &= \dot{\theta} = x_2, \\ \dot{x}_2 &= \ddot{\theta} = -\frac{g}{\ell}\sin\theta = -\frac{g}{\ell}\sin x_1. \end{aligned} \tag{6.3}$$

Therefore, our single second-order nonlinear equation has been converted to a system of two first-order nonlinear equations.

Equilibrium positions are where the angular velocity $\dot{\theta}$ and angular acceleration $\ddot{\theta}$ are zero; hence,

$$\{\dot{\theta},\ddot{\theta}\} = \{\dot{x}_1,\dot{x}_2\} = \{0,0\}.$$

Thus, from the system (6.3)

$$x_2 = 0, \quad -\frac{g}{\ell}\sin x_1 = 0,$$

and the stationary points are given by

$$s_1 = x_1 = n\pi, \quad s_2 = x_2 = 0, \quad n = 0,1,2,\ldots.$$

The equilibrium states of the system $\mathbf{s} = [s_1,s_2]^T$ are then

$$\mathbf{s} = [s_1,s_2]^T = [n\pi,0]^T, \quad n = 0,1,2,\ldots.$$

As before, there are two equilibrium points corresponding to n even and n odd.

In order to evaluate stability of the equilibrium states, let us impose small perturbations about the equilibrium positions according to

$$x_1(t) = s_1 + \epsilon u_1(t) = n\pi + \epsilon u_1(t),$$
$$x_2(t) = s_2 + \epsilon u_2(t) = \epsilon u_2(t),$$

where $\epsilon \ll 1$. Substituting into equations (6.3) leads to

$$\begin{aligned} \epsilon\dot{u}_1 &= \epsilon u_2, \\ \epsilon\dot{u}_2 &= -\frac{g}{\ell}\sin(n\pi + \epsilon u_1). \end{aligned} \tag{6.4}$$

Linearizing the second equation as before leads to the linear system of equations for the perturbations $u_1(t)$ and $u_2(t)$ written in matrix form $\dot{\mathbf{u}} = \mathbf{A}\mathbf{u}$ as

$$\begin{bmatrix} \dot{u}_1 \\ \dot{u}_2 \end{bmatrix} = \begin{bmatrix} 0 & 1 \\ \mp\frac{g}{\ell} & 0 \end{bmatrix} \begin{bmatrix} u_1 \\ u_2 \end{bmatrix}, \tag{6.5}$$

where the minus sign corresponds to the equilibrium point with n even, and the plus sign to n odd. Let us consider each case in turn.

For n even, the system (6.5) is

$$\begin{bmatrix} \dot{u}_1 \\ \dot{u}_2 \end{bmatrix} = \begin{bmatrix} 0 & 1 \\ -\frac{g}{\ell} & 0 \end{bmatrix} \begin{bmatrix} u_1 \\ u_2 \end{bmatrix}.$$

In order to diagonalize the system, determine the eigenvalues of $\mathbf{A}$. We write

$$(\mathbf{A} - \lambda\mathbf{I}) = \mathbf{0}.$$

For a nontrivial solution, the determinant must be zero, such that

$$\begin{vmatrix} -\lambda & 1 \\ -\frac{g}{\ell} & -\lambda \end{vmatrix} = 0,$$

or

$$\lambda^2 + \frac{g}{\ell} = 0.$$

Factoring gives the eigenvalues

$$\lambda_{1,2} = \pm\sqrt{-\frac{g}{\ell}} = \pm\sqrt{\frac{g}{\ell}}i \quad (\pm \text{ imaginary}).$$

Hence, the solution near the equilibrium point corresponding to n even is of the form

$$u(t) = u_1(t) = c_1 \cos\left(\sqrt{\frac{g}{\ell}}t\right) + c_2 \sin\left(\sqrt{\frac{g}{\ell}}t\right),$$

which is the same as that obtained previously. This oscillatory solution is *linearly stable* in the form of a *stable center* (see the next section) as the solution remains bounded for all time.

In the case of n odd, the system (6.5) is

$$\begin{bmatrix} \dot{u}_1 \\ \dot{u}_2 \end{bmatrix} = \begin{bmatrix} 0 & 1 \\ \frac{g}{\ell} & 0 \end{bmatrix} \begin{bmatrix} u_1 \\ u_2 \end{bmatrix}.$$

Finding the eigenvalues produces

$$\begin{vmatrix} -\lambda & 1 \\ \frac{g}{\ell} & -\lambda \end{vmatrix} = 0$$

$$\lambda^2 - \frac{g}{\ell} = 0$$

$$\lambda_{1,2} = \pm\sqrt{\frac{g}{\ell}} \quad (\pm \text{ real}).$$

Then the solution near the equilibrium point corresponding to n odd is of the form

$$u(t) = c_1 e^{\sqrt{\frac{g}{\ell}}t} + c_2 e^{-\sqrt{\frac{g}{\ell}}t},$$

which again is the same as before. Observe that while the second term decays exponentially as $t \to \infty$, the first term grows exponentially, eventually becoming unbounded. Therefore, this equilibrium point is *linearly unstable* in the form of an *unstable saddle* (see the next section).

6.3 Linear, Second-Order, Autonomous Systems

As we have seen for the simple pendulum, the stability of at least some dynamical systems is determined from the nature of the eigenvalues of the coefficient matrix in the equations of motion expressed as a system of first-order linear equations $\dot{\mathbf{u}} = \mathbf{A}\mathbf{u}$. This is the case when the coefficient matrix $\mathbf{A}$ is comprised of constants, that is, it does not change with time. Such systems are called *autonomous*. We consider an example of a *nonautonomous* system in the next section. Stability analysis that is based on the behavior of the eigenvalues, or modes, of the system is called *modal stability analysis*.

In order to illustrate the possible types of stable and unstable equilibrium points, we evaluate stability of the general second-order system expressed as a system of first-order equations in the form

$$\begin{bmatrix} \dot{u}_1 \\ \dot{u}_2 \end{bmatrix} = \begin{bmatrix} A_{11} & A_{12} \\ A_{21} & A_{22} \end{bmatrix} \begin{bmatrix} u_1 \\ u_2 \end{bmatrix}.$$

The original system may be nonlinear; however, it has been linearized about the equilibrium point(s) as illustrated for the simple pendulum. For autonomous systems, the equilibrium points are steady and do not change with time. Therefore, $\dot{\mathbf{u}} = 0$, and the steady equilibrium points are given by $\mathbf{Au} = \mathbf{0}$.

In order to characterize the stability of the system, we obtain the eigenvalues of $\mathbf{A}$ as follows:

$$|\mathbf{A} - \lambda \mathbf{I}| = 0$$

$$\begin{vmatrix} A_{11} - \lambda & A_{12} \\ A_{21} & A_{22} - \lambda \end{vmatrix} = 0$$

$$(A_{11} - \lambda)(A_{22} - \lambda) - A_{12}A_{21} = 0$$

$$\lambda^2 - (A_{11} + A_{22})\lambda + A_{11}A_{22} - A_{12}A_{21} = 0$$

$$\lambda^2 - \text{tr}(\mathbf{A})\lambda + |\mathbf{A}| = 0,$$

where from the quadratic formula

$$\lambda_{1,2} = \frac{1}{2}\left[\text{tr}(\mathbf{A}) \pm \sqrt{\text{tr}(\mathbf{A})^2 - 4|\mathbf{A}|}\right].$$

Many references summarize the various possibilities graphically by demarcating the various stable and unstable regions on a plot of $|\mathbf{A}|$ versus $\text{tr}(\mathbf{A})$. It is also informative to plot solution trajectories in the *phase plane*, which is a plot of $u_2(t)$ versus $u_1(t)$. This can be interpreted as a plot of velocity versus position of the solution trajectory. In the phase plane, steady solutions are represented by a single point, and periodic, or limit cycle, solutions are closed curves.

Each type of stable or unstable point is considered in turn by plotting several solution trajectories on the phase plane. The dots on each trajectory indicate the initial condition for that trajectory. In each case, the origin of the phase plane is the equilibrium point around which the system has been linearized for consideration of stability. In general, complex systems may have numerous equilibrium points, each of which exhibits one of the stable or unstable behaviors illustrated below:

1. Unstable Saddle Point: $\text{tr}(\mathbf{A})^2 - 4|\mathbf{A}| > 0, |\mathbf{A}| < 0$
   If the eigenvalues of $\mathbf{A}$ are positive and negative real values, the equilibrium point is an *unstable saddle* as shown in Figure 6.2. Observe that while some trajectories move toward the equilibrium point, all trajectories eventually move away from the origin and become unbounded. Therefore, it is called an *unstable saddle point*. This is the behavior exhibited by the simple pendulum with n odd ($\theta = \pi$).
2. Unstable Node or Source: $\text{tr}(\mathbf{A})^2 - 4|\mathbf{A}| > 0, \text{tr}(\mathbf{A}) > 0, |\mathbf{A}| > 0$
   If the eigenvalues of $\mathbf{A}$ are both positive real values, the equilibrium point is

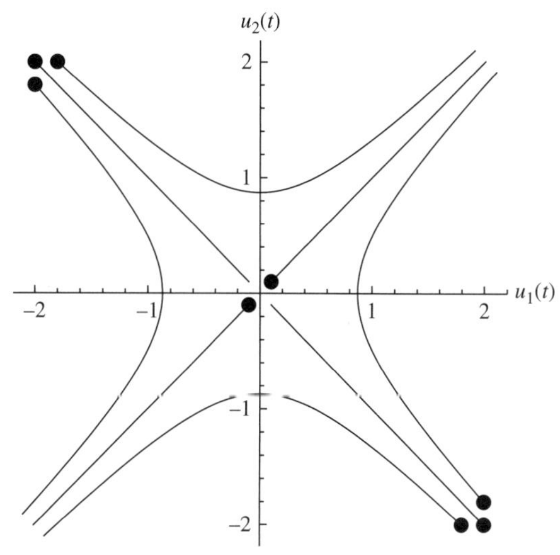

Figure 6.2. Phase plane plot of an unstable saddle point.

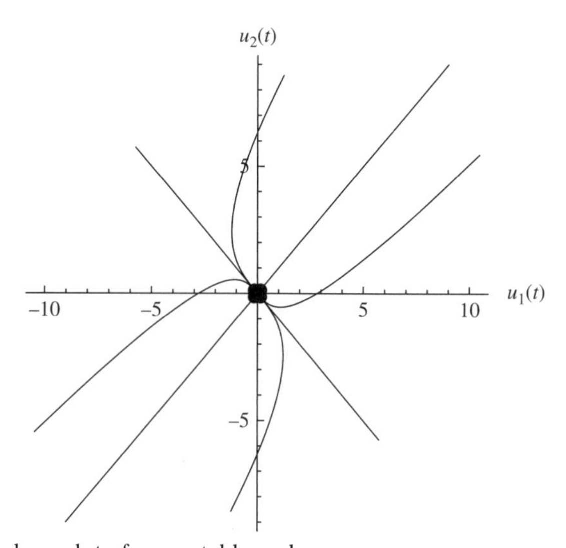

Figure 6.3. Phase plane plot of an unstable node or source.

an *unstable node* as shown in Figure 6.3. In the case of an unstable node, all trajectories move away from the origin to become unbounded.

3. Stable Node or Sink: $\text{tr}(\mathbf{A})^2 - 4|\mathbf{A}| > 0, \text{tr}(\mathbf{A}) < 0, |\mathbf{A}| > 0$
   If the eigenvalues of **A** are both negative real values, the equilibrium point is a *stable node* as shown in Figure 6.4. In the case of a stable node, all trajectories move toward the equilibrium point and remain there.
4. Unstable Spiral or Focus: $\text{tr}(\mathbf{A})^2 - 4|\mathbf{A}| < 0, \text{Re}[\lambda_{1,2}] > 0\ (\text{tr}(\mathbf{A}) \neq 0)$
   If the eigenvalues of **A** are complex, but with positive real parts, the equilibrium point is an *unstable focus*. A plot of position versus time is shown in Figure 6.5 and the phase plane plot is shown in Figure 6.6. As can be seen from both plots, the trajectory begins at the origin, that is, the equilibrium point, and spirals away, becoming unbounded.

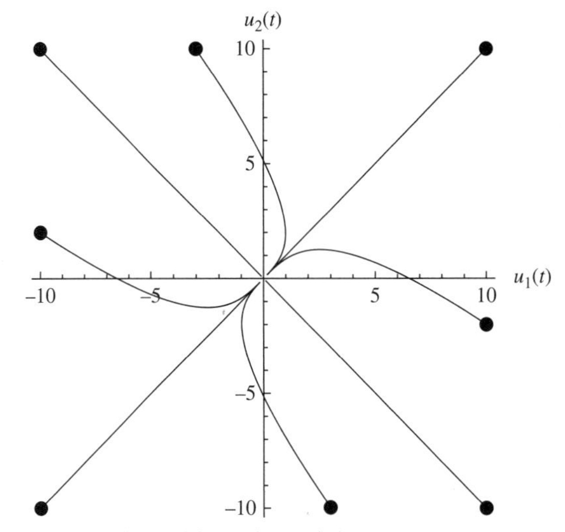

Figure 6.4. Phase plane plot of a stable node or sink.

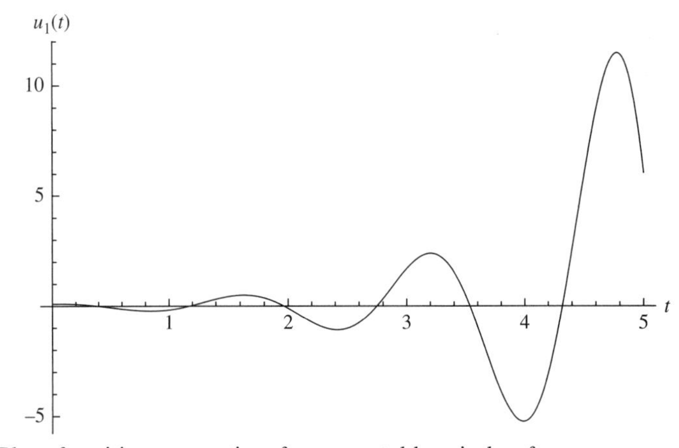

Figure 6.5. Plot of position versus time for an unstable spiral or focus.

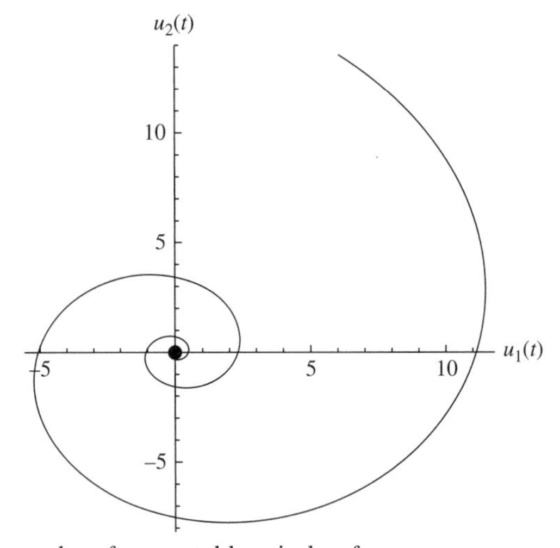

Figure 6.6. Phase plane plot of an unstable spiral or focus.

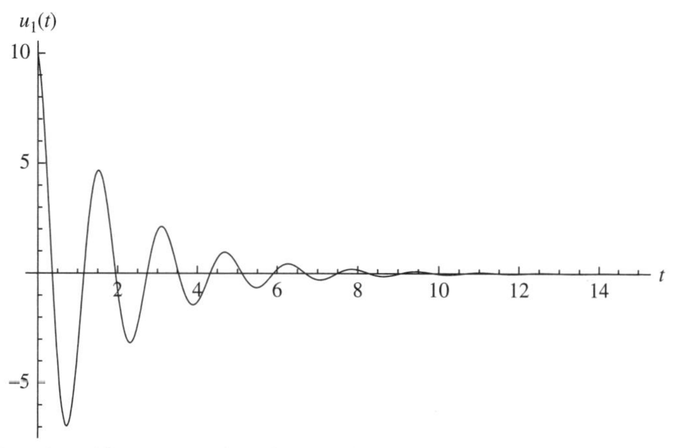

Figure 6.7. Plot of position versus time for a stable spiral or focus.

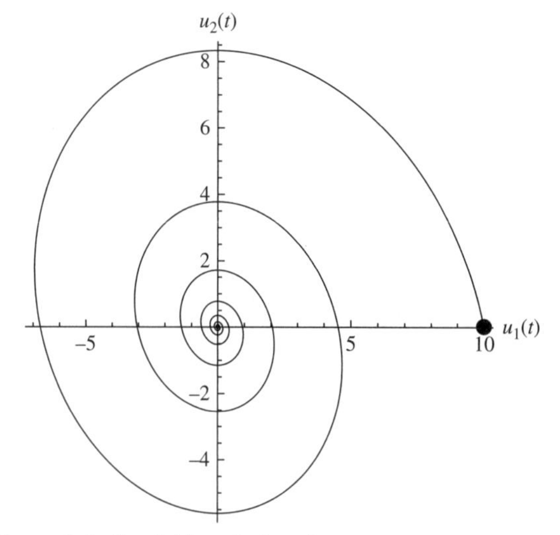

Figure 6.8. Phase plane plot of a stable spiral or focus.

5. Stable Spiral or Focus: $\mathrm{tr}(\mathbf{A})^2 - 4|\mathbf{A}| < 0, \mathrm{Re}[\lambda_{1,2}] < 0$ $(\mathrm{tr}(\mathbf{A}) \neq 0)$
   If the eigenvalues of $\mathbf{A}$ are complex, but with negative real parts, the equilibrium point is a *stable focus*. A plot of position versus time is shown in Figure 6.7 and the phase plane plot is shown in Figure 6.8. In the case of a stable focus, the trajectory spirals in toward the equilibrium point from any initial condition.
6. Stable Center: $\mathrm{tr}(\mathbf{A})^2 - 4|\mathbf{A}| < 0, \mathrm{tr}(\mathbf{A}) = 0$
   If the eigenvalues of $\mathbf{A}$ are purely imaginary, the equilibrium point is a *stable center*. A plot of position versus time is shown in Figure 6.9 and the phase plane plot is shown in Figure 6.10. A stable center is comprised of a periodic limit cycle solution centered at the equilibrium point. Because the trajectory remains bounded, the solution is stable. Recall that the simple pendulum with n even $(\theta = 0)$ has a stable center.

There are situations for which such a modal analysis is not possible, as for nonautonomous systems illustrated in the next section, or incomplete, as for non-normal systems illustrated in Section 6.5.

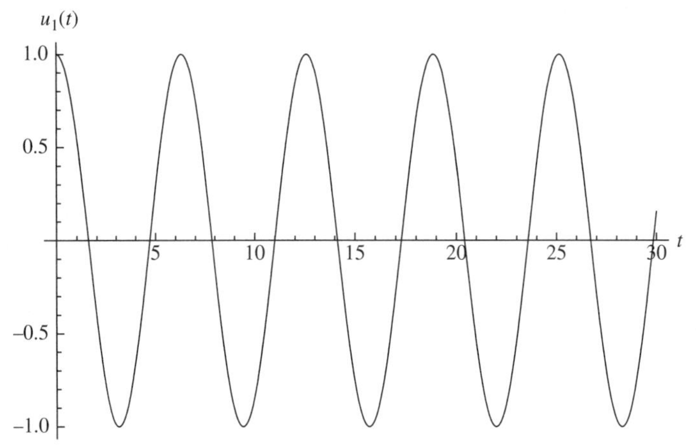

Figure 6.9. Plot of position versus time for a stable center.

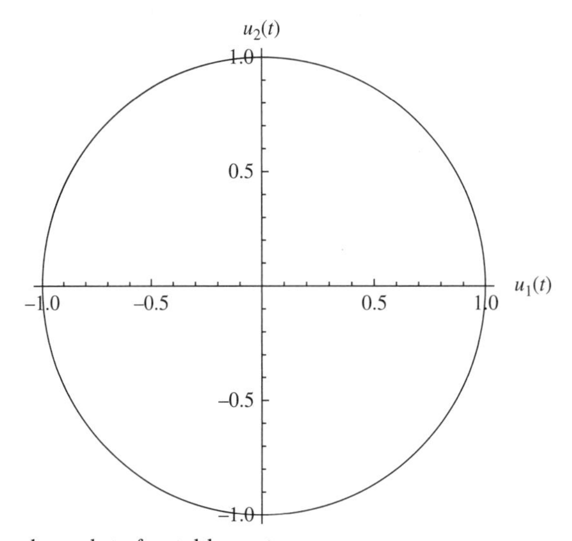

Figure 6.10. Phase plane plot of a stable center.

6.4 Nonautonomous Systems – Forced Pendulum

Let us make a small change to the simple pendulum considered in Section 6.2 by adding vertical forcing of the pivot point. Consider the motion of a pendulum of mass m that is forced vertically according to $y_O = A\cos(\omega t)$ as shown in Figure 6.11. From the geometry, the coordinates of the mass are

$$x = \ell \sin\theta, \quad y = A\cos(\omega t) - \ell\cos\theta.$$

In terms of x and y, the kinetic energy is

$$T = \frac{1}{2}m(\dot{x}^2 + \dot{y}^2).$$

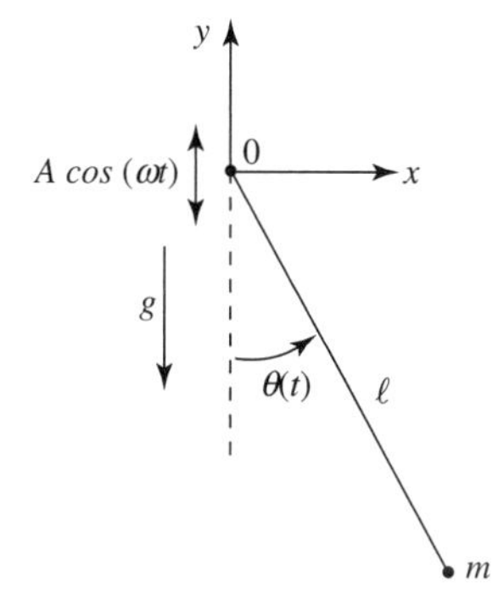

Figure 6.11. Schematic of the forced pendulum.

Because the system has one degree of freedom, we can write this in terms of the angle θ as

$$T = \frac{1}{2}m\left\{\left[\ell\dot{\theta}\cos\theta\right]^2 + \left[-A\omega\sin(\omega t) + \ell\dot{\theta}\sin\theta\right]^2\right\},$$

or

$$T = \frac{1}{2}m\left[\ell^2\dot{\theta}^2 + A^2\omega^2\sin^2(\omega t) - 2A\ell\omega\sin(\omega t)\dot{\theta}\sin\theta\right],$$

where the fact that $\cos^2\theta + \sin^2\theta = 1$ has been used in simplifying the last expression. The potential energy is

$$V = mgy = mg\left[A\cos(\omega t) - \ell\cos\theta\right].$$

Note that the external forcing does work on the system, continuously supplying additional energy to the pendulum. Therefore, mechanical energy is not conserved for such externally forced systems.

For the generalized coordinate $q = \theta$, the Euler-Lagrange equation of motion is

$$\frac{\partial L}{\partial \theta} - \frac{d}{dt}\left(\frac{\partial L}{\partial \dot{\theta}}\right) = 0,$$

which leads to the following equation of motion for the forced pendulum

$$\ddot{\theta} + \left[\frac{g}{\ell} - \frac{A\omega^2}{\ell}\cos(\omega t)\right]\sin\theta = 0.$$

Observe that this reduces to the equation of motion (6.1) for the simple pendulum when the amplitude of forcing vanishes ($A = 0$). For convenience, let us nondimensionalize according to the forcing frequency by setting $\tau = \omega t$, in which case $d^2/dt^2 = \omega^2 d^2/d\tau^2$, and the equation of motion becomes

$$\ddot{\theta} + (\Omega + \alpha\cos\tau)\sin\theta = 0, \tag{6.6}$$

where $\Omega = g/\ell\omega^2$ and $\alpha = -A/\ell$. Note that the natural frequency of the unforced pendulum is $\sqrt{g/\ell}$; therefore, Ω is the square of the ratio of the natural frequency of the pendulum to the forcing frequency ω. Forcing at the natural frequency, $\omega = \sqrt{g/\ell}$, corresponds to $\Omega = 1$.

As for the simple pendulum, the equilibrium positions s, for which $\ddot{q}_i = \dot{q}_i = 0$, require that $\sin s = 0$; thus, $s = n\pi, n = 0, 1, 2, \ldots$. The cases for which n is even correspond to $s = 0$, in which case the pendulum is hanging vertically down. Likewise, the cases for which n is odd correspond to $s = \pi$, for which the pendulum is located vertically above the pivot point.

Now let us consider stability of the equilibrium position $s = 0$, which is stable for the simple pendulum ($A = 0$). Introducing small perturbations about the equilibrium position

$$\theta(\tau) = s + \epsilon u(\tau) = \epsilon u(\tau),$$

equation (6.6) leads to the perturbation equation

$$\ddot{u} + (\Omega + \alpha \cos \tau)\, u = 0, \tag{6.7}$$

where again we linearize using the Taylor series $\sin(\epsilon u) = \epsilon u - \frac{\epsilon^3 u^3}{3!} + \cdots$. Equation (6.7) is known as *Mathieu's equation*. As with the simple pendulum, we can convert this to a system of first-order equations in matrix form using the transformation

$$u_1(\tau) = u(\tau),$$
$$u_2(\tau) = \dot{u}(\tau).$$

Differentiating and substituting Mathieu's equation leads to the system

$$\begin{bmatrix} \dot{u}_1 \\ \dot{u}_2 \end{bmatrix} = \begin{bmatrix} 0 & 1 \\ -(\Omega + \alpha \cos \tau) & 0 \end{bmatrix} \begin{bmatrix} u_1 \\ u_2 \end{bmatrix}. \tag{6.8}$$

Observe that the matrix $\mathbf{A}$ in (6.8) for Mathieu's equation is time-dependent owing to the forcing. Such systems are called *nonautonomous*, and we cannot evaluate stability by examining the eigenvalues. Instead, we directly solve Mathieu's equation numerically for the perturbations $u(\tau)$.

Whereas the simple pendulum is stable near the equilibrium position corresponding to $s = 0$, surprisingly, the forced pendulum allows for regions of instability for $s = 0$ and certain values of the parameters Ω and α, which are related to the forcing frequency and amplitude, respectively. To illustrate this behavior, we will set the initial conditions to be $u(0) = 1$ and $\dot{u}(0) = 0$, and the amplitude parameter is set to $\alpha = 0.01$. Figure 6.12 shows results obtained by setting the forcing frequency parameter to $\Omega = 0.24$ and $\Omega = 0.25$. For $\Omega = 0.24$, the oscillation amplitude remains bounded and is stable. For just a small change in the forcing frequency, however, the case with $\Omega = 0.25$ becomes unbounded and is unstable. The case with $\Omega = 0.26$ (not shown) is stable and looks very similar to Figure 6.12(a) for $\Omega = 0.24$. Recall that all of these cases correspond to the equilibrium state for which the pendulum is hanging down, which is stable when there is no forcing ($A = 0$). Thus, when the forcing is at a certain frequency, we can observe unstable behavior of the forced pendulum. In fact, there is a very narrow range of forcing frequencies for which the solution becomes unstable. Observe also that $\Omega = 0.25$ does not correspond to forcing at the natural frequency of the unforced pendulum, which would be $\Omega = 1$. Therefore, the instability is not a result of some resonance phenomena, nor is it due to nonlinear effects as Mathieu's equation is linear. Based on further analysis, it can be shown

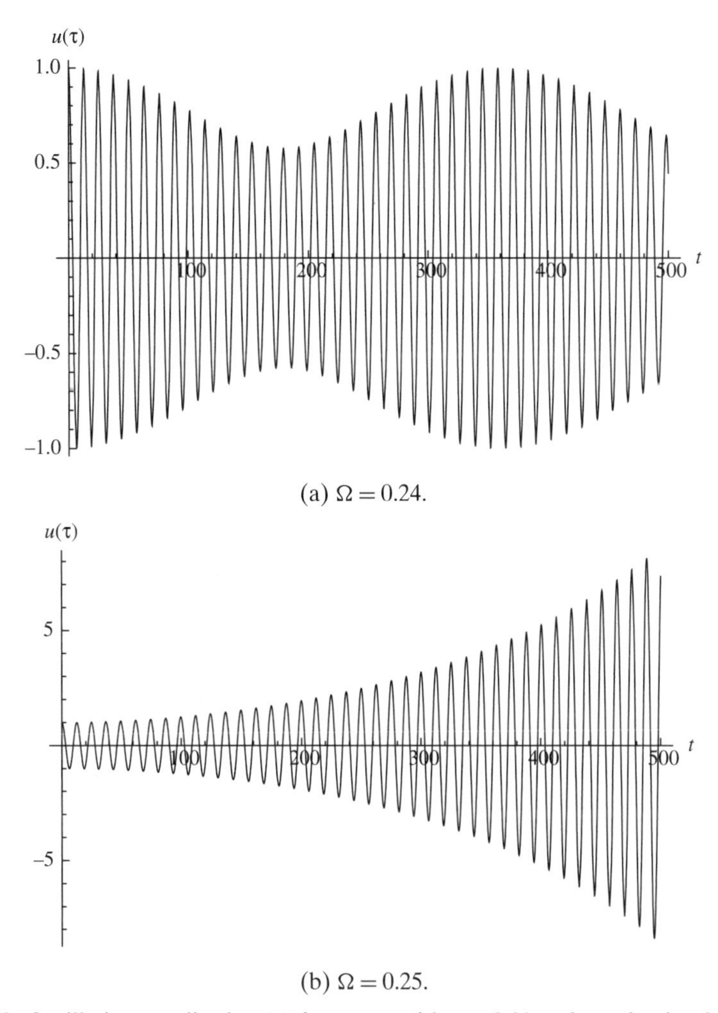

(a) $\Omega = 0.24$.

(b) $\Omega = 0.25$.

Figure 6.12. Oscillation amplitude $u(\tau)$ for a case with $\alpha = 0.01$ and two forcing frequencies.

that there are an infinity of "unstable tongues" similar to the one near $\Omega = 1/4$ at $\Omega = \frac{1}{4}n^2, n = 0, 1, 2, \ldots$.

Even more surprising than the fact that there are unstable regions near the $\theta = 0$ equilibrium point is that there is a small region for negative Ω, which corresponds to the $\theta = \pi$ equilibrium point, for which the system is stable. An example is shown in Figure 6.13. As the pendulum begins to fall off to one side or the other, which would be unstable without the forcing, the small amplitude forcing acts to attenuate the pendulum's motion and force it back toward the equilibrium position, keeping the pendulum's motion stable. This unusual behavior is only observed for small magnitudes of negative Ω, that is, large forcing frequencies.

6.5 Non-Normal Systems – Transient Growth

A conventional stability analysis considers each individual stability mode separately, with the overall stability being determined by the growth rate of the most unstable (or

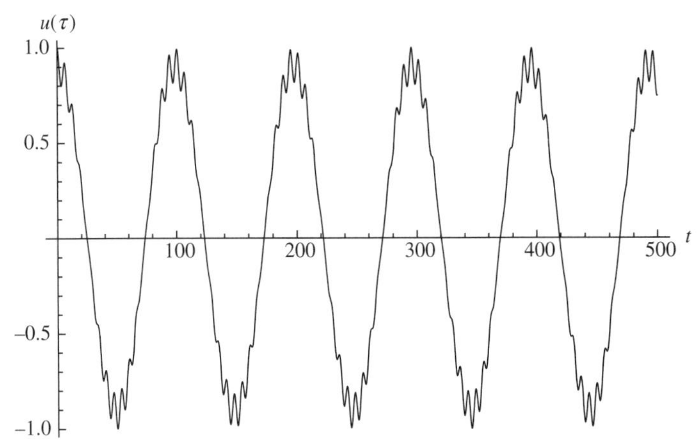

Figure 6.13. Oscillation amplitude $u(\tau)$ for case with $\alpha = 0.1$ and $\Omega = -0.001$.

least stable) mode. Therefore, it is called a *modal stability analysis*, and the stability characteristics as $t \to \infty$ are determined from each corresponding eigenvalue. This is known as *asymptotic stability* as it applies for large times, and the eigenvalues provide the growth rate of the infinitesimal perturbations. A modal analysis is sufficient for systems governed by *normal matrices*, for which the eigenvectors are mutually orthogonal and, therefore, do not interact with one another, allowing for each mode to be considered individually. A normal matrix is one for which the coefficient matrix commutes with its conjugate transpose according to

$$\mathbf{A}\overline{\mathbf{A}}^T = \overline{\mathbf{A}}^T\mathbf{A}.$$

A real, symmetric matrix, for example, is normal, but not all normal matrices are symmetric.

For systems governed by non-normal matrices, the asymptotic stability behavior for large times is still determined by the eigenvalues of the system's coefficient matrix. However, owing to exchange of energy between the nonorthogonal modes, it is possible that the system will exhibit a different behavior for $O(1)$ times as compared to the asymptotic behavior observed for large times. This is known as *transient growth*. For example, a system may be asymptotically stable, meaning that none of the eigenvalues lead to a growth of their corresponding modes (eigenvectors) as $t \to \infty$, but it displays a transient growth for finite times in which the amplitude of a perturbation grows for some period of time before yielding to the large-time asymptotic behavior.

Consider a second-order example given in Farrell and Ioannou (1996) in order to illustrate the effect of non-normality of the system on the transient growth behavior for finite times before the solution approaches its asymptotically stable equilibrium point for large time. The system is governed by the following first-order ordinary differential equations

$$\begin{bmatrix} \dot{u}_1 \\ \dot{u}_2 \end{bmatrix} = \begin{bmatrix} -1 & -\cot\theta \\ 0 & -2 \end{bmatrix} \begin{bmatrix} u_1 \\ u_2 \end{bmatrix}. \tag{6.9}$$

The coefficient matrix $\mathbf{A}$ is normal if $\theta = \pi/2$ and non-normal for all other θ. The system has an equilibrium point at $(u_1, u_2) = (0,0)$.

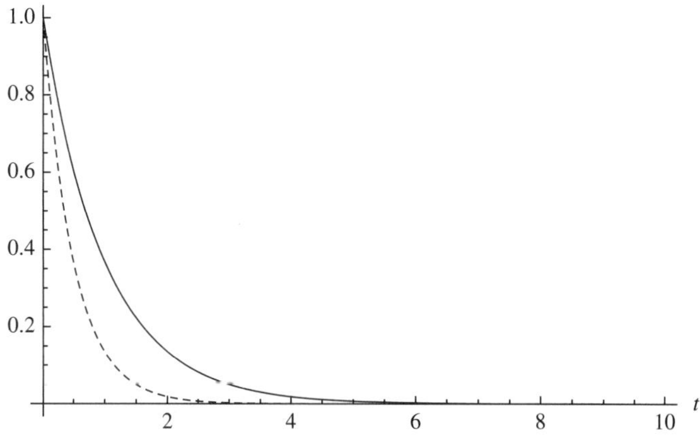

Figure 6.14. Solution of the normal system with $\theta = \pi/2$ for $u_1(t)$ (solid line) and $u_2(t)$ (dashed line).

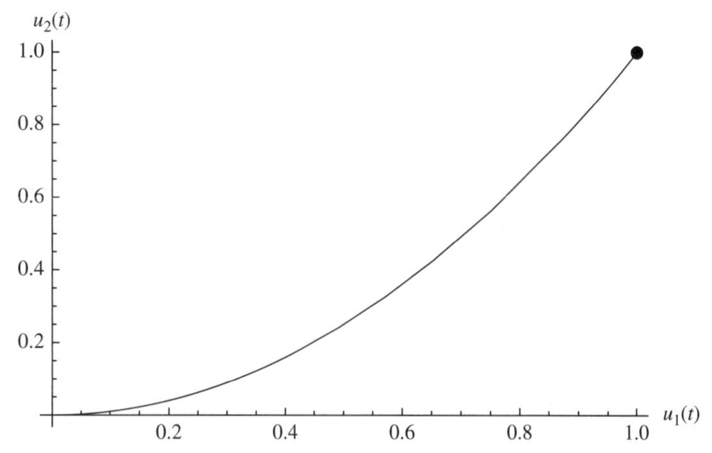

Figure 6.15. Solution of the normal system with $\theta = \pi/2$ for $u_1(t)$ and $u_2(t)$ in phase space (the dot indicates the initial condition).

Let us first consider the behavior of the normal system with $\theta = \pi/2$. In this case, the eigenvalues are $\lambda_1 = -2$ and $\lambda_2 = -1$. Because the eigenvalues are negative and real, the system is asymptotically stable in the form of a stable node. The corresponding eigenvectors are given by

$$\mathbf{v}_1 = \begin{bmatrix} 0 \\ 1 \end{bmatrix}, \quad \mathbf{v}_2 = \begin{bmatrix} 1 \\ 0 \end{bmatrix},$$

which are mutually orthogonal as expected for a normal system. The solution for the system (6.9) with $\theta = \pi/2$ is shown in Figures 6.14 and 6.15 for the initial conditions $u_1(0) = 1$ and $u_2(0) = 1$. For such a stable normal system, the perturbed solution decays exponentially toward the stable equilibrium point $(u_1, u_2) = (0,0)$ according to the rates given by the eigenvalues. In phase space, the trajectory starts at the initial condition $(u_1(0), u_2(0)) = (1,1)$ and moves progressively toward the equilibrium point $(0,0)$.

Now consider the system (6.9) with $\theta = \pi/100$, for which the system is no longer normal. While the eigenvalues $\lambda_1 = -2$ and $\lambda_2 = -1$ remain the same, the

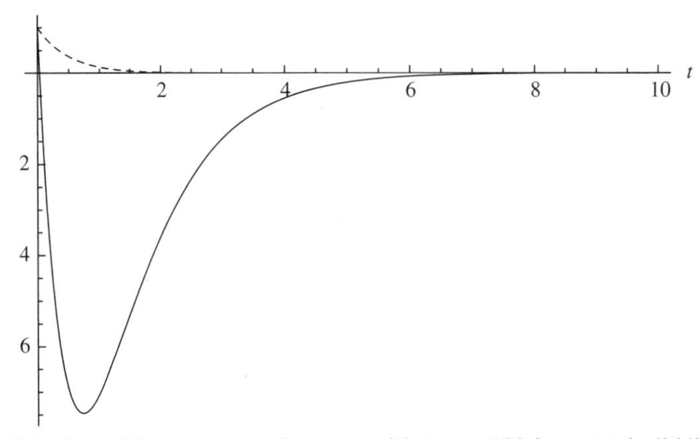

Figure 6.16. Solution of the non-normal system with $\theta = \pi/100$ for $u_1(t)$ (solid line) and $u_2(t)$ (dashed line).

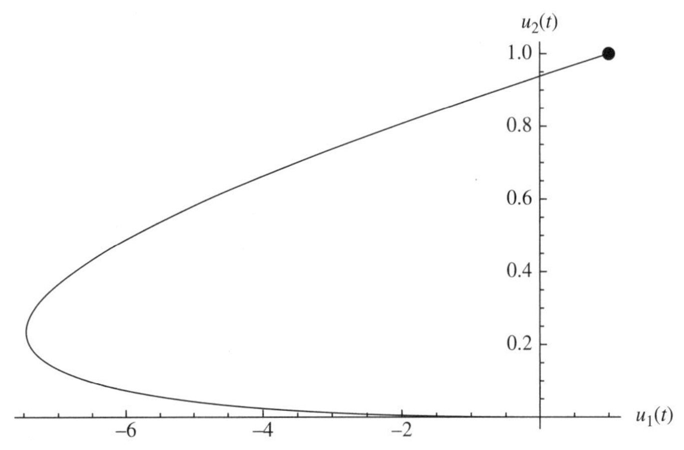

Figure 6.17. Solution of the non-normal system with $\theta = \pi/100$ for $u_1(t)$ and $u_2(t)$ in phase space (the dot indicates the initial condition).

corresponding eigenvectors are now

$$\mathbf{v}_1 = \begin{bmatrix} \cot\left(\frac{\pi}{100}\right) \\ 1 \end{bmatrix}, \quad \mathbf{v}_2 = \begin{bmatrix} 1 \\ 0 \end{bmatrix},$$

which are no longer orthogonal giving rise to the potential for transient growth behavior. The solution for the system (6.9) with $\theta = \pi/100$ is shown in Figures 6.16 and 6.17 for the same initial conditions $u_1(0) = 1$ and $u_2(0) = 1$ as before. Recall that the system is asymptotically stable as predicted by its eigenvalues, and indeed the perturbations do decay toward zero for large times. Whereas $u_2(t)$ again decays monotonically[1] toward the equilibrium point, however, $u_1(t)$ first grows to become quite large before decaying toward the stable equilibrium point. In phase space, the trajectory starts at the initial condition $(u_1(0), u_2(0)) = (1,1)$ but moves farther away from the equilibrium point at the origin before being attracted to it, essentially succumbing to the asymptotically stable large-t behavior.

[1] If a function decays monotonically, it never increases over any interval.

The transient growth behavior can occur for non-normal systems because the nonorthogonal eigenvectors lead to the possibility that the individual modes can exchange energy leading to transient growth. In other words, even though each eigenvalue leads to exponential decay of their respective modes, the corresponding nonorthogonal eigenvectors may interact to induce a temporary growth of the solution owing to the different rates of decay of each mode.

Note that for a nonlinear system, the linearized behavior represented by the perturbation equations only governs in the vicinity of the equilibrium point. Therefore, transient growth behavior could lead to the solution moving too far from the equilibrium point for which the linearized behavior is valid, thereby bringing in nonlinear effects and altering the stability properties before the eventual decay to the asymptotic stability state can be realized.

Not all non-normal systems exhibit such a transient growth behavior. For example, try the system (6.9) with $\theta = \pi/4$. Also, observe that the coefficient matrices are non-normal for both the simple and forced pendulum examples considered previously; however, neither exhibits transient growth behavior. For non-normal systems that do exhibit transient growth, the question becomes, "What is the initial perturbation that leads to the maximum possible transient growth in the system within finite time before succumbing to the asymptotic stability behavior as $t \to \infty$?"

The initial perturbation that experiences maximum growth over a specified time range is called the *optimal perturbation* and is found using variational methods. Specifically, we seek to determine the initial perturbation $\mathbf{u}(0)$ for the system

$$\dot{\mathbf{u}} = \mathbf{A}\mathbf{u}, \tag{6.10}$$

for which the energy growth over some time interval $0 \leq t \leq t_f$ is a maximum. We define the energy of a disturbance $\mathbf{u}(t)$ at time t using the norm operator as follows:

$$E(t) = ||\mathbf{u}(t)||^2 = \langle \mathbf{u}(t), \mathbf{u}(t) \rangle,$$

where $\langle \cdot, \cdot \rangle$ is the inner product. We seek the initial perturbation $\mathbf{u}(0)$ that produces the greatest relative energy gain during the time interval $0 \leq t \leq t_f$ defined by

$$G(t_f) = \max_{\mathbf{u}(0)} \frac{E\left[\mathbf{u}(t_f)\right]}{E\left[\mathbf{u}(0)\right]}.$$

In other words, we seek the initial disturbance that maximizes the gain functional

$$G[\mathbf{u}] = \frac{||\mathbf{u}(t_f)||^2}{||\mathbf{u}(0)||^2} = \frac{\langle \mathbf{u}(t_f), \mathbf{u}(t_f) \rangle}{\langle \mathbf{u}(0), \mathbf{u}(0) \rangle} = \frac{\mathbf{u}^T(t_f)\mathbf{u}(t_f)}{\mathbf{u}^T(0)\mathbf{u}(0)}, \tag{6.11}$$

subject to the differential constraint that the governing equation $\dot{\mathbf{u}} = \mathbf{A}\mathbf{u}$ is satisfied for all time in the range $0 \leq t \leq t_f$. As in Section 2.7.3, we define the augmented functional

$$\tilde{G}[\mathbf{u}, \mathbf{\Lambda}] = \frac{\mathbf{u}^T(t_f)\mathbf{u}(t_f)}{\mathbf{u}^T(0)\mathbf{u}(0)} - \int_0^{t_f} \mathbf{\Lambda}^T (\dot{\mathbf{u}} - \mathbf{A}\mathbf{u}) dt, \tag{6.12}$$

where $\mathbf{\Lambda}^T(t) = [\Lambda_1(t), \Lambda_2(t)]$ are the Lagrange multipliers required to enforce satisfaction of the governing equation. Note that the negative sign is introduced for the constraint in order to ensure a maximum in the gain. For each terminal time t_f, a different optimal initial perturbation $\mathbf{u}(0)$ will result from maximizing the functional $G[\mathbf{u}]$. Therefore, by altering the terminal time t_f over which the optimization of the initial condition is determined, one can produce a maximum growth curve $G(t_f)$ versus the terminal time t_f.

The augmented cost functional (6.12) is a *dual functional* as in Section 2.2 with a *terminal-time performance measure* as in Section 10.4.1. Hence, transient growth analysis has a great deal in common with the optimal control approach addressed in Chapter 10. In fact, we are essentially optimizing the initial perturbation in order to produce the maximum relative gain.

Keeping in mind that $\mathbf{u}(t)$, $\mathbf{\Lambda}(t)$, and $\mathbf{u}(t_f)$ vary in the augmented cost functional (6.12), let us take the variation of $\tilde{G}$ and set it equal to zero according to

$$\delta\tilde{G} = \frac{\delta\left[\mathbf{u}^T(t_f)\mathbf{u}(t_f)\right]}{\mathbf{u}^T(0)\mathbf{u}(0)} - \frac{\mathbf{u}^T(t_f)\mathbf{u}(t_f)\delta\left[\mathbf{u}^T(0)\mathbf{u}(0)\right]}{\left[\mathbf{u}^T(0)\mathbf{u}(0)\right]^2} - \delta\int_0^{t_f}\mathbf{\Lambda}^T(\dot{\mathbf{u}} - \mathbf{A}\mathbf{u})dt = 0. \quad (6.13)$$

In order to evaluate the variations in the first two terms, observe that for a general $n \times 1$ vector $\mathbf{v}$

$$\delta\left[\mathbf{v}^T\mathbf{v}\right] = \delta\left[v_1^2 + v_2^2 + \ldots + v_n^2\right] = 2\left[v_1\delta v_1 + v_2\delta v_2 + \ldots + v_n\delta v_n\right] = 2\mathbf{v}^T\delta\mathbf{v}.$$

Integrating the third term of equation (6.13) by parts leads to

$$\int_0^{t_f}\mathbf{\Lambda}^T\delta\dot{\mathbf{u}}dt = \mathbf{\Lambda}^T(t_f)\delta\mathbf{u}(t_f) - \mathbf{\Lambda}^T(0)\delta\mathbf{u}(0) - \int_0^{t_f}\dot{\mathbf{\Lambda}}^T\delta\mathbf{u}dt.$$

Substituting these results into equation (6.13) and collecting terms yields

$$\left\{-\frac{2\mathbf{u}^T(t_f)\mathbf{u}(t_f)\mathbf{u}^T(0)}{\left[\mathbf{u}^T(0)\mathbf{u}(0)\right]^2} + \mathbf{\Lambda}^T(0)\right\}\delta\mathbf{u}(0) + \left[\frac{2\mathbf{u}^T(t_f)}{\mathbf{u}^T(0)\mathbf{u}(0)} - \mathbf{\Lambda}^T(t_f)\right]\delta\mathbf{u}(t_f)$$
$$+ \int_0^{t_f}\left(\dot{\mathbf{\Lambda}}^T + \mathbf{\Lambda}^T\mathbf{A}\right)\delta\mathbf{u}dt = 0.$$

Because $\mathbf{u}(t)$, $\mathbf{u}(0)$, and $\mathbf{u}(t_f)$ vary, the expressions multiplied by each variation must vanish. We have the Euler equation

$$\dot{\mathbf{\Lambda}}^T = -\mathbf{\Lambda}^T\mathbf{A},$$

or taking the transpose of both sides yields

$$\dot{\mathbf{\Lambda}} = -\mathbf{A}^T\mathbf{\Lambda}, \quad (6.14)$$

and the initial conditions

$$\mathbf{u}(0) = \frac{\left[\mathbf{u}^T(0)\mathbf{u}(0)\right]^2}{2\mathbf{u}^T(t_f)\mathbf{u}(t_f)}\mathbf{\Lambda}(0), \quad (6.15)$$

$$\mathbf{\Lambda}(t_f) = \frac{2}{\mathbf{u}^T(0)\mathbf{u}(0)}\mathbf{u}(t_f). \quad (6.16)$$

The differential operator in the Euler equation (6.14) is the adjoint of that in the original governing equation $\dot{\mathbf{u}} = \mathbf{A}\mathbf{u}$ (see Section 1.8); therefore, we call the Lagrange multiplier $\mathbf{\Lambda}(t)$ the *adjoint variable* of the perturbation $\mathbf{u}(t)$. Note that the negative sign requires equation (6.14) to be integrated backward in time from $t = t_f$ to $t = 0$.

Observe that because the Lagrange multiplier $\mathbf{\Lambda}(t)$ varies, taking the variation of the augmented functional also produces the term

$$\int_0^{t_f} (\dot{\mathbf{u}} - \mathbf{A}\mathbf{u})\delta\mathbf{\Lambda}^T dt.$$

However, this simply returns the original governing equation (6.10). Therefore, it need not be included.

In order to determine the optimal initial perturbation that maximizes the gain for a given terminal time t_f, we solve the governing equation (6.10) forward in time using the initial condition obtained from equation (6.15). The solution at the terminal time $\mathbf{u}(t_f)$ is then used in equation (6.16) to determine the initial condition $\mathbf{\Lambda}(t_f)$ for the backward-time integration of the *adjoint equation* (6.14). The resulting solution for the adjoint variable $\mathbf{\Lambda}(t)$ provides $\mathbf{\Lambda}(0)$ for use in equation (6.15), which is used to obtain the initial condition for the governing equation (6.10). This procedure is repeated iteratively until a converged solution for $\mathbf{u}(t)$ and $\mathbf{\Lambda}(t)$ is obtained, from which the optimal initial perturbation $\mathbf{u}(0)$ is determined.

Let us return to the system governed by equation (6.9). Recall that the system is non-normal for $\theta = \pi/100$ and leads to transient growth for finite times before succumbing to the asymptotically stable behavior for large times predicted by a modal analysis. After carrying out a series of optimal perturbation calculations as described above for a range of terminal times t_f, we can calculate the gain $G(t_f)$ as defined by equation (6.11) and shown in Figure 6.18 for a range of terminal times. The optimal initial perturbation $\mathbf{u}(0)$ that produces these maximum gains for each terminal time t_f is plotted in Figure 6.19. The maximum gain is found to be $G = 63.6$ for a terminal time of $t_f = 0.69$, which means that the transient growth can result in a perturbation energy growing to nearly 64 times its initial energy at $t = 0$. The initial perturbation that produces this maximum gain is $(u_1(0), u_2(0)) = (-0.134, 2.134)$. As

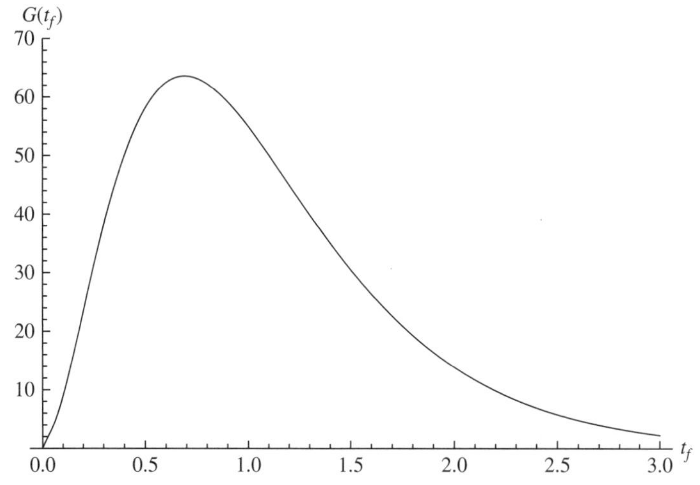

Figure 6.18. Gain $G(t_f)$ for a range of terminal times t_f for the non-normal system with $\theta = \pi/100$.

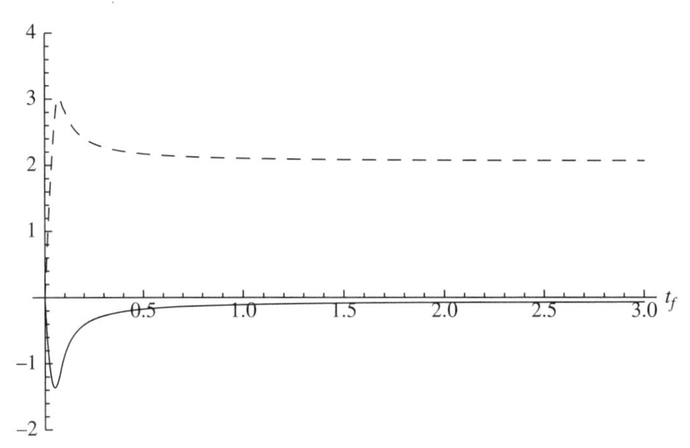

Figure 6.19. Initial perturbations $u_1(0)$ (solid line) and $u_2(0)$ (dashed line) that produce the maximum gain for a range of terminal times t_f for the non-normal system with $\theta = \pi/100$.

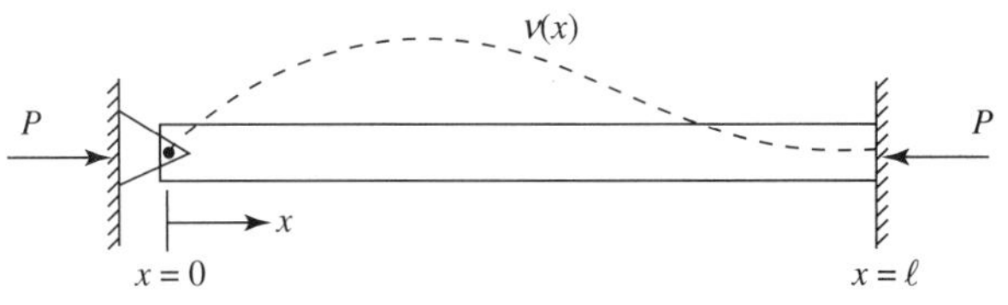

Figure 6.20. Schematic of an axially-loaded beam-column.

shown by the plot of gain $G(t_f)$, the gain eventually decays to zero as the terminal time increases, consistent with the fact that the system is asymptotically stable for large times.

For the linear stability analysis of small perturbations about stationary equilibrium points considered here, the transient growth analysis for non-normal systems can be reduced to performing a singular-value decomposition (SVD) to obtain the optimal disturbance. The primary virtue of the variational approach presented here is that it can be extended to evaluate stability of nonlinear state equations and/or stability of perturbations about time-evolving (unsteady) solutions. The nonlinear case corresponds to analysis of nonlinear systems subject to perturbations that are not infinitesimally small. Stability of non-normal, continuous fluid dynamical systems will be encountered in Section 9.5.4.

6.6 Continuous Systems – Beam-Column Buckling

As illustrated for the simple pendulum, asymptotic stability of discrete systems may be evaluated in terms of an *algebraic eigenproblem* $\mathbf{Au} = \lambda\mathbf{u}$ with n degrees of freedom. In contrast, stability of continuous systems, such as deformable bodies, is governed by *differential eigenproblems* of the form $\mathcal{L}u = \lambda u$, where $\mathcal{L}$ is a differential operator with infinite degrees of freedom.

As an example of stability of a continuous system, let us consider stability of a beam-column under an axial load as shown in Figure 6.20. Here, P is the axial compressive force, $u(x)$ is the longitudinal displacement, and $v(x)$ is the lateral deflection of the beam-column. For a beam-column subject to an axial load P, the potential energy per unit volume in the x-direction is (see longitudinal vibration of a rod in Section 5.4.1)

$$\mathcal{V}_x = \frac{1}{Adx} dU_x = \frac{1}{2} E \left(\frac{du}{dx} \right)^2,$$

where A is the cross-sectional area, and E is Young's modulus. The total potential energy of the beam-column in the x-direction is then

$$V_x[u] = \int_0^\ell \mathcal{V}_x dx = \frac{1}{2} \int_0^\ell E \left(\frac{du}{dx} \right)^2 dx,$$

and the associated Euler-Lagrange equation is

$$\frac{d}{dx} \left(E \frac{du}{dx} \right) = 0.$$

Integrating leads to the following expression for the longitudinal displacement

$$u(x) = \int_0^x \frac{C}{E(\bar{x})} d\bar{x},$$

where the integration constant C depends upon P, ℓ, and the boundary conditions. The static equilibrium solution for the longitudinal displacement $u(x)$ can then be obtained for a given $E(x)$ distribution. Note that if E is constant, then $u(x)$ is a linear function of x. The lateral deflection under the compressive force P is simply zero. That is, the equilibrium solution for lateral deflection is $v(x) = 0$.

We seek to investigate the stability of this solution when a small lateral perturbation $\epsilon \bar{v}(x)$ is introduced. This small lateral deflection will introduce additional strain energy due to elongation and bending of the beam-column. The strain energy due to elongation is as in the case of a laterally loaded string, and the strain energy due to bending is taken from Section 5.3.2 for a beam. Thus, the potential energy per unit volume in the y-direction is

$$\mathcal{V}_y = \frac{1}{Adx} dU_y = \frac{1}{Adx} \frac{1}{2} \left[EI(\epsilon \bar{v}'')^2 - P(\epsilon \bar{v}')^2 \right] dx,$$

and the potential energy is

$$V_y[\bar{v}] = \int_0^\ell \mathcal{V}_y dx = \frac{1}{2A} \int_0^\ell \left[EI(\epsilon \bar{v}'')^2 - P(\epsilon \bar{v}')^2 \right] dx. \tag{6.17}$$

Observe that the strain energy due to bending is related to the average magnitude of the curvature of the laterally deflected beam-column, and that the strain energy due to elongation is related to the average magnitude of the slope.

The Euler equation for a functional of the form (6.17), with $F = F(x, \bar{v}', \bar{v}'')$, is

$$-\frac{d}{dx} \left(\frac{\partial \mathcal{V}_y}{\partial \bar{v}'} \right) + \frac{d^2}{dx^2} \left(\frac{\partial \mathcal{V}_y}{\partial \bar{v}''} \right) = 0,$$

leading to the *buckling equation*

$$\frac{d^2}{dx^2} \left(EI \frac{d^2 \bar{v}}{dx^2} \right) + P \frac{d^2 \bar{v}}{dx^2} = 0. \tag{6.18}$$

If the bending stiffness EI is a constant, this is a differential eigenproblem with eigenvalues $\lambda = P/EI$. Observe that the trivial solution $\bar{v}(x) = 0$ is obtained for the majority of values of the parameter $\lambda = P/EI$. These solutions are stable as the lateral deflection returns to zero after a small perturbation.

The eigenfunctions corresponding to the eigenvalues of the differential equation are critical buckling loads for which the lateral deflection is nontrivial. In the case of buckling instabilities, we are interested in the smallest load P (for given stiffness EI) that produces buckling, which is called the *Euler buckling load*. The corresponding smallest eigenvalue $\lambda = P/EI$ can be estimated using the Rayleigh-Ritz methods as illustrated in Section 3.1.3. Notice from the potential energy given by equation (6.17) that increasing the bending stiffness EI increases the potential energy of the beam-column, thereby stabilizing it, while increasing P decreases the potential energy owing to the minus sign in the functional, thereby destabilizing it. Once again, we find that the variational form provides additional physical insight that is not evident from the differential form.

The analog of a discrete normal system in the continuous case is a system governed by a self-adjoint differential operator (see Section 1.8) as is the beam-column buckling equation. Similar to the eigenvectors of a normal matrix, the eigenfunctions of a self-adjoint operator are all mutually orthogonal, and asymptotic stability of the system is governed by the (now infinite) set of eigenvalues of the operator. Systems that are governed by non-normal operators, that is, those that are not self-adjoint, can exhibit the same transient growth behavior observed for non-normal discrete systems. Stability of non-normal, continuous systems will be considered in Section 9.5.4 as applied to fluid dynamics.

Stability of dynamical systems is a rich and multi-faceted topic for which we have only hinted at some of the basic approaches and results. Much more could be said about stability of discrete and continuous, autonomous and nonautonomous, and normal and non-normal systems.

7 Optics and Electromagnetics

> The miracle of the appropriateness of the language of mathematics to the formulation of the laws of physics is a wonderful gift which we neither understand nor deserve. We should be grateful for it and hope that it will remain valid in future research and that it will extend, for better or worse, to our pleasure, even though perhaps also to our bafflement, to wide branches of learning.
>
> (Eugene Wigner)

Electromagnetic wave radiation is remarkable for two reasons: 1) it spans a vast spectrum of wavelengths providing for a wide range of physics and applications, and 2) it does not require a medium within which to travel. While all electromagnetic waves travel at the speed of light if traveling through a vacuum, they span ten orders of magnitude in wavelength, essentially corresponding to the full range of sizes that exist in the universe. Amazingly, the governing equations of electromagnetic waves, which do not require a medium, and physical waves, which do, are essentially the same. Rather than a guitar string or gas molecules, however, electromagnetic waves travel via massless photons. While the photons do not have mass, they do carry energy; it is this energy that determines their wavelength. A photon's energy is directly proportional to its frequency (and inversely proportional to the wavelength).

From low frequency to high, the electromagnetic spectrum includes radio waves, microwave radiation, infrared radiation, visible light, ultraviolet radiation, X-ray radiation, and gamma radiation. Until two centuries ago, visible light was the only portion of the electromagnetic spectrum that we were aware of – although we unknowingly experienced the ultraviolet radiation from the Sun. Since that time, however, man has invented ways to use nearly every portion of the spectrum for an amazing variety of purposes. They are used to transmit information, image the human body and the universe, heat our food, control electronic devices remotely, explore the subatomic world, and to treat cancer. Before providing a general treatment of electromagnetics, we discuss the classical subject of geometric optics.

7.1 Optics

Optics has been considered briefly in Section 1.3.1 as a means to motivate the need for the calculus of variations and to contrast it with differential calculus. Having established the basic principles of the calculus of variations, we can now revisit optics in a broader context and treat some typical cases. Recall that the path of light traveling through a medium is governed by Fermat's principle, which states that the path of light between two points is the one that requires the minimum travel time. We apply Fermat's principle in a two-dimensional plane by seeking to minimize the travel time $T[u(x)]$ of light along all the possible paths $u(x)$ that the light could take through a medium between the points (x_0,u_0) and (x_1,u_1), where the travel time is given by the functional

$$T[u(x)] = \int_{x_0}^{x_1} \frac{ds}{v(x,u)} = \frac{1}{c}\int_{x_0}^{x_1} n(x,u)ds = \frac{1}{c}\int_{x_0}^{x_1} n(x,u)\sqrt{1+\left(\frac{du}{dx}\right)^2}dx. \tag{7.1}$$

Here, $v(x,u)$ is the speed of light in the medium, c is the speed of light in a vacuum, $n(x,u)$ is the index of refraction in the medium, and ds is a differential element along the path $u(x)$. The index of refraction is the ratio of the speed of light in a vacuum to that in the medium, that is, $n(x,u) = c/v(x,u)$. Note that because $v(x,u) \leq c$, the index of refraction must be such that $n(x,u) \geq 1$.

If the index of refraction is a constant, then equation (7.1) is simply the length functional considered in Section 2.1.4, for which the stationary function is a straight line. Accordingly, light travels along straight paths in homogeneous media. This fact has been used in Section 1.3.1 to derive *Snell's law* for light traveling through two homogeneous media that each have constant index of refraction. Snell's law states that the ratio of the sines of the angles of incidence and refraction is equal to the ratio of the indices of refraction in the two adjacent homogeneous media. Likewise, the same considerations lead to the *law of reflection*, which states that the angles of incidence and reflection are equal in a homogeneous medium.

In the general case of nonhomogeneous media, it is necessary to reconsider the functional (7.1) representing Fermat's principle for the given index of refraction distribution. Notice, however, that the integrand in the functional (7.1) is of the form $F = F(x,u,u')$. Therefore, the corresponding Euler equation is of the usual form

$$\frac{\partial F}{\partial u} - \frac{d}{dx}\left(\frac{\partial F}{\partial u'}\right) = 0. \tag{7.2}$$

Because of the obvious similarity to the Euler-Lagrange equations of motion (4.21), one can draw analogies between optics and the dynamics of discrete particles.

EXAMPLE 7.1 Consider Fermat's principle applied to a medium in which the index of refraction only depends upon the vertical coordinate such that $n = n(u)$. (a) Determine the path of light if $n = n_0 = \text{constant}$. (b) Determine the path of light if $n(u) = n_0/u$.

Solution: If the index of refraction only depends on the vertical coordinate u, then the integrand in Fermat's principle (7.1) is independent of the independent variable x, in which case we can utilize special case II from Section 2.1.4. From equation (7.1), the integrand is

$$F = \frac{n(u)}{c}\sqrt{1+(u')^2}.$$

According to special case II, we switch the roles of the independent and dependent variables by writing

$$\bar{F} = x'F\left[u,(x')^{-1}\right] = x'\frac{n(u)}{c}\sqrt{1+(x')^{-2}} = \frac{n(u)}{c}\sqrt{(x')^2+1}.$$

Then the Euler equation for special case II is

$$\frac{\partial \bar{F}}{\partial x'} = \frac{A}{c},$$

where A is a constant. Substituting F yields

$$\frac{n(u)}{c}\frac{1}{2}\left[(x')^2+1\right]^{-1/2}(2x') = \frac{A}{c}.$$

Squaring both sides and solving for x' leads to

$$\frac{dx}{du} = \frac{1}{\sqrt{\left(\frac{n}{A}\right)^2-1}},$$

or

$$\frac{du}{dx} = \sqrt{\left(\frac{n}{A}\right)^2-1}. \tag{7.3}$$

This first order equation can be solved for the path $u(x)$ given the starting and ending points (x_0,u_0) and (x_1,u_1), respectively.

(a) Consider the case when $n(u) = n_0 =$ constant. With the index of refraction being a constant, the right-hand side of equation (7.3) is a constant, say B. Integrating gives

$$u(x) = Bx + C,$$

which is a straight line as expected for a homogeneous medium having a constant index of refraction. Substitution of the end conditions would provide the constants B and C.

(b) For the case with $n(u) = n_0/u$, equation (7.3) yields

$$\frac{du}{dx} = \frac{1}{u}\sqrt{\left(\frac{n_0}{A}\right)^2-u^2}.$$

Separating variables and integrating produces

$$-\sqrt{\left(\frac{n_0}{A}\right)^2-u^2} = x+B.$$

Squaring both sides and solving for $u(x)$ yields

$$u(x) = \sqrt{\left(\frac{n_0}{A}\right)^2-(x+B)^2}.$$

For example, if the beginning and ending points are given by $u(0) = 1$ and $u(1) = 0$, then $A = n_0$ and $B = 0$, which gives the solution for the path of light as

$$u(x) = \sqrt{1 - x^2},$$

which is a circle of unit radius centered at the origin, that is, $u^2 + x^2 = 1$.

The scenario considered in the previous example, in which the index of refraction only depends upon the vertical coordinate, can be used to explain the occurrence of mirages on hot surfaces on a sunny day. The index of refraction being only a function of the vertical coordinate u is a reasonable approximation for the air above a heated surface on a day with very little wind. The temperature affects the density of the air, which in turn influences the index of refraction. For the hot air near the surface, the density is smaller than that of the cooler air above. Likewise, the index of refraction of the hot air is smaller corresponding to a faster speed of the light in the hot air as compared to the cool air. The changing index of refraction causes the light to bend, or refract. Because the index of refraction is close to unity at the hot surface and increases slowly with height above the surface, the light from the sky is bent as it approaches the heated surface. Hence, it can appear that the light is coming from near the surface some distance ahead of the observer, thereby producing a mirage.

We can easily extend Fermat's principle to three dimensions as follows:

$$T[u(x), v(x)] = \frac{1}{c} \int_{x_0}^{x_1} n(x,u,v) \sqrt{1 + \left(\frac{du}{dx}\right)^2 + \left(\frac{dv}{dx}\right)^2}\, dx, \tag{7.4}$$

where the three-dimensional path of the light is given by $u(x)$ and $v(x)$.

As with so many other areas of classical mechanics, for example, the principle of least action, Fermat's principle as historically stated is somewhat misleading. Just as the action is not always a minimum in Hamilton's principle, the actual path of light does not always correspond to a minimum in Fermat's principle. In some cases, the path may correspond to a maximum or a stationary function. Therefore, Fermat's principle should more accurately be stated as the path of light between two points is the one that requires the travel time to be stationary. An example in which the actual path produces a maximum in travel time is the case of reflection of light by a concave spherical mirror (Smith 1974).

Fermat's principle captures the approach to optics in which the light is imagined to travel along a continuous path given by $u(x)$ [and possibly $v(x)$]. This is known as *geometric optics* and explains reflection and refraction of light rays. In this context we regard light as a particle, that is, a photon, traveling along a path, much as we do in dynamics of particles, giving rise to the analogy between Fermat's principle and the Euler-Lagrange equations of motion. Although many optical phenomena can be explained using such a framework, light is known to have a dual character in which it behaves as a particle in some respects and as a wave in other respects. For example, the phenomena of diffraction, interference, and polarization require a wave propagation approach to light. In this context, light is governed by the wave equation (see Section 5.4). If the wavelength of light is small compared to the spatial scale over which the index of refraction changes, then the geometric optics approach using Fermat's principle is a good approximation. If these spatial scales become comparable, then it is necessary to account for the wave nature of light. This is called *physical*, or *wave*, *optics*. Fundamentally, visible light is electromagnetic waves within a narrow range of the full electromagnetic spectrum. Electromagnetic waves are governed by Maxwell's equations, which are discussed in the next section.

7.2 Maxwell's Equations of Electromagnetics

An important topic that leads naturally to some of the essential aspects of modern physics to be considered in the next chapter is electromagnetics (or electrodynamics). We consider this important topic in the context of both discrete and continuous systems, involving charged particles and distributed charges, respectively.

Electromagnetism is governed by Maxwell's equations relating electric charges and currents to the coupled electric and magnetic fields, and they account for how an unsteady electric field generates an unsteady magnetic field and vice versa. The four Maxwell equations[1] are

$$\nabla \cdot (\epsilon \mathbf{E}) = \rho_e, \tag{7.5}$$

$$\nabla \cdot \mathbf{B} = 0, \tag{7.6}$$

$$\nabla \times \mathbf{E} = -\frac{\partial \mathbf{B}}{\partial t}, \tag{7.7}$$

$$\nabla \times \left(\frac{1}{\mu}\mathbf{B}\right) = \frac{\partial}{\partial t}(\epsilon \mathbf{E}) + \mathbf{J}. \tag{7.8}$$

Here, $\mathbf{E}(\mathbf{r},t)$ is the electric field intensity, $\mathbf{B}(\mathbf{r},t)$ is the magnetic field intensity, $\rho_e(\mathbf{r},t)$ is the charge density (charge per unit volume), $\mathbf{J}(\mathbf{r},t)$ is the total current density, $\mu(\mathbf{r})$ is the permeability of the medium, and $\epsilon(\mathbf{r})$ is the permittivity of the medium. If the medium is a vacuum, then the electromagnetic properties take on their reference values μ_0 and ϵ_0, which are universal constants such that $c = 1/\sqrt{\mu_0 \epsilon_0}$, where c is the speed of light in a vacuum. Note that the index of refraction in the previous section on optics is related to the electromagnetic properties by $n = \sqrt{\mu\epsilon/\mu_0\epsilon_0}$. The charge density and current density are related through a continuity equation

$$\frac{\partial \rho_e}{\partial t} + \nabla \cdot \mathbf{J} = 0$$

that enforces conservation of charge. Equation (7.5) is called *Gauss's law* and accounts for the relationship between electric charges and the electric fields they produce, namely that the divergence (expansion) of the electric field equals the charge density. Equation (7.6) is called *Gauss's law of magnetism*, which indicates that the magnetic field is divergence free as there is no "magnetic charge" to alter the magnetic field. A divergence free field is such that there is no expansion or contraction of the field. Equation (7.7) is *Faraday's law* and accounts for how an unsteady magnetic field induces an electric field. Equation (7.8) is *Ampere's law*,

[1] In electromagnetics, an electric displacement field $\mathbf{D} = \epsilon \mathbf{E}$ and a magnetic induction field $\mathbf{H} = \mathbf{B}/\mu$ are typically defined in order to highlight the similarities in the two divergence and two curl equations that constitute Maxwell's equations. These are known as *constitutive relations*. We forgo this formulation in order to keep the number of variables to a minimum.

which indicates that a magnetic field can be produced by an unsteady electric field or an electrical current $\mathbf{J}(\mathbf{r},t)$. An alternative form of Maxwell's equations will be derived from Hamilton's variational principle in Section 7.5.

In most applications, the permeability and permittivity of the material can be regarded as known and independent of time. It is only when the electromagnetic fields are very strong, such as in a focused laser, that these properties would be coupled with the field strengths. In general, the permeability and permittivity of the material must be expressed as tensors. However, if the material is isotropic, for which the electromagnetic behavior of the material does not change locally with direction, then μ and ϵ are scalar functions. If, in addition, the medium is homogeneous, that is, the electromagnetic properties do not change with location, then the permeability and permittivity are constants. Under these conditions, Maxwell's equations become

$$\nabla \cdot \mathbf{E} = \frac{\rho_e}{\epsilon}. \tag{7.9}$$

$$\nabla \cdot \mathbf{B} = 0, \tag{7.10}$$

$$\nabla \times \mathbf{E} = -\frac{\partial \mathbf{B}}{\partial t}, \tag{7.11}$$

$$\nabla \times \mathbf{B} = \mu\epsilon \frac{\partial \mathbf{E}}{\partial t} + \mu \mathbf{J}. \tag{7.12}$$

Because the electric and magnetic fields are both vectors, six components are required to specify an electromagnetic field in three dimensions. However, the formulation can be simplified somewhat by defining potential functions. To do so, we take advantage of the fact that the divergence of the curl of any vector is zero. Hence, we can define a vector magnetic potential $\mathbf{A}(\mathbf{r},t)$ according to

$$\mathbf{B} = \nabla \times \mathbf{A}, \tag{7.13}$$

such that Gauss's law of magnetism (7.10) is identically satisfied. Substituting the definition for the magnetic potential (7.13) into Faraday's law (7.11) gives

$$\nabla \times \left(\mathbf{E} + \frac{\partial \mathbf{A}}{\partial t} \right) = 0.$$

Because the curl of the gradient of any scalar is zero, we can define an electric scalar potential $\phi(\mathbf{r},t)$ such that

$$\mathbf{E} + \frac{\partial \mathbf{A}}{\partial t} = -\nabla \phi,$$

where the minus sign is included for convenience. Then, we have

$$\mathbf{E} = -\nabla \phi - \frac{\partial \mathbf{A}}{\partial t}. \tag{7.14}$$

Defining the electric and magnetic potentials according to (7.13) and (7.14) causes Gauss's law of magnetism (7.10) and Faraday's law (7.11) to be identically satisfied, such that they no longer need to be considered. Substitution of these potentials into Gauss's law (7.9) and Ampere's law (7.12) produce the following equivalent form of the Maxwell equations

$$-\nabla^2 \phi - \frac{\partial}{\partial t}(\nabla \cdot \mathbf{A}) = \frac{\rho_e}{\epsilon}, \tag{7.15}$$

and

$$\nabla\left(\nabla\cdot\mathbf{A}+\mu\epsilon\frac{\partial\phi}{\partial t}\right)-\nabla^2\mathbf{A}+\mu\epsilon\frac{\partial^2\mathbf{A}}{\partial t^2}=\mu\mathbf{J}, \tag{7.16}$$

where the vector identity

$$\nabla\times(\nabla\times\mathbf{A})=\nabla(\nabla\cdot\mathbf{A})-\nabla^2\mathbf{A}$$

has been employed. Observe that we have converted the four first-order Maxwell equations into two coupled second-order partial differential equations.

Equations (7.15) and (7.16) can be uncoupled using the following procedure that takes advantage of the *gauge invariance*[2] of Maxwell's equations. Recall that $\mathbf{A}$ has been defined such that its curl is equal to the magnetic field $\mathbf{B}$; however, this only determines its derivatives uniquely, not $\mathbf{A}$ itself. Therefore, we have the freedom to define the divergence of $\mathbf{A}$ in any convenient form. We do so such that the expression in $(\cdot)$ in equation (7.16) vanishes according to

$$\nabla\cdot\mathbf{A}=-\mu\epsilon\frac{\partial\phi}{\partial t}.$$

Substituting into equations (7.15) and (7.16) leads to the alternative uncoupled form of Maxwell's equations

$$\mu\epsilon\frac{\partial^2\phi}{\partial t^2}-\nabla^2\phi=\frac{\rho_e}{\epsilon}, \tag{7.17}$$

$$\mu\epsilon\frac{\partial^2\mathbf{A}}{\partial t^2}-\nabla^2\mathbf{A}=\mu\mathbf{J}. \tag{7.18}$$

Hence, the electric and magnetic potentials satisfy nonhomogeneous wave equations with wave speed $1/\sqrt{\mu\epsilon}$. This formulation of Maxwell's equations only has four components in three dimensions as compared to six for the original form. If the electric and magnetic fields do not change with time, that is, they are steady, then the problem reduces to solving Poisson equations for $\phi(\mathbf{r})$ and $\mathbf{A}(\mathbf{r})$.

Given a specified electric charge density $\rho_e(\mathbf{r},t)$ and total current density $\mathbf{J}(\mathbf{r},t)$, Maxwell's equations (7.9)–(7.12) are partial differential equations that can be solved for the electric field intensity $\mathbf{E}(\mathbf{r},t)$ and the magnetic field intensity $\mathbf{B}(\mathbf{r},t)$. More commonly, the alternative form of Maxwell's equations (7.17) and (7.18) are wave equations that can be solved for the scalar electric potential $\phi(\mathbf{r},t)$ and vector magnetic potential $\mathbf{A}(\mathbf{r},t)$. The relationships (7.13) and (7.14) can then be used to obtain $\mathbf{E}(\mathbf{r},t)$ and $\mathbf{B}(\mathbf{r},t)$.

[2] A *gauge transformation* is of the form

$$\mathbf{A}'=\mathbf{A}-\nabla\chi, \quad \phi'=\phi+\frac{1}{c}\frac{\partial\chi}{\partial t},$$

where χ is a solution of the wave equation. Maxwell's equations are gauge invariant if the transformed equations for $\mathbf{A}'$ and ϕ' are the same as those for $\mathbf{A}$ and ϕ, respectively, which is indeed the case. This gauge invariance allows us to choose the divergence of the magnetic potential $\mathbf{A}$, or $\nabla\cdot\mathbf{A}$, in any form.

7.3 Electromagnetic Wave Equations

If no electric charges or currents are present, in which case $\rho_e = 0$ and $\mathbf{J} = 0$, then the Maxwell equations (7.17) and (7.18) reduce to wave equations for the electric and magnetic potentials. It can also be shown that the electric and magnetic fields $\mathbf{E}$ and $\mathbf{B}$ must satisfy wave equations as well. To do so, observe that Gauss's law (7.9) and Gauss's law of magnetism (7.10) reduce to

$$\nabla \cdot \mathbf{E} = 0, \quad \nabla \cdot \mathbf{B} = 0,$$

which indicates that both the electric and magnetic fields are divergence free. Taking the curl of Faraday's law (7.11) gives

$$\nabla \times \nabla \times \mathbf{E} = -\frac{\partial}{\partial t}(\nabla \times \mathbf{B}).$$

From Ampere's law (7.12), this becomes

$$\nabla \times \nabla \times \mathbf{E} = -\mu\epsilon \frac{\partial^2 \mathbf{E}}{\partial t^2}.$$

Using a vector identity, the left-hand side can be written

$$\nabla \times \nabla \times \mathbf{E} = \nabla(\nabla \cdot \mathbf{E}) - \nabla^2 \mathbf{E}.$$

However, the electric field is divergence free; therefore, the electric field satisfies the wave equation

$$\mu\epsilon \frac{\partial^2 \mathbf{E}}{\partial t^2} - \nabla^2 \mathbf{E} = 0. \tag{7.19}$$

A similar procedure applied to Ampere's law (7.12) leads to the wave equation for the magnetic field

$$\mu\epsilon \frac{\partial^2 \mathbf{B}}{\partial t^2} - \nabla^2 \mathbf{B} = 0. \tag{7.20}$$

The wave speed for both fields is the speed of light in the medium, which is $1/\sqrt{\mu\epsilon}$.

If in addition to there being no electric charges or currents, the fields are steady, then the wave equations become Laplace equations

$$\nabla^2 \mathbf{E} = 0, \quad \nabla^2 \mathbf{B} = 0. \tag{7.21}$$

Similarly, in the alternate formulation (7.17) and (7.18), the electric and magnetic potentials each also satisfy the Laplace equation

$$\nabla^2 \phi = 0, \quad \nabla^2 \mathbf{A} = 0. \tag{7.22}$$

In this context, $\phi(\mathbf{r})$ is known as the electrostatic potential. We then have the Dirichlet problem considered in Section 2.5.3. In this situation, the electric and magnetic field lines are mutually orthogonal throughout the domain.

7.4 Discrete Charged Particles in an Electromagnetic Field

If the charge density $\rho_e(\mathbf{r},t)$ is zero except in small regions of the domain, we can regard each of these regions as a charged particle with charge e such that

$$e = \int_{\bar{V}} \rho_e d\bar{V},$$

where $\bar{V}$ is the volume of the small region with non-zero charge density. In this case, we have a combined discrete and continuous system as the motion of the discrete charged particles are considered within a continuous electromagnetic field.

If the charged particles are free to move throughout an electromagnetic field, we need to supplement Maxwell's equations with equations of motion for the charged particles. Consider a particle of mass m with charge e moving with velocity $\dot{\mathbf{r}}(t)$ in an electric field $\mathbf{E}(\mathbf{r},t)$ and magnetic field $\mathbf{B}(\mathbf{r},t)$, where $\mathbf{r}(t)$ is the position vector of the particle with respect to some coordinate system. The force acting on the charged particle in an electromagnetic field is given by the Lorentz force

$$\mathbf{f}(\mathbf{r},t) = e\left[\mathbf{E}(\mathbf{r},t) + \dot{\mathbf{r}}(t) \times \mathbf{B}(\mathbf{r},t)\right]. \tag{7.23}$$

The first term is the electric force, and the second term is the magnetic force. Note that the electric force on a charged particle is in the same direction as the electric field $\mathbf{E}(\mathbf{r},t)$ and that the magnetic force is perpendicular to both the velocity $\dot{\mathbf{r}}(\mathbf{r},t)$ and the magnetic field $\mathbf{B}(\mathbf{r},t)$.

Observe that because the force owing to the magnetic field depends on the velocity of the particle, it cannot be expressed in terms of a potential energy $V(q_i)$, which can only depend on the position of the particle. Whereas the force owing to the electric field is conservative, therefore, that owing to the magnetic field is nonconservative. As such we must use the expression (4.25) to obtain the generalized force according to

$$Q_i = \mathbf{f} \cdot \frac{\partial \mathbf{r}}{\partial q_i} = e\left[\mathbf{E} + \dot{\mathbf{r}} \times \mathbf{B}\right] \cdot \frac{\partial \mathbf{r}}{\partial q_i}, \quad i = 1, \ldots, n. \tag{7.24}$$

Because we account for all of the forces in the generalized force,[3] the Lagrangian is then

$$L = T - \cancelto{0}{V} = \frac{1}{2} m \dot{\mathbf{r}}^2.$$

The Euler-Lagrange equation of motion for nonconservative forces (4.27) then becomes

$$\frac{\partial L}{\partial q_i} - \frac{d}{dt}\left(\frac{\partial L}{\partial \dot{q}_i}\right) = -e\left[\mathbf{E} + \dot{\mathbf{r}} \times \mathbf{B}\right] \cdot \frac{\partial \mathbf{r}}{\partial q_i}, \quad i = 1, \ldots, n. \tag{7.25}$$

The Lorentz force (7.23) expressed with respect to the electric and magnetic potentials (7.13) and (7.14) is

$$\mathbf{f}(\mathbf{r},t) = e\left[-\nabla\phi - \frac{\partial \mathbf{A}}{\partial t} + \dot{\mathbf{r}} \times (\nabla \times \mathbf{A})\right]. \tag{7.26}$$

[3] Note that we are not really saying that the potential energy is zero; it is not. We are simply accounting for all the forces – even those that contribute to the potential energy – within the generalized force.

With this alternative form of the Lorentz force, the Euler-Lagrange equations are

$$\frac{\partial L}{\partial q_i} - \frac{d}{dt}\left(\frac{\partial L}{\partial \dot{q}_i}\right) = e\left[\nabla\phi + \frac{\partial \mathbf{A}}{\partial t} - \dot{\mathbf{r}} \times (\nabla \times \mathbf{A})\right] \cdot \frac{\partial \mathbf{r}}{\partial q_i}, \quad i = 1,\ldots,n. \tag{7.27}$$

The Euler-Lagrange equations in the form (7.25) or (7.27) can be employed to obtain the equations of motion for a moving charged particle.

Because the magnetic force is nonconservative owing to its dependence on velocity, we would not expect that a Lagrangian of the form $L(t,q_i,\dot{q}_i) = T(q_i,\dot{q}_i) - V(q_i)$ could be defined that fully accounts for all of the forces on a charged particle (this is why we have used the generalized force above). However, it turns out that if we allow for the potential energy to depend on velocity, such that $V = V(q_i,\dot{q}_i)$, it is possible to define a Lagrangian that produces the correct Euler-Lagrange equations of motion. This Lagrangian is

$$L(\mathbf{r},t) = T(\mathbf{r},t) - V(\mathbf{r},t) = \frac{1}{2}m\dot{\mathbf{r}}^2 - e\left[\phi(\mathbf{r},t) - \dot{\mathbf{r}} \cdot \mathbf{A}(\mathbf{r},t)\right], \tag{7.28}$$

where $V = e\phi$ is the potential owing to the electric field, and $V = -e\dot{\mathbf{r}} \cdot \mathbf{A}$ is the velocity-dependent potential owing to the magnetic field. Allowing for a velocity-dependent potential, sometimes called a *generalized potential*, makes it possible to frame electromagnetics within the usual Lagrangian (or Hamiltonian) framework.[4]

Before we show that the Lagrangian (7.28) leads to the correct Euler-Lagrange equations of motion, it is necessary to address a subtle issue that arises because of the fact that we have a mixed discrete and continuous representation of the problem. The framework considered throughout the text thus far is known as a *Lagrangian description* in which the motion of each discrete particle is followed in time as it moves throughout space. Alternatively, it is sometimes advantageous to consider continuous (field) phenomena using what is known as an *Eulerian description*, in which we focus on fixed points in space and describe the motion of the particles as they pass by these fixed points (see, for example, fluid mechanics in Chapter 9). In the case of electromagnetics, the discrete charged particles are moving within a continuous electromagnetic field. Therefore, it is necessary to have a means to translate information between the two descriptions. Consider a generic scalar $\Phi = \Phi(\mathbf{r},t)$ that depends on both position and time. The total differential of the scalar is

$$d\Phi = \frac{\partial \Phi}{\partial t}dt + (\nabla\Phi) \cdot d\mathbf{r}.$$

However, the velocity is $\dot{\mathbf{r}}(t) = d\mathbf{r}/dt$, in which case $d\mathbf{r} = \dot{\mathbf{r}}dt$. Accordingly, we can write the total differential as

$$d\Phi = \frac{\partial \Phi}{\partial t}dt + \dot{\mathbf{r}} \cdot \nabla\Phi dt.$$

Thus, the total derivative of the scalar function, which is the rate of change following the particle, is

$$\frac{d\Phi}{dt} = \frac{\partial \Phi}{\partial t} + \dot{\mathbf{r}} \cdot \nabla\Phi. \tag{7.29}$$

[4] Observe, however, that it is not possible to follow our preferred route in which we form the Lagrangian directly from first principles. Instead, we deduce the Lagrangian that produces the known form of the differential Euler-Lagrange equations.

Essentially, the left-hand side of (7.29) is the Lagrangian derivative, whereas the right-hand side is the Eulerian derivative. The $\partial\Phi/\partial t$ term, called the *local derivative*, accounts for the explicit change of the scalar function with time, and the $\dot{\mathbf{r}}\cdot\nabla\Phi$ term, called the *convective derivative*, accounts for the change in the scalar function owing to the motion of the particle throughout the continuous field. Similarly, the scalar function Φ can be replaced by a vector; for example, if Φ is replaced by the velocity vector, then the total derivative of the velocity with respect to time is the acceleration of the particle. This *total derivative* is sometimes referred to as a *substantial derivative* or *material derivative*. For example, in Cartesian coordinates, with $\mathbf{r} = x\mathbf{i} + y\mathbf{j} + z\mathbf{k}$, the substantial derivative is

$$\frac{d}{dt}\Phi(x,y,z,t) = \frac{\partial\Phi}{\partial t} + \dot{x}\frac{\partial\Phi}{\partial x} + \dot{y}\frac{\partial\Phi}{\partial y} + \dot{z}\frac{\partial\Phi}{\partial z}.$$

Now, let us show that the Lagrangian (7.28) produces the equations of motion for a charged particle moving with velocity $\dot{\mathbf{r}}(t)$ in an electromagnetic field. To illustrate, we consider the x-component of the equation of motion (7.27) in Cartesian coordinates. Thus, the position vector is $\mathbf{r} = x\mathbf{i} + y\mathbf{j} + z\mathbf{k}$, and the magnetic potential is $\mathbf{A} = A_x\mathbf{i} + A_y\mathbf{j} + A_z\mathbf{k}$. Recall that the kinetic energy is $T = \frac{1}{2}m\dot{\mathbf{r}}^2$; therefore, the equation of motion (7.27) with $q_1 = x$, for which $\partial\mathbf{r}/\partial q_1 = \mathbf{i}$, is

$$m\ddot{x} = e\left\{-\frac{\partial\phi}{\partial x} - \frac{\partial A_x}{\partial t} + [\dot{\mathbf{r}}\times(\nabla\times\mathbf{A})]_x\right\},$$

where $[\cdot]_x$ indicates the x-component. Evaluating the x-component of the cross-product yields

$$m\ddot{x} = e\left[-\frac{\partial\phi}{\partial x} - \frac{\partial A_x}{\partial t} + \dot{y}\left(\frac{\partial A_y}{\partial x} - \frac{\partial A_x}{\partial y}\right) - \dot{z}\left(\frac{\partial A_x}{\partial z} - \frac{\partial A_z}{\partial x}\right)\right]. \tag{7.30}$$

We now show that this is the equation of motion that results from application of the Lagrangian (7.28) for electromagnetic fields with $q_1 = x$. The Euler-Lagrange equation of motion for conservative forces with $q_1 = x$ is

$$\frac{\partial L}{\partial x} - \frac{d}{dt}\left(\frac{\partial L}{\partial \dot{x}}\right) = 0.$$

Consider the Lagrangian (7.28), where in Cartesian coordinates

$$\dot{\mathbf{r}}^2 = \dot{\mathbf{r}}\cdot\dot{\mathbf{r}} = \dot{x}^2 + \dot{y}^2 + \dot{z}^2, \quad \dot{\mathbf{r}}\cdot\mathbf{A} = \dot{x}A_x + \dot{y}A_y + \dot{z}A_z.$$

Substituting the Lagrangian into the above Euler-Lagrange equation yields

$$e\left[-\frac{\partial\phi}{\partial x} + \dot{x}\frac{\partial A_x}{\partial x} + \dot{y}\frac{\partial A_y}{\partial x} + \dot{z}\frac{\partial A_z}{\partial x}\right] - \frac{d}{dt}[m\dot{x} + eA_x] = 0.$$

From the total derivative of A_x with respect to time [see equation (7.29) with $\Phi = A_x$], this becomes

$$m\ddot{x} = e\left[-\frac{\partial\phi}{\partial x} - \frac{\partial A_x}{\partial t} + \dot{x}\frac{\partial A_x}{\partial x} + \dot{y}\frac{\partial A_y}{\partial x} + \dot{z}\frac{\partial A_z}{\partial x} - \dot{x}\frac{\partial A_x}{\partial x} - \dot{y}\frac{\partial A_x}{\partial y} - \dot{z}\frac{\partial A_x}{\partial z}\right].$$

Canceling the $\dot{x}\partial A_x/\partial x$ terms and rearranging leads to

$$m\ddot{x} = e\left[-\frac{\partial \phi}{\partial x} - \frac{\partial A_x}{\partial t} + \dot{y}\left(\frac{\partial A_y}{\partial x} - \frac{\partial A_x}{\partial y}\right) - \dot{z}\left(\frac{\partial A_x}{\partial z} - \frac{\partial A_z}{\partial x}\right)\right].$$

Hence, it is confirmed that the equation of motion resulting from the Lagrangian (7.28) does indeed produce the appropriate form (7.30). The same is true for the $q_2 = y$ and $q_3 = z$ equations of motion as well.

EXAMPLE 7.2 Determine the equations of motion of a particle having mass m and charge e in a horizontal plane with a fixed charge e_0 at the origin.

Solution: Let us first solve the problem using the Lagrangian (7.28). With no magnetic field ($\mathbf{A} = 0$), we have

$$L = \frac{1}{2}m\dot{\mathbf{r}}^2 - e\phi. \tag{7.31}$$

In polar coordinates, the position vector and velocity of the charged particle are

$$\mathbf{r} = r\mathbf{e}_r, \quad \dot{\mathbf{r}} = \dot{r}\mathbf{e}_r + r\dot{\theta}\mathbf{e}_\theta. \tag{7.32}$$

From Coulomb's law, the electric field $\mathbf{E}$ owing to a charge e_0 at the origin is

$$\mathbf{E} = \frac{e_0}{4\pi\epsilon_0 r^2}\mathbf{e}_r, \tag{7.33}$$

which like gravity is an inverse-square law. For a steady electric field, equation (7.14) requires that

$$\mathbf{E} = -\nabla\phi.$$

or

$$\mathbf{E} = -\left(\frac{\partial \phi}{\partial r}\mathbf{e}_r + \frac{1}{r}\frac{\partial \phi}{\partial \theta}\mathbf{e}_\theta\right).$$

Because $\mathbf{E} = E(r)\mathbf{e}_r$, we must have

$$\frac{\partial \phi}{\partial r} = -E(r) = -\frac{e_0}{4\pi\epsilon_0 r^2}, \quad \frac{\partial \phi}{\partial \theta} = 0,$$

which when integrated gives the electric potential

$$\phi(r) = \frac{e_0}{4\pi\epsilon_0 r}. \tag{7.34}$$

Substituting equations (7.32) and (7.34) into (7.31), the Lagrangian becomes

$$L = \frac{1}{2}m\left[\dot{r}^2 + (r\dot{\theta})^2\right] - \frac{e_0 e}{4\pi\epsilon_0 r}.$$

Then from the Euler-Lagrange equation of motion for $q_1 = r$

$$\frac{\partial L}{\partial r} - \frac{d}{dt}\left(\frac{\partial L}{\partial \dot{r}}\right) = 0$$

$$mr\dot{\theta}^2 + \frac{e_0 e}{4\pi\epsilon_0 r^2} - \frac{d}{dt}(m\dot{r}) = 0$$

$$m\ddot{r} - mr\dot{\theta}^2 - \frac{e_0 e}{4\pi\epsilon_0 r^2} = 0. \tag{7.35}$$

Likewise, for $q_2 = \theta$

$$\frac{\partial L}{\partial \theta} - \frac{d}{dt}\left(\frac{\partial L}{\partial \dot{\theta}}\right) = 0$$

$$-\frac{d}{dt}(mr^2\dot{\theta}) = 0$$

$$p_\theta = mr^2\dot{\theta} = \text{constant}. \tag{7.36}$$

Thus, the angular momentum p_θ is a constant and is conserved. Note that based on Noether's theorem in Section 4.7, this is expected because θ does not appear in the Lagrangian. Rewriting equation (7.35) in terms of p_θ yields the equation of motion in the form

$$m\ddot{r} - \frac{p_\theta^2}{mr^3} - \frac{e_0 e}{4\pi\epsilon_0 r^2} = 0. \tag{7.37}$$

Hence, given initial conditions for $r(0)$, $\dot{r}(0)$, $\theta(0)$, and $\dot{\theta}(0)$, we can calculate the constant angular momentum from equation (7.36) and solve the equation of motion (7.37) for $r(t)$. Note that $\theta(0)$ merely fixes the starting location of the charge; it does not influence the subsequent dynamics of the charged particle. Charges e_0 and e of the same sign repel one another, while charges of opposite sign will attract.

Alternatively, we can use the equation of motion in the form (7.25) or (7.27). In polar coordinates, the kinetic energy is

$$T = \frac{1}{2}m\dot{\mathbf{r}}^2 = \frac{1}{2}m\left[\dot{r}^2 + (r\dot{\theta})^2\right].$$

From equation (7.25) with no magnetic field ($\mathbf{B} = 0$), we have

$$\frac{\partial T}{\partial q_i} - \frac{d}{dt}\left(\frac{\partial T}{\partial \dot{q}_i}\right) = -e\mathbf{E}\cdot\frac{\partial \mathbf{r}}{\partial q_i}.$$

For $q_1 = r$ with $\mathbf{r}(t)$ given by equation (7.32) and $\mathbf{E}$ given by (7.33), this yields

$$mr\dot{\theta}^2 - \frac{d}{dt}(m\dot{r}) = -e\left(\frac{e_0}{4\pi\epsilon_0 r^2}\mathbf{e}_r\right)\cdot\mathbf{e}_r,$$

or

$$m\ddot{r} - mr\dot{\theta}^2 - \frac{e_0 e}{4\pi\epsilon_0 r^2} = 0,$$

which is the same as equation (7.35). For $q_2 = \theta$, we have

$$-\frac{d}{dt}(mr^2\dot{\theta}) = 0,$$

which results in equation (7.36) as before.

7.5 Continuous Charges in an Electromagnetic Field

Let us now return to the more general case in which the charge is distributed continuously throughout the medium with charge density $\rho_e(\mathbf{r},t)$. The Lorentz force is then

$$\mathbf{f}(\mathbf{r},t) = \rho_e(\mathbf{r},t)\left[\mathbf{E}(\mathbf{r},t) + \dot{\mathbf{r}}(t)\times\mathbf{B}(\mathbf{r},t)\right], \tag{7.38}$$

where now $\mathbf{f}(\mathbf{r},t)$ is the force per unit volume caused by the charge. Similar to equation (7.28), the Lagrangian density is then

$$\mathcal{L}(\mathbf{r},t) = \frac{1}{2}\rho\dot{\mathbf{r}}^2 - \rho_e\left[\phi(\mathbf{r},t) - \dot{\mathbf{r}}\cdot\mathbf{A}(\mathbf{r},t)\right], \tag{7.39}$$

where ρ is the density (mass per unit volume) of the distributed charge. In the previous section, we show how the equations of motion for a moving charged particle follow from application of the Euler-Lagrange equations. The same equations of motion hold for the continuous case with the mass m and charge e replaced by their density counterparts ρ and ρ_e, respectively.

We can also show how the Maxwell equations themselves result from application of Hamilton's principle to a continuous system. Recall that Gauss's law of magnetism (7.10) and Faraday's law (7.11) essentially serve to define the electric field $\mathbf{E}$ and magnetic field $\mathbf{B}$ in terms of the scalar and vector potentials ϕ and $\mathbf{A}$ according to (7.13) and (7.14), respectively. We can then obtain Gauss's law (7.9) and Ampere's law (7.12) from the Lagrangian density

$$\mathcal{L}(\mathbf{r},t) = \frac{1}{2}\left[\mu\epsilon\mathbf{E}(\mathbf{r},t)^2 - \mathbf{B}(\mathbf{r},t)^2\right] - \mu\rho_e\phi(\mathbf{r},t) + \mu\mathbf{J}(\mathbf{r},t)\cdot\mathbf{A}(\mathbf{r},t). \tag{7.40}$$

From the definitions of the scalar and vector potentials (7.13) and (7.14), observe that

$$\mathbf{E}^2 = \mathbf{E}\cdot\mathbf{E} = \left(\nabla\phi + \frac{\partial\mathbf{A}}{\partial t}\right)\cdot\left(\nabla\phi + \frac{\partial\mathbf{A}}{\partial t}\right) = (\nabla\phi)^2 + 2\nabla\phi\cdot\dot{\mathbf{A}} + \dot{\mathbf{A}}^2,$$
$$\mathbf{B}^2 = \mathbf{B}\cdot\mathbf{B} = (\nabla\times\mathbf{A})^2.$$

Although it is not important which terms we regard as the kinetic and potential energies within the Lagrangian density, only that the correct Euler-Lagrange equations result from Hamilton's principle for continuous systems, observe the similarities between the electromagnetic Lagrangian density and that which produces the wave equation in Section 5.4. The square of the electric field $\mathbf{E}$ contains $\dot{\mathbf{A}}^2$, which is analogous to the kinetic energy term in the Lagrangian density for the wave equation. The square of the magnetic field $\mathbf{B}$ contains spatial derivatives of $\mathbf{A}$; therefore, the square of $\mathbf{B}$ resembles the potential energy term in the wave equation Lagrangian density.

Let us obtain the Euler-Lagrange equations for the Lagrangian density (7.40) and show that they lead to Gauss's law and Ampere's law. Substituting into Hamilton's principle for continuous systems

$$\delta\int_{t_1}^{t_2} L dt = \delta\int_{t_1}^{t_2}\int_{\bar{V}} \mathcal{L} d\bar{V} dt = 0$$

gives

$$\delta\int_{t_1}^{t_2}\int_{\bar{V}}\left[\frac{\mu\epsilon}{2}(\nabla\phi)^2 + \mu\epsilon\nabla\phi\cdot\dot{\mathbf{A}} + \frac{\mu\epsilon}{2}\dot{\mathbf{A}}^2 - \frac{1}{2}(\nabla\times\mathbf{A})^2\right.$$
$$\left. - \mu\rho_e\phi + \mu\mathbf{J}\cdot\mathbf{A}\right] d\bar{V} dt = 0,$$

where $\bar{V}$ is the volume of the domain. Taking the variation inside the integral and assuming that ρ_e and $\mathbf{J}$ are specified fields yields

$$\int_{t_1}^{t_2}\int_{\bar{V}}\left[\mu\epsilon\nabla\phi\cdot\delta(\nabla\phi)+\mu\epsilon\dot{\mathbf{A}}\cdot\delta(\nabla\phi)+\mu\epsilon\nabla\phi\cdot\delta\dot{\mathbf{A}}+\mu\epsilon\dot{\mathbf{A}}\cdot\delta\dot{\mathbf{A}}\right.$$
$$\left.-(\nabla\times\mathbf{A})\cdot(\nabla\times\delta\mathbf{A})-\mu\rho_e\delta\phi+\mu\mathbf{J}\cdot\delta\mathbf{A}\right]d\bar{V}dt=0.$$

Integrating the first four terms by parts (for example, see the Dirichlet problem in Section 2.5.3 for the first term) and applying a vector identity to the fifth term leads to

$$\int_{t_1}^{t_2}\int_{\bar{V}}\left\{-\mu\epsilon\nabla^2\phi\delta\phi-\mu\epsilon\frac{\partial}{\partial t}(\nabla\cdot\mathbf{A})\delta\phi-\mu\epsilon\nabla\left(\frac{\partial\phi}{\partial t}\right)\cdot\delta\mathbf{A}-\mu\epsilon\frac{\partial^2\mathbf{A}}{\partial t^2}\cdot\delta\mathbf{A}\right.$$
$$\left.-\left[\nabla(\nabla\cdot\mathbf{A})-\nabla^2\mathbf{A}\right]\cdot\delta\mathbf{A}-\mu\rho_e\delta\phi+\mu\mathbf{J}\cdot\delta\mathbf{A}\right\}d\bar{V}dt=0.$$

Collecting terms, we have

$$\int_{t_1}^{t_2}\int_{\bar{V}}\left\{\left[\mu\epsilon\nabla^2\phi+\mu\epsilon\frac{\partial}{\partial t}(\nabla\cdot\mathbf{A})+\mu\rho_e\right]\delta\phi\right.$$
$$\left.+\left[\nabla(\nabla\cdot\mathbf{A})+\mu\epsilon\nabla\left(\frac{\partial\phi}{\partial t}\right)-\nabla^2\mathbf{A}+\mu\epsilon\frac{\partial^2\mathbf{A}}{\partial t^2}-\mu\mathbf{J}\right]\cdot\delta\mathbf{A}\right\}d\bar{V}dt=0.$$

Therefore, the Euler-Lagrange equations are

$$\nabla^2\phi+\frac{\partial}{\partial t}(\nabla\cdot\mathbf{A})=-\frac{\rho_e}{\epsilon}$$

and

$$\nabla\left(\nabla\cdot\mathbf{A}+\mu\epsilon\frac{\partial\phi}{\partial t}\right)-\nabla^2\mathbf{A}+\mu\epsilon\frac{\partial^2\mathbf{A}}{\partial t^2}=\mu\mathbf{J},$$

which are Gauss's law and Ampere's law expressed in terms of potentials (7.15) and (7.16) as expected. We can, in fact, combine the Lagrangian densities (7.39) and (7.40) such that Gauss's law, Faraday's law, and the equations of motion are all produced from the Euler-Lagrange equations.

Maxwell's equations govern a wide variety of phenomena involving electromagnetic waves throughout the electromagnetic spectrum, including light and optics considered in the previous section. Moreover, they govern phenomena over a wide range of spatial scales ranging from the atomic scale to the galactic scale. Maxwell's equations play an essential role in special relativity theory as they served as the starting point for Einstein's derivation. Therefore, electromagnetism is the bridge between classical mechanics and modern physics to be discussed in the next chapter.

8 Modern Physics

> The mathematician plays a game in which he himself invents the rules, while the physicist plays a game in which the rules are provided by nature, but as time goes on, it becomes increasingly evident that the rules which the mathematician finds interesting are the same as those which nature has chosen.
>
> (Paul A. M. Dirac)

As described in Chapter 5, *classical (Newtonian) mechanics* is centered around Hamilton's principle and the associated Newton's second law, particularly as applied to statics and dynamics of nondeformable (discrete) and deformable (continuous) bodies. Such systems are typically of $O(1)$ size operating at $O(1)$ speeds. *Modern physics* generally refers to branches of physics that have arisen since Einstein's relativity theory in the early part of the twentieth century, including special relativity, general relativity, and quantum mechanics, for example. Modern physics addresses physical phenomena that occur at very small sizes, such as atoms, or very large sizes, such as galaxies, and/or very high speeds close to the speed of light. We focus on the theory of relativity and quantum mechanics, both of which reduce to the classical mechanics limit that has occupied our considerations thus far. Relativity reduces to classical mechanics when $v \ll c$, where c is the speed of light in a vacuum, and v is a characteristic velocity of the system; quantum mechanics reduces to classical mechanics when length scales dominant within the system are much larger than Planck's universal constant $\hbar$.

We distinguish between classical mechanics and modern physics because of the dramatically different methods required to explain and analyze such familiar things as light, matter, and energy and the relations between them in the context of relativity and quantum theories. It is a remarkable example of how scientific observations and the striving for mathematical beauty and simplicity coupled to produce fundamental – and often counterintuitive – theories of the most basic building blocks of our existence. Given the breakdown of the classical approach in explaining certain phenomena observed in electromagnetics, for example, indisputable mathematics was required to unseat traditional thinking through the exposition of mathematical theories of Einstein and others that were so elegant and beautiful that they couldn't be wrong. This being the case despite the fact that many of the consequences of relativity and quantum mechanics have taken decades to confirm.

In view of the fact that the scientific implications of relativity and quantum mechanics are incredibly profound, and they represent extraordinary advances in our understanding of how the physical laws of nature operate, it is remarkable that they can be formulated naturally within the context of the same Hamilton's variational principle that forms the basis for classical mechanics. We explore this connection in the present chapter. As with the other applications of the calculus of variations in this book, many of the details and implications of modern physics are left to dedicated texts on the subject, and we focus on formulating these topics within the context of a unified variational framework.

8.1 Relativistic Mechanics

There is no more recognizable portrait of a scientist than that of Albert Einstein, and there is perhaps no more publicly recognizable equation in physics than his $E = mc^2$. The former is due to the fact that he was a very public figure both in Europe and in the United States – not to mention that he looks like the quintessential scientist; the latter is because this simple algebraic relationship caused a revolution in our understanding of fundamental physics.

In a span of six months during the year 1905, Einstein published four seminal papers, including the formulation of special relativity, early developments in quantum theory, and the basis for nuclear energy – and weapons. The remaining paper was on Brownian motion in statistical physics. While he continued to be a prolific scientist for many years to come, his contributions in 1905 alone led to a new era in physics – *modern physics* – and to a Nobel Prize in Physics.

Einstein's theory of relativity is a wonderful example of the mathematical intuition of a genius leading to a revolutionary physical law. His notions and expectations of mathematical symmetry in physics led to a new physical law in order to rectify mathematical shortcomings of previous laws. This is the case for both *special* and *general relativity*. Special relativity addresses the changes required to treat physics that occurs near the speed of light, and general relativity incorporates the effect of gravity on space and time. Relativity requires a rethinking of the way we view space and time and the transformations between reference frames.

8.1.1 Special Relativity

When we speak of dimensionality in classical mechanics, we are talking about the one, two, or three spatial dimensions. We then consider time dependence separately as necessary. This approach is very much second nature for most scientists and engineers as we have adapted mathematical constructs, such as vectors, to handle the spatial aspects of position, velocity, and so on. We then imagine these vectors to be functions of time for time-dependent systems. In this way, we regard time and space to be fundamentally different. For example, whereas time is always an independent variable, spatial coordinates are considered dependent variables in the case of discrete systems and independent variables in the case of continuous systems.

Similarly, we often execute elaborate transformations on the spatial coordinates in order to simplify problems and facilitate their solution, whereas we typically only shift time. Time is viewed as a fixed independent variable that is common to all systems, and there is no particular need to transform time (other than to shift the start point $t = 0$ to some convenient point in time). With the arrival of electromagnetic theory, however, hints of the limitations of this approach began to be revealed, and Einstein's relativity theory showed that in fact transforming time has fundamental importance in certain settings. Specifically, when velocities approach the speed of light, as in electromagnetic waves, relative times between different coordinate systems, or reference frames, become of paramount importance.

Because coordinate systems are arbitrary and chosen by us for convenience, physical laws should not be dependent on our choice of coordinate system. In other words, the physics of the system remains the same no matter what our vantage point is. For physics of systems having $O(1)$ velocities, this requires that such laws must be invariant to *Galilean transformations*. As velocities approach the speed of light, as in electromagnetics and relativity, this requires the more stringent *Lorentz transformation*.

A Galilean transformation is one in which one reference frame moves with a constant velocity relative to the other. Such coordinate systems are called *inertial frames*. Because the laws of motion in classical mechanics are invariant to a Galilean transformation between two coordinate systems, the behavior and properties of systems do not change under such transformations. For example, this invariance allows one to apply Newton's second law in its usual form to a passenger walking with constant speed down the aisle of a moving airplane and the airplane itself, which is cruising at constant speed.

More generally, consider the transformation from the coordinate system $A(x,y,z,t)$ to $A'(x',y',z',t')$, where A' is moving with constant velocity v in the x-direction relative to A (with the x-axes being collinear and the y- and z-axes remaining parallel). The Galilean transformation for this case is given by

$$x' = x - vt, \quad y' = y, \quad z' = z, \quad t' = t. \tag{8.1}$$

Because the coordinate systems only differ by a constant velocity, the accelerations are the same in both. Therefore, Newton's second law remains the same in both coordinate systems. The Galilean transformation is consistent with our experience with objects that move with $O(1)$ velocities. Note that the transformation from Cartesian to cylindrical coordinates is not a Galilean transformation. Therefore, the governing equations, such as Newton's second law, change form and include "new" inertial terms in cylindrical coordinates (see Section 5.1.1). As velocities increase and become of the same order as the speed of light c, however, problems arise with this viewpoint. The first hints of this came with the formulation of Maxwell's equations in 1864 (see Chapter 7), for which the electromagnetic fields move at speeds close to the speed of light (or at the speed of light in a vacuum). Before this discovery, Galilean invariance, or symmetry, was thought to be a fundamental property of all physical laws; however, Maxwell's equations turned out not to be invariant to such transformations.

Order was restored when it was realized that Maxwell's equations are invariant to a *Lorentz transformation*. Whereas, the Galilean transformation involves a constant velocity between spatial coordinates, the Lorentz transformation is four-dimensional, involving both space and time, so called *spacetime*. The key insight to formulating the Lorentz transformation is that the speed of light does not depend upon the reference frame used, that is, it is a constant that does not depend upon whether the observer is stationary or moving with constant velocity. Because the speed of light is the inherent wave speed found in Maxwell's equations, therefore, the equations remain the same (invariant) as long as the reference frames are not accelerating with respect to one another. This is the case under a Lorentz transformation, but not a Galilean transformation. Although this seems like a subtle point, it signifies that Newton's laws, which are not invariant to a Lorentz transformation, are only approximate and must be adjusted to be symmetric under a Lorentz transformation as is the case with Maxwell's equations.

Einstein took the experimentally verified fact that the speed of light was independent of whether a coordinate system is stationary or moving with constant velocity a step further. He postulated that this meant that all of the laws of physics must be invariant to a Lorentz transformation. In this way, neither of any two coordinate systems moving with constant speed relative to each other can be favored over the other; the laws of physics are precisely the same in both. For example, Maxwell's equations (and their variants) formulated in Chapter 7 are the same in both coordinate systems. This is not the case with Newton's laws of motion, however. Therefore, they do not apply for motions traveling close to the speed of light and must be generalized to account for such relativistic settings.

On our way to obtaining the Lorentz transformation, let us first consider light in two coordinate systems A and A' that are moving with a constant relative velocity to one another. Imagine a light source at the origin of both coordinate systems. The speed of light is the same in both cases; therefore, the light will be observed to travel outward in all directions with speed c. The spherical wave front of the light in A will satisfy

$$(ict)^2 + x^2 + y^2 + z^2 = 0,$$

where $i = \sqrt{-1}$ is the imaginary number. Likewise, the spherical wave front of the light in A' will satisfy

$$(ict')^2 + x'^2 + y'^2 + z'^2 = 0.$$

Just as we define the *metric* $ds^2 = dx^2 + dy^2 + dz^2$ to be an infinitesimal arc length in three-dimensional (x,y,z)-space in classical mechanics, we can define the metric

$$ds^2 = (icdt)^2 + dx^2 + dy^2 + dz^2 = -c^2dt^2 + dx^2 + dy^2 + dz^2$$

to be the arc length in four-dimensional (x,y,z,t)-space in relativity theory. As usual, the metric is written in terms of the space-like arc length ds. In terms of the time-like arc length $d\tau$, where $c^2d\tau^2 = -ds^2$, we have

$$c^2d\tau^2 = -(icdt)^2 - dx^2 - dy^2 - dz^2 = c^2dt^2 - dx^2 - dy^2 - dz^2.$$

The coordinate τ increases along a trajectory of the system, and $t(\tau)$ becomes a dependent variable along with the positions $x(\tau)$, $y(\tau)$, and $z(\tau)$. The variable τ is called the *proper time*, and the curve defined in four-dimensional space by (x,y,z,t) is called the *world line*. The proper time as measured along a world line through spacetime is the actual time that an observer on that trajectory would measure.

For the differential arc length of a world line in the coordinate system A, we have

$$cd\tau = \sqrt{c^2dt^2 - dx^2 - dy^2 - dz^2}.$$

Factoring c^2dt^2 out of the square root yields

$$d\tau = \sqrt{1 - \frac{dx^2 + dy^2 + dz^2}{c^2dt^2}}\,dt.$$

Because the velocity of the particle is given by

$$v^2 = \frac{dx^2 + dy^2 + dz^2}{dt^2},$$

our arc length expression along the world line becomes

$$d\tau = \sqrt{1 - \frac{v^2}{c^2}}\,dt. \tag{8.2}$$

This provides the relationship between proper time and time.

We are now ready to formulate the Lagrangian for the relativistic case. Because the potential energy is only a function of position, it remains the same as in the nonrelativistic case. However, the kinetic energy, which involves velocities, must be reformulated for the relativistic setting. Whereas we denote kinetic energy by T in the classical case, let us use E_k in the relativistic case.

Hamilton's principle is of the usual form

$$\delta \int_{t_1}^{t_2} L dt = 0, \tag{8.3}$$

or

$$\delta \left[\int_{\tau_1}^{\tau_2} E_k d\tau - \int_{t_1}^{t_2} V dt \right] = 0,$$

where τ_1 and τ_2 are the values of τ corresponding to t_1 and t_2, respectively. Observe that we integrate the relativistic kinetic energy with respect to the proper time τ, rather than time t. Because the kinetic energy is a system's energy owing to its motion, it is the difference between its total energy and *rest energy*. This is

$$\delta \left[\int_{\tau_1}^{\tau_2} \left(\frac{mc^2}{\sqrt{1 - \frac{v^2}{c^2}}} - mc^2 \right) d\tau - \int_{t_1}^{t_2} V dt \right] = 0.$$

From the relationship (8.2) between proper time and time, this becomes

$$\delta \left[\int_{t_1}^{t_2} mc^2 \left(\frac{1}{\sqrt{1 - \frac{v^2}{c^2}}} - 1 \right) \sqrt{1 - \frac{v^2}{c^2}}\,dt - \int_{t_1}^{t_2} V dt \right] = 0,$$

or

$$\delta \int_{t_1}^{t_2} \left[mc^2 \left(1 - \sqrt{1 - \frac{v^2}{c^2}} \right) - V \right] dt = 0.$$

Thus, the Lagrangian for relativistic mechanics is

$$L = mc^2\left(1 - \sqrt{1 - \frac{v^2}{c^2}}\right) - V. \tag{8.4}$$

Writing the Lagrangian in terms of generalized coordinates, we have

$$L(t,q_i,\dot{q}_i) = mc^2\left[1 - \sqrt{1 - \frac{1}{c^2}\left(\dot{q}_1^2 + \dot{q}_2^2 + \dot{q}_3^2\right)}\right] - V(q_1,q_2,q_3). \tag{8.5}$$

In order to encompass relativistic mechanics, along with classical mechanics as a special case, Hamilton's principle (8.3) remains of the same form, but the Lagrangian is no longer $L = T - V$. Specifically, the kinetic energy is generalized for relativistic effects. The potential energy is applied for the given setting as before. For example, if the relativistic particle has a charge and is being acted upon by an electromagnetic field, then the potential energy is as given in equation (7.28).

Because Hamilton's principle for the Lagrangian $L(t,q_i,\dot{q}_i)$ is the same as for classical mechanics, the Euler-Lagrange equations of motion are the same as well

$$\frac{\partial L}{\partial q_i} - \frac{d}{dt}\left(\frac{\partial L}{\partial \dot{q}_i}\right) = 0.$$

Evaluating the partial derivatives for the relativistic Lagrangian (8.5), we have

$$\frac{\partial L}{\partial q_i} = -\frac{\partial V}{\partial q_i},$$

and with mc^2 being constant

$$\frac{\partial L}{\partial \dot{q}_i} = -\frac{1}{2}mc^2\left(1 - \frac{v^2}{c^2}\right)^{-1/2}\left(-\frac{2}{c^2}\dot{q}_i\right) = \frac{m\dot{q}_i}{\sqrt{1 - \frac{v^2}{c^2}}}.$$

Hence, the equations of motion for a single particle are

$$\frac{d}{dt}\left(\frac{m\dot{q}_i}{\sqrt{1 - \frac{v^2}{c^2}}}\right) = -\frac{\partial V}{\partial x_i}, \quad i = 1,2,3. \tag{8.6}$$

Alternatively, we can write the equations of motion in terms of the generalized forces $Q_i = -\partial V/\partial x_i$. This is the relativistic form of Newton's second law. Clearly, for $v \ll c$, this reduces to the classical form of Newton's second law.

It is of interest to determine the Hamiltonian for a relativistic particle. Recall from equation (4.46) that the Hamiltonian for an autonomous system, in which time does not appear explicitly in the Lagrangian, is related to the Lagrangian by

$$H = \sum_{i=1}^{n} \dot{q}_i \frac{\partial L}{\partial \dot{q}_i} - L,$$

which still holds in a relativistic context. It is also still the case that the generalized momentum is given by $p_i = \partial L/\partial \dot{q}_i$. From equation (8.4) [or equation (8.5)], we then obtain

$$\begin{aligned} H &= \dot{q}_1 \frac{\partial L}{\partial \dot{q}_1} + \dot{q}_2 \frac{\partial L}{\partial \dot{q}_2} + \dot{q}_3 \frac{\partial L}{\partial \dot{q}_3} - L \\ &= \frac{mv^2}{\sqrt{1-\frac{v^2}{c^2}}} + mc^2\sqrt{1-\frac{v^2}{c^2}} + V \\ &= \frac{m}{\sqrt{1-\frac{v^2}{c^2}}}\left[v^2 + c^2\left(1-\frac{v^2}{c^2}\right)\right] + V \\ H &= \frac{mc^2}{\sqrt{1-\frac{v^2}{c^2}}} + V. \end{aligned}$$

Recalling that the Hamiltonian H is simply the total energy of the system, this is Einstein's famous *mass-energy relation*. In the nonrelativistic case $v \ll c$, and with vanishing potential energy, this gives the familiar form $E = H = mc^2$. Recall that in classical mechanics, the kinetic energy of a particle is only due to its motion. In the relativistic case, however, the particle also has energy simply owing to its mass. In fact, because of the very large c^2 factor, the energy of a particle owing to its mass is typically far larger than that owing to its $O(1)$ velocity if $v \ll c$. In his usual understated fashion, Lanczos (1970) recounts:

> Considering the ease with which the various forms of energy may be transformed into each other, the possibility of converting the latent energy, associated with a given mass, into other energy forms appeared on the horizon. The successful construction of the atom bomb tragically confirmed this conclusion of relativity.

There is a great deal of unnecessary confusion surrounding use of the term *mass* in relativistic settings. The mass m that we use here, often referred to as the *rest mass*, is the same mass that we have used throughout our discussion of classical mechanics. Many authors denote the rest mass by m_0 and talk in terms of a *relativistic mass* given by

$$m = \frac{m_0}{\sqrt{1-\frac{v^2}{c^2}}}.$$

While $m = m_0$ in the classical case when $v \ll c$, the relativistic mass would become very large as the velocity approaches the speed of light c. In fact, it is the energy $E = H$ of the body that becomes very large as v approaches c, not the mass itself, which does not change. Therefore, we use mass (or rest mass) m in the same sense throughout the text, and we do not appeal to a relativistic mass.

Having obtained the relativistic equations of motion, we return to the matter of transformations and reference frames. Recall that nonrelativistic equations of motion are invariant to Galilean transformations, whereas in the relativistic case, a more general Lorentz transformation is required. The Lorentz transformation for the same scenario as considered in equation (8.1) for the Galilean transformation is given by

$$x' = \frac{1}{\sqrt{1-\frac{v^2}{c^2}}}(x - vt), \quad y' = y, \quad z' = z, \quad t' = \frac{1}{\sqrt{1-\frac{v^2}{c^2}}}\left(t - \frac{vx}{c^2}\right).$$

[See Goldstein (1959) for a detailed derivation of the Lorentz transformation.] Observe that if $v \ll c$, then the Lorentz transformation reduces to a Galilean transformation, in which case we are back to the familiar Newtonian (classical) mechanics. The Lorentz transformation reflects the fact that in relativity theory, time is relative such that it passes at different rates depending upon the observer's (coordinate system's) velocity. Inherent in our formulation of relativistic mechanics and the Lorentz transformation is the fact that the velocity v can never be larger than the speed of light c.

Recall that the Euler-Lagrange equations are invariant to any transformation. Therefore, the primary contribution of the more limited Lorentz transformation is to recognize the effect of different reference frames on the advancement of time and relative lengths. Specifically, relativity theory indicates that there is a *time dilation* effect between two moving reference frames, such that a clock on the faster moving reference frame will advance more slowly relative to the slower moving reference frame.[1] Similarly, the length of an object will appear shorter in the direction of its motion, which is known as *length contraction*. These effects are imperceptible at normal speeds.

In the preceding developments, we have taken the approach of Goldstein (1959) and Lanczos (1970) in which we have regarded time as a parameter distinct from the spatial coordinates. Thus, we have not truly taken into account the four-dimensionality inherent in special relativity. Although this approach is not completely faithful to the principles of special relativity, it is more palatable for those of us who primarily deal with classical mechanics applications as we are accustomed to thinking of time as the independent variable, and the spatial coordinates as the dependent variables for systems of discrete particles.

8.1.2 General Relativity

Special relativity represented a significant advance in our understanding of relativistic effects on objects moving at very large velocities. However, a significant issue remained unresolved that occupied Einstein for more than a decade. Forces acting at a distance, such as gravity, rather than by contact proved unsatisfactorily accounted for in special relativity. Einstein extended his theory of special relativity to include gravitational fields leading to the theory of *general relativity* in which the effect of gravity on masses results in a geometric warping of spacetime, which is envisioned to be flat in special relativity. Whereas special relativity unifies space and time into one entity, general relativity unifies space, time, and matter into one entity by providing a geometric theory of gravitation. That is, rather than imagining gravity to be a force acting between masses within a system, it is simply a property of spacetime.

[1] Although his wife does not buy this argument, this is why the author likes to drive fast so as to slow down the aging process little by little: "But dear, my driving so fast makes you look so young!"

In classical mechanics, Newton's laws tell us that masses move along straight lines until acted upon by an external force causing the mass to adjust its path accordingly, and gravity is thought of as just another force. Rather than representing gravity by a force, general relativity represents the effect of gravity on matter through a curving of spacetime. As in special relativity, the trajectory of an object in spacetime is called a world line, but because spacetime is curved, the trajectories in spacetime are *geodesics*. Recall that a geodesic is a curve on a surface having minimum length. In the general relativity context, masses move along geodesics in curved spacetime until acted upon by a force (other than gravity).

As with special relativity, general relativity hinges upon generalizing the transformations with respect to which the laws of physics are invariant. Whereas only uniformly moving reference frames are equivalent through a Lorentz transformation in special relativity, all reference frames are equivalent in general relativity. That is, the laws of physics are the same for all observers. In the parlance of tensors, this requires that general relativity display *covariance*. To fully appreciate the independence of systems to all coordinate system transformations requires the use of tensors, and the curved nature of spacetime demands differential geometry. Because we have not emphasized tensors or differential geometry in this text, we cannot fully exploit these mathematical constructs in articulating general relativity. Without such machinery, we will simply state the basic result and proceed to an example.

General relativity is governed by the *Einstein-Hilbert action* in four-dimensional spacetime

$$S = \iiiint \mathcal{L}\, dxdydzdt = \iiiint \sqrt{g}R\, dxdydzdt, \tag{8.7}$$

where $g = \frac{1}{c^2}\det(g_{\mu\nu})$ is the determinant of the metric tensor $g_{\mu\nu}$, and R is the scalar curvature. Note that double subscripts indicate tensors. The corresponding Euler-Lagrange equation, known as the *Einstein equation*, to be solved for the metric tensor is

$$R_{\mu\nu} - \frac{1}{2}g_{\mu\nu}R = \frac{8\pi G}{c^4}T_{\mu\nu}, \tag{8.8}$$

where $R_{\mu\nu}$ is the Ricci tensor, $T_{\mu\nu}$ is the energy-momentum tensor, and G is Newton's gravitational constant. This tensor equation corresponds to ten coupled nonlinear partial differential equations in the most general case. Einstein's field equations indicate how the geometry of spacetime, given by the left-hand side, is influenced by matter and energy, represented by the right-hand side, by enforcing energy-momentum conservation.

The cosmological consequences of general relativity are vast. They encompass the deflection of light owing to gravity, gravitational redshift, the prediction of black holes, in which spacetime is so distorted that light cannot escape, and the Big Bang and the expanding universe among other remarkable phenomena.

Before considering some of the consequences of general relativity, let us recall the metric introduced in the previous section. In differential geometry, the *metric* relates physical distances within the chosen coordinate system. In classical mechanics, the metric allows us to represent the geometry of space. For example, in three-dimensional Euclidean space, the metric is given by

$$ds^2 = dx^2 + dy^2 + dz^2 = dr^2 + r^2\left(d\theta^2 + \sin^2\theta\, d\phi^2\right),$$

where ds is the differential length along a curve in three-dimensional space. The first representation is in Cartesian coordinates (x,y,z), and the second is in spherical coordinates (r,θ,ϕ). In four-dimensional special relativity, the metric allows us to represent the geometry of spacetime, with the metric being given by

$$ds^2 = (icdt)^2 + dx^2 + dy^2 + dz^2 = -c^2dt^2 + dx^2 + dy^2 + dz^2$$

in Cartesian coordinates, or

$$ds^2 = -c^2dt^2 + dr^2 + r^2\left(d\theta^2 + \sin^2\theta d\phi^2\right)$$

in spherical coordinates.

When applied to general relativity, in which spacetime is curved, the metric becomes much more complicated. We focus on the common case where spacetime is static, in which it does not change with time, and spherically symmetric or isotropic. Under these assumptions, the metric can be written in the general form

$$ds^2 = -A(r)c^2dt^2 + B(r)dr^2 + r^2\left(d\theta^2 + \sin^2\theta d\phi^2\right), \tag{8.9}$$

where $A(r)$ and $B(r)$ are to be found by solving the Einstein equations. For a static, spherically symmetric metric, the Einstein equations (8.8) simplify to

$$R_{\mu\nu} = 0,$$

where the energy-momentum tensor $T_{\mu\nu}$ vanishes if the space outside the mass distribution is empty, for example.

Very soon after Einstein put forth his governing equations for general relativity (8.8), the first exact solution was obtained by Karl Schwarzschild in 1916 for the gravitational field surrounding a nonrotating spherical body.[2] This famous solution, which follows from the static, spherically symmetric form of the equations of general relativity, leads to the prediction of the presence of black holes. By analyzing the components of the Ricci tensor $R_{\mu\nu}$, it can be shown that the unknown functions in the metric (8.9) are of the form

$$A(r) = 1 + \frac{C}{r}, \quad B(r) = \frac{1}{A(r)},$$

where C is an unknown constant. The constant C is determined by considering the limit for which Newtonian mechanics applies, which gives

$$C = -\frac{2Gm}{c^2},$$

where m is the total mass of the body,[3] say a star, and G is the gravitational constant. This leads to the *Schwarzschild metric*

$$ds^2 = -\left(1 - \frac{2Gm}{c^2r}\right)c^2dt^2 + \left(1 - \frac{2Gm}{c^2r}\right)^{-1}dr^2 + r^2\left(d\theta^2 + \sin^2\theta d\phi^2\right). \tag{8.10}$$

[2] The Schwarzschild solution is a good approximation for slowly rotating bodies, such as planets and stars.

[3] Many authors use M for mass in this context because the object is very large (and massive). However, it is the same mass used throughout this chapter.

The corresponding metric tensor for the Schwarzschild metric is then of the form

$$g_{\mu\nu} = \begin{bmatrix} -\left(1-\frac{2Gm}{c^2 r}\right) & 0 & 0 & 0 \\ 0 & \left(1-\frac{2Gm}{c^2 r}\right)^{-1} & 0 & 0 \\ 0 & 0 & r^2 & 0 \\ 0 & 0 & 0 & r^2\sin^2\theta \end{bmatrix}.$$

One of the primary simplifying features of the Schwarzschild solution is that it is not necessary to account for the particular distribution of mass and energy in the body in the form of the energy-momentum tensor in order to determine the spacetime field surrounding the mass.

The first thing we notice about the Schwarzschild metric (8.10) is that the first term becomes singular at $r = 0$, and the second term becomes singular at $r = r_s$, where

$$r_s = \frac{2Gm}{c^2},$$

which is called the *Schwarzschild radius* or *event horizon*. It turns out that the singularity at $r = 0$ is a physical singularity, whereas the singularity at $r = r_s$ is a coordinate singularity. That is, the latter is an artifact of our choice of coordinate system and is not physical. Nevertheless, the Schwarzschild radius has a great deal of physical significance. At $r = 0$, the curvature becomes singular (infinite), and at this point the metric, and spacetime itself, is no longer well-defined. Such solutions are now believed to exist and correspond to black holes.

The Schwarzschild radius r_s only becomes large in comparison to the radius r_0 of the mass close to black holes and other extremely dense objects, such as neutron stars. In all other cases, the effects of general relativity lead to only small corrections to the classical Newtonian mechanics of celestial bodies. For example, the Sun has radius $r_0 = 7 \times 10^5$ km, while its Schwarzschild radius is only $r_s = 3$ km. Similarly, the Earth has radius $r_0 = 6 \times 10^3$ km, while its Schwarzschild radius is only $r_s = 0.44$ cm. Even the radius of a tiny proton is many times larger than its Schwarzschild radius. Despite this apparent small effect of general relativity, it explains important phenomena, such as the gravitational deflection of light rays and the slight precession of the perihelion[4] of planets.

The Lagrangian density in the Einstein-Hilbert action (8.7) for the Schwarzschild metric is given by

$$\mathcal{L}(t,r,\theta,\phi) = -\left(1-\frac{2Gm}{c^2 r}\right)c^2\dot{t}^2 + \left(1-\frac{2Gm}{c^2 r}\right)^{-1}\dot{r}^2 + r^2\dot{\theta}^2. \tag{8.11}$$

Observe that ϕ does not appear in the Lagrangian; therefore, by Noether's theorem, angular momentum $p_\phi = mr^2\dot{\phi}$ is conserved (constant). From the Lagrangian density, one can determine the gravitational potential energy surrounding the body of mass m to be

$$V(r) = -\frac{Gm}{r} - \frac{D}{r^3}, \tag{8.12}$$

[4] Perihelion is the point in a planet's orbit at which it is closest to the Sun.

where the constant D is related to the mass and constant angular momentum. The first term is the familiar gravitational potential owing to classical Newtonian mechanics, and the second term is a small correction owing to general relativistic effects. Note that for large r, r^3 becomes very large rendering the second term extremely small. For example, based on Newtonian mechanics, the planets undergo elliptic orbits around the Sun. However, the effect of the general relativistic term is to modify these orbits such that there is a small shift in the perihelion of the orbit. For the planet Mercury, for example, this results in an adjustment of 43.03″ of arc per century.[5] This small, but observable, effect had been known to astronomers prior to its explanation via general relativity.

When the radius of a mass is much larger than its Schwarzschild radius, that is, $r_0 \gg r_s$, general relativity theory leads to small corrections to classical Newtonian mechanics. When $r_0 < r_s$, however, things get very interesting. Essentially, whenever the radius of a mass is smaller than its Schwarzschild radius, it will form a black hole. When this is the case, the Schwarzschild radius, also known as the event horizon, acts as a barrier between the two regions $r < r_s$ and $r > r_s$, such that nothing occurring within the event horizon can influence happenings outside the event horizon. For example, no light emanating from inside the event horizon can propagate beyond the Schwarzschild radius. This is why it is called a *black hole* as there is no way to "see" what is happening within the event horizon.

It is believed that when a star's fuel is exhausted, the inward gravitational pull of the massive object causes it to collapse to form a *white dwarf*, *neutron star*, or a black hole in the case of sufficiently massive stars. Observe from the gravitational potential (8.12) that the correction for general relativity, that is, the second term, becomes much larger than the classical Newtonian term as r becomes very small. Consequently, the gravitational pull of the collapsing mass is sufficient to crush the mass of an entire star into the size of the proverbial period at the end of this sentence. It is not yet clear what physics prevails as such singularities are approached. One would expect that quantum effects must become important at such small scales.

8.2 Quantum Mechanics

Whereas relativity primarily addresses the very fast and the very large, quantum mechanics addresses the very small. Just as geometric optics is the small wavelength limit of physical (or wave) optics, classical Newtonian mechanics is the limiting case of quantum mechanics when the system dimensions and properties are much larger than Planck's constant. Being the "pixel size" of all matter and processes, Planck's constant is the smallest unit in which trajectories – or world lines in the relativistic case – are traced out. Because it addresses the fundamental nature and behavior of matter on the smallest of scales, quantum mechanics has broad applications in numerous fields in physics and chemistry at the atomic and molecular scales. Relativity and quantum mechanics are independent but not mutually exclusive. Therefore, we can speak of nonrelativistic quantum mechanics or relativistic quantum mechanics, which would account for the very small at very high speeds.

[5] One second, 1″, is one degree divided by 3,600.

The name quantum mechanics refers to the fact that many aspects of matter on the atomic scale are found to exist in *quanta*, or discrete quantities of energy, wavelengths, and so forth. For example, photons and electrons exist in discrete (as opposed to continuous) wavelengths and contain discrete amounts of energy, each of which being related to Planck's universal constant $\hbar$. The primary objective of quantum mechanics is to determine the probabilities that a system, such as the constituent parts of an atom, occupies its possible states and that a transition between states will occur. There are two primary peculiarities of quantum mechanics relative to classical mechanics: 1) the particle-wave duality of light and electromagnetic waves owing to their simultaneous discrete particle and continuous wave behavior, and 2) the Heisenberg uncertainty principle.

In some respects, photons of electromagnetic energy exhibit the properties of particles, while in other respects they behave as waves. For example, electromagnetic energy is carried by discrete photons that have total energy $E = \hbar\omega$, where ω is the frequency of the radiation field. On the other hand, light displays interference and diffraction properties as with other continuous waves. Owing to this particle-wave duality, photons display certain probabilistic characteristics in that the continuous wave nature indicates the probability that a photon will travel along a particular path. That is, the continuous wave solution provides the average behavior of a large number of photons. The probability that a photon is at a particular location is proportional to the intensity of the wave at that location, and the intensity of an electromagnetic wave is proportional to the square of the magnitude of the local electric field. The particle-wave duality of photons of light carries over to electrons, protons, neutrons, and the other subatomic particles that comprise matter.

Heisenberg's uncertainty principle asserts that it is not possible to measure both position and momentum simultaneously to arbitrary precision. This is because any attempt to measure one or the other of these quantities will alter the system such that the outcome of subsequent measurements would be different. In other words, the act of measuring position or momentum will disturb the system in such a way as to lead to different results for other measurements. It is this uncertainty due to measurement that leads to the probabilistic nature of quantum mechanics. Morse and Feshbach (1953) explain that, "The very fact that matter (and electromagnetic radiation, and so on) comes only in finite packages produces an inherent clumsiness in all our experiments which makes the uncertainty principle inevitable."

We focus our attention on nonrelativistic quantum mechanics and leave relativistic quantum mechanics to the more ambitious. We begin with the famous Schrödinger equation, which governs the probability wave function that allows us to determine the probability of a particle being in a certain location. We then briefly discuss *density-functional theory*, which allows for calculation of the electron states for large systems of atoms and molecules, and *Feynman's path-integral formulation of quantum mechanics*.

8.2.1 Schrödinger's Equation

The motion of electrons is described by the Schrödinger equation discovered in 1926, which is the analog of the wave equation obtained from Maxwell's equations that governs photons. The Schrödinger equation governs the *wave function* $\psi(x,y,z,t)$ that provides the continuous distribution of the intensity of the electron wave. The probability that there is an electron at a given point is then the square of the magnitude of the wave function, or equivalently the product of the wave function ψ and its complex conjugate $\bar{\psi}$, evaluated at that position. The wave function comes in the form of eigenfunctions, where the eigenvalues provide the discrete frequencies to which the energy of the electron is proportional, where Planck's constant provides the constant of proportionality.

The wave function is complex, and because the Schrödinger equation will turn out to be linear, the real and imaginary parts are each solutions of the equation independently. For the special case of only one spatial dimension, the Lagrangian density is

$$\mathcal{L} = -\frac{\hbar^2}{2\mu}\frac{\partial \bar{\psi}}{\partial x}\frac{\partial \psi}{\partial x} + \frac{1}{2}i\hbar\left(\bar{\psi}\frac{\partial \psi}{\partial t} - \frac{\partial \bar{\psi}}{\partial t}\psi\right) - V\bar{\psi}\psi, \tag{8.13}$$

where μ is the *reduced mass* of the system, which is the equivalent mass of a point located at the center of gravity of the system and is constant in the nonrelativistic case considered here, and $V(x)$ is the potential energy as usual. Once we have solved for the wave functions $\psi(x,t)$ and $\bar{\psi}(x,t)$, the probability P of an electron being located in the range $a \leq x \leq b$ at time t is then given by

$$P = \int_a^b |\psi(x,t)|^2 dx = \int_a^b \psi(x,t)\bar{\psi}(x,t)dx.$$

Because the total probability for all locations must be unity, we have that the norm of the wave function is unity according to

$$\int_{-\infty}^{\infty} \psi(x,t)\bar{\psi}(x,t)dx = 1.$$

Applying this normalization as a constraint, the augmented Lagrangian density is

$$\tilde{\mathcal{L}}(x,t,\psi,\bar{\psi},\psi_x,\bar{\psi}_x) = -\frac{\hbar^2}{2\mu}\frac{\partial \bar{\psi}}{\partial x}\frac{\partial \psi}{\partial x} + \frac{1}{2}i\hbar\left(\bar{\psi}\frac{\partial \psi}{\partial t} - \frac{\partial \bar{\psi}}{\partial t}\psi\right) - V\bar{\psi}\psi + E\bar{\psi}\psi, \tag{8.14}$$

where $E(x,t)$ is the Lagrange multiplier. The Euler-Lagrange equations for $L = \int_{-\infty}^{\infty}\tilde{\mathcal{L}}dx$ are

$$\frac{\partial L}{\partial \psi} - \frac{\partial}{\partial x}\left(\frac{\partial L}{\partial \psi_x}\right) - \frac{\partial}{\partial t}\left(\frac{\partial L}{\partial \psi_t}\right) = 0, \tag{8.15}$$

and

$$\frac{\partial L}{\partial \bar{\psi}} - \frac{\partial}{\partial x}\left(\frac{\partial L}{\partial \bar{\psi}_x}\right) - \frac{\partial}{\partial t}\left(\frac{\partial L}{\partial \bar{\psi}_t}\right) = 0. \tag{8.16}$$

Consider equation (8.16) with the Lagrangian density (8.14). Evaluating the derivatives gives

$$\frac{\partial L}{\partial \bar{\psi}} = \frac{1}{2}i\hbar\frac{\partial \psi}{\partial t} - V\psi + E\psi, \quad \frac{\partial L}{\partial \bar{\psi}_x} = -\frac{\hbar^2}{2\mu}\frac{\partial \psi}{\partial x}, \quad \frac{\partial L}{\partial \bar{\psi}_t} = -\frac{1}{2}i\hbar\psi,$$

and substituting into equation (8.16) yields

$$\frac{1}{2}i\hbar\frac{\partial \psi}{\partial t} - V\psi + E\psi - \frac{\partial}{\partial x}\left(-\frac{\hbar^2}{2\mu}\frac{\partial \psi}{\partial x}\right) - \frac{\partial}{\partial t}\left(-\frac{1}{2}i\hbar\psi\right) = 0,$$

or

$$-\frac{\hbar^2}{2\mu}\frac{\partial^2\psi}{\partial x^2}+V(x)\psi(x,t)=E\psi(x,t)+i\hbar\frac{\partial\psi}{\partial t}. \tag{8.17}$$

Similarly, equation (8.15) results in

$$-\frac{\hbar^2}{2\mu}\frac{\partial^2\bar{\psi}}{\partial x^2}+V(x)\bar{\psi}(x,t)=E\bar{\psi}(x,t)-i\hbar\frac{\partial\bar{\psi}}{\partial t}. \tag{8.18}$$

Equations (8.17) and (8.18) are each Schrödinger equations. We only need to solve one of these as the other is simply its complex conjugate. The first term of Schrödinger's equation results from the kinetic energy. Observe that we have an unsteady eigenproblem in which the Lagrange multiplier E is the eigenvalue and $\psi(x,t)$ [or $\bar{\psi}(x,t)$] is the eigenfunction.

In the general three-dimensional case, the Lagrangian density becomes

$$\mathcal{L}=-\frac{\hbar^2}{2\mu}(\nabla\bar{\psi})\cdot(\nabla\psi)+\frac{1}{2}i\hbar\left(\bar{\psi}\frac{\partial\psi}{\partial t}-\frac{\partial\bar{\psi}}{\partial t}\psi\right)-V\bar{\psi}\psi, \tag{8.19}$$

for which the Euler-Lagrange, or Schrödinger, equation is

$$-\frac{\hbar^2}{2\mu}\nabla^2\psi+V\psi=E\psi+i\hbar\frac{\partial\psi}{\partial t}. \tag{8.20}$$

The most compact form of Schrödinger's equation is in terms of the Hamiltonian, which is

$$H(p_i,q_i)=T+V=\frac{1}{2}\mu\left(\dot{x}^2+\dot{y}^2+\dot{z}^2\right)+V=\frac{1}{2\mu}\left(p_1^2+p_2^2+p_3^2\right)+V(q_1,q_2,q_3),$$

where $p_1=\partial L/\partial\dot{q}_1=\mu\dot{q}_1$, for example. The Schrödinger equation is then

$$H\left(\frac{\hbar}{i}\frac{\partial}{\partial q_i},q_i\right)\psi(q_i)=E\psi(q_i)+i\hbar\frac{\partial\psi}{\partial t}.$$

This notation indicates that p_i is replaced by the operator $\frac{\hbar}{i}\frac{\partial}{\partial q_i}$ in the Hamiltonian, and it operates on $\psi(q_i)$. Substituting the Hamiltonian yields

$$\left[-\frac{\hbar^2}{2\mu}\left(\frac{\partial^2}{\partial {q_1}^2}+\frac{\partial^2}{\partial {q_2}^2}+\frac{\partial^2}{\partial {q_3}^2}\right)+V\right]\psi=E\psi+i\hbar\frac{\partial\psi}{\partial t},$$

or

$$-\frac{\hbar^2}{2\mu}\left(\frac{\partial^2\psi}{\partial {q_1}^2}+\frac{\partial^2\psi}{\partial {q_2}^2}+\frac{\partial^2\psi}{\partial {q_3}^2}\right)+(V-E)\psi=i\hbar\frac{\partial\psi}{\partial t}.$$

Finally, in terms of the Laplacian operator, we have

$$-\frac{\hbar^2}{2\mu}\nabla^2\psi+(V-E)\psi=i\hbar\frac{\partial\psi}{\partial t},$$

which is the same as equation (8.20). By deriving Schrödinger's equation from a variational principle, it is straightforward to incorporate other physics, such as electromagnetic fields, by inserting the appropriate expression for the potential energy V.

In order to illustrate solutions of the Schrödinger equation, let us consider the case for which it was originally developed – the hydrogen atom. Because there is only a single electron orbiting a single proton, the hydrogen atom can be modeled using the Coulomb potential

$$V(r) = -\frac{e^2}{4\pi\epsilon_0 r}, \tag{8.21}$$

where e is the charge of an electron. Recall that ϵ_0 is the permittivity of a vacuum. Because of the spherical geometry, we write the Schrödinger equation in spherical coordinates (r,θ,ϕ) with the proton located at the origin. The Laplacian operator in spherical coordinates is

$$\nabla^2 = \frac{1}{r^2}\frac{\partial}{\partial r}\left(r^2\frac{\partial}{\partial r}\right) + \frac{1}{r^2\sin\theta}\frac{\partial}{\partial\theta}\left(\sin\theta\frac{\partial}{\partial\theta}\right) + \frac{1}{r^2\sin^2\theta}\frac{\partial^2}{\partial\phi^2}.$$

The steady Schrödinger equation (8.20) in spherical coordinates, with the potential energy given by equation (8.21), is then

$$-\frac{\hbar^2}{2\mu}\frac{1}{r^2\sin\theta}\left[\sin\theta\frac{\partial}{\partial r}\left(r^2\frac{\partial\psi}{\partial r}\right) + \frac{\partial}{\partial\theta}\left(\sin\theta\frac{\partial\psi}{\partial\theta}\right) + \frac{1}{\sin\theta}\frac{\partial^2\psi}{\partial\phi^2}\right] - \frac{e^2}{4\pi\epsilon_0 r}\psi = E\psi. \tag{8.22}$$

This form of the Schrödinger equation can be solved using the method of separation of variables, which is beyond the scope of this text, so we will simply summarize the results. Separation of variables allows us to convert the single partial differential equation (8.22) into three ordinary differential eigenproblems for the three coordinate directions.

We find that the state of the hydrogen atom is determined by three integer quantum numbers known as the *principal*, *orbital* or *angular momentum*, and *magnetic quantum numbers*. The corresponding eigenfunctions ψ_n that result from the method of separation of variables provide the probability density functions from which the probability P is obtained for the location of the electron. The discrete energy levels of the hydrogen atom E_n only depend upon the principal quantum number n and are given by

$$E_n = \frac{1}{n^2}E_1, \quad n = 1,2,\ldots,\infty,$$

where

$$E_1 = -\frac{\mu}{32}\left(\frac{e^2}{\pi\epsilon_0\hbar}\right)^2 = -13.598 \text{ eV},$$

where the units are electron volts. Here, the reduced mass μ is related to the masses of the proton and electron, m_e and m_p, respectively, as follows

$$\mu = \frac{m_e m_p}{m_e + m_p}.$$

When the electron moves from an energy level E_l to E_m, a photon of energy $E = E_l - E_m$ is given off. The lowest energy E_1 is known as the *ionization energy*, and the theoretical value provided above has been confirmed by experiments. Observe that the energy levels are all negative, and the difference between respective energy levels decreases until reaching $E_\infty = 0$. For more complex atoms, the energy levels depend upon all three quantum numbers, not just the principal one as for the hydrogen atom.

The knowledge gained from solutions of the Schrödinger equation for the hydrogen atom or more complex elements can be used to identify the elements comprising distant stars, for example. When a series of photons having a range of frequencies passes through an atom, photons of particular frequencies can be absorbed by the atom causing the electron(s) to be excited to a higher energy level. When the electron(s) returns to a lower energy state, a photon of energy is released with a frequency determined by the difference in energy states. Because each element has a different sequence of discrete energy levels that electrons occupy, the frequency of the photons released by a substance uniquely identifies the elements that comprise the substance. This is called the *emission and absorption spectrum*. This procedure allows us to deduce the chemical make up of substances at great distances simply by examining the electromagnetic spectrum emitted by the object. For example, our Sun is found to contain primarily hydrogen (91.2%) and helium (8.7%) atoms.

In so far as all matter is comprised of atoms and their constituent subatomic particles, all matter behaves according to quantum theory. Thankfully, however, for most $O(1)$-sized systems that are of everyday interest, these quantum effects are extremely small and can be neglected. Attempts to articulate a relativistic theory of quantum mechanics led to replacement of the Schrödinger equation with the Klein-Gordon and Dirac equations. However, these were not entirely consistent leading to the development of *quantum field theory*. Unlike relativity theory for particles, this approach applies quantum mechanics to continuous fields. Eventually, quantum field theories for the strong and weak nuclear forces were developed. However, gravity has not been fully explained within a quantum mechanical framework pointing to the need for an even deeper explanation.

8.2.2 Density-Functional Theory

The Schrödinger equation has been very successful in providing solutions for systems involving small numbers of electrons. However, it does not scale efficiently to large systems of electrons, such as required for determining the atomic structure of condensed matter and solid states. This limitation of the Schrödinger equation led to development of *density-functional theory*, which is based on two Hohenberg-Kohn theorems articulated in 1964 that established a variational principle governing the electron density. Density-functional theory is used for first-principles calculations of the electronic structure of many-body systems of atoms, molecules, and condensed phases in physics and chemistry. It is commonly used in calculations of solid state and condensed matter physics.

Rather than solving for $3N$ three-dimensional spatial coordinates of N electrons as in the Schrödinger equation, density-functional theory is formulated in terms of a nonlinear field equation in three dimensions for the continuous electron density. To do so, the wave function ψ in the Schrödinger equation is integrated to form a spatially dependent electron density. The total energy of a condensed system is a functional of its electron density that is to be minimized. This leads to a nonlinear differential equation for the optimal arrangement of atoms in a solid that provides a more efficient alternative than solving the Schrödinger equation.

Density-functional theory remains an active area of research, and its scope of applications continues to expand in physics, chemistry, and materials science. For in-depth treatments of this important tool in modern physics and chemistry, see Parr and Yang (1989), Dreizler and Gross (1990), and Koch and Holthausen (2001).

8.2.3 Feynman Path-Integral Formulation of Quantum Mechanics

As applied in classical mechanics, Hamilton's principle gives one the impression that a particle must consider all possible paths between the initial and final states before determining which one is the stationary path. Clearly, this is not what actually happens as the particle somehow seems to know which path is stationary without trying all of the alternatives. In 1948, Richard Feynman reformulated quantum mechanics in such a way as to show that this really is what happens at the quantum level. *Feynman's path-integral* is an elegant and concise reformulation of quantum mechanics in terms of an integral that sums over the classical Lagrangians of all possible paths to determine a probability amplitude K.

The Feynman path-integral is given by

$$K(b,a) = \int_a^b \exp\left(\frac{i}{\hbar}S[x(t)]\right)\mathcal{D}x(t),$$

where $K(b,a)$ is called the *propagator* from the point (x_a,t_a) to the point (x_b,t_b), and the action is given by

$$S[x(t)] = \int_0^{t_f} L[x(t),\dot{x}(t)]dt,$$

where $t=0$ and $t=t_f$ correspond to the starting and ending points a and b, respectively. In the path-integral, it is necessary to employ a new notation, $\mathcal{D}x(t)$, to indicate that the integration is to be carried out over all possible paths $x(t)$ between a and b. The Lagrangian $L[x(t),\dot{x}(t)]$ is the classical Lagrangian for the scenario under consideration. The exponential involving the action acts as a weight function such that paths near the stationary action provide a larger contribution to the path-integral.

We imagine each path contributing to a complex probability amplitude that is related to the integrated action along the corresponding paths. The square of the modulus of the complex probability amplitude gives the probability of the system following that path. Summed up through integration, the distribution of probability amplitudes then provides the likelihood of each outcome occurring. In a quantum mechanical sense, therefore, the system really does "try" all possible paths in order to determine the likelihood of each actually occurring. Thus, in a sense, Hamilton's principle has anticipated the more fundamental quantum explanation of how systems evolve.

Because the probability amplitudes are complex, they consist of a modulus (length) and a phase. While the modulus of each path does not differ significantly, the phase for each path can, and it is this phase difference that most contributes to the probability amplitude for each possible path. The phase is determined by the action along each path. Whereas actions for very different paths have vastly different phases that tend to cancel each other through destructive interference, actions for paths that are very close to one another constructively interfere and contribute the most to the probability amplitude. Because the action is normalized by Planck's constant, $\hbar$ is the "pixel size" that determines how different the actions of various paths are.

The nature of the probability amplitude distribution is determined by the magnitude of the action $S[x]$ relative to Planck's universal constant $\hbar$. When $S[x]$ is extremely small and of the same order as Planck's constant, the contribution from a set of paths near the stationary path to the amplitude is significant leading to a broader range of paths with comparable amplitude. When $S[x] \gg \hbar$, we reach the classical limit in which the path having stationary action dominates the path-integral. This occurs because any path away from the stationary one leads to rapid fluctuations in the exponential, and the contributions to the path-integral cancel out (except for that owing to the stationary path).

Although the Feynman path-integral is not a variational principle, it does involve the Lagrangian, and it reduces to Hamilton's variational principle in the classical limit. In addition, the Schrödinger equation can be derived formally from the path-integral formulation of quantum mechanics; hence, the two formulations are equivalent, and the wave function $\psi(x,t)$ in the Schrödinger equation can be related to the probability amplitude $K(b,a)$ in the path-integral formulation. Not only is the path-integral formulation equivalent to other formulations of quantum mechanics, it provides a very intuitive framework for treating quantum theory. For more on path-integrals, see Feynman and Hibbs (1965), Schulman (1981), and Basdevant (2007).

We have only touched on the revolutionary theories in modern physics developed throughout the twentieth century. For more, see Morse and Feshback (1953), Dirac (1958), Yourgrau and Mandelstam (1968), Nesbet (2003), Basdevant (2007), and the numerous other texts on modern physics and its relativistic and quantum theories.

9 Fluid Mechanics

> I am an old man now, and when I die and go to heaven there are two matters on which I hope for enlightenment. One is quantum electrodynamics, and the other is the turbulent motion of fluids. And about the former I am rather optimistic.
>
> (Sir Horace Lamb)

Fluid mechanics is one of the fields that is contributing significantly to the current resurgence of interest in variational methods. Historically, calculus of variations has not been taught or utilized as a core mathematical tool in the arsenal of the fluid mechanics practitioner; it is a rare fluid mechanics textbook that even mentions variational calculus. However, recent research developments are revolutionizing the field by reframing certain fluid mechanics phenomena within a variational framework. This is particularly the case in the areas of *flow control* and *hydrodynamic stability*.

Traditionally, flow control has been implemented in an ad hoc manner involving a significant amount of trial and error within an empirical (experimental or numerical) framework. Beginning in the 1990s, flow control is increasingly being framed within the context of optimal control theory. As will be seen in Chapter 10, the solution to optimal control formulations, particularly for large problems involving many degrees of freedom, is very computationally intensive. It is only recently that the computational resources have become available that are capable of solving realistic fluid mechanics scenarios within such an optimal control framework.

Similarly, significant advances are being made in our understanding of stability of fluid flows through the application of *transient growth analysis* that seeks the "optimal," or most unstable, initial perturbation (disturbance) that results in the greatest growth of the instability. This approach follows from framing the stability problem within a variational framework. Transient growth stability analysis is taken up in Section 9.5.4 (see also Section 6.5).

Before undertaking a development of these more recent topics in the application of variational methods to fluid mechanics, we derive the governing equations of incompressible fluid mechanics from Hamilton's principle. Bernoulli's equation, which is discussed in Section 9.2.1, follows from treating a single fluid particle as a discrete, nondeformable system and applying Hamilton's principle accordingly. The inviscid Euler and viscous Navier-Stokes equations follow from treating the fluid as a continuous (deformable) medium and again applying the appropriate form of Hamilton's principle. The Euler equations for inviscid flow and the Navier-Stokes equations for viscous flow are derived in Sections 9.2.2 and 9.3, respectively. The chapter closes with an introduction to transient growth stability analysis in Section 9.5.4 and a brief discussion of flow control.

Variational methods are also utilized in aerodynamic shape optimization. Shape optimization and optimal control theory will be addressed in a more general context in Chapter 10. Furthermore, numerical grid generation, which is treated in Chapter 12, is an important issue in computational fluid dynamics (CFD).

9.1 Introduction

Fluid mechanics is a fascinating topic that encompasses numerous applications in many flow regimes. Aerodynamics, for example, spans the full set of flow regimes from incompressible at low speeds, such as automobiles, compressible at moderate speeds, such as commercial aircraft, supersonic at high speeds, such as fighter aircraft, to hypersonic at extremely high speeds, such as space shuttle reentry. The inherently nonlinear nature of fluids leads to a wide variety of physical phenomena and stability behaviors. In this section we introduce some essential concepts and terminology that are unique to fluid mechanics and related fields.

Fluids consist of liquids and gases, which are both governed by the same underlying physics. Whereas solids experience a fixed deformation, or strain, under the action of a shear stress, a fluid cannot support such a shear stress. Instead, a fluid experiences a continuous deformation, or strain rate.

While it is possible to treat a fluid in terms of the discrete atoms and molecules fundamental to all matter, it is advantageous under most conditions to consider a fluid as a continuously varying deformable substance. This viewpoint is known as *continuum mechanics* and has been implicit in our description of elasticity in Section 5.3.3 and electromagnetics in Sections 7.2 to 7.5. In the microscopic approach, the motion of each individual discrete molecule is considered using Newtonian mechanics. The advantage to this approach, which is the basis for molecular dynamics simulations, is that it leads to solving linear equations for the motion of the particles. The primary difficulties are that a collision model is required, and it is necessary to follow each of the prohibitively large number of individual molecules. This approach is used for rarefied gas dynamics, in which the gas molecules are far apart, and modeling of nano- and micro-flows, in which the size of the system is comparable to the distance between molecules. The macroscopic continuum approach considers the motion of fluid "particles" that each contain many microscopic molecules having properties, such as density, viscosity, velocity, and pressure, that are averaged over many such microscopic particles to produce a smoothly varying, or continuous, distribution. In this way, properties are represented as continuous functions of space and time. While it is no longer necessary to consider the motion of each and every atom and/or molecule, the equations are nonlinear and very difficult to solve. The continuum approach is valid for situations in which the size of the system is very large compared to the distance between molecules, which is the case for the vast majority of engineering applications. Therefore, this is the approach taken here.

Within the continuum approximation, there are two different approaches to describing the motion of the fluid particles. The *Lagrangian description* requires following the motion of each individual fluid particle (but not the many atoms and molecules that comprise each particle). This is the approach taken throughout this text with its emphasis on Lagrangian mechanics and the associated Euler-Lagrange equations of motion. In the Lagrangian formulation, the dependent variables are functions of the initial positions of the particles and time, and all flows are unsteady within the context of the Lagrangian description. In the *Eulerian description*, the focus is on fixed points in space, and the motion of the fluid particles are described as they move past these fixed locations in space. In the Eulerian description, the dependent variables, such as velocity and pressure, are functions of the current locations of the fluid particles and time.

Recall from Section 7.4 that the time derivative of a scalar or vector quantity in a Lagrangian reference frame is related to that in the Eulerian description through the *substantial*, or *material*, *derivative*. In the context of fluid mechanics, we apply the substantial derivative to the velocity vector, which in three-dimensional Cartesian coordinates is

$$\dot{\mathbf{r}} = \mathbf{v} = u\mathbf{i} + v\mathbf{j} + w\mathbf{k},$$

where $u(x,y,z,t)$, $v(x,y,z,t)$, and $w(x,y,z,t)$ are the velocity components in the x-, y-, and z-direction, respectively. Confusingly, the positions $x(t)$, $y(t)$, and $z(t)$ are dependent variables in the Lagrangian description, whereas they are independent variables indicating the fixed spatial locations in the Eulerian reference frame. Similarly, in the Lagrangian reference frame, $\dot{x} = u$, $\dot{y} = v$, and $\dot{z} = w$ are the time rates of change of the positions, that is, the velocities. The derivative of the velocity vector with respect to time is the acceleration of the Lagrangian fluid particle. Applying equation (7.29) accordingly yields the substantial derivative

$$\frac{d\mathbf{v}}{dt} = \frac{\partial \mathbf{v}}{\partial t} + (\mathbf{v} \cdot \nabla)\mathbf{v}. \tag{9.1}$$

The $\partial\mathbf{v}/\partial t$ term is called the *local acceleration* and accounts for the explicit change of the velocity with time at a fixed position, and the $(\mathbf{v} \cdot \nabla)\mathbf{v}$ term is the *convective acceleration* and accounts for the change in the velocity owing to the change of position of the fluid particle as it moves throughout the continuous velocity field. In Cartesian coordinates, the substantial derivative is

$$\frac{d\mathbf{v}}{dt} = \frac{\partial \mathbf{v}}{\partial t} + u\frac{\partial \mathbf{v}}{\partial x} + v\frac{\partial \mathbf{v}}{\partial y} + w\frac{\partial \mathbf{v}}{\partial z}. \tag{9.2}$$

Note that in fluid mechanics, the notation $D\mathbf{v}/Dt$ is often used for the substantial derivative instead of $d\mathbf{v}/dt$ in order to prevent confusion regarding what is meant by the Lagrangian derivative following the fluid particle.

The expansion or compression of a vector field is quantified by the vector divergence operator. A fluid is regarded as *incompressible* if the divergence of the velocity field is zero, that is,

$$\nabla \cdot \mathbf{v} = 0. \tag{9.3}$$

If a fluid is incompressible, then its density ρ, or mass per unit volume, will be a constant throughout the flow field. Liquids and low speed gases can accurately be assumed to be incompressible, which will be the focus of our discussion here. In Cartesian coordinates, the zero divergence condition becomes

$$\frac{\partial u}{\partial x}+\frac{\partial v}{\partial y}+\frac{\partial w}{\partial z}=0. \tag{9.4}$$

Equations (9.3) and (9.4) are known as the *continuity equation* and enforce conservation of mass for each fluid particle.

In the two-dimensional case, it is convenient to define the *streamfunction* $\psi(x,y)$ by

$$u=\frac{\partial \psi}{\partial y}, \quad v=-\frac{\partial \psi}{\partial x}, \tag{9.5}$$

such that the continuity equation for incompressible flow is identically satisfied. In other words, defining the streamfunction in this manner effectively replaces the continuity equation. *Streamlines* are contours of constant streamfunction. Consider the total differential of the constant streamfunction $\psi(x,y)$ along a streamline

$$d\psi=\frac{\partial \psi}{\partial x}dx+\frac{\partial \psi}{\partial y}dy=0;$$

therefore, the slope of the streamline is

$$\frac{dy}{dx}=\frac{-\partial\psi/\partial x}{\partial\psi/\partial y}=\frac{v}{u},$$

where the ratio of velocity components is the slope of the velocity vector. Thus, the streamlines of a flow field are everywhere tangent to the velocity vector, thereby indicating local flow direction. It can also be shown that the difference between the values of ψ for any two streamlines is equal to the volume flow rate between them. Therefore, one can infer the relative magnitudes of velocity between streamlines as they converge or diverge in the flow.

9.2 Inviscid Flow

All fluids are *viscous* as they are subject to frictional forces between the moving fluid particles. However, the property that indicates how viscous a fluid is, the dynamic viscosity μ, is typically very small for most fluids. Hence, it seems reasonable to neglect such viscous effects. We call such fluids *inviscid*. We first consider each fluid particle to be a nondeformable body moving throughout a steady, incompressible, inviscid fluid, which leads to the *Bernoulli equation*. Allowing for deformation of the fluid particle under the action of the surrounding fluid pressure leads to the *Euler equations* for inviscid flow. Finally, we reformulate the steady, incompressible, inviscid flow problem in terms of a potential and a streamfunction analogous to the formulation of electromagnetics in Section 7.3 in terms of potential functions. This leads to *potential flow theory*.

9.2.1 Fluid Particles as Nondeformable Bodies – Bernoulli Equation

We begin by considering a steady, nondeformable differential fluid element. If the fluid behaves in an inviscid manner with zero viscosity (friction), the fluid element is only subject to normal stresses owing to pressure, which is taken as positive acting on the fluid element. In two-dimensional Cartesian coordinates, the position vector of the fluid element is

$$\mathbf{r} = x\mathbf{i} + y\mathbf{j},$$

and the velocity is

$$\dot{\mathbf{r}} = \dot{x}\mathbf{i} + \dot{y}\mathbf{j} = u\mathbf{i} + v\mathbf{j}.$$

The kinetic energy density is then

$$\mathcal{T} = \frac{1}{2}\rho\dot{\mathbf{r}}^2 = \frac{1}{2}\rho\mathbf{v}^2,$$

where $\dot{\mathbf{r}}^2 = \mathbf{v}^2 = (u\mathbf{i} + v\mathbf{j}) \cdot (u\mathbf{i} + v\mathbf{j}) = u^2 + v^2$ is the square of the magnitude of the velocity of the fluid element, and ρ is the density of the fluid, which is constant for incompressible flows. The potential energy density is

$$\mathcal{V} = p + \rho g y.$$

The first term is due to *flow work*, which is the work per unit volume done by the pressure in moving a fluid particle, and the second term is due to gravitational acceleration in the y-direction. The Hamiltonian density for the fluid particle is

$$\mathcal{H} = \mathcal{T} + \mathcal{V} = \frac{1}{2}\rho\mathbf{v}^2 + p + \rho g y.$$

Recall that for conservative systems, the Hamiltonian is a constant of the particle's motion. Therefore, we have

$$\frac{\mathbf{v}^2}{2} + \frac{p}{\rho} + gy = \frac{\mathcal{H}}{\rho} = \text{constant}. \tag{9.6}$$

This is the famous *Bernoulli equation*, which enforces conservation of mechanical energy of a nondeformable fluid particle in a steady, inviscid, incompressible flow, in which the movement of the fluid particle is due to the action of gravitational and pressure forces. Because we are following a fluid particle in a Lagrangian sense, Bernoulli's equation applies along a *streamline*, which is the path of a fluid particle in a steady flow. We can apply Bernoulli's equation to any fluid particle moving along its streamline, but the constant $\mathcal{H}/\rho$ will be different for each streamline.

The simplicity of Bernoulli's equation belies its usefulness. For example, observe that a fluid particle at constant elevation y that experiences an increase in velocity must undergo a corresponding decrease in pressure. This fact alone explains numerous phenomena in incompressible fluid mechanics.

9.2.2 Fluid Particles as Deformable Bodies – Euler Equations

Considering the fluid particle to be nondeformable as in the previous section means that pressure only serves to move the fluid particles around in the flow field. If we also allow the fluid particles to deform owing to the action of the local pressure field, then the pressure moves the particles and deforms them.

Recall from Section 4.2 that Hamilton's principle applied to continuous systems is typically applied to the system as a whole. The Lagrangian density $\mathcal{L} = \mathcal{T} - \mathcal{V}$ is formed for each differential element of the system and summed through integration. This approach is preferred as it has the advantage that it is not necessary to explicitly account for internal forces acting between differential elements (see, for example, Seliger and Whitham 1968 and Salmon 1988). However, in some instances, particularly continuous systems with nonconservative forces acting on them, it is advantageous to consider a single generic differential element of the overall system being careful to include all internal forces acting on the differential element owing to surrounding elements. This latter approach is applied here to fluid mechanics.

In anticipation of including shear forces owing to viscous effects, which are nonconservative, in Section 9.3 we will include all surface forces as generalized forces in Hamilton's principle applied to a differential element having volume $d\bar{V}$. Recalling Hamilton's principle of the form (4.27) with (4.25), and writing in terms of the Lagrangian density, gives

$$\frac{\partial \mathcal{L}}{\partial q_i} - \frac{d}{dt}\left(\frac{\partial \mathcal{L}}{\partial \dot{q}_i}\right) = -\frac{Q_i}{d\bar{V}} = -\mathbf{f} \cdot \frac{\partial \mathbf{r}}{\partial q_i}, \quad i = 1, \ldots n, \tag{9.7}$$

where here $\mathbf{f}$ is the force per unit volume acting on the differential element. The Lagrangian density is

$$\mathcal{L} = \mathcal{T} - \mathcal{V} = \frac{1}{2}\rho v^2 - \rho g y,$$

where we have not included the pressure as it will be included with the other surface forces in $\mathbf{f}$ and allowed to deform the fluid element.

For the sake of simplicity, let us consider two-dimensional Cartesian coordinates, in which case the Lagrangian density is

$$\mathcal{L} = \frac{1}{2}\rho\left(\dot{x}^2 + \dot{y}^2\right) - \rho g y.$$

Then with $q_1 = x$, and $q_2 = y$, we have $\partial \mathbf{r}/\partial q_1 = \mathbf{i}$ and $\partial \mathbf{r}/\partial q_2 = \mathbf{j}$. With the net surface force per unit volume acting on the fluid element given by $\mathbf{f} = f_x\mathbf{i} + f_y\mathbf{j}$, the Euler-Lagrange equation (9.7) for $q_1 = x$ is

$$0 - \frac{d}{dt}(\rho \dot{x}) = -f_x,$$

or with $\dot{x} = u$

$$\rho \frac{du}{dt} = f_x. \tag{9.8}$$

Similarly, for $q_2 = y$

$$-\rho g - \frac{d}{dt}(\rho \dot{y}) = -f_y,$$

or with $\dot{y} = v$

$$\rho \frac{dv}{dt} = f_y - \rho g. \tag{9.9}$$

In order to write equations (9.8) and (9.9) in the Eulerian frame of reference, we apply the substantial derivative (9.2) for the acceleration on the left-hand sides yielding

$$\rho\left(\frac{\partial u}{\partial t}+u\frac{\partial u}{\partial x}+v\frac{\partial u}{\partial y}\right)=f_x, \tag{9.10}$$

$$\rho\left(\frac{\partial v}{\partial t}+u\frac{\partial v}{\partial x}+v\frac{\partial v}{\partial y}\right)=f_y-\rho g. \tag{9.11}$$

The ρg term acts on the entire volume of the fluid element; therefore, it is referred to as a body force. The f_x and f_y terms, then, will be used to treat surface forces acting on the boundaries of the element. Observe that whereas equations (9.8) and (9.9) expressed in the Lagrangian viewpoint are linear, equations (9.10) and (9.11) expressed in the Eulerian viewpoint are nonlinear owing to the convective acceleration terms. This is the primary disadvantage of using the Eulerian formulation of the equations of fluid mechanics.

In deriving the Bernoulli equation, we account for the pressure acting to move the nondeformable fluid particles. Here, we also allow for the deformation of the differential fluid particles owing to the pressure distribution throughout the domain. Note that for inviscid (frictionless) flow, only normal stresses owing to pressure act on the fluid elements, and no shear stresses are present. Therefore, the differential element in two-dimensions is given in Figure 9.1, where we use the same subscript notation as in Section 5.3.3 for elasticity, in which the first subscript denotes the face on which the stress acts, designated by the normal to the surface, and the second subscript indicates the direction in which the stress is acting. Because the fluid element is infinitesimally small, truncated Taylor series are used to account for the change in each stress across the width and height of the differential element. Summing the forces per unit volume acting on the differential fluid element in the x-direction (assuming unit depth in the z-direction) gives

$$f_x=\frac{1}{dxdy}\left[\left(\tau_{xx}+\frac{\partial\tau_{xx}}{\partial x}dx\right)dy-\tau_{xx}dy\right]=\frac{\partial\tau_{xx}}{\partial x},$$

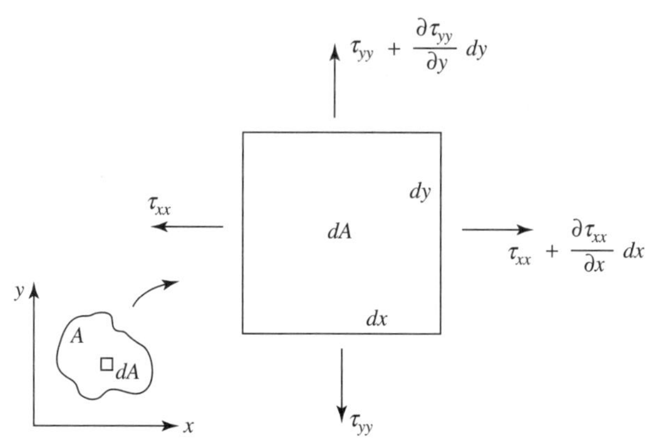

Figure 9.1. Normal stresses acting on a two-dimensional differential fluid element in an inviscid flow.

and in the y-direction

$$f_y = \frac{1}{dxdy}\left[\left(\tau_{yy} + \frac{\partial \tau_{yy}}{\partial y}dy\right)dx - \tau_{yy}dx\right] = \frac{\partial \tau_{yy}}{\partial y}.$$

In inviscid flow, the only normal stress is that due to the pressure acting on the fluid element. Although the pressure changes with position and time, such that $p = p(x,y,t)$, it is the same in all directions at each point and acts inward on the element. Therefore,

$$\tau_{xx} = \tau_{yy} = -p.$$

Hence, equations (9.10) and (9.11) become

$$\rho\left(\frac{\partial u}{\partial t} + u\frac{\partial u}{\partial x} + v\frac{\partial u}{\partial y}\right) = -\frac{\partial p}{\partial x}, \tag{9.12}$$

$$\rho\left(\frac{\partial v}{\partial t} + u\frac{\partial v}{\partial x} + v\frac{\partial v}{\partial y}\right) = -\frac{\partial p}{\partial y} - \rho g. \tag{9.13}$$

We also have the continuity equation for incompressible flow (9.4), which is now

$$\frac{\partial u}{\partial x} + \frac{\partial v}{\partial y} = 0. \tag{9.14}$$

These are the *Euler equations for inviscid, incompressible flow* in two-dimensional Cartesian coordinates expressed in the Eulerian reference frame. In vector form, these equations are

$$\rho\left[\frac{\partial \mathbf{v}}{\partial t} + (\mathbf{v}\cdot\nabla)\mathbf{v}\right] = -\nabla p - \rho g\mathbf{j}, \quad \nabla\cdot\mathbf{v} = 0. \tag{9.15}$$

The expression in $[\cdot]$ on the left-hand side is the acceleration, with the first term being the local acceleration owing to unsteadiness of the flow, and the second term representing the convective acceleration owing to the movement of fluid particles among regions of varying local velocity. The terms on the right-hand side are the forces (per unit volume) owing to the local pressure and the body force owing to gravity, respectively.

Note that Bernoulli's equation (9.6) and the above Euler equations both apply for inviscid, incompressible flow. The Bernoulli equation in the form provided further assumes that the flow is steady, that is, $\partial\mathbf{v}/\partial t = 0$. The primary difference is that Bernoulli's equation follows a single fluid particle in a Lagrangian sense, that is, along a streamline, and considers it to be a nondeformable body. On the other hand, the Euler equations apply at every point in a flow domain viewed from an Eulerian viewpoint and accounts for deformation of the fluid elements owing to local pressure forces.

9.2.3 Potential Flow

The nonlinear partial differential Euler equations (9.15) that govern inviscid, incompressible flow are difficult to solve analytically and in general must be solved numerically. However, if we make one additional assumption, that the flow is *irrotational*, the formulation can be greatly simplified in terms of a potential similar to that in electromagnetics and affords treatment using complex variable theory with conformal mapping. Although not all inviscid flows are also irrotational, it is true that the primary mechanism for producing rotation of fluid particles is through the action of viscous forces. Therefore, the assumption of irrotationality is often appropriate for inviscid flows.

For any two-dimensional flow, the *vorticity* $\omega(x,y)$ is defined by the curl of the velocity field

$$\boldsymbol{\omega} = \nabla \times \mathbf{v} = \frac{\partial v}{\partial x} - \frac{\partial u}{\partial y}, \tag{9.16}$$

which is a measure of the local rotation rate of the fluid particles. If there is no such rotation, the flow is *irrotational* and

$$\boldsymbol{\omega} = \nabla \times \mathbf{v} = 0.$$

As we saw in electromagnetics, because the curl of the gradient of any scalar is zero, that is, $\nabla \times (\nabla\phi) = 0$, we can define a scalar potential $\phi(x,y)$ such that we can write the velocity vector as the gradient of a scalar function $\phi(x,y)$ according to

$$\mathbf{v} = \nabla\phi. \tag{9.17}$$

Because $\mathbf{v} = u\mathbf{i} + v\mathbf{j}$ and $\nabla\phi = \frac{\partial\phi}{\partial x}\mathbf{i} + \frac{\partial\phi}{\partial y}\mathbf{j}$, we have that

$$u = \frac{\partial\phi}{\partial x}, \quad v = \frac{\partial\phi}{\partial y}, \tag{9.18}$$

and the scalar function $\phi(x,y)$ is called the *velocity potential*. Substituting into the continuity equation (9.14) gives

$$\frac{\partial^2\phi}{\partial x^2} + \frac{\partial^2\phi}{\partial y^2} = 0; \tag{9.19}$$

accordingly, the velocity potential satisfies the Laplace equation. Substituting the definitions of the streamfunction (9.5) into that of the vorticity (9.16) yields

$$\frac{\partial^2\psi}{\partial x^2} + \frac{\partial^2\psi}{\partial y^2} = -\omega. \tag{9.20}$$

This is known as a Poisson equation. Because the flow is assumed to be irrotational, $\psi(x,y)$ satisfies Laplace's equation $\nabla^2\psi = 0$. We call steady, two-dimensional, incompressible, inviscid, and irrotational flow *ideal flow* or *potential flow*. Observe that both the velocity potential $\phi(x,y)$ and the streamfunction $\psi(x,y)$ satisfy Laplace's equation. Remarkably, the potential flow formulation, which requires the assumption of irrotationality, is fully linear as compared to the Euler equations for inviscid, but rotational, flow, which are nonlinear. Therefore, the potential flow formulation admits a much larger number of exact solutions than those given by the Euler equations, and numerical solutions are much easier to obtain if exact solutions are not available. Notwithstanding its seemingly restrictive assumptions, potential flow is in common use today for flows in which viscous (friction) effects are negligible or in regions of flow fields away from areas where viscous effects are important, such as near solid surfaces.

Because the streamfunction and velocity potential are each solutions of Laplace equations, we may define the *complex potential function* by

$$\Phi(z) = \phi(x,y) + i\psi(x,y),$$

where $\Phi(z)$ is an analytic function of $z = x + iy$. In this way, we can take advantage of complex variable theory in solving the potential flow problem in the two-dimensional case (see, for example, Kreyszig 2011, Jeffrey 2002, or Greenberg 1998). Note that owing to the properties of analytic functions, the lines of constant $\phi(x,y)$ (equipotential lines) and lines of constant $\psi(x,y)$ (streamlines) are mutually orthogonal throughout the domain. We can envision the flow being driven from regions of high to low equipotential along the streamlines. Observe also that the pressure does not appear explicitly in the potential flow formulation; however, it can be obtained from the Bernoulli equation. Bernoulli's equation applied to steady, inviscid, incompressible, *and* irrotational flow is of the form (9.6) except that the constant is the same throughout the flow field, not just along each streamline. Three-dimensional potential flow is a natural extension to the above formulation except that the elegance of complex variable theory no longer applies.

Because the potential flow formulation leads to Laplace equations governing the velocity potential and streamfunction, we are led to recall Section 2.5.3 where we discuss the Dirichlet problem. There we find that the functional that produces the Laplace equation (9.19) is of the form

$$\delta \iint_A \left[\left(\frac{\partial \phi}{\partial x} \right)^2 + \left(\frac{\partial \phi}{\partial y} \right)^2 \right] dx\, dy = 0. \tag{9.21}$$

Observe that the kinetic energy density for inviscid flow when modified for irrotational flow, in which the velocity can be expressed in terms of the velocity potential, is

$$\mathcal{T} = \frac{1}{2}\rho \mathbf{v}^2 = \frac{1}{2}\rho(\nabla\phi)^2 = \frac{1}{2}\rho \left[\left(\frac{\partial \phi}{\partial x} \right)^2 + \left(\frac{\partial \phi}{\partial y} \right)^2 \right].$$

Similarly, we may produce the Poisson equation for streamfunction in a rotational flow (9.20) using the functional

$$\delta \iint_A \left[\left(\frac{\partial \psi}{\partial x} \right)^2 + \left(\frac{\partial \psi}{\partial y} \right)^2 - 2\omega\psi \right] dx\, dy = 0.$$

Once again, we may interpret the first two terms in the functional as the kinetic energy density

$$\mathcal{T} = \frac{1}{2}\rho \mathbf{v}^2 = \frac{1}{2}\rho \left[u^2 + v^2 \right] = \frac{1}{2}\rho \left[\left(\frac{\partial \psi}{\partial y} \right)^2 + \left(-\frac{\partial \psi}{\partial x} \right)^2 \right] = \frac{1}{2}\rho \left[\left(\frac{\partial \psi}{\partial x} \right)^2 + \left(\frac{\partial \psi}{\partial y} \right)^2 \right].$$

In the irrotational case corresponding to potential flow, the Poisson equation for streamfunction becomes a Laplace equation, which once again only involves the kinetic energy.

9.3 Viscous Flow – Navier-Stokes Equations

Viscosity is a macroscopic property accounting for friction between adjacent moving fluid particles. It arises from the microscopic interactions between molecules in a fluid. Although viscosity is small for most fluids, there are situations where the influence of viscosity is not small and is essential to explain relevant phenomena. This may occur throughout the domain in internal flows or locally in certain regions of a flow field. The latter is the case in the regions near solid surfaces of external flows where the velocity gradients may become large enough to overcome the small viscosity to render viscous effects essential in such *boundary layers*.

The presence of viscous effects in a fluid causes shear stresses to occur between adjacent fluid elements in addition to the normal stresses considered in the inviscid case. The stresses acting on a two-dimensional differential fluid element are shown in Figure 9.2 (compare with elasticity in Section 5.3.3). We follow the same approach used in Section 9.2.2 in deriving Euler's equation for inviscid flow, except that now we must include the shear stresses acting on the differential element. For example, summing the forces per unit volume acting on the differential fluid element in the x-direction (assuming unit depth) leads to

$$f_x = \frac{1}{dxdy}\left[\left(\tau_{xx} + \frac{\partial \tau_{xx}}{\partial x}dx\right)dy - \tau_{xx}dy + \left(\tau_{yx} + \frac{\partial \tau_{yx}}{\partial y}dy\right)dx - \tau_{yx}dx\right].$$

Simplifying yields

$$f_x = \frac{\partial \tau_{xx}}{\partial x} + \frac{\partial \tau_{yx}}{\partial y}.$$

Similarly, summing forces in the y-direction and simplifying gives

$$f_y = \frac{\partial \tau_{xy}}{\partial x} + \frac{\partial \tau_{yy}}{\partial y}.$$

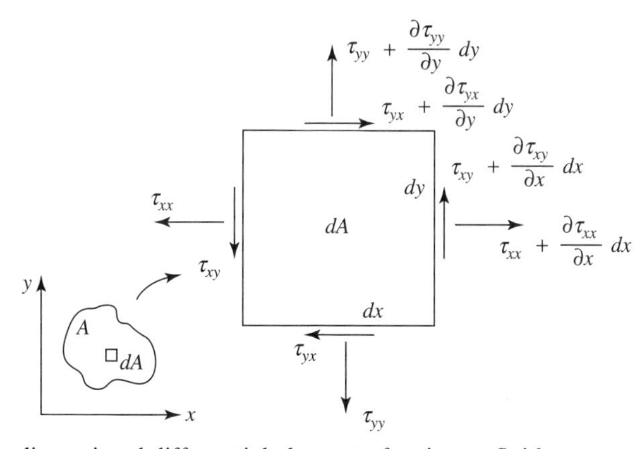

Figure 9.2. Two-dimensional differential element of a viscous fluid.

Substituting these forces per unit volume into the general forms of the Euler-Lagrange equations (9.10) and (9.11) gives

$$\rho\left(\frac{\partial u}{\partial t}+u\frac{\partial u}{\partial x}+v\frac{\partial u}{\partial y}\right)=\frac{\partial \tau_{xx}}{\partial x}+\frac{\partial \tau_{yx}}{\partial y}, \tag{9.22}$$

$$\rho\left(\frac{\partial v}{\partial t}+u\frac{\partial v}{\partial x}+v\frac{\partial v}{\partial y}\right)=\frac{\partial \tau_{xy}}{\partial x}+\frac{\partial \tau_{yy}}{\partial y}-\rho g. \tag{9.23}$$

Equations (9.22), (9.23), and the continuity equation (9.14) involve two unknown velocity components and three stress components (note that $\tau_{xy}=\tau_{yx}$). Therefore, the three equations are insufficient to solve for all five variables. We need to relate stresses and velocities in order to complete the formulation. Recall that in elasticity, we have kinematic relationships between displacements and strains and the constitutive relations (Hooke's law) that relate stresses to strain in a linear manner. In a similar fashion, in fluid mechanics, we have kinematic relationships between velocities and strain rates, and constitutive relations between stress and strain rate. Most common fluids, such as air and water, behave as *Newtonian fluids* in which the stresses are linearly proportional to strain rates. The constitutive relations for Newtonian fluids are expressed by the *Stokes relations*, which for two-dimensional Cartesian coordinates are

$$\tau_{xx}=-p+\lambda\nabla\cdot\mathbf{v}+2\mu\epsilon_{xx},\quad \tau_{yy}=-p+\lambda\nabla\cdot\mathbf{v}+2\mu\epsilon_{yy},\quad \tau_{xy}=\tau_{yx}=2\mu\epsilon_{xy},$$

where μ is the dynamic viscosity, and λ is the bulk viscosity. For incompressible flow, as considered here, the continuity equation requires that $\nabla\cdot\mathbf{v}=0$, and the corresponding terms vanish in the Stokes relations. Because we are not considering heat transfer effects as the flow is considered isothermal, the dynamic viscosity μ will be taken as constant. From kinematic considerations, the strain rates are related to the velocities by

$$\epsilon_{xx}=\frac{\partial u}{\partial x},\quad \epsilon_{yy}=\frac{\partial v}{\partial y},\quad \epsilon_{xy}=\frac{1}{2}\left(\frac{\partial v}{\partial x}+\frac{\partial u}{\partial y}\right),$$

where we use the same subscript convention for strain rates as for stresses. Substituting Stokes relations for Newtonian fluids and the kinematic relationships into the right-hand side of equation (9.22) gives

$$\begin{aligned} f_x&=\frac{\partial \tau_{xx}}{\partial x}+\frac{\partial \tau_{yx}}{\partial y}\\ &=\frac{\partial}{\partial x}(-p+2\mu\epsilon_{xx})+2\mu\frac{\partial \epsilon_{xy}}{\partial y}\\ &=\frac{\partial}{\partial x}\left(-p+2\mu\frac{\partial u}{\partial x}\right)+\mu\frac{\partial}{\partial y}\left(\frac{\partial v}{\partial x}+\frac{\partial u}{\partial y}\right) \end{aligned}$$

$$= -\frac{\partial p}{\partial x} + 2\mu\frac{\partial^2 u}{\partial x^2} + \mu\frac{\partial^2 v}{\partial x \partial y} + \mu\frac{\partial^2 u}{\partial y^2}$$

$$= -\frac{\partial p}{\partial x} + \mu\frac{\partial^2 u}{\partial x^2} + \mu\frac{\partial}{\partial x}\left(\cancelto{0}{\frac{\partial u}{\partial x} + \frac{\partial v}{\partial y}}\right) + \mu\frac{\partial^2 u}{\partial y^2}$$

$$f_x = -\frac{\partial p}{\partial x} + \mu\left(\frac{\partial^2 u}{\partial x^2} + \frac{\partial^2 u}{\partial y^2}\right),$$

where the term canceled in the last step vanishes owing to the continuity equation. A similar procedure applied to the right-hand side of equation (9.23) leads to

$$f_y = \frac{\partial \tau_{xy}}{\partial x} + \frac{\partial \tau_{yy}}{\partial y} - \rho g = -\frac{\partial p}{\partial y} + \mu\left(\frac{\partial^2 v}{\partial x^2} + \frac{\partial^2 v}{\partial y^2}\right) - \rho g.$$

Substitution into equations (9.22) and (9.23) yields the final form of the governing equations for incompressible, viscous flow as

$$\rho\left(\frac{\partial u}{\partial t} + u\frac{\partial u}{\partial x} + v\frac{\partial u}{\partial y}\right) = -\frac{\partial p}{\partial x} + \mu\left(\frac{\partial^2 u}{\partial x^2} + \frac{\partial^2 u}{\partial y^2}\right), \tag{9.24}$$

$$\rho\left(\frac{\partial v}{\partial t} + u\frac{\partial v}{\partial x} + v\frac{\partial v}{\partial y}\right) = -\frac{\partial p}{\partial y} + \mu\left(\frac{\partial^2 v}{\partial x^2} + \frac{\partial^2 v}{\partial y^2}\right) - \rho g. \tag{9.25}$$

Because they essentially arise from Newton's second law, which enforces conservation of momentum, they are sometimes referred to as the momentum equations. The continuity equation for incompressible flow (9.14) remains unchanged and is

$$\frac{\partial u}{\partial x} + \frac{\partial v}{\partial y} = 0. \tag{9.26}$$

Equations (9.24)–(9.26) are known as the *Navier-Stokes equations* and govern incompressible, viscous flow of a Newtonian fluid. The vector form of the Navier-Stokes equations is

$$\rho\left[\frac{\partial \mathbf{v}}{\partial t} + (\mathbf{v}\cdot\nabla)\mathbf{v}\right] = -\nabla p + \mu\nabla^2\mathbf{v} - \rho g\mathbf{j}, \quad \nabla\cdot\mathbf{v} = 0. \tag{9.27}$$

This vector form applies to three-dimensional flows, not simply the two-dimensional case considered here, and it can be used to write the governing equations in other than the Cartesian coordinates used here. One simply needs to use the appropriate forms of the vector operations in the desired coordinate system. The *convection*, or acceleration, terms on the left-hand side of the Navier-Stokes equations account for the movement of fluid owing to the unsteady velocity field and the local velocity. The $\mu\nabla^2\mathbf{v}$ terms on the right-hand side account for *diffusion* owing to the viscous (frictional) effects between adjacent fluid particles. The remaining terms on the right-hand side account for the forces owing to pressure and gravity. Thus, the momentum equation enforces that the mass per unit volume (density) times acceleration is equal to the sum of the forces per unit volume.

It is convenient to nondimensionalize the Navier-Stokes equations in order to combine all of the relevant physical parameters into a single nondimensional parameter. To do so, we identify a characteristic length scale L and velocity scale U and nondimensionalize equations (9.24)–(9.26), or equivalently (9.27), according to

$$x^* = \frac{x}{L}, \quad y^* = \frac{y}{L}, \quad t^* = \frac{t}{L/U}, \quad u^* = \frac{u}{U}, \quad v^* = \frac{v}{U}, \quad p^* = \frac{p + \rho g y}{\rho U^2}.$$

Observe that the nondimensionalization for the pressure includes the body force term owing to gravity. Substituting into the Navier-Stokes equations (9.24)–(9.26), and dropping the asterisks for clarity[1], leads to the nondimensional form of the Navier-Stokes equations

$$\frac{\partial u}{\partial t} + u\frac{\partial u}{\partial x} + v\frac{\partial u}{\partial y} = -\frac{\partial p}{\partial x} + \frac{1}{Re}\left(\frac{\partial^2 u}{\partial x^2} + \frac{\partial^2 u}{\partial y^2}\right), \tag{9.28}$$

$$\frac{\partial v}{\partial t} + u\frac{\partial v}{\partial x} + v\frac{\partial v}{\partial y} = -\frac{\partial p}{\partial y} + \frac{1}{Re}\left(\frac{\partial^2 v}{\partial x^2} + \frac{\partial^2 v}{\partial y^2}\right), \tag{9.29}$$

$$\frac{\partial u}{\partial x} + \frac{\partial v}{\partial y} = 0, \tag{9.30}$$

where

$$Re = \frac{\rho U L}{\mu}$$

is the *Reynolds number*. In vector form, the nondimensional Navier-Stokes equations are

$$\frac{\partial \mathbf{v}}{\partial t} + (\mathbf{v}\cdot\nabla)\mathbf{v} = -\nabla p + \frac{1}{Re}\nabla^2\mathbf{v}, \quad \nabla\cdot\mathbf{v} = 0. \tag{9.31}$$

The Reynolds number is a nondimensional parameter that encompasses all of the physical parameters into one. Not only is the nondimensional form more convenient to deal with as it only has one physical parameter, it also reveals useful insights into the relative effects of these parameters. For example, we see from the Reynolds number that the effect of doubling the characteristic velocity is the same as halving the viscosity of the fluid. That is, solutions of the Navier-Stokes equations for a given Reynolds number are the same; however, there are an infinite number of combinations of ρ, μ, L, and U that would produce such a value of the Reynolds number.

The Reynolds number can be thought of as the ratio of the magnitude of the inertial forces, owing to the convective terms, to the viscous forces. Low Reynolds numbers correspond to flows in which the viscous diffusion dominates over convection, while high Reynolds numbers correspond to cases in which the nonlinear convection dominates over diffusion. In the limit as the Reynolds number goes to infinity, the Navier-Stokes equations reduce to the Euler equations for inviscid flow with the caveat that boundary conditions for the two sets of equations differ owing to the fact that the Navier-Stokes equations are second order, while the Euler equations are first order. As a result, only one boundary condition can be imposed on the Euler equations for inviscid flow. At a solid surface, for example, this one condition would be that the surface is impermeable, such that there is no flow through

[1] It is easy to see whether we are dealing with the dimensional or nondimensional form based on the presence of the parameters ρ and μ or Re in the equations.

the boundary. For the Navier-Stokes equations, on the other hand, two boundary conditions are required. In the case of a solid surface, the impermeability condition is applied along with the so called *no-slip* boundary condition, which requires that the fluid immediately adjacent to the surface move at the same tangential velocity as the surface. Not surprisingly, the no-slip condition had been met with significant controversy in the early days of fluid mechanics until it could be confirmed with careful experimental measurements. Essentially, it assumes that the microscopic interactions between the fluid and the solid wall are at least as strong as those between the fluid particles. Owing to the no-slip boundary condition, solutions to the Navier-Stokes equations as the Reynolds number goes to infinity are not the same as solutions of the Euler equations for the same flow. This is because in order to satisfy the no-slip boundary condition, a thin *boundary layer* is required immediately adjacent to the surface, with the flow away from the surface governed by the inviscid Euler equations. The boundary layer serves to adjust the non-zero relative tangential velocity from the solutions to the Euler equations to zero at the solid surface.

Previously, we had introduced the streamfunction and vorticity via equations (9.5) and (9.16), respectively. We can reframe the so called *primitive-variables formulation* of the Navier-Stokes equations given above, which are solved for the two velocity components and pressure, with the *vorticity-streamfunction formulation* as follows. Recall that the streamfunction $\psi(x,y,t)$ is defined for two-dimensional, incompressible flow by

$$u = \frac{\partial \psi}{\partial y}, \quad v = -\frac{\partial \psi}{\partial x}, \tag{9.32}$$

such that the continuity equation (9.30) is identically satisfied. In three dimensions, the vorticity is a vector with three components. In two dimensions, however, there is only one component in the direction normal to the xy-plane. The vorticity $\omega(x,y,t)$ in two-dimensions is defined by

$$\boldsymbol{\omega} = \nabla \times \mathbf{v} = \left(\frac{\partial v}{\partial x} - \frac{\partial u}{\partial y} \right) \mathbf{k} = \omega \mathbf{k}. \tag{9.33}$$

Taking the curl of the momentum equation (9.31) in vector form leads to

$$\frac{\partial \omega}{\partial t} + (\mathbf{v} \cdot \nabla)\omega = \frac{1}{Re} \nabla^2 \omega, \tag{9.34}$$

which is the *vorticity-transport equation*. Observe that it is of the same form as in the primitive-variables formulation except that the pressure has been eliminated. In the two-dimensional case considered here, we have

$$\frac{\partial \omega}{\partial t} + u\frac{\partial \omega}{\partial x} + v\frac{\partial \omega}{\partial y} = \frac{1}{Re}\left(\frac{\partial^2 \omega}{\partial x^2} + \frac{\partial^2 \omega}{\partial y^2} \right). \tag{9.35}$$

As before, the Poisson equation for streamfunction (9.20) is

$$\frac{\partial^2 \psi}{\partial x^2} + \frac{\partial^2 \psi}{\partial y^2} = -\omega, \tag{9.36}$$

or in vector form $\nabla^2 \psi = -\omega$. Equations (9.35) and (9.36), along with the definitions of streamfunction (9.32), are two equations for the vorticity $\omega(x,y,t)$ and streamfunction

$\psi(x,y,t)$ in two-dimensional, incompressible flow. Recall that in the primitive-variables formulation, we have three equations for $u(x,y,t)$, $v(x,y,t)$, and $p(x,y,t)$. Therefore, there is some economy in using the vorticity-streamfunction formulation for two-dimensional, incompressible flows.

There are relatively few exact solutions of the nonlinear Navier-Stokes equations (see, for example, Drazin and Riley 2006); therefore, several techniques have been developed for obtaining approximate solutions. It is often the case that we consider flows with parameters, such as Reynolds number, being large or small. In such situations, *asymptotic*, or *perturbation*, *methods* can be utilized to great effect in determining the dominant physics in such situations and leading to reduced sets of equations to be solved. See, for example, van Dyke (1975) and Hinch (1991). More generally, *computational fluid dynamics* (CFD) has progressed dramatically in the last half century along with computational capacity and resources to become a tool used commonly in research and industry. See, for example, Fletcher (1991a and b), Ferziger and Peric (1996), Tannehill, Anderson, and Pletcher (1997), and Chung (2010). For much more on the fundamentals of potential, inviscid, viscous, incompressible, and compressible flows, see Batchelor (1967), White (1974), and Panton (2005), and see Yourgrau and Mandelstam (1968) for an alternative derivation of the Euler and Navier-Stokes equations from variational principles. For a recent review of extensions to the variational theory to handle complex fluids, including ionic solutions, see Eisenberg, Hyon, and Liu (2010).

EXAMPLE 9.1 Consider the two-dimensional, viscous flow between two infinite parallel plates a distance $2H$ apart and driven by a constant streamwise pressure gradient in the x-direction as shown in Figure 9.3. This is known as *plane-Poiseuille flow*.

Solution: If we nondimensionalize based on the half-channel height H and maximum velocity U, the Reynolds number is then $Re = \rho UH/\mu$. Because the channel is infinitely long in the x-direction, the velocity components will not change in this direction; we say that the flow is *fully developed*. Note, however, that the pressure does change with x in order to maintain a constant pressure gradient $\partial p/\partial x = \text{constant} < 0$ that drives the flow through the channel.

Let us first consider the continuity equation (9.30) for fully developed flow, for which $\partial u/\partial x = 0$. We are then left with $\partial v/\partial y = 0$. Because v does not change with x for fully developed flow, integrating requires that $v = \text{constant}$. However, owing to the impermeability boundary condition at the lower wall, $v = 0$ at $y = 0$; therefore, $v = 0$ throughout the flow. From the y-momentum equation (9.29), we then conclude

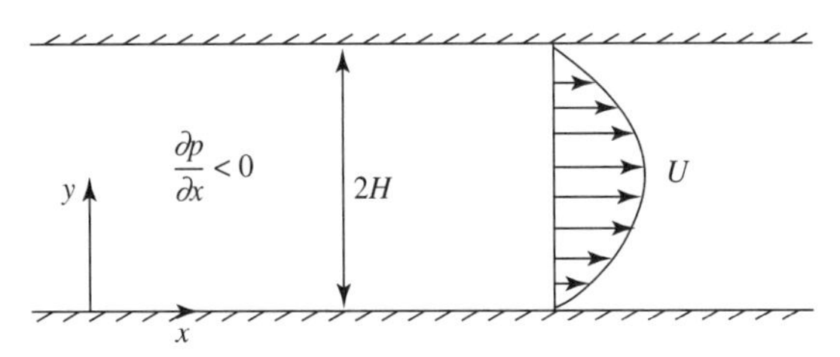

Figure 9.3. Schematic of plane-Poiseuille flow.

that $\partial p/\partial y = 0$, in which case $p = p(x)$. Finally, from the x-momentum equation (9.28), with fully developed flow and $v = 0$, we are left with

$$\frac{d^2u}{dy^2} = Re\frac{dp}{dx} = \text{constant}. \tag{9.37}$$

Integrating twice results in

$$u(y) = \frac{1}{2}Re\frac{dp}{dx}y^2 + c_0 y + c_1,$$

where c_0 and c_1 are constants of integration. Applying the no-slip boundary condition at the lower wall, which is $u = 0$ at $y = 0$, requires that $c_1 = 0$. Similarly, applying the no-slip boundary condition at the upper wall, which is $u = 0$ at $y = 2$, requires that $c_0 = -Re\, dp/dx$. Therefore, the velocity profile across the channel is

$$u(y) = -\frac{1}{2}Re\frac{dp}{dx}y(2-y). \tag{9.38}$$

By definition of the nondimensional velocity, the maximum velocity is $u = 1$, which occurs along the channel midplane at $y = 1$. Let us then determine the specified pressure gradient that is required to produce this maximum velocity. Applying the solution (9.38) with $u = 1$ at $y = 1$ leads to

$$\frac{dp}{dx} = -\frac{2}{Re}.$$

Substituting back into the solution (9.38) yields

$$u(y) = y(2-y), \tag{9.39}$$

and substituting into the governing equation for plane-Poiseuille flow (9.37) provides the form

$$\frac{d^2u}{dy^2} = -2. \tag{9.40}$$

Observe that the plane-Poiseuille flow solution (9.39) is independent of the Reynolds number. This is unusual for viscous flows in general, but often occurs for parallel flows, in which the velocity only varies with one direction, in this case y.

Note that for use in Chapter 10, we can write the nondimensional, unsteady version of equation (9.40) in the form

$$\frac{\partial u}{\partial t} = \frac{1}{Re}\frac{\partial^2 u}{\partial y^2} + \frac{2}{Re} + X, \tag{9.41}$$

where $X(y)$ is the nondimensional body force in the x-direction, and $2/Re$ is the specified pressure gradient.

9.4 Multiphase and Multicomponent Flows

Many applications involve flow of multiple phases and/or multiple chemical components. Such flows pose a considerable challenge in order to track the moving interfaces between phases or components and to account for phase change and other physical processes that may take place at these interfaces. Two closely related methods have been developed to handle these types of flows within a variational framework, namely *level-set methods* and *phase-field models*. Level-set methods are used to handle circumstances in which the flow of multiple phases or multiple immiscible[2] fluids is considered on the macroscopic scale of the overall problem. The interfaces between phases and/or immiscible fluids are considered sharp, with the properties changing suddenly across the interface. On the mesoscopic scale,[3] however, such interfaces reveal that, in fact, there is a thin region at the interface across which more complicated physics and chemistry may be taking place. Phase-field models treat the macroscopic and mesoscopic scales as a continuum, with the interfaces being handled by a *diffuse-interface model*. Normally, fluid mechanics is considered on the macroscopic scale; fluids in which multiple scales must be considered are sometimes referred to as *complex fluids*.

Both level-set methods and phase-field models track the interfaces implicitly by calculating a field function throughout the domain rather than tracking the interfaces explicitly. Level-set methods are used for scenarios in which the physics on the scale of the interface can be ignored; phase-field models are used for situations in which such physics cannot be ignored. For example, a gas bubble in a liquid has molecules of each component moving across the interface; however, this does not affect the overall macroscopic evolution of the two-phase flow, and the level-set method is used. On the other hand, if phase change is taking place at the interface, such as in solidification processes, then phase-field models are used.

9.4.1 Level-Set Methods

Originally pioneered by Osher and Sethian (1988) to treat interfaces in multiphase fluid flows, level-set methods provide a conceptually simple and computationally convenient framework for representing and tracking moving boundaries, interfaces, or edges in many settings – we will encounter them again when discussing image processing in Chapter 11. The primary advantage of the level-set method is that sharp interfaces can be defined simply by computing an unsteady field equation for the level-set function viewed from an Eulerian point of view. The level set is a nonphysical, continuous function, typically denoted by $\phi(x,y,t)$, defined throughout a domain for which the $\phi = 0$ level set corresponds to a set of internal interfaces within the domain and with the respective phases corresponding to opposing signs of the level-set function. Although it may seem to be overkill to define a new function throughout the entire domain where only one of its values is physically relevant, it provides an approach with several important advantages over other front-tracking methods. More will be highlighted later, but the primary advantage is that topological changes to the interface, such as splittings or mergings, are handled naturally. Such

[2] Immiscible substances are such that there is no mass transfer between substances.

[3] Mesoscale means "intermediate scale." In the context of phase-field models, the macroscopic scale refers to the overall scale of the system; the microscopic scale is on the scale of atoms and molecules, and the mesoscopic scale is the intermediate scale of the interface between phases or components.

events are notoriously problematic to deal with in alternative interface-tracking methods.

Having the interfaces be defined implicitly by a continuous function throughout the domain allows for straightforward calculation of the level-set function on a fixed Eulerian grid corresponding to the discretized differential equations governing a physical problem, such as fluid mechanics. The evolution of the level-set function $\phi(x,y,t)$ is determined from calculations of a partial differential equation given by

$$\frac{\partial \phi}{\partial t} + v_n |\nabla \phi| = 0, \tag{9.42}$$

where $v_n(x,y,t)$ is the local velocity normal to the level sets, which are contours of constant $\phi(x,y,t)$, and is determined from a separate calculation. The normal velocity in physical applications, such as fluid mechanics, comes from the solution of the relevent evolution equations. Where a velocity vector field $\mathbf{v}(x,y,t)$ is defined throughout the domain from such a physical solution of the Navier-Stokes equations, for example, the level-set equation can be written

$$\frac{\partial \phi}{\partial t} + (\mathbf{v} \cdot \nabla)\phi = 0, \tag{9.43}$$

where the scalar dot product provides the desired normal component of the velocity vector.

In many applications of level sets, such as image processing segmentation (see Section 11.1.4), the level-set framework can be incorporated directly into the variational framework to create a very natural and computationally efficient formulation. Additional virtues of the level-set method include:

1. Topological changes to the interface are handled naturally by the level-set calculation on a fixed Eulerian grid.
2. Geometric quantities can be computed directly from the level-set function.
3. Extension from two- to three-dimensions is straightforward.

The first point is important in multiphase flows as the level-set method is capable of treating merging or splitting of bubbles, for example, without any additional complications. The second point is also essential in multiphase flows in order to account for surface tension at the interface between two phases, which is directly proportional to the curvature of the interface. Once the level-set function $\phi(x,y,t)$ is calculated, the outward facing normal unit vector is given by

$$\mathbf{n} = \frac{\nabla \phi}{|\nabla \phi|},$$

because the gradient of a function is in the direction of the normal to the function. In two dimensions, the normal vector is

$$\mathbf{n} = \left[\left(\frac{\partial \phi}{\partial x} \right)^2 + \left(\frac{\partial \phi}{\partial y} \right)^2 \right]^{-\frac{1}{2}} \left(\frac{\partial \phi}{\partial x}\mathbf{i} + \frac{\partial \phi}{\partial y}\mathbf{j} \right).$$

The local curvature of the interface is then given by

$$\kappa = \nabla \cdot \mathbf{n} = \nabla \cdot \left(\frac{\nabla \phi}{|\nabla \phi|} \right),$$

which in two dimensions is

$$\kappa = \left[\left(\frac{\partial \phi}{\partial x} \right)^2 + \left(\frac{\partial \phi}{\partial y} \right)^2 \right]^{-\frac{3}{2}} \left[\left(\frac{\partial \phi}{\partial y} \right)^2 \frac{\partial^2 \phi}{\partial x^2} - 2 \frac{\partial \phi}{\partial x} \frac{\partial \phi}{\partial y} \frac{\partial^2 \phi}{\partial x \partial y} + \left(\frac{\partial \phi}{\partial x} \right)^2 \frac{\partial^2 \phi}{\partial y^2} \right].$$

Although level-set methods treat interfaces as sharp discontinuities within the domain, they are approximated numerically using a thin region across which the properties and variables vary rapidly from one phase to the other.

In addition to fluid mechanics, level-set methods have found application in image processing, computational geometry, computer vision, computer graphics, materials science, shape and topology optimization, and numerous other fields. In shape and topology optimization, the level-set method is used to define the interface between different materials or regions of a domain (see Burger and Osher 2005). The unknown variable to be optimized is the domain shape and/or topology, and the normal interface velocity moves the shape toward its optimum. For a review of level-set methods as applied to fluid mechanics, see Sethian and Smereka (2003). For level-set methods applied to a variety of applications, see Sethian (1999) and Osher and Fedkiw (2003).

9.4.2 Phase-Field Models

Level-set methods are closely related to phase-field models that are often used in solidification applications to track the interface between different phases of the same substance where phase change is taking place or the interface between different chemical components. Essentially, phase-field models are physically motivated level-set methods. Although following the same basic approach as level sets, phase-field models are defined such that the model parameter is zero for one phase or component and unity for the other with a smooth adjustment across a narrow interface. Such models are known as *diffuse-interface models* and recognize that the seemingly sharp interface in such applications is really a thin layer across which the phase change or mixing of components occurs. In the level-set method, the interfacial physics is encapsulated in the surface tension, which is included as a body force. In the phase-field model, the interfacial physics is contained in a stress tensor throughout the thin interfacial region. Consequently, surface tension at a sharp interface is the macroscopic manifestation of a distributed stress across a mesoscopic-scale diffuse interface. Note that level-set methods treat the interface as having small, but finite, thickness for numerical reasons, whereas phase-field models do so for physical reasons.

Phase-field models involve determining the equation for the stationary function of a functional for the appropriate form of thermodynamic mixing energy in the diffuse interface. This results in an equation for evolution of the phase-field function, which is often called the *order parameter*. Phase-field models are a popular approach for modeling of dendrite solidification and alloy solidification processes as well as other materials science applications including certain crystal growth and thin film growth processes. The microstructure of a material may consist of various phases of the same and different components, a range of grain orientations, as well as other interfacial inhomogeneities. Understanding such microstructures is essential to prediction of macroscopic material properties. The primary difficulty in such applications is the presence of a network of interfaces within the system that move, deform, and change topology. In order to treat such scenarios, it is necessary to account for the macroscopic fluid mechanics of the bulk regions as well as the mesoscale physics in the interfacial regions.

Just as with level-set methods, an additional field equation is required throughout the domain to define the phase-field function. Once again we use $\phi(x,y,t)$; however, the phase field is defined such that it is constant within each phase or component, say $\phi = 0$ and $\phi = 1$, and varies smoothly across a thin region at the interface between phases or components. The interface is then defined where $\phi = \frac{1}{2}$. In a single-component, multiphase setting, $\phi(x,y,t)$ is the density of the substance, while in two-component (binary) fluids, it would be the concentration of one component.

The phase-field function $\phi(x,y,t)$ is governed by an equation that is obtained from an appropriate energy functional for the thermodynamics within the interface. Let us illustrate this for the case of a two-dimensional, isothermal (constant temperature), single-component, multiphase system with diffusion but no convection. In this case, the molecular interaction between the two phases is given by the *total free energy*, which is the energy within the system that can be converted to do work. The free energy of the system is given in the form of the functional

$$E[\phi] = \iint_A \left[f(\phi) + \frac{1}{2} K |\nabla \phi|^2 \right] dx dy, \tag{9.44}$$

where $f(\phi)$ is the local free energy density function, K is the gradient energy coefficient related to the interfacial length scale, and A is the domain of the system. The first term accounts for the bulk energy density of the substances or phases, and the second term accounts for the surface energy owing to mixing within the interface. The local free energy density $f(\phi)$ is such that it has two minimums. A typical expression for the local free energy density function takes the form

$$f(\phi) = \frac{1}{4} \left(\phi^2 - 1 \right)^2.$$

This is known as a double-well function with the two minimums corresponding to the bulk energy densities of the two phases. The free energy density function can become rather complicated when more components are added and additional physics is incorporated at the interface. Physically, the two terms represent the tendency of phases to want to separate owing to the double-well potential in the local free energy density function and the weakly nonlocal interactions between the different phases in the thin, but finite, interface that seek complete mixing. The first seeks to promote a thin interface and is referred to as a *phobic* effect, while the second seeks a thick interface and is a *philic* effect. The actual interface profile results from the competition between these two opposing effects. When the thickness of the diffuse interface collapses to zero, that is, in the sharp-interface limit, this formally leads to the macroscopic notion of interfacial surface tension, usually denoted by σ.

From the second law of thermodynamics, the free energy of the system $E[\phi]$ must decrease monotonically[4] with time in order to ensure a nonnegative entropy

[4] If a function decreases monotonically, it never increases over any interval.

production. This requires that the time rate of change of $\phi(x,y,t)$ be related to the functional derivative of the free energy functional with respect to the phase field according to

$$\frac{\partial \phi}{\partial t} = -M\frac{\delta E}{\delta \phi} = M\left(K\nabla^2\phi - \frac{\partial f}{\partial \phi}\right), \tag{9.45}$$

where $M(\phi) > 0$ is the mobility coefficient. This is known as the *Allen-Cahn equation*.

Recall that we are considering the evolution of a single-component system in which the phase-field function (order parameter) $\phi(x,y,t)$ corresponds to a normalized density. We say that $\phi(x,y,t)$ is *conserved* if

$$\frac{d}{dt}\iint_A \phi(x,y,t)dxdy = 0;$$

that is, the total integrated value of the phase-field function $\phi(x,y,t)$ throughout the domain does not change with time. Because this is not the case for density in a system involving phase change, such single-component systems are *nonconservative* and the Allen-Cahn equation is used.

In the case of conserved phase-field functions, we can reformulate the problem as follows. Consider, for example, a binary system comprised of two components. In this case, the phase-field function is typically defined to be the concentration of one of the components such that $0 \leq \phi \leq 1$, with the other component having the concentration $1-\phi$. The system is said to be conserved because the total amounts of each of the two components remain the same despite their diffusion and possible convection. The equation governing the interface dynamics is obtained from a generalization of Fick's law for diffusion, which requires that the mass flux across the interface be proportional to the gradient of the chemical potential denoted by $\gamma(\phi)$. This gives rise to the *Cahn-Hilliard equation*

$$\frac{\partial \phi}{\partial t} = \nabla\cdot(M(\phi)\nabla\gamma), \tag{9.46}$$

where again $M(\phi) > 0$ is the mobility coefficient. The chemical potential is given by the first variation of the free energy, which is

$$\gamma(\phi) = \frac{\delta E}{\delta \phi} = -K\nabla^2\phi + \frac{\partial f}{\partial \phi}.$$

If the mobility is a constant, then the Cahn-Hilliard equation (9.46) becomes

$$\frac{\partial \phi}{\partial t} = M\nabla^2\gamma. \tag{9.47}$$

For an excellent description of the Cahn-Hilliard equation, including some of its applications and mathematical properties, see Chapter 4 of Dafermos and Pokorny (2008). Convection can be included in the Allen-Cahn and Cahn-Hilliard equations by adding a term of the form $(\mathbf{v}\cdot\nabla)\phi$ as in the Navier-Stokes equation (9.31), where $\mathbf{v}$ is the velocity field. The Allen-Cahn or Cahn-Hilliard equation for $\phi(x,y,t)$ would then be coupled with the Navier-Stokes equations modified to include variable fluid properties and surface forces that depend on the phase field.

Some representative references on phase-field models include Anderson, McFadden, and Wheeler (1998) who review research on single-component and binary fluids. Anderson, McFadden, and Wheeler (2000) and Yue, Feng, Liu, and Shen (2004) consider the nonisothermal case with convection, and Chen (2002) reviews the application of phase-field models in microstructure evolution. Singer-Loginova and Singer (2008) also review phase-field models and provide a number of fascinating examples of computational simulations of various phenomena. In an interesting application of Noether's theorem, Anderson, McFadden, and Wheeler (1998) show that a particular stress tensor must be conserved as a consequence of a symmetry in the energy functional.

Note that the sharp-interface model can be recovered formally from the phase-field model using asymptotic methods, and phase-field models can be shown to be a consequence of density-functional theory discussed in Section 8.2.2 (Pruessner and Sutton 2008). Therefore, density-functional theory, phase-field models, and sharp-interface models are formally consistent as we move from the microscopic, to mesoscopic, to macroscopic viewpoints, respectively.

9.5 Hydrodynamic Stability Analysis

9.5.1 Introduction

Solutions of the Navier-Stokes equations (9.28)–(9.30) are highly susceptible to instability, particularly at high Reynolds numbers where convection dominates over diffusion. Therefore, it is of interest to consider stability analysis in the spirit of Chapter 6, where it is applied to discrete systems with few degrees of freedom and continuous systems only involving one spatial coordinate. When applied to fluid mechanics, such stability analysis is often referred to as *hydrodynamic stability*. Although the basic approach is the same as in Chapter 6, there are a number of issues that arise in the fluid mechanics setting.

Recall that in the context of dynamical systems expressed in the Lagrangian frame of reference, as in Chapter 5, the system does not change with time and is static (stationary) at its equilibrium positions. Therefore, we consider stability owing to small perturbations about the static equilibrium points of the system. In fluid mechanics, where we utilize the Eulerian frame of reference, we consider stability of flows about their *steady* equilibrium solutions. A steady solution is one for which the velocity and pressure, or vorticity and streamfunction, do not change with time. The flow is not typically static as convection still occurs. That is, although there is no local acceleration $\partial \mathbf{v}/\partial t$, there is still convective acceleration $(\mathbf{v} \cdot \nabla)\mathbf{v}$. Therefore, rather than perturbing about stationary equilibrium states, we perturb about steady solutions of the inviscid Euler or viscous Navier-Stokes equations. As such, the equilibrium points are steady, but not static.

The ultimate goal of stability analysis of fluid flows is to illuminate the process by which a flow transitions from being *laminar* to *turbulent* at high Reynolds numbers. Laminar flow is a state in which the flow behaves in an orderly fashion, while turbulent flow displays what appears to be random fluctuations as the flow becomes very disordered, involving very small-scale fluctuations.

We begin by considering the traditional linear approach to hydrodynamic stability using modal analysis. Such a modal analysis can be performed locally at a particular streamwise location or globally to account for two- or three-dimensional effects. We then introduce transient-growth (optimal disturbance) analysis to account for non-normal effects on the short-time and long-time stability properties of a system. Transient-growth analysis is based on a variational principle and is applicable to steady and unsteady base flows, parallel and nonparallel base flows, and linear and nonlinear perturbation equations. We close the section on hydrodynamic stability with a brief discussion of energy methods.

9.5.2 Linear Stability of the Navier-Stokes Equations

Similar to Chapter 6, we apply small perturbations ($\epsilon \ll 1$) to the steady, two-dimensional base flow according to

$$\begin{aligned} u(x,y,t) &= u_0(x,y) + \epsilon \hat{u}(x,y,t), \\ v(x,y,t) &= v_0(x,y) + \epsilon \hat{v}(x,y,t), \\ p(x,y,t) &= p_0(x,y) + \epsilon \hat{p}(x,y,t), \end{aligned} \tag{9.48}$$

where u_0, v_0, and p_0 are the steady base flow, and $\hat{u}$, $\hat{v}$, and $\hat{p}$ are the unsteady perturbations acting on the base flow. Substituting these small perturbations into the Navier-Stokes equations (9.28)–(9.30) and neglecting quadratic terms in ϵ results in two sets of equations. The $O(1)$ equations are simply the steady Navier-Stokes equations for the base flow about which we want to consider stability of the small perturbations as follows:

$$u_0 \frac{\partial u_0}{\partial x} + v_0 \frac{\partial u_0}{\partial y} = -\frac{\partial p_0}{\partial x} + \frac{1}{Re}\left(\frac{\partial^2 u_0}{\partial x^2} + \frac{\partial^2 u_0}{\partial y^2}\right), \tag{9.49}$$

$$u_0 \frac{\partial v_0}{\partial x} + v_0 \frac{\partial v_0}{\partial y} = -\frac{\partial p_0}{\partial y} + \frac{1}{Re}\left(\frac{\partial^2 v_0}{\partial x^2} + \frac{\partial^2 v_0}{\partial y^2}\right), \tag{9.50}$$

$$\frac{\partial u_0}{\partial x} + \frac{\partial v_0}{\partial y} = 0. \tag{9.51}$$

The second set of equations are the unsteady perturbation equations that govern the $O(\epsilon)$ disturbances. They are

$$\frac{\partial \hat{u}}{\partial t} + u_0 \frac{\partial \hat{u}}{\partial x} + v_0 \frac{\partial \hat{u}}{\partial y} + \hat{u} \frac{\partial u_0}{\partial x} + \hat{v} \frac{\partial u_0}{\partial y} = -\frac{\partial \hat{p}}{\partial x} + \frac{1}{Re}\left(\frac{\partial^2 \hat{u}}{\partial x^2} + \frac{\partial^2 \hat{u}}{\partial y^2}\right), \tag{9.52}$$

$$\frac{\partial \hat{v}}{\partial t} + u_0 \frac{\partial \hat{v}}{\partial x} + v_0 \frac{\partial \hat{v}}{\partial y} + \hat{u} \frac{\partial v_0}{\partial x} + \hat{v} \frac{\partial v_0}{\partial y} = -\frac{\partial \hat{p}}{\partial y} + \frac{1}{Re}\left(\frac{\partial^2 \hat{v}}{\partial x^2} + \frac{\partial^2 \hat{v}}{\partial y^2}\right), \tag{9.53}$$

$$\frac{\partial \hat{u}}{\partial x} + \frac{\partial \hat{v}}{\partial y} = 0. \tag{9.54}$$

Observe that the nonlinear convective terms have been linearized owing to the small disturbances. Therefore, these equations are called the *linearized Navier-Stokes equations*.

Similar to equation (9.31), the vector form of the steady base-flow equations (9.49)–(9.51) are

$$(\mathbf{v}_0 \cdot \nabla)\mathbf{v}_0 = -\nabla p_0 + \frac{1}{Re}\nabla^2 \mathbf{v}_0, \quad \nabla \cdot \mathbf{v}_0 = 0. \tag{9.55}$$

Likewise, the linearized Navier-Stokes equations (9.52)–(9.54) are

$$\frac{\partial \hat{\mathbf{v}}}{\partial t} + (\mathbf{v}_0 \cdot \nabla)\hat{\mathbf{v}} + (\hat{\mathbf{v}} \cdot \nabla)\mathbf{v}_0 = -\nabla \hat{p} + \frac{1}{Re}\nabla^2 \hat{\mathbf{v}}, \quad \nabla \cdot \hat{\mathbf{v}} = 0. \tag{9.56}$$

In compact form, these perturbation equations are written as

$$\frac{\partial \hat{\mathbf{v}}}{\partial t} = \mathcal{L}_{LNS}(\hat{\mathbf{v}}), \tag{9.57}$$

where the differential operator $\mathcal{L}_{LNS}$ contains the spatial derivatives.

In Chapter 6 for discreet systems, we solve the linear(ized) ordinary differential equation(s) for the disturbances to determine the stability characteristics of the system. Because the solution is determined by the eigenvalues of the coefficient matrix, the stability characteristics are established by the finite spectrum of eigenvalues (modes). Stability of continuous systems, such as the column-buckling problem in Section 6.6, results in a differential eigenproblem having an infinite number of eigenvalues. Nevertheless, the stability properties are again established by the nature of the eigenvalues.

In fluid mechanics, we could imagine solving equations (9.49)–(9.51) for the steady base flow $u_0(x,y)$, $v_0(x,y)$, and $p_0(x,y)$, and then using $u_0(x,y)$ and $v_0(x,y)$ in equations (9.52)–(9.54) to solve for the perturbation quantities $\hat{u}(x,y,t)$, $\hat{v}(x,y,t)$, and $\hat{p}(x,y,t)$. If, given some initial conditions on the perturbation quantities, we find that they grow with time to become unbounded, then we say that the base flow is unstable. The linearized Navier-Stokes equations for the disturbances must, in general, be solved numerically, and it is not practical to consider all possible initial conditions (disturbances) to evaluate their stability, so it is advantageous to simplify the analysis by making assumptions about the behavior of the base flow and/or the nature of the disturbances. Both of these simplifications are addressed in the following section.

9.5.3 Modal Analysis

In *modal (eigenvalue) analysis*, the linearized Navier-Stokes equations for the perturbations (9.52)–(9.54) are simplified to the form of a differential eigenproblem. This is done by assuming that the temporal behavior of the disturbance is either exponential growth or decay. In *local (parallel) analysis*, we further assume the form of the disturbances to be infinitesimally small sinusoidal perturbations. The primary advantage of modal analysis is that each mode can be analyzed independently of the others, that is, they are uncoupled. This is the case because the modes are mutually orthogonal (normal) and cannot influence each other. Such an approach is frequently referred to as *normal-mode analysis*. As will be seen in Section 9.5.4, such a separation of modes is not always possible; however, it greatly simplifies the analysis and is often a good starting point for stability analysis.

We first consider the scenario in which the base flow can be assumed to be parallel, that is, only evolving in one coordinate direction. We refer to this as a *local stability analysis* as it only takes into account the local behavior of the base flow in the modal analysis. We then consider the influence of two-dimensionality in the base flow on the modal analysis using *global stability analysis*.

As in Chapter 6, we seek the response of the system to infinitesimally small disturbances. If the disturbances remain small for all time and do not grow in amplitude, then the flow is stable. If at least one (of the infinity) of the modes grows to become large, then the flow is unstable. Asymptotic stability or instability refers to the behavior of the modes as $t \to \infty$.

Local (Parallel) Stability Analysis

A local stability analysis applies when the flow varies slowly in the streamwise direction as compared to the wavelength of instability waves. Here, we consider the situation in which the streamwise velocity distribution changes with the normal coordinate y, but not the streamwise coordinate x. In this case, the base-flow velocity in the streamwise direction is of the form $u_0 = u_0(y)$. From continuity (9.51), if u_0 does not change with x, then the normal base-flow velocity must be such that $v_0 = v_0(x)$ (if a solid wall parallel to the x-axis is present, then $v_0 = 0$). The pressure, which drives the flow in the streamwise direction, can depend upon the streamwise coordinate only; therefore, $p_0 = p_0(x)$. See Example 9.1 for the case of plane-Poiseuille flow, which is an example of parallel flow.

Consideration of a local modal analysis involves imposing sinusoidal disturbances of all possible frequencies on the base flow and determining their stability as time proceeds. Thus, we impose disturbances of the form

$$\begin{aligned}\hat{u}(x,y,t) &= u_1(y)e^{i(\alpha x - \alpha c t)},\\ \hat{v}(x,y,t) &= v_1(y)e^{i(\alpha x - \alpha c t)},\\ \hat{p}(x,y,t) &= p_1(y)e^{i(\alpha x - \alpha c t)},\end{aligned} \tag{9.58}$$

where the real part of the right-hand side is assumed to be taken in order to produce the sinusoidal disturbance. The wavenumber α is real and positive, and the complex wave speed is $c = c_r + ic_i$, where c_r and c_i are the real and imaginary parts of the complex wave speed, respectively.

Representing the sinusoidal disturbances as a complex exponential in this fashion is permissible because if a complex function satisfies a linear differential equation, then the real and imaginary parts each satisfy the same equation individually. To see how the sinusoidal disturbance results from (9.58), consider, for example,[5]

$$\begin{aligned}\hat{u}(x,y,t) &= \mathrm{Re}\left[u_1(y)e^{i(\alpha x - \alpha c t)}\right]\\ &= \mathrm{Re}\left[u_1(y)e^{\alpha c_i t}e^{\alpha i(x - c_r t)}\right]\\ &= \mathrm{Re}\left[u_1(y)e^{\alpha c_i t}\left\{\cos[\alpha(x - c_r t)] + i\sin[\alpha(x - c_r t)]\right\}\right]\\ \hat{u}(x,y,t) &= u_1(y)e^{\alpha c_i t}\cos[\alpha(x - c_r t)].\end{aligned}$$

[5] Note that Re corresponds to the real part of a complex function, while *Re* is the nondimensional Reynolds number. Although this distinction is rather subtle, this is the only location where it must be made.

As a result, the perturbation is in the form of a sine wave with wavenumber α and phase velocity c_r in the x-direction; the wavelength of each disturbance is proportional to $1/\alpha$. The y-dependence of the perturbation is determined by the eigenfunctions $u_1(y)$, $v_1(y)$, and $p_1(y)$ corresponding to each eigenvalue c. If $c_i > 0$ for a given eigenvalue, the amplitude of the corresponding perturbation grows exponentially to become unbounded as $t \to \infty$ with growth rate αc_i. Because equations (9.52)–(9.54) are linear, each mode with wavenumber α may be considered independently of one another. Therefore, we seek the least stable mode, which is the one with the fastest growth rate given by $\alpha \max(c_i)$. The presupposition in modal analysis is that any infinitesimally small initial perturbation can be expressed as a superposition of the necessary individual modes, such as in the form of a Fourier series, but that because the modes are uncoupled, we can evaluate stability of each mode with wavenumber α individually.

For a steady, parallel base flow with $u_0 = u_0(y)$, $v_0 = 0$, and $p_0 = p_0(x)$, the Navier-Stokes equations (9.49)–(9.51) reduce to simply

$$\frac{d^2u_0}{dy^2} = Re\frac{dp_0}{dx}, \tag{9.59}$$

where $Re p_0'(x)$ is a constant for Poiseuille flow (see Example 9.1). Similarly, the disturbance equations (9.52)–(9.54) become

$$\frac{\partial \hat{u}}{\partial x} + \frac{\partial \hat{v}}{\partial y} = 0, \tag{9.60}$$

$$\frac{\partial \hat{u}}{\partial t} + u_0\frac{\partial \hat{u}}{\partial x} + \hat{v}\frac{\partial u_0}{\partial y} = -\frac{\partial \hat{p}}{\partial x} + \frac{1}{Re}\left(\frac{\partial^2 \hat{u}}{\partial x^2} + \frac{\partial^2 \hat{u}}{\partial y^2}\right), \tag{9.61}$$

$$\frac{\partial \hat{v}}{\partial t} + u_0\frac{\partial \hat{v}}{\partial x} = -\frac{\partial \hat{p}}{\partial y} + \frac{1}{Re}\left(\frac{\partial^2 \hat{v}}{\partial x^2} + \frac{\partial^2 \hat{v}}{\partial y^2}\right). \tag{9.62}$$

Substitution of the ansatz[6] (9.58) into the disturbance equations (9.60)–(9.62) leads to

$$\alpha i u_1 + \frac{dv_1}{dy} = 0, \tag{9.63}$$

$$\alpha i(u_0 - c)u_1 + \frac{du_0}{dy}v_1 = -\alpha i p_1 + \frac{1}{Re}\left(\frac{d^2u_1}{dy^2} - \alpha^2 u_1\right), \tag{9.64}$$

$$\alpha i(u_0 - c)v_1 = -\frac{dp_1}{dy} + \frac{1}{Re}\left(\frac{d^2v_1}{dy^2} - \alpha^2 v_1\right), \tag{9.65}$$

respectively. Solving equation (9.63) for u_1 and substituting into equation (9.64) results in

$$-(u_0 - c)\frac{dv_1}{dy} + \frac{du_0}{dy}v_1 = -\alpha i p_1 - \frac{1}{\alpha i}\frac{1}{Re}\left(\frac{d^3v_1}{dy^3} - \alpha^2\frac{dv_1}{dy}\right), \tag{9.66}$$

[6] An *ansatz* is an assumed form of a mathematical statement that is not based on any underlying theory or principle, that is, it is an educated guess.

leaving equations (9.65) and (9.66) for $v_1(y)$ and $p_1(y)$. Solving equation (9.66) for $p_1(y)$, differentiating with respect to y and substituting into equation (9.65) leads to

$$(u_0 - c)\left(\frac{d^2v_1}{dy^2} - \alpha^2 v_1\right) - \frac{d^2u_0}{dy^2}v_1 = \frac{1}{\alpha i}\frac{1}{Re}\left(\frac{d^4v_1}{dy^4} - 2\alpha^2\frac{d^2v_1}{dy^2} + \alpha^4 v_1\right), \tag{9.67}$$

which is the *Orr-Sommerfeld equation*. For a given base flow velocity profile $u_0(y)$, Reynolds number Re, and wavenumber α, the Orr-Sommerfeld equation is a fourth-order ordinary differential generalized eigenproblem of the form

$$\mathcal{L}_1 v_1 = c\mathcal{L}_2 v_1,$$

where the wave speeds c are the (complex) eigenvalues, the disturbance velocities $v_1(y)$ are the eigenfunctions, and $\mathcal{L}_1$ and $\mathcal{L}_2$ are one-dimensional ordinary differential operators in y.

Recall that the Orr-Sommerfeld equation applies for steady, parallel, viscous flow perturbed by infinitesimally small disturbances corresponding to a local analysis. Based on solutions of the Orr-Sommerfeld equation for a range of Reynolds numbers and wavenumbers, we can determine the smallest Reynolds number for which some mode is unstable. We call this the critical Reynolds number and denote it by Re_{cr}. For $Re \leq Re_{cr}$, the real part of c_i is less than or equal to zero for all eigenvalues such that all modes are stable and decay exponentially (or at least do not grow) as $t \to \infty$. For $Re > Re_{cr}$ there is at least one eigenvalue for which the real part is greater than zero, and the flow is asymptotically unstable.

The analysis above is a *temporal stability analysis* for which α is real and c is complex. An instability that results from a temporal analysis is called an *absolute instability* in which infinitesimally small perturbations grow to become unbounded as $t \to \infty$ at a fixed streamwise location. It is also possible to perform a *spatial stability analysis*, for which the wavenumber α is complex and the wave speed c is real. Instabilities that result from a spatial analysis are called *convective instabilities*, in which infinitesimally small perturbations grow to become unbounded at a point moving (convecting) with the flow. At a given streamwise location, the convective instability will appear to grow and then decay as it convects past the point.

If the parallel flow is inviscid, that is, if Reynolds number Re is infinite, then the Orr-Sommerfeld equation (9.67) reduces to

$$(u_0 - c)\left(\frac{d^2v_1}{dy^2} - \alpha^2 v_1\right) - \frac{d^2u_0}{dy^2}v_1 = 0, \tag{9.68}$$

which is a second-order ordinary differential eigenproblem known as the *Rayleigh equation*. For a given parallel base flow $u_0(y)$ and wavenumber α, we determine the complex wave speeds (eigenvalues) c of the Rayleigh equation. Once again, if $c_i > 0$, then the corresponding mode with wavenumber α is unstable.

For the inviscid case governed by the Rayleigh equation, two important theorems can be proven. The first is *Rayleigh's inflection point theorem*, which states that for an inviscid flow to be unstable, the existence of an inflection point in the base velocity profile $u_0 = u_0(y)$ is a necessary condition. *Fjørtoft's theorem* extends Rayleigh's theorem by proving that to be unstable, not only must an inflection point be present, but the velocity profile must be such that

$$\frac{d^2u_0}{dy^2}\left[u_0(y)-u_0(y_{ip})\right]<0$$

somewhere in the flow, where y_{ip} is the location of the inflection point. Essentially, Fjørtoft's theorem places requirements on the behavior of the curvature of the velocity profile above and below the inflection point. Because Rayleigh's and Fjørtoft's theorems are necessary, but not sufficient, for an instability, they can be used to determine whether a parallel inviscid flow is stable (if the theorems are not satisfied), but only whether it might be unstable (if the theorems are satisfied). There are some situations in which Rayleigh's and Fjørtoft's theorems together provide a necessary and sufficient condition for an instability. This is the case, for example, for symmetric flows in a channel.

EXAMPLE 9.2 Consider a parallel modal analysis of plane-Poiseuille flow (see Example 9.1).

Solution: Plane-Poiseuille flow is parallel; therefore, the base flow is a solution of equation (9.59). From Example 9.1, the solution is a parabolic velocity profile given by

$$u_0(y)=y(2-y), \quad 0\leq y\leq 2. \tag{9.69}$$

Note that the base flow is independent of the Reynolds number. Given this velocity profile, we calculate numerically the complex wave speeds (eigenvalues) for a given wavenumber α and Reynolds number Re from the Orr-Sommerfeld equation (9.67) in order to evaluate stability. By performing a large number of such calculations for a range of wavenumbers and Reynolds numbers, we can plot the *marginal stability curve* for plane-Poiseuille flow. This shows the curve in α-Re parameter space for which $\max(c_i)=0$ delineates the regions of parameter space in which the flow is stable and unstable to infinitesimally small perturbations. For plane-Poiseuille flow, the marginal stability curve is shown in Figure 9.4. The critical Reynolds number is $Re_{cr}=5{,}772$, which occurs for a wavenumber of $\alpha=1.02$ (Orszag 1971). Observe

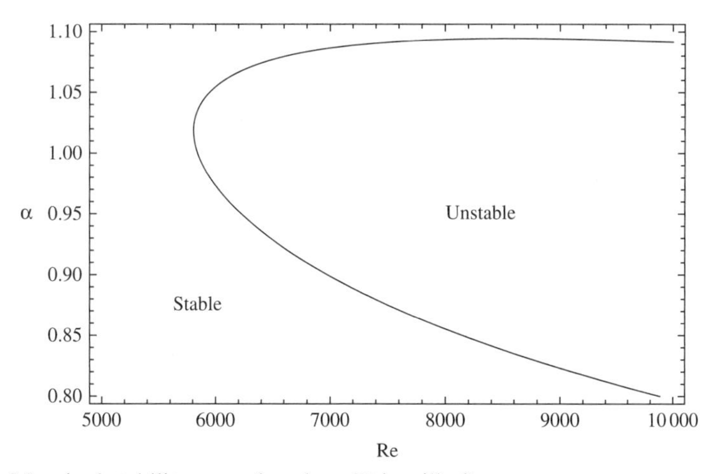

Figure 9.4. Marginal stability curve for plane-Poiseuille flow.

that the velocity profile for plane-Poiseuille flow does not have any inflection points; therefore, it is stable in the context of an inviscid stability analysis based on the Rayleigh equation in the $Re \to \infty$ limit. In contrast, the solutions to the Orr-Sommerfeld equation, which includes viscous effects, are unstable for $Re > Re_{cr}$. Therefore, viscous effects are destabilizing for large, but finite, Reynolds numbers and stabilizing for $Re < Re_{cr}$ when viscous effects tend to damp out the disturbances.

The perceptive reader may be wondering, "Even though the base flow only depends on y, and we only consider disturbances in the x and y directions, disturbances also could occur in the z-direction." This possibility is addressed by a beautiful result known as *Squire's theorem*, which states that each three-dimensional mode has a corresponding two-dimensional mode that is more unstable. Therefore, it is only necessary to perform a two-dimensional stability analysis for parallel flows. It is not necessarily the case, however, that the most unstable two-dimensional modes occur in the (x,y)-plane.

The vast majority of hydrodynamic stability analyses carried out in the twentieth century are local (parallel) modal analysis. Given its limitations, it is remarkable that such a wide variety of stability phenomena have been explained using modal analysis (see, for example, Drazin and Reid 1981 and Drazin 2002). Notwithstanding the many notable successes of local modal analysis, a number of important shortcomings have also come to light. For example, there are several examples in which the critical Reynolds numbers from a local analysis differ significantly from the experimentally observed transitional Reynolds numbers. This is the case, for example, in wall-bounded shear layers, pipe flow, and plane-Couette flow. In fact, in some cases, the linear modal stability analysis predicts that the flow is stable for *all* finite Reynolds numbers, which is certainly not the case in real flows. Some of these shortcomings have been resolved using more general analyses that account for two- and three-dimensional base flows as well as short-time stability behavior in the form of global stability and transient growth analyses, respectively. These are described in the following sections.

Global Stability Analysis

The parallel-flow assumption enlisted in the previous section to obtain the Orr-Sommerfeld and Rayleigh equations is very restrictive. Although much has been learned from this classical approach, the majority of base flows vary substantially in the streamwise direction. Therefore, a more general approach is necessary, but this will come at a substantial cost.

We once again assume that the perturbations are infinitesimally small ($\epsilon \ll 1$) of the form (9.48) and consider the linearized Navier-Stokes equations (9.52)–(9.54) for the perturbations of the base-flow equations (9.49)–(9.51). As such, a linear stability analysis is being considered. In general, the base flow must be computed numerically by solving the coupled Navier-Stokes equations (9.49)–(9.51) to obtain $u_0(x,y)$, $v_0(x,y)$, and $p_0(x,y)$. We then impose the following perturbations

$$\begin{aligned} \hat{u}(x,y,t) &= u_1(x,y)e^{-ict}, \\ \hat{v}(x,y,t) &= v_1(x,y)e^{-ict}, \\ \hat{p}(x,y,t) &= p_1(x,y)e^{-ict}, \end{aligned} \tag{9.70}$$

in place of equation (9.58). Observe that we are no longer assuming the specific form of the disturbances in the streamwise x-direction. Instead, the eigenfunctions $u_1(x,y)$, $v_1(x,y)$, and $p_1(x,y)$ representing the mode shapes are allowed to be two-dimensional. The real part of the disturbances, which involves c_i, are associated with the growth rate of the global modes in time, while the imaginary part, which involves c_r, is the phase velocity or angular frequency. As for parallel flows, substituting the global modes (9.70) leads to the differential eigenproblem

$$\mathcal{L}_1\boldsymbol{\phi}_1 = c\mathcal{L}_2\boldsymbol{\phi}_1.$$

However, the operators $\mathcal{L}_1$ and $\mathcal{L}_2$ are now two-dimensional partial differential operators in x and y rather than one-dimensional ordinary differential operators in y alone, and the eigenfunctions $\boldsymbol{\phi}_1$ contain all of the unknowns in the perturbation equations, including $u_1(x,y)$, $v_1(x,y)$, and $p_1(x,y)$. Whereas the eigenproblem for parallel flows is in the form of an ordinary differential equation, that in a global analysis is a partial differential equation. If the imaginary part of the least stable eigenvalue is positive, that is, $c_i > 0$, then the flow is globally unstable. Note that some authors refer to such an analysis as *BiGlobal* when the base flow is two-dimensional or *TriGlobal* when the base flow is three-dimensional.

After discretizing the partial differential operators $\mathcal{L}_1$ and $\mathcal{L}_2$ using a spectral or finite-difference method, for example, a generalized matrix eigenproblem results that is much larger than that obtained for local analysis. Global stability calculations are extremely costly in terms of computer memory requirements. However, we can take advantage of the fact that we typically care only about the least stable mode, not the full spectrum of eigenvalues. In such scenarios, methods based on Krylov subspaces, for example, the Arnoldi method, are often employed. Even so, it is only relatively recently that the computer power and memory have been available to begin performing such global stability calculations, and there is significant room for improvements in speed and memory resulting from further algorithm development.

Global stability analysis is advancing rapidly and has greatly enhanced our understanding of hydrodynamic stability in a number of settings. This is particularly true for base flows that involve separation, in which there is a region adjacent to a surface where the flow is moving upstream against the balance of the flow.

9.5.4 Nonmodal Transient Growth (Optimal Perturbation) Analysis

Paralleling our developments for stability analysis of discrete systems in Chapter 6, the modal analysis considered in the previous section often provides an incomplete picture of the stability characteristics of continuous fluid flows. Traditional modal analysis only considers the behavior of the least stable mode as $t \to \infty$. Recall from Section 6.6, however, that the analog of a discrete normal system in the continuous case is a system governed by a self-adjoint differential operator (see Section 1.8). Similar to the eigenvectors of a normal matrix, the eigenfunctions of a self-adjoint operator are all mutually orthogonal, and asymptotic stability of the system is governed by the (now infinite) set of eigenvalues of the operator. Systems that are governed by non-normal operators, that is, those that are not self-adjoint, can exhibit the same transient growth behavior observed for non-normal discrete systems (see Section 6.5). Thus, owing to the non-normal character of the linearized Navier-Stokes equations, the full spectrum of unstable modes must be considered. In such cases, the non-normal behavior may lead to significant amplification over a finite-time interval, even for cases in which the large-t asymptotic behavior is stable.

In modal analysis of parallel flows, we assume that the growth or decay of the disturbances is exponential in time and that the streamwise "shape" of the perturbations are prescribed. In global stability analysis, we allow for two- or three-dimensional base flows, but we still consider only exponential growth or decay in time. As in Section 6.5, however, it is sometimes the case that a disturbance can grow for some finite period of time before succumbing to the asymptotic stability behavior as $t \to \infty$ given by the modal analysis. If this growth is sufficient, it is possible that nonlinear effects may come into play prior to the eventual exponential decay for large times and lead to transition to turbulence. Such a mechanism is often referred to as *bypass transition* as it bypasses the traditionally postulated mechanism in which the infinitesimally small and asymptotically unstable linear modes are imagined to grow to a sufficient extent to induce nonlinear effects and transition.

Recall that transient growth (sometimes called algebraic growth in contrast to the exponential growth assumed in modal analysis) is a possibility for discrete systems when the coefficient matrix of the governing equation(s) is non-normal. A matrix is normal if it commutes with its conjugate transpose. A commonly encountered subclass of normal matrices are symmetric matrices.

In the continuous case, the analog of a discrete system governed by a symmetric matrix is a continuous system governed by a self-adjoint differential operator, which is a special case of a normal differential operator. Similar to the eigenvectors of a normal matrix, the eigenfunctions of a normal operator are all mutually orthogonal, and asymptotic stability of the system is governed by the (now infinite) set of eigenvalues of the differential operator, each of which can be considered independently. Non-normal matrices and operators, on the other hand, have nonorthogonal eigenvectors and eigenfunctions, and the coupling of all modes must be considered as they can interact to produce transient-growth behavior. Therefore, continuous systems that are governed by non-normal operators can exhibit the same transient growth behavior observed for non-normal discrete systems. Because the differential operator in the linearized Navier-Stokes equations (9.57) is non-normal, transient growth is possible, even for parameter ranges for which the flow is asymptotically stable according to a modal analysis as described in Section 9.5.3. A comparison of modal and nonmodal transient-growth analysis is summarized in Table 9.1. In particular, note that we no longer assume the form of the disturbance or its exponential growth in time; instead, we determine the initial condition that leads to the maximum growth rate over some finite time interval.

Notwithstanding the differences between the manner in which discrete and continuous systems are represented, the transient-growth analysis presented here for a continuous fluid flow will closely parallel that for discrete systems in Section 6.5. Not only does transient-growth analysis provide a quantitative description of the disturbance behavior over short time periods, it leads to the same asymptotic behavior for $t \to \infty$ predicted by a local or global modal analysis for one- or two-dimensional base flows, respectively. Whereas local or global modal analysis leads to an eigenproblem to evaluate stability, we will find that nonmodal transient-growth

Table 9.1. *A summary of the limitations of modal (local and global) analysis as compared to nonmodal (transient-growth) analysis.*

Modal Analysis	Nonmodal Analysis
A particular perturbation shape is assumed for local analysis.	No perturbation shape or time horizon is assumed.
Exponential growth or decay in time of the disturbances is assumed.	No prescribed time evolution of disturbances.
Only addresses asymptotic stability characteristics of a system as $t \to \infty$.	Addresses finite-time and asymptotic stability ($t \to \infty$) properties.
Only applies to stability of steady base flows, that is, steady equilibrium states.	Can apply to steady and unsteady base flows.
Only takes into account growth or decay of each individual mode.	Takes into account the interaction between non-normal modes.
Applies to linear (or linearized) perturbation equations.	Can treat linear stability or be extended to treat nonlinear effects.

analysis requires solution of an initial-value problem in order to determine the optimal initial disturbance that leads to the maximum amplification of its kinetic energy within a finite-time horizon.

The initial perturbation that experiences maximum growth over a specified time range is called the *optimal perturbation* and is found using variational methods. As has been the case throughout this chapter, we treat the two-dimensional case for simplicity with straightforward extension to three dimensions. We define the energy $E(t)$ of a disturbance velocity $\hat{\mathbf{v}}(x,y,t)$ at time t using the norm operator as follows:

$$E(t) = ||\hat{\mathbf{v}}(t)||^2 = \langle \hat{\mathbf{v}}(t), \hat{\mathbf{v}}(t) \rangle = \iint_A \hat{\mathbf{v}}(t) \cdot \hat{\mathbf{v}}(t)\, dA,$$

where $\langle \cdot, \cdot \rangle$ is the inner product, and A is the area of the domain. Therefore, $E(t)$ is related to the kinetic energy of the disturbance integrated over the domain at time t. Note that the outcome of the subsequent analysis depends on the choice of norm used to define the energy. We seek the initial perturbation $\hat{\mathbf{v}}(x,y,0)$ that produces the greatest relative energy gain during the time interval $0 \le t \le t_f$ defined by

$$G(t_f) = \max_{\hat{\mathbf{v}}(0)} \frac{E\left[\hat{\mathbf{v}}(t_f)\right]}{E\left[\hat{\mathbf{v}}(0)\right]},$$

which is the ratio of the perturbation energy at $t = t_f$ to that at $t = 0$. In other words, we seek the initial disturbance that maximizes the gain functional

$$G[\hat{\mathbf{v}}] = \frac{||\hat{\mathbf{v}}(t_f)||^2}{||\hat{\mathbf{v}}(0)||^2} = \frac{\langle \hat{\mathbf{v}}(t_f), \hat{\mathbf{v}}(t_f) \rangle}{\langle \hat{\mathbf{v}}(0), \hat{\mathbf{v}}(0) \rangle}. \tag{9.71}$$

Transient growth occurs when the maximum value of the gain is $G > 1$ for finite time, and it is not unusual for the maximum gain to be several orders of magnitude. If the flow is asymptotically unstable according to a linear modal analysis, then the maximum gain will be such that $G \to \infty$ as $t \to \infty$ (with or without transient growth).

As in Sections 2.7.3 and 6.5, we define the augmented functional such that (9.71) is maximized subject to the differential constraint that the linearized Navier-Stokes equations for the disturbances (9.57), which are $\partial\hat{\mathbf{v}}/\partial t = \mathcal{L}_{LNS}(\hat{\mathbf{v}})$, are satisfied for all space and time in the range $0 \le t \le t_f$. The augmented functional is then

$$\tilde{G}[\hat{\mathbf{v}},\mathbf{v}^*] = \frac{\langle\hat{\mathbf{v}}(t_f),\hat{\mathbf{v}}(t_f)\rangle}{\langle\hat{\mathbf{v}}(0),\hat{\mathbf{v}}(0)\rangle} - \int_0^{t_f}\left[\iint_A \mathbf{v}^{*T}\left(\frac{\partial\hat{\mathbf{v}}}{\partial t} - \mathcal{L}_{LNS}(\hat{\mathbf{v}})\right)dA\right]dt,$$

where $\mathbf{v}^*(x,y,t)$ is a vector of Lagrange multipliers, which are the adjoint variables of the perturbation equations.[7] Taking the variation of the augmented functional and setting equal to zero, performing the necessary integration by parts, and collecting common variations as in Section 6.5, leads to the following Euler equations

$$-\frac{\partial\mathbf{v}^*}{\partial t} + (\mathbf{v}_0\cdot\nabla)\mathbf{v}^* - \mathbf{v}^*\cdot(\nabla\mathbf{v}_0)^T = -\nabla p^* + \frac{1}{Re}\nabla^2\mathbf{v}^*, \quad \nabla\cdot\mathbf{v}^* = 0, \tag{9.72}$$

and initial conditions of the form

$$\hat{\mathbf{v}}(0) = \frac{\langle\hat{\mathbf{v}}(0),\hat{\mathbf{v}}(0)\rangle^2}{2\langle\hat{\mathbf{v}}(t_f),\hat{\mathbf{v}}(t_f)\rangle}\mathbf{v}^*(0), \tag{9.73}$$

$$\mathbf{v}^*(t_f) = \frac{2}{\langle\hat{\mathbf{v}}(0),\hat{\mathbf{v}}(0)\rangle}\hat{\mathbf{v}}(t_f), \tag{9.74}$$

which are analogous to equations (6.14)–(6.16). Equations (9.72), which can be written in the compact form

$$-\frac{\partial\mathbf{v}^*}{\partial t} = \mathcal{L}^*_{LNS}(\mathbf{v}^*),$$

are the *adjoint equations* of the linearized Navier-Stokes equations (9.56), and the starred variables are the adjoint variables (Lagrange multipliers) of $\mathbf{v}$ and p. The operator $\mathcal{L}^*_{LNS}$ is the adjoint of that in the linearized Navier-Stokes equations denoted by $\mathcal{L}_{LNS}$. Observe the negative sign on the time derivative indicating that the adjoint equations are to be integrated backward in time from $t = t_f$ to $t = 0$.

In order to determine the optimal initial perturbation $\hat{\mathbf{v}}(0)$ that maximizes the gain G for a given terminal time t_f, we solve the linearized Navier-Stokes equations (9.56) forward in time using the initial condition (9.73), which converts the adjoint solution at $t = 0$ into an initial condition for the perturbations at $t = 0$. The solution at the terminal time $\hat{\mathbf{v}}(t_f)$ is then used in equation (9.74) to determine the initial condition for the backward-time integration of the adjoint equations (9.72). This procedure is repeated iteratively until a converged solution for the linearized Navier-Stokes equations and their adjoints are obtained, from which the optimal initial perturbation $\hat{\mathbf{v}}(x,y,0)$ is determined. Note that for each terminal time t_f, the gain is maximized over all possible initial conditions. Therefore, the optimal initial perturbation for each terminal time will be different, and the gain curve represents the envelope of individual growth curves for each possible initial condition.

If Re_{cr} is the lowest Reynolds number for which at least one of the modes is asymptotically unstable, and Re_{ener} is the lowest Reynolds number for which

[7] Previously, we denoted Lagrange multipliers by Λ. However, we found in Section 6.5 that the Lagrange multiplier(s) turn out to be the adjoint variable(s) of the governing equation, which we take advantage of here from the start.

transient growth occurs, then the following possibilities can arise. For normal systems, $Re_{ener} = Re_{cr}$, and for non-normal systems $Re_{ener} \leq Re_{cr}$, that is, the system may experience transient growth for Reynolds numbers that would otherwise be asymptotically stable.

For the linear stability analysis of small perturbations about steady base flows considered here, the transient growth analysis for non-normal systems can be reduced to performing a singular-value decomposition (SVD) to obtain the optimal disturbance. For unsteady base flows $\mathbf{v}_0 = \mathbf{v}_0(x,y,t)$, however, it is necessary to use the variational (or adjoint) approach to evaluate transient growth.

In order to extend transient-growth analysis to the full nonlinear Navier-Stokes equations (9.28)–(9.30), it is necessary to use the variational approach outlined here. Recall that in the linear case, the forward and backward integrations of the linearized Navier-Stokes and adjoint equations, respectively, can be carried out with information communicated between equations only at the initial ($t = 0$) and terminal ($t = t_f$) times of the calculations. In the nonlinear case, however, it becomes necessary to utilize information from the adjoint solution throughout the interval $0 \leq t \leq t_f$ while computing the nonlinear disturbance equations forward in time, and vice versa. This requires that solutions for both the forward and backward problems be saved throughout the time horizon considered for use in the other calculation. Therefore, there is additional computational and computer memory requirements for transient-growth analysis of nonlinear disturbance equations.

Squire's theorem does not apply in the context of transient-growth analysis. Therefore, it is often necessary to consider three-dimensional disturbances even for one- or two-dimensional base flows. The results reflected in the next example for plane-Poiseuille flow are for a two-dimensional transient-growth analysis. Maximum gains many times higher can be obtained by considering three-dimensional disturbances as shown in Butler and Farrell (1992).

EXAMPLE 9.3 Consider a transient-growth analysis of plane-Poiseuille flow.
Solution: Recall from Example 9.2 that a local modal analysis leads to a critical Reynolds number for instability of $Re_{cr} = 5{,}772$. For $Re < Re_{cr}$, the flow is asymptotically stable to small perturbations. Now consider a transient-growth analysis to determine the optimal disturbance for the parallel base-flow profile

$$\mathbf{v}_0 = u_0(y)\mathbf{i} = y(2-y)\mathbf{i}, \quad 0 \leq y \leq 2. \tag{9.75}$$

Farrell (1988) and Reddy and Henningson (1993) performed such an analysis as summarized here.

Plots of the maximum gain G versus terminal time t_f for three cases are shown in Figure 9.5, with α being the streamwise wavenumber of the optimal initial disturbance. Transient-growth results are reflected in the gain curves labeled "Stable" and "Unstable." The stable case is for $\alpha = 1$ and $Re = 5{,}000 < Re_{cr}$. Observe that there is a transient growth for finite times before decaying exponentially with time according to the asymptotic stability characteristics predicted by the linear modal analysis. The maximum gain is more than 30 times the magnitude of the initial perturbation. Note that it is not unusual for the maximum gain to be several orders of magnitude larger than the energy of the initial perturbation. The unstable case is for $\alpha = 1$ and

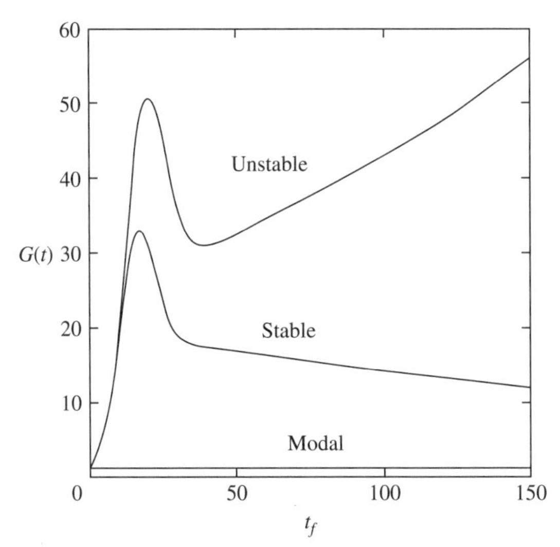

Figure 9.5. Maximum gain versus terminal time from a transient-growth analysis for plane-Poiseuille flow. The unstable case is for $\alpha = 1$ and $Re = 8{,}000$; the stable case is for $\alpha = 1$ and $Re = 5{,}000$; the modal case is for $\alpha = 1$ and $Re = 8{,}000$. (From Reddy and Henningson 1993, Figure 9).

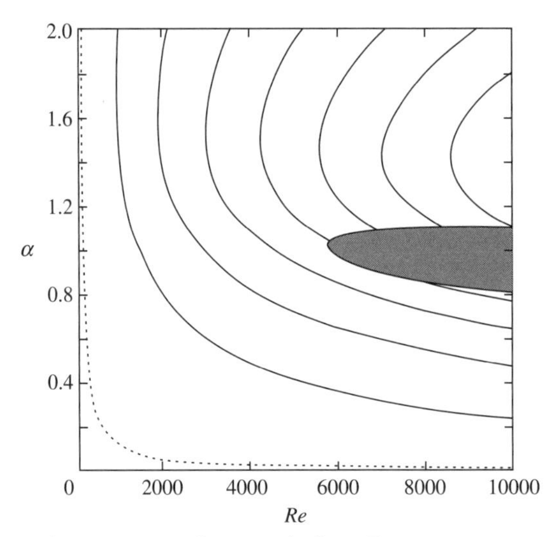

Figure 9.6. Contours of constant maximum gain in α-Re parameter space, with the dashed curve being $G = 1$, that is, marginal stability for transient growth. The shaded region is asymptotically unstable from a linear modal analysis and is the same as the marginal stability curve in Figure 9.4. (From Reddy and Henningson 1993, Figure 8).

$Re = 8{,}000 > Re_{cr}$. Once again there is substantial transient growth for finite times, with the maximum gain being upward of 50 times the initial energy. In this case, however, the flow is asymptotically unstable; therefore, the gain $G \to \infty$ as $t \to \infty$ consistent with the linear modal analysis. The curve labeled "Modal" is the gain computed for the single mode corresponding to $\alpha = 1$ and $Re = 8{,}000$. Although this mode is asymptotically unstable, the transient growth for the same set of parameters is significantly larger owing to the non-normal superposition of many eigenfunctions.

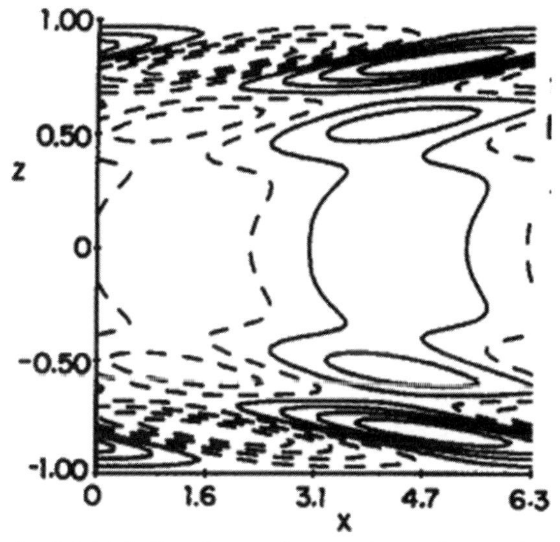

Figure 9.7. The optimal initial disturbance that produces the maximum gain for $\alpha = 1$, $Re = 10{,}000$, and $t_f = 20$. (From Farrell 1988, Figure 8).

Based on a series of transient-growth calculations for a range of Reynolds numbers and streamwise wavenumbers α, a neutral stability curve is shown in Figure 9.6. The neutral stability curve from modal analysis demarcating the asymptotically unstable shaded region is reproduced from Figure 9.4 revealing the critical Reynolds number for asymptotic instability of $Re_{cr} = 5{,}772$. The additional contours represent curves of constant maximum gain in the α-Re parameter space from transient-growth computations, with the dashed curve being the $G = 1$ contour representing neutral gain of the initial disturbance. Allowing for transient growth indicates that $Re_{ener} = 87.7$, which is significantly lower than the critical asymptotic value. For $Re_{ener} \leq Re < Re_{cr}$, a substantial transient growth is observed for short times prior to succumbing to asymptotic decay corresponding to the eigenvalues from a modal analysis. Although each of the modes decays exponentially, the nonorthogonal superposition of these exponentially decaying modes can give rise to short-term transient growth. While the asymptotic fate of the modes is determined by their respective eigenvalues, they do not account for short-term transient growth of the initial disturbances. For each point on this parameter map that experiences transient growth, there is an optimal initial perturbation that produces this maximum gain. For example, the optimal initial perturbation that produces the maximum gain for $\alpha = 1$ and $Re = 10{,}000$ is shown in Figure 9.7. Observe that the optimal initial disturbances are elongated in the streamwise direction and are adjacent to the upper and lower walls of the channel.

The approach used in Section 6.5 and in the present section to analyze transient growth of optimal initial perturbations foreshadows the methods used in optimal control of discrete and continuous systems in Chapter 10. For recent reviews of global stability and transient-growth analysis in fluid mechanics, see Schmid and Henningson (2001), Theofilis (2003), Chomaz (2005), Schmid (2007), Theofilis (2010).

9.5.5 Energy Methods

The *energy method* provides an alternative, and complementary, approach to stability analysis. As stated by Drazin (2002):

> The energy method is complementary to the linear theory of stability in the sense that the linear theory may show that a given flow is unstable to some perturbations but not that the flow is stable to all, whereas the energy method may show that a flow is stable to all perturbations but not that the flow is unstable to some.

The energy method also proves to be complementary to nonmodal transient-growth analysis as well because it does not assume anything about the nature of the disturbance. Rather than providing *necessary* (and sometimes *sufficient*) criteria for instability of fluid flows as is the case with modal analysis, the energy method provides *sufficient* conditions for stability. In other words, a flow can be determined to be stable to any disturbance below some Reynolds number, but we can only say that the flow *might* be stable or unstable for higher Reynolds numbers. As a consequence, the sufficient criteria for stability from an energy analysis rarely coincides with the necessary condition for instability from a linear modal analysis.

The evolution of the kinetic energy of the disturbances in a flow field is the focus of the energy method. The time rate of change of the integrated kinetic energy per unit mass of the disturbances over the entire domain is given by the *Reynolds-Orr energy equation*, which can be derived from the perturbation equations. If the total (integrated) kinetic energy of the disturbances decreases with time for a given base flow, then the base flow is stable in the mean. Variational techniques can be used to rigorously determine sufficient conditions for stability with respect to *any* disturbance. Therefore, the results of an energy analysis apply to disturbances of any magnitude and are not limited to the linear regime with its infinitesimally small perturbations.

The primary feature of energy methods is their ability to lead to rigorous and provable criteria for stability. Because these criteria are only sufficient for stability, however, they often under predict critical Reynolds numbers for instability, sometimes significantly. For example, recall that plane-Poiseuille flow has a critical Reynolds number from a linear modal analysis of $Re_{cr} = 5,772$, and that a nonmodal analysis reveals that transient growth occurs for $Re_{ener} \geq 89$. In contrast, an energy analysis for plane-Poiseuille flow shows that the flow is stable to all (even nonlinear) disturbances for $Re < 49.6$ (Joseph and Carmi 1969). Although this provides a lower bound for Re_{ener}, it is an overly restrictive one that provides little practical insight. More generally, Serrin (1959) proved that *any* steady or unsteady viscous flow in a bounded domain is stable to arbitrary disturbances of any magnitude for $Re = 5.71$. Mathematically, this result is remarkable for its generality; practically, however, it provides a very poor estimate of the critical Reynolds number for the onset of instability in most real flows.

Before transient growth analysis became computationally tractable, the use of energy methods displayed some success in rigorously obtaining sufficient conditions for stability in simple, primarily parallel, flows. It has found limited application to more complex base flows, and the sufficient criteria obtained are often very conservative; therefore, it has largely been subsumed by transient-growth analysis, which is becoming increasingly feasible for flows in complex geometries. Although it is based on variational methods, the energy method will not be considered further owing to its limited applicability. See Sections 26.2 and 53 of Drazin and Reid (1981) and Sections 5.3 and 8.6.1 of Drazin (2002) for an introduction to energy methods, and see Joseph (1976) for a more comprehensive treatment of energy methods. Reddy and Henningson (1993) consider linear stability, energy, and transient growth analyses for plane-Poiseuille and Couette flows.

9.6 Flow Control

There are significant overlaps and commonalities between transient-growth stability analysis and that for flow control. Certainly, the transient growth analysis bears strong similarities to the optimal control methodology outlined in Chapter 10 with regard to the forward- and backward-time integrations of the state and adjoint equations. It is also the case that the objective is often to control the instabilities that are elucidated through stability analysis. For example, the transient growth analysis discussed in Section 9.5.4 provides valuable information for control as it reveals the initial perturbation (location and modes) that leads to the strongest transient growth of the instability. Therefore, these optimal perturbations are the ones that most logically should be the target of control owing to the flow's sensitivity to such disturbances. Therefore, stability analysis can provide qualitative information and increased understanding of the flows for which control is to be implemented, and it provides quantitative information that gives guidance concerning the location and type of actuation that would be most effective. In addition, control is being implemented to reduce the effects of turbulence, eliminate separation, and reduce acoustical noise, for example.

There is a long history of flow control research and practice, particularly in aerodynamic applications where the objective typically is to manage aerodynamic forces, such as increasing lift and decreasing drag. Flow control techniques can broadly be classified into *active* and *passive* depending upon whether external energy is introduced into the flow or not, respectively. The goal of *active flow control* is to leverage small energy inputs via the control mechanism into large changes in the overall flow field in such a way as to improve performance and/or efficiency.

The typical heuristic approach to active flow control is as follows: 1) choose or develop an actuator or approach that modifies a flow field typically through the boundary conditions or a body force, 2) test the approach experimentally or computationally on a prescribed problem, and 3) compare the control and no control behavior to determine the effectiveness of the control strategy. Such an ad hoc approach to flow control has developed into an art form combining fundamental physical understanding of the underlying flow physics and trial and error. Although much has been learned through this heuristic approach, there are several issues with its execution: 1) the approach requires a detailed physical understanding of the underlying flow in order to devise effective means of flow control, 2) the measures of effectiveness are typically too qualitative and subjective, appealing to an imprecise notion of how the flow should behave, and 3) there is often little effort to quantitatively compare various control schemes with regard to their effectiveness and energy input requirements.

A more formal approach to active flow control is to utilize *optimal control theory* to determine the best means of accomplishing a desired control objective. Modern computational fluid dynamics (CFD) techniques are being combined with optimal control theory to develop a formal and systematic framework with which to investigate active flow control in a variety of settings. This systematic approach will enable us to determine the optimal, that is "best," active flow control approach for specific flow configurations. According to Kim and Bewley (2007), "The incorporation of model-based control theory into many open problems in fluid mechanics presents a host of new opportunities." The approach is as follows: 1) specify a cost functional to be minimized, 2) determine the optimal solution for the state and control variables required to minimize the cost functional and satisfy the state (Navier-Stokes) equations, and 3) choose/develop control actuators that most closely emulate the optimal control solution. Selection of the cost functional requires a clearly articulated prescription of the desired controlled flow properties and conditions – in other words what is meant by *optimal* control.

Optimal control theory is a standalone field with widespread applications in many fields, and it will be introduced in Chapter 10. Its application to fluid mechanics has only recently been pioneered by Abergel and Temam (1990). There are a number of investigations that illustrate how model-based optimal control has been successfully applied to active flow control problems in which an instability can be exploited by the control mechanism to significantly alter the flow. See, for example, the review articles by Bewley (2001), Collis, Joslin, Seifert, and Theofilis (2004), and Kim and Bewley (2007) and the book by Gunzburger (2003).

Previous research utilizing optimal control theory in fluid dynamics has primarily been limited to consideration of linear state and adjoint equations, often to control instabilities for which such a framework can be justified. As we see in Section 9.3, however, the governing state equations of fluid mechanics, the Navier-Stokes equations, are nonlinear. Therefore, one must choose how the fundamentally nonlinear nature of the governing equations of fluid mechanics are to be addressed relative to the well-developed linear control theory. Four possibilities exist: 1) linearize the governing equations about a steady or mean flow, 2) retain the nonlinear state equations, but linearize the control problem at each time for which control is to be determined, 3) apply suboptimal control in which the control calculation is performed in a quasi-steady manner at each time, or 4) implement optimal control based on the fully nonlinear state and control (adjoint) equations throughout. Which choice is appropriate depends upon the underlying behavior of the flow to be controlled. In many cases in which a hydrodynamic instability is the target for control, a steady or mean flow exists about which to linearize the state equations, justifying use of approach (1). In scenarios for which the nonlinear nature of the state equations cannot be avoided, approach (2) is utilized such as in Bewley, Moin, and Temam (2001). Min and Choi (1999) applied approach (3), sub-optimal control, to the flow about a circular cylinder. The fully nonlinear approach (4) has received relatively little attention owing to the complexity and enormity of the resulting numerical problem to be solved in the context of control of fluid flows.

The most critical step in obtaining an effective control methodology is selection of an appropriate cost functional as it defines what is meant by *optimal* control. Making the selection, particularly in fluid dynamical applications, requires a clear understanding of the physics of the flow to be controlled as it is important to target the cause, rather than the effect, of the phenomena. It is common to choose cost functionals that are formulated in terms of the squares (or products) of energies per unit volume in order to minimize their magnitudes, ensure dimensional consistency, and aid in physical interpretation.

The various approaches to flow control can be either classified as those that seek to modify the flow through changes in boundary conditions, such as blowing, suction, moving wall, or shape change, and those that seek to modify the flow through a body force internal to the flow, such as buoyancy effects, magnetohydrodynamics (MHD), electrohydrodynamics (EHD), Lorentz forces, plasma actuators, and so on. In the optimal control approach, however, we do not need to decide on a specific method of actuation, only on whether it is domain or boundary based. However, the optimal control approach does require one to decide explicitly how the flow should behave through specification of an appropriate cost functional. Just as with the actuation methods, the cost functional, which in practice would be linked to measurements from sensors, can be based on quantities throughout the domain, such as velocity, or only at the boundaries, such as wall shear stress. The domain- or boundary-based performance measures may be applied in one of two ways: 1) over a specified time interval $0 \leq t \leq t_f$, in which case the control is determined that will minimize the cost functional over the entire time interval, or 2) at a terminal time $t = t_f$, in which case the control is determined over the time interval $0 \leq t \leq t_f$ in order to minimize the cost functional at $t = t_f$. Although any of the control approaches may be combined with any of the cost functionals to provide an optimal control methodology, boundary-based performance measures are more practical as they only require surface-based sensors to determine the state of the flow.

The variational approach to optimal control of discrete and continuous systems will be addressed in Chapter 10 in a broad context. The example in Section 10.5.3 illustrates how these methods can be applied to optimal control of fluid flows specifically, and partial differential equations generally, using a simple example based on the plane-Poiseuille flow that has served as a unifying example throughout this chapter.

PART III

OPTIMIZATION

10 Optimization and Control

> Catbert: Wally, you can't float through life with no goals and no ambition.
> Wally: You misjudge me. I have my entire career planned out. My five-year plan is to avoid any sort of work in which my individual accomplishments can be measured. I'll hoard knowledge about one of our legacy systems so I seem indispensable. When I get to within four years of retirement, I'll only work on projects that have a five-year payback. I'll protect my cardiovascular system by getting plenty of naps and not caring about the quality of my work. Then I'll stick a straw in our pension fund and suck on it for the next forty years.
>
> (*Dilbert* Comic by Scott Adams)

The focus of Chapters 4 through 9 has been to develop physical variational principles and their associated differential equations of mathematical physics. The number and variety of physical phenomena captured in Hamilton's principle are truly remarkable and are a testament to the fundamental nature of the law of conservation of energy on which it is based. Whereas our emphasis has been to elucidate the laws of physics using calculus of variations, we now turn our attention to optimization and control of these systems. Interestingly, the same variational methods used to formulate the governing Euler equations provide the framework for performing their optimization and control. The governing Euler-Lagrange equations of motion obtained in the previous chapters become the differential constraints for control of such dynamical, fluid mechanical, and electromagnetic systems.

We have already encountered the following examples of optimization thus far: optimization of a river-crossing trajectory in Section 1.3.3, brachistochrone problem in Section 2.2, and transient growth (optimal perturbation) stability analysis in Sections 6.5 and 9.5.4. In fact, two of the first problems tackled during the development of the calculus of variations were optimization problems. Johann Bernoulli proposed and solved the brachistochrone problem (optimize the path in order to minimize the time of travel of a falling object), and Isaac Newton solved a minimum drag problem (optimize the body shape in order to minimize drag) (see Section 10.2).

Engineers, economists, and, indeed, each one of us have always had as our objective to optimize the systems, processes, and entities that we design, manage, and operate. Traditionally, however, such "optimization" has been based on intuition, trial and error, and experience. Each field has developed its own tools and methods that have become accepted practice in each respective arena. Over time these become increasingly mathematically based, and to varying degrees, these heuristic optimization techniques are yielding to the unified theory of optimization and control based on variational principles presented in this chapter. Because optimization and control are being applied so broadly, including financial optimization, shape optimization, grid generation, and image processing, it is instructive to view these topics within a common framework.

Optimization and control are areas in which the calculus of variations find some of its most far-reaching applications and where the mathematics and the applications are very tightly integrated. What we have considered thus far in this text is regarded as *analysis*. Analysis of systems governed by differential equations requires: 1) specification of the system geometry, 2) prescription of the initial and/or boundary conditions on the state variable(s), and 3) application of the appropriate governing (state) equations. Given these ingredients, analysis determines the state of the system by solving the resulting problem. Optimization and control may be regarded as "analysis in reverse." Given a desired target state of the system, we determine the geometry, boundary and/or initial conditions, or control input to the system that most closely produces the target state. If it is the "best" means of producing the desired objective, then it is referred to as *optimal control* (see Sections 10.4–10.6). In the case of discrete systems, the state is adjusted through the action of an external control input, or forcing, incorporated into the state equations. For continuous systems, optimal control can be implemented through forcing acting throughout the domain or optimizing the boundary conditions in order to accomplish the desired objective. Optimizing the geometry of the domain to achieve an objective is called *shape optimization* (see Section 10.2).

If the optimization or control objective is defined by specification of an algebraic *objective*, or *cost*, *function*, then it is addressed using *differential calculus* as in Section 1.4. If the optimization or control objective is defined by specification of an integral *objective*, or *cost*, *functional*, then it is addressed using *variational calculus*. The latter is the focus of this chapter in the context of both discrete and continuous systems. We emphasize general approaches based on variational methods rather than specific results for narrow classes of systems and their optimization or control so as to emphasize the versatility of the variational approach applied to optimization and control. We will also supplement the classical variational approach with a brief discussion of Pontryagin's principle, which allows for piecewise continuous control inputs, unlike the continuously differentiable control required by the classical approach. This is necessary because even Wally from the *Dilbert* comic recognizes that there are times when discontinuous changes in the control input provide the optimal solution to an objective.

It is helpful at this point to clarify what we mean when we use the terms *discrete* and *continuous* as they have related, but subtly different, connotations in various settings. Beginning in Chapter 4 on Hamilton's principle, we have used the terms discrete and continuous in their usual classical mechanics context to describe the system under consideration. Discrete systems are those made of up discrete nondeformable bodies, while continuous systems are comprised of continuous deformable bodies. All of the systems considered, including the discrete systems, are continuous in time; thus, the distinction between a discrete or continuous system indicates whether it is discrete or continuous in space, that is, whether the mass is concentrated in discrete objects or distributed continuously throughout the domain. In mathematics generally, and in the theory of differential equations in particular, the terms discrete and continuous are used in reference to the variables being considered. If they are continuous functions of time and/or space, then we regard it as a continuous problem; that is, the variables vary continuously with time and/or space. The problem is discrete if the variables are broken up into discrete intervals, in which they are only evaluated at a finite number of points in the temporal and/or spatial domain. Thus, we make a distinction between the actual *system* and the mathematical *model* that we use to express its physics. All mathematical *models* in this text are continuous; however, we apply these models to both discrete and continuous *systems*. Discrete models are formulated in terms of vectors and matrices; continuous models are formulated in terms of functions and differential operators. Because all of our models in this text are continuous, the continuous models of discrete systems are finite dimensional and comprised of ordinary differential equations, whereas the continuous models of continuous systems are infinite dimensional and typically consist of partial differential equations. If we convert our continuous model into a discrete one to solve it numerically, for example, we say that it has been *discretized*. When reading this or other texts, then, take note of whether it is the actual system being described or its mathematical model.

10.1 Optimization and Control Examples

In order to give a flavor for the remarkable generality of the variational approach to optimization and control, here are some examples that can be addressed using the techniques to be developed in this chapter:

1. One of the largest operating expenses in the commercial aircraft industry is the cost of fuel. Even small reductions in the total drag, which is the resistance to air flow owing to friction, of an aircraft can have a measurable impact on total fuel consumption. With this in mind, an aerospace engineer would like to determine the optimal shape of the wing of a new aircraft in order to minimize the drag under cruise conditions. At the same time, the engineer would like to maximize the wing's lift, which keeps the aircraft aloft. See Section 10.2 for a classical example of aerodynamic shape optimization first carried out by Isaac Newton.
2. A company generates a profit each year that is the difference between its revenues and expenses. Two of the most important financial questions a company must answer are, "How do we maximize our current profit, that is, maximize revenue and minimize expenses?" and "What is the best use of that profit in order to maximize future profitability?" See the *profit reinvestment* example in Sections 10.3 and 10.4.4 for a simple treatment of the second question.
3. A dynamical system has been outfitted with a set of sensors and actuators. The sensors are used to determine the current state of the system, and the actuators are to be used to provide control inputs into the system in order to force the system into a particular state. A controls engineer is asked to determine the optimal actuator configuration and input that produces the desired state

while minimizing the total control input to the system. See the examples in Sections 10.4–10.6 for the application of *optimal control theory* to discrete and continuous systems.

4. A computational fluid dynamics (CFD) expert is using a computer code that takes a very long time to complete a single run for one set of input parameters (several days, or even weeks, is not unusual). In order to minimize the computational time, the CFD expert would like to determine the best computational grid on which to discretize the problem that minimizes the total number of grid points required to accurately perform the calculations. See Chapter 12 for an introduction to grid generation based on variational principles.
5. A space agency seeks to launch a probe that will rendezvous with the planet Saturn. They would like to reach this objective using a balance between minimizing the amount of fuel required (to save on weight) and minimizing the time of travel from Earth to Saturn (to complete the program before the project manager retires).

You will observe the frequent use of terms such as *optimal*, *best*, *minimize*, and *maximize* in these examples. These are questions that are tailor-made for the calculus of variations to answer in a formal way. In the following sections we show how shape optimization, financial optimization, and control theory are implemented in a variational calculus context and how this variational approach to controls is related to the adjoint approach to differential operators introduced in Section 1.8.

10.2 Shape Optimization

We have already found that shapes observed in physical settings often correspond to some optimization principle. For example, the shape of a hanging cable is found to be that which minimizes its potential energy (see Section 2.7.1), and the shape of a soap film is such that the surface energy (area) is minimized (see Section 2.1.4). In addition, we consider minimal surfaces in a more general context in Section 2.5.2, and the brachistochrone problem treated in Section 2.1.4 can be framed as a shape optimization problem.

Motivated by such physical shape optimization principles, engineers in a wide variety of fields have sought to develop techniques for determining the best – or optimal – shapes to accomplish some objective. In the previous section, for example, we refer to shape optimization in the context of designing airfoil shapes in order to minimize drag and maximize lift. In fact, one of the first problems solved using the calculus of variations was the minimum-drag nose shape for a body of revolution traveling through a fluid. This problem was solved by Isaac Newton in 1687; however, the model that he used for aerodynamic forces is found to be applicable for hypersonic (very high speed) flows[1] only, for which Newton could not have even conceived. This problem is treated in the following example.

[1] Hypersonic flow corresponds to velocities that are at least five to six times that of the speed of sound in air. For example, the Space Shuttle experiences flows of twenty-five times the speed of sound during re-entry into the Earth's atmosphere.

EXAMPLE 10.1 Determine the body of revolution about the x-axis that produces the least pressure drag when moving through a uniform hypersonic fluid with speed U as illustrated in Figure 10.1. We assume that $x(r)$ is monotonic, such that no part of the surface shields another part.

Solution: From the geometry, we see that $dr = \cos\theta ds$, in which case $\cos\theta = dr/ds$. As usual, the arc length is given by

$$ds = \sqrt{1 + [x'(r)]^2} dr.$$

Newton introduced the following expression for the approximate pressure distribution in the x-direction for a body traveling through a fluid. This expression turns out to be a good approximation only for hypersonic flows. The approximation for the pressure component in the x-direction p_x is

$$p_x = -\frac{1}{2}\rho U^2 \cos^2\theta,$$

where ρ is the density of the fluid.

In order to obtain the pressure drag F_D, we integrate the x-component of the pressure over the entire surface of revolution, with the projected area in the x-direction denoted by dA_x, as follows

$$\begin{aligned}
F_D &= -\int_A p_x dA_x \\
&= \int \left(\frac{1}{2}\rho U^2 \cos^2\theta\right)(2\pi r dr) \\
&= \pi\rho U^2 \int \cos^2\theta \,(r\cos\theta ds) \\
&= \pi\rho U^2 \int r\left(\frac{dr}{ds}\right)^3 ds \\
F_D &= \pi\rho U^2 \int_0^a \frac{r dr}{1 + x'^2}.
\end{aligned}$$

Alternatively, it is convenient to nondimensionalize the drag force to form the coefficient of drag C_D. We nondimensionalize the stress using the dynamic pressure ρU^2 and the lengths using a, such that $\bar{r} = r/a$ and $\bar{x} = x/a$ with $0 \leq \bar{r} \leq 1$ and

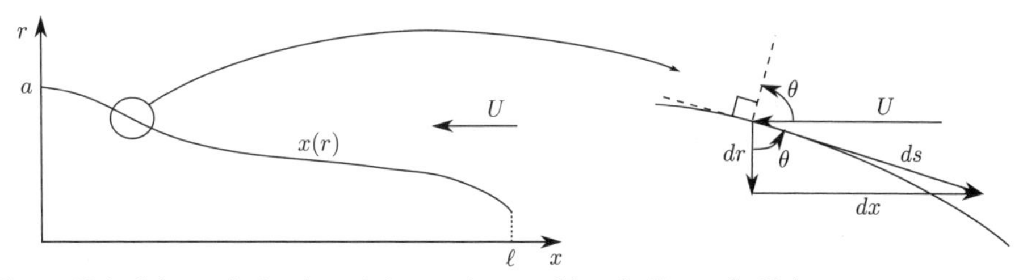

Figure 10.1. Schematic for the minimum-drag problem in Example 10.1.

$0 \leq \bar{x} \leq \ell/a$, as follows

$$\begin{aligned}
C_D &= \frac{F_D}{\rho U^2 A_x} \\
&= \left(\frac{1}{\rho U^2 \pi a^2}\right) \pi \rho U^2 a^2 \int_0^1 \frac{\bar{r} d\bar{r}}{1+\bar{x}'^2} \\
C_D &= \int_0^1 \frac{\bar{r} d\bar{r}}{1+\bar{x}'^2}.
\end{aligned}$$

We seek the shape $\bar{x}(\bar{r})$ that minimizes the functional for the coefficient of drag $C_D[\bar{x}(\bar{r})]$. Observe that the integrand is of the form $F = F(\bar{r}, \bar{x}')$, which corresponds to special case I in Section 2.1.4 as the dependent variable $\bar{x}$ is missing; therefore, the Euler equation is

$$\cancelto{0}{\frac{\partial F}{\partial \bar{x}}} - \frac{d}{dr}\left(\frac{\partial F}{\partial \bar{x}'}\right) = 0.$$

Integrating leads to

$$\begin{aligned}
\frac{\partial F}{\partial \bar{x}'} &= c \\
\frac{-2\bar{r}\bar{x}'}{(1+\bar{x}'^2)^2} &= c \\
\bar{r}\bar{x}' &= -c_1\left(1+\bar{x}'^2\right)^2.
\end{aligned}$$

Typically, we would seek to solve for $\bar{x}'$ and integrate to obtain $\bar{x}(\bar{r})$; however, this is not possible in this case. Instead, we observe that $\bar{x}$ does not appear explicitly in the above expression; therefore, we seek to parameterize the solution with respect to $u = -\bar{x}' > 0$, in which case we have

$$\bar{r}u = c_1\left(1+u^2\right)^2,$$

or

$$\frac{\bar{r}}{c_1} = \frac{\left(1+u^2\right)^2}{u}, \tag{10.1}$$

which is a parametric equation for $\bar{r}$ with respect to u. In order to obtain a parametric equation for $\bar{x}$, let us rewrite equation (10.1) in the form

$$\bar{r} = c_1 \frac{1}{u}(1+2u^2+u^4) = c_1\left(\frac{1}{u} + 2u + u^3\right);$$

as a result,

$$d\bar{r} = c_1\left(-\frac{1}{u^2} + 2 + 3u^2\right) du.$$

Table 10.1. *Drag coefficient for cases with $\ell/a = 1, 2,$ and 4.*

ℓ/a	C_D
1	0.146
2	0.076
4	0.024

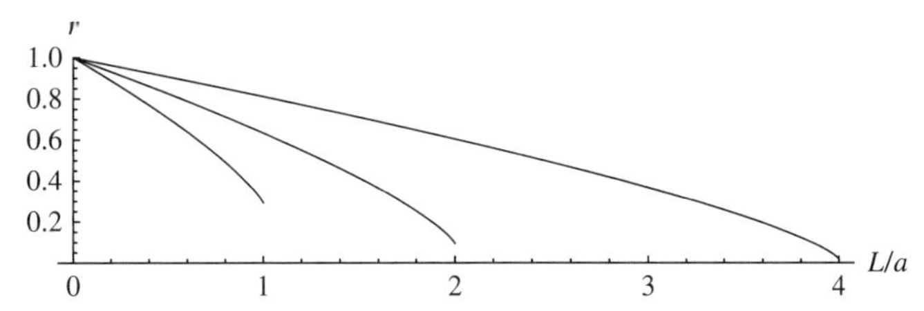

Figure 10.2. Minimum-drag shapes for $\ell/a = 1, 2,$ and 4.

Integration of the relationship $u = -d\bar{x}/d\bar{r}$ yields

$$\begin{aligned}
\frac{\bar{x}}{c_1} &= -\frac{1}{c_1}\int u\,d\bar{r} \\
&= -\int u\left(-\frac{1}{u^2} + 2 + 3u^2\right)du \\
&= \int \left(\frac{1}{u} - 2u - 3u^3\right)du \\
\frac{\bar{x}}{c_1} &= \ln u - u^2 - \frac{3}{4}u^4 + c_2,
\end{aligned}$$

which is a parametric equation for $\bar{x}$ with respect to $u = -\bar{x}'$.

Through differentiation of equation (10.1), it can be shown that a minimum occurs for $u_{min} = 1/\sqrt{3}$. This provides a bound on the slope of the body at the tip $\bar{x} = \ell/a$. The constants c_1, c_2, and u_{max} can be found from boundary conditions and limitations on the shape of the surface of revolution. The parametric equations for $\bar{r}(u)$ and $\bar{x}(u)$ are plotted in Figure 10.2 for $\ell/a = 1, 2,$ and 4.

Observe that the coefficient of drag is given by

$$\begin{aligned}
C_D &= \int_0^1 \frac{\bar{r}\,d\bar{r}}{1 + \bar{x}'^2} \\
&= \int_{u_{min}}^{u_{max}} \frac{c_1^2(1+u^2)^2}{u(1+u^2)}\left(-\frac{1}{u^2} + 2 + 3u^2\right)du \\
C_D &= c_1^2 \int_{u_{min}}^{u_{max}} \frac{1+u^2}{u}\left(3u^2 - \frac{1}{u^2} + 2\right)du.
\end{aligned}$$

The coefficients of drag are given in Table 10.1 for the cases considered in Figure 10.2.

Shape optimization has advanced significantly beyond Newton's humble beginnings. Modern shape optimization is based on the principles of optimal control theory, which is the subject of the remainder of this chapter. Essentially, shape optimization is a constrained extremum problem. An objective functional is devised that defines what is to be optimized about the domain shape, and the constraints are imposed using Lagrange multipliers in the usual manner. Typically, the constraints are partial differential equations that govern the physics within the domain whose shape is to be optimized. The objective could be based on material cost, weight, or strength, for example. The shape is optimized using optimal control theory with the boundary shape acting as the control.

In fluid mechanics, the Navier-Stokes equations (9.31), or some reduced set of these equations, provide the differential constraints to finding an extremum of the objective functional. Optimal shape design in fluid mechanics applications has been pioneered by Pironneau (1974) with numerous applications to aerodynamic shape optimization (see, for example, Jameson, Martinelli, and Pierce 1998). Additional references include Pironneau (1984) and the recent review by Mohammadi and Pironneau (2004). In solid mechanics and structures, in addition to optimizing the shape of an object or structural component, it is also possible to optimize material properties (through coefficients in the differential equations) and/or load distributions (through forcing terms in the differential equations).

A significant amount of current research is devoted to *multiobjective optimization*, in which a system is optimized according to several, often competing, objectives. In *multidisciplinary design optimization* (MDO) these competing objectives span multiple disciplines. For example, an aircraft wing could be optimized to both maximize lift (fluid mechanics) and to minimize structural loads within the wing components (solid mechanics). *Topology optimization* is also of increased interest as an extension of shape optimization. In traditional shape optimization, the topology of the structure or system is predetermined, and the specific parameters that define the geometry are optimized. In topology optimization, the optimization algorithm is tasked with determining the best topology of the material or geometry to achieve the objective. For example, shape optimization of a truss structure would involve sizing the members of the truss given a prescribed connectivity between the fixed number of members. Topology optimization would simply start with a distribution of material and seek the optimum number of members and their connectivity as part of the optimization procedure. Clearly, this is much more challenging than traditional shape optimization, but it holds the promise of significantly enhancing design capabilities for minimum-weight/maximum-strength design, for example.

10.3 Financial Optimization

In the shape optimization example considered in the previous section, we minimize a functional in which all of the physics is captured in the functional itself. More typically in optimization and control, we seek to minimize or maximize an *objective* (or *cost*) *functional* subject to constraints that the governing equation is satisfied. In order to illustrate the approach that will be used throughout the remainder of the chapter, let us consider a simple model to determine the optimal profit reinvestment over some period of time for a company.

EXAMPLE 10.2 A company wants to know what rate $\phi(t)$, expressed as a percentage with $0 \leq \phi \leq 1$, that its profit rate $u(t)$ should be reinvested in order to maximize its total retained profit J over a specified time period $0 \leq t \leq t_f$. We assume that there is a linear relationship between profit reinvested, which is $\phi(t)u(t)$, and rate of change of profit with time of the form

$$\dot{u} = \alpha\phi u, \tag{10.2}$$

where $\alpha > 0$ is the proportionality constant.

Solution: For a short time period dt, the retained profit, or profit not reinvested, is $(1-\phi)udt$. Hence, over the time interval $0 \leq t \leq t_f$, the total retained profit is then

$$J[u(t),\phi(t)] = \int_0^{t_f} (1-\phi)udt. \tag{10.3}$$

Therefore, we seek to maximize the functional[2] (10.3) subject to the constraints that the differential constraint (10.2) and the inequality constraint $0 \leq \phi \leq 1$ are satisfied. We call (10.3) the *cost*, or *objective*, *functional* as it specifies the objective of the optimization problem (maximize accumulated profit), and the differential equation (10.2) is referred to as the *state equation* as it determines the state of the system, which in this case is the company's profit and loss statement. Correspondingly, we call $u(t)$ the *state variable* and $\phi(t)$ the *control variable*.

As in Section 2.7.3, we write the augmented integrand

$$\tilde{F}(u,\dot{u},\phi,\Lambda) = [1-\phi(t)]u(t) + \Lambda(t)[\dot{u}(t) - \alpha\phi(t)u(t)],$$

where we recall that the Lagrange multiplier $\Lambda(t)$ is a function of time in order to impose the differential constraint at each time in the interval $0 \leq t \leq t_f$. There are two dependent variables $u(t)$ and $\phi(t)$; accordingly, we have two Euler equations of the form

$$\frac{\partial \tilde{F}}{\partial u} - \frac{d}{dt}\left(\frac{\partial \tilde{F}}{\partial \dot{u}}\right) = 0, \quad \frac{\partial \tilde{F}}{\partial \phi} - \frac{d}{dt}\left(\frac{\partial \tilde{F}}{\partial \dot{\phi}}\right) = 0,$$

which lead to

$$1 - \phi(1+\alpha\Lambda) - \dot{\Lambda} = 0, \quad -u(1+\alpha\Lambda) = 0, \tag{10.4}$$

respectively. In optimization and control settings, we call the first of these equations an *adjoint differential equation* for reasons that will become clear in Section 10.5.2, and the second is the *optimality condition*, an algebraic relationship between the state $u(t)$ and the Lagrange multiplier $\Lambda(t)$, which is also called the *adjoint variable*.

If $u(t) \neq 0$, that is, there is a profit, then the optimality condition requires that the adjoint variable be $\Lambda = -1/\alpha$. Because α is a constant, $\dot{\Lambda}(t) = 0$, and the adjoint equation leads to the contradiction $1 = 0$. This contradiction would seem to suggest

[2] As is customary in optimization and control theory, $J[\cdot]$ is used to designate objective or cost functionals rather than the generic notation $I[\cdot]$ used for functionals thus far.

that the only possibility is that the profit be zero ($u = 0$), which is not the answer that the company wants to hear.

It may seem somewhat surprising (and disappointing) that the classical variational methods that have served us so well do not produce an optimal solution for such a simple problem. In fact, it merely points to the need for an even more general theory in order to handle certain circumstances. The issue in the present case is that there is no optimal solution among the admissible solutions of continuously differentiable functions. Instead, the optimal solution must allow for discontinuous (specifically, piecewise continuous) solutions for the control $\phi(t)$. This possibility arises owing to the inequality constraint $0 \leq u \leq 1$, which will be addressed in Section 10.4.4 using *Pontryagin's principle*, at which point we will revisit this profit reinvestment problem.

There are numerous opportunities for optimization theory to be applied to financial issues, such as production planning, distribution and transportation of goods, financial and capital investments, resource utilization, and optimization of the economics of individual organizations and entire economies. For example, the 1996 Nobel Memorial Prize in Economic Sciences was awarded to James A. Mirrlees for his work on optimal income tax theory, for which he used variational methods to optimize the level of taxation such that total societal utility is maximized while conserving total wealth. For more on financial optimization, see Seierstad and Sydsaeter (1987), Sydsaeter and Hammond (2012), and Sydsaeter, Hammond, Seierstad, and Strom (2008).

10.4 Optimal Control of Discrete Systems

When optimization theory is applied to dynamical systems, such that the motion or behavior of the system is optimized in some way, it is referred to as *optimal control theory*. The state of a real system may be determined (or *estimated*) using a series of sensors that measure appropriate quantities that fix the state of the system. Alternatively, the system (real or imagined) may be modeled using some state equation that governs the evolution of the system, which is referred to as *model-based control*. The state equations are based on the equations of motion derived in Chapters 4 through 9. For discrete systems, such state equations are ordinary differential equations, while for continuous systems, they are typically partial differential equations. We consider discrete systems in this section and continuous systems in Section 10.5. For discrete systems, we first present a simple example before considering the more general framework.

10.4.1 Example: Control of an Undamped Harmonic Oscillator

In order to introduce the basic approach to optimal control, we consider the simple example of controlling an undamped harmonic oscillator. This will allow us to highlight some of the features that arise in the variational approach to optimal control theory. We consider two versions of this example with different cost (objective) functionals.

Terminal-Time Performance Measure
Consider control of an undamped harmonic oscillator governed by the state equation

$$\ddot{u} + u = 0,$$

where $u(t)$ is the state variable and represents the amplitude of oscillation. This ordinary differential equation governs the linear spring-mass system considered in Section 4.1 (with $k/m = 1$) and the simple pendulum linearized about the stable equilibrium point $\theta = 0$ considered in Section 6.2 (with $g/\ell = 1$). We will consider the case with initial conditions

$$u(0) = 1, \quad \dot{u}(0) = 0. \tag{10.5}$$

The exact solution for the uncontrolled case with these initial conditions is $u(t) = \cos(t)$.

Let us now seek to control the amplitude of the oscillations through some forcing function $\phi(t)$ that we place on the right-hand side of the state equation as follows:

$$\ddot{u} + u = \phi, \quad 0 \leq t \leq t_f. \tag{10.6}$$

We will refer to $\phi(t)$ as the control variable.[3] The state variable $u(t)$ and the control variable $\phi(t)$ are sought such that the cost functional

$$J[u,\phi] = \frac{1}{2}\gamma^2 u(t_f)^2 + \frac{1}{2}\int_0^{t_f} \phi^2 dt \tag{10.7}$$

is minimized over the time interval $0 \leq t \leq t_f$. Note that equation (10.7) is a *dual functional* as described in Section 2.2. The first term is the *performance measure* and seeks to minimize the magnitude of the oscillation amplitude at the terminal time $t = t_f$. The second term is the *penalty function* that minimizes the magnitude of the control input over the entire range $0 \leq t \leq t_f$. It is because we want to minimize the magnitudes of these two quantities, not their actual values, that they are quadratics in the cost functional. Choosing the weight coefficient γ^2 determines the relative importance of the performance measure and the penalty function; setting $\gamma^2 = 0$ corresponds to the uncontrolled case.

As shown in Sections 2.7.3, 6.5, and 9.5.4, in order to minimize the functional (10.7) subject to the differential constraint that the state (governing) equation (10.6) is satisfied, we introduce the Lagrange multiplier $\Lambda(t)$, integrate to form the augmented functional, take the variation, and set it equal to zero as follows:

$$\delta\tilde{J} = \delta\left\{\frac{1}{2}\gamma^2 u(t_f)^2 + \int_0^{t_f}\left[\frac{1}{2}\phi^2 + \Lambda(t)(\ddot{u} + u - \phi)\right]dt\right\} = 0.$$

In this way, we are seeking the state and control variables that best meet the control objective while satisfying the constraint that the dynamics of the system

[3] Typically, texts that emphasize control of discrete systems use $x(t)$ [or $\mathbf{x}(t)$] to represent the state variable(s) and $u(t)$ [or $\mathbf{u}(t)$] to represent the control variable(s). Here, however, we use $u(t)$ [or $\mathbf{u}(t)$] for the state variable(s) to be consistent with the balance of the text. We then reserve x to be used as an independent variable representing the spatial coordinate when treating control of continuous system, which involves partial differential equations.

must be faithfully adhered to. Recall that the Lagrange multiplier is a function of the independent variable, here time, for differential constraints that apply throughout the domain. It is sometimes referred to as an *influence*, or *sensitivity*, *function* that indicates how the cost functional J is "influenced by" or "sensitive to" changes in the corresponding variable. Evaluating the variation gives

$$\gamma^2 u(t_f)\delta u(t_f) + \int_0^{t_f} [\phi\delta\phi + \Lambda(\delta\ddot{u} + \delta u - \delta\phi)]\, dt = 0, \tag{10.8}$$

where we note that the amplitude at the terminal time $u(t_f)$ does vary. Observe that because the Lagrange multiplier $\Lambda(t)$ varies, taking the variation of the augmented functional also produces the term

$$\int_0^{t_f} (\ddot{u} + u - \phi)\delta\Lambda dt,$$

but this simply returns the governing (state or constraint) equation (10.6) and has not been included in (10.8).

Integrating the required term by parts in equation (10.8) twice leads to

$$\begin{aligned}\int_0^{t_f} \Lambda\delta\ddot{u}dt &= \Lambda(t_f)\delta\dot{u}(t_f) - \Lambda(0)\delta\dot{u}(0) - \int_0^{t_f} \dot{\Lambda}\delta\dot{u}dt \\ &= \Lambda(t_f)\delta\dot{u}(t_f) - \Lambda(0)\delta\dot{u}(0) \\ &\quad -\dot{\Lambda}(t_f)\delta u(t_f) + \dot{\Lambda}(0)\delta u(0) + \int_0^{t_f} \ddot{\Lambda}\delta u dt.\end{aligned}$$

Owing to the specified initial conditions, we have $\delta u(0) = \delta\dot{u}(0) = 0$, and the corresponding terms vanish. Substituting into equation (10.8) and collecting terms yields

$$[\gamma^2 u(t_f) - \dot{\Lambda}(t_f)]\delta u(t_f) + \Lambda(t_f)\delta\dot{u}(t_f) + \int_0^{t_f} \{[\ddot{\Lambda} + \Lambda]\delta u + [\phi - \Lambda]\delta\phi\}\, dt = 0.$$

Because $u(t)$, $\phi(t)$, $u(t_f)$, and $\dot{u}(t_f)$ vary, the expressions multiplied by each variation lead to the following Euler equations and natural boundary conditions

$$\ddot{\Lambda} + \Lambda = 0, \tag{10.9}$$

$$\Lambda = \phi, \tag{10.10}$$

$$\dot{\Lambda}(t_f) = \gamma^2 u(t_f), \tag{10.11}$$

$$\Lambda(t_f) = 0. \tag{10.12}$$

Equation (10.9) governs the Lagrange multiplier, which is also called the *adjoint variable*, and is sometimes referred to as the *adjoint equation* for reasons that will become apparent in Section 10.5.2. Equation (10.10) is called the *optimality condition* and relates the control variable $\phi(t)$ to the Lagrange multiplier $\Lambda(t)$. Substituting the optimality condition into equation (10.9) gives the control equation

$$\ddot{\phi} + \phi = 0. \tag{10.13}$$

Alternatively, we could substitute $\phi = \Lambda$ from the optimality condition into the right-hand side of the state equation (10.7) in order to formulate the optimal control problem in terms of the state variable $u(t)$ and the Lagrange multiplier $\Lambda(t)$. Equations (10.11) and (10.12) provide the initial conditions for the backward-time integration of the control equation (10.13) from $t = t_f$ to $t = 0$. The terminal-time initial conditions on the control variable are

$$\phi(t_f) = 0, \quad \dot{\phi}(t_f) = \gamma^2 u(t_f). \tag{10.14}$$

Note that the control equation (10.13) for $\phi(t)$ is coupled to the state equation (10.6) for $u(t)$ through the initial conditions (10.14). Specifically, the amplitude at the terminal time t_f is required as an initial condition for the backward-time integration of the control equation (10.13). One can view the coupled state and control equations as two initial-value problems to be integrated forward and backward in time, respectively, or as a single boundary-value problem in time with boundary conditions at $t = 0$ and $t = t_f$.

Because the control equation does not involve the state variable throughout the time interval $0 \leq t \leq t_f$, but only at the terminal time t_f, we can solve it exactly. The general solution to (10.13) is

$$\phi(t) = c_1 \cos t + c_2 \sin t, \tag{10.15}$$

and application of the terminal-time initial conditions (10.14) leads to the following system of equations for the integration constants:

$$\begin{bmatrix} \cos t_f & \sin t_f \\ -\sin t_f & \cos t_f \end{bmatrix} \begin{bmatrix} c_1 \\ c_2 \end{bmatrix} = \begin{bmatrix} 0 \\ -\gamma^2 u(t_f) \end{bmatrix}.$$

The solution to this system results in

$$c_1 = -\gamma^2 u(t_f) \sin t_f, \quad c_2 = \gamma^2 u(t_f) \cos t_f.$$

Substituting into (10.15) provides the following solution for the control equation

$$\phi(t) = \gamma^2 u(t_f)[\cos t_f \sin t - \sin t_f \cos t],$$

or

$$\phi(t) = \gamma^2 u(t_f) \sin(t - t_f). \tag{10.16}$$

Observe that this solution depends upon the final amplitude $u(t_f)$, which is unknown at this point. Placing $\phi(t)$ on the right-hand side of equation (10.6), the homogeneous solution to the state equation is

$$u_H(t) = c_3 \cos t + c_4 \sin t. \tag{10.17}$$

Using the method of undetermined coefficients with the particular solution

$$u_P(t) = c_5 t \cos t + c_6 t \sin t, \tag{10.18}$$

the constants are found to be

$$c_5 = -\frac{1}{2}\gamma^2 u(t_f) \cos t_f, \quad c_6 = -\frac{1}{2}\gamma^2 u(t_f) \sin t_f.$$

Therefore, the general solution is

$$u(t) = u_H(t) + u_P(t) = c_3 \cos t + c_4 \sin t - \frac{1}{2}\gamma^2 u(t_f) t \left[\cos t_f \cos t + \sin t_f \sin t\right].$$

Applying the initial conditions (10.5) leads to

$$c_3 = 1, \quad c_4 = \frac{1}{2}\gamma^2 u(t_f) \cos t_f.$$

Therefore, the solution to the state equation is

$$u(t) = \cos t + \frac{1}{2}\gamma^2 u(t_f) \left[\cos t_f \sin t - t(\cos t_f \cos t + \sin t_f \sin t)\right]. \tag{10.19}$$

Note that the uncontrolled solution $u(t) = \cos(t), \phi(t) = 0$ results when $\gamma^2 = 0$ as expected.

Because the solution for the state variable $u(t)$ involves the amplitude at the terminal time t_f, let us set $t = t_f$ in equation (10.19) and solve for $u(t_f)$. This leads to the result

$$u(t_f) = \frac{\cos t_f}{1 - \frac{1}{2}\gamma^2 \left(\cos t_f \sin t_f - t_f\right)}.$$

This result can then be substituted into equation (10.19) for the state variable $u(t)$ and into equation (10.16) for the control variable $\phi(t)$.

The solution for terminal time $t_f = 5\pi$ and weight coefficient $\gamma^2 = 1{,}000$ is shown in Figure 10.3. Note that the control damps the oscillation such that the amplitude is close to zero at the terminal time. The corresponding control input consists of out-of-phase forcing. If the control objective is for the state of the system to be exactly equal to some target value, say $u(t_f) = \bar{u}$, rather than simply close to zero, then it would be applied as a constraint on the cost functional rather than as a terminal-time performance measure.

In the general scenario, where exact solutions of the state and control equations are not available, the following iterative procedure would be implemented:

1. Set initial guess for control variable, say $\phi^n(t) = 0, n = 0$.
2. Set iteration number $n = 1$.
3. Integrate state equation (10.6) forward in time for the amplitude $u^n(t)$.
4. Obtain the terminal-time initial conditions (10.14) for the control variable at $t = t_f$ from $u^n(t)$.
5. Integrate control equation (10.13) backward in time subject to (10.14) for $\hat{\phi}^n(t)$.
6. Use a gradient-based method or Newton's method to determine the next iterate for the control input.
7. Increment the iteration number $n = n + 1$.
8. Repeat steps (3)–(7) until convergence.

$0 \le t \le t_f$ Performance Measure

Let us now reconsider control of the same undamped harmonic oscillator governed by the state equation (10.6) with the initial conditions (10.5), but with a revised cost functional

$$J[u,\phi] = \frac{1}{2}\int_0^{t_f} \left(\gamma^2 u^2 + \phi^2\right) dt, \tag{10.20}$$

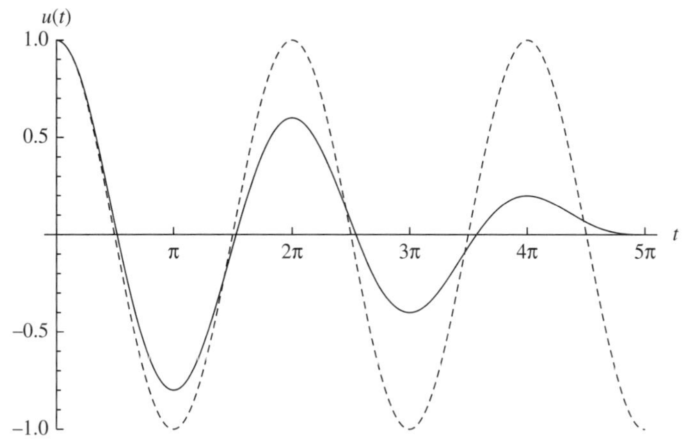

(a) Oscillation amplitude $u(t)$: dashed line represents the uncontrolled solution, $u(t) = \cos(t)$, and the solid line is the controlled solution.

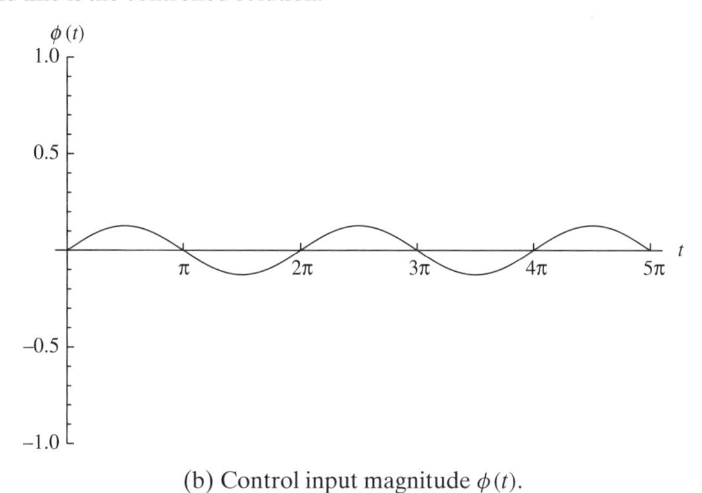

(b) Control input magnitude $\phi(t)$.

Figure 10.3. Optimal control solution of the undamped harmonic oscillator for the case with $t_f = 5\pi$ and weight coefficient $\gamma^2 = 1{,}000$.

where now the performance measure seeks to minimize the magnitude of the oscillation amplitude over the entire range $0 \leq t \leq t_f$, rather than only at the terminal time $t = t_f$ as in the previous case. Following the same procedure as in the previous case, which is left as an exercise, we obtain the control equation

$$\ddot{\phi} + \phi = -\gamma^2 u, \tag{10.21}$$

with the initial conditions applied at the terminal time for the backward-time integration being

$$\phi(t_f) = 0, \quad \dot{\phi}(t_f) = 0. \tag{10.22}$$

Note that in this case, the coupling of the control equation (10.21) to the state equation (10.6) is through the forcing term, which is related to the amplitude of the oscillation, and the initial conditions are homogeneous.

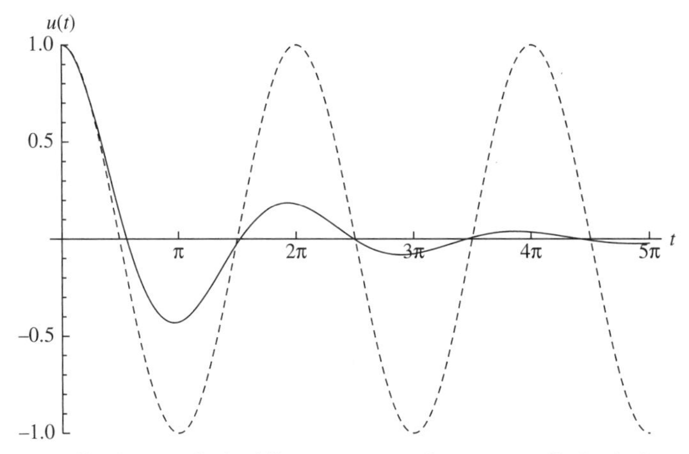

(a) Oscillation amplitude $u(t)$: dashed line represents the uncontrolled solution, $u(t) = \cos(t)$, and the solid line is the controlled solution.

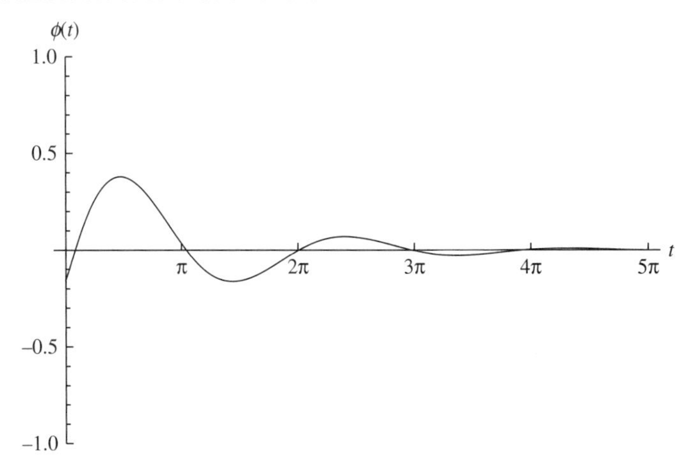

(b) Control input magnitude $\phi(t)$.

Figure 10.4. Optimal control solution of the undamped harmonic oscillator for the case with $t_f = 5\pi$ and weight coefficient $\gamma^2 = 1/3$.

In this case, the coupled state and control equations can be solved simultaneously, that is, without iteration, as a boundary-value problem. We show solutions for two cases, both with $t_f = 5\pi$. First is the case with $\gamma^2 = 1/3$ shown in Figure 10.4. For this case the value of the penalty function, which is indicative of the energy input required for control, is

$$\frac{1}{2}\int_0^{t_f} \phi^2 dt = 0.1308.$$

Figure 10.5 shows the results for a larger weight coefficient $\gamma^2 = 1$, which increases the emphasis on the performance measure as compared to the penalty function. Note that the amplitude of oscillation throughout the range $0 \leq t \leq t_f$, including near the terminal time, has decreased as compared to the case with $\gamma^2 = 1/3$, but that the required control input has a slightly larger magnitude. The corresponding penalty

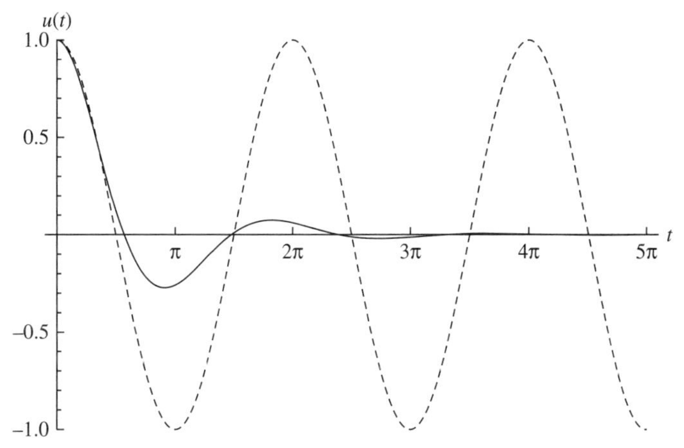

(a) Oscillation amplitude $u(t)$: dashed line represents the uncontrolled solution, $u(t) = \cos(t)$, and the solid line is the controlled solution.

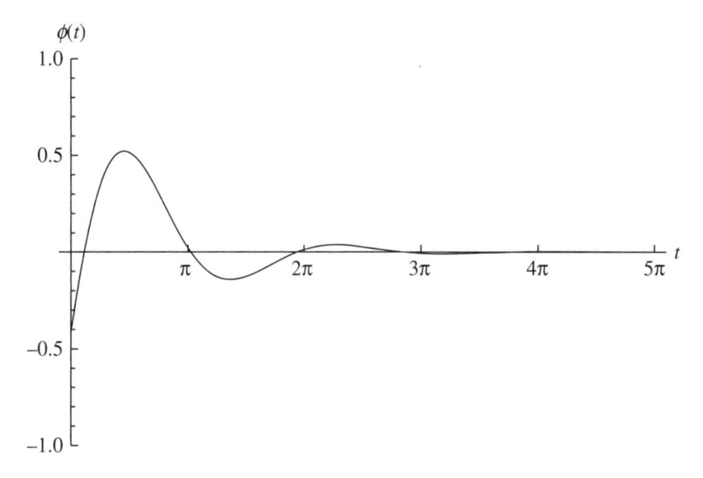

(b) Control input magnitude $\phi(t)$.

Figure 10.5. Optimal control solution of the undamped harmonic oscillator for the case with $t_f = 5\pi$ and weight coefficient $\gamma^2 = 1$.

function is

$$\frac{1}{2}\int_0^{t_f} \phi^2 dt = 0.2080,$$

which is 60% larger than that for $\gamma^2 = 1/3$.

REMARKS:

1. The performance measure and penalty function in the cost functional must be dimensionally consistent. For example, it is often the case that they represent forms of energy. If they are not dimensionally consistent, then a coefficient, say d, having appropriate dimensions may be included in the penalty function.
2. As we see from the previous examples, optimal control theory essentially consists of minimizing a cost functional, which involves a performance measure and

penalty function that define what is meant by optimal control, subject to the state equation being satisfied as a differential constraint.

10.4.2 Riccati Equation for the LQ Problem

Linear Quadratic (LQ) Problems

In this section we generalize the case considered in the previous example to that with n state variables and m control variables. Suppose that the state equations may be expressed in terms of the first-order, linear system of ordinary differential equations

$$\dot{\mathbf{u}} = \mathbf{A}\mathbf{u} + \mathbf{B}\boldsymbol{\phi}, \tag{10.23}$$

where $\mathbf{u}(t)$ is the $n \times 1$ vector of state variables, $\boldsymbol{\phi}(t)$ is the $m \times 1$ vector of control variables, and $\mathbf{A}$ and $\mathbf{B}$ are $n \times n$ and $n \times m$ matrices, respectively. Recall that any higher-order, linear ordinary differential equation (or system of equations) can be recast as a system of first-order ordinary differential equations. Therefore, any linear system, such as the undamped harmonic oscillator considered in the previous section, can be expressed in the form (10.23). Furthermore, nonlinear systems often can be linearized about a steady solution (equilibrium state) so as to produce local state equations of the form (10.23). If the matrices $\mathbf{A}$ and $\mathbf{B}$ are not functions of time, the system is *autonomous*; if they are functions of time, then it is *nonautonomous*. The state variables are subject to the initial conditions

$$\mathbf{u}(0) = \mathbf{u}_0 \quad \text{at} \quad t = 0. \tag{10.24}$$

Generalizing (10.20), we seek to minimize the cost functional

$$J[\mathbf{u},\boldsymbol{\phi}] = \frac{1}{2}\int_0^{t_f} \left(\gamma^2 \mathbf{u}^T \mathbf{Q}\mathbf{u} + \boldsymbol{\phi}^T \mathbf{R}\boldsymbol{\phi}\right) dt, \tag{10.25}$$

where $\mathbf{Q}$ is an $n \times n$ matrix, and $\mathbf{R}$ is $m \times m$. Both $\mathbf{Q}$ and $\mathbf{R}$ are symmetric and positive definite, such that, for example, $\mathbf{u}^T\mathbf{Q}\mathbf{u} \geq 0$ for all $\mathbf{u}$. As before, the first term in the cost functional is the performance measure that depends upon the state of the system $\mathbf{u}(t)$. The second term, the penalty function, regulates the control input $\boldsymbol{\phi}(t)$. It is the cost functional that formally defines what the objective of the control is. The entries in the $\mathbf{Q}$ and $\mathbf{R}$ matrices determine the relative weights of the quadratic quantities in the performance measure and penalty function, respectively. For example, if $\mathbf{Q}$ is the identity matrix, then $u_1^2, u_2^2, \ldots, u_i^2, \ldots, u_n^2$ are equally weighted in the performance measure (and $u_i u_j, i \neq j$ carry zero weight).

The cost functional seeks to minimize the magnitude of the performance measure as well as the control input over the time interval $0 \leq t \leq t_f$. It is because we want to minimize the magnitudes of these two quantities, not their actual values, that they are expressed as quadratics in the cost functional. The weight coefficient γ^2 determines the relative importance of the performance measure and the penalty function. Alternatively, the inverse of the weight coefficient may be viewed as the "cost" of the control. If the control is "expensive," then one would increase the relative weight of the penalty function by decreasing γ^2. Setting $\gamma^2=0$ corresponds to the no-control case, in which case the cost functional minimizes the magnitude of the control input such that $J[\mathbf{u},\boldsymbol{\phi}]=0$ corresponding to $\boldsymbol{\phi}(t)=0$. Alternatively, if $\gamma^2\to\infty$, thereby penalizing the control input, then the performance measure is minimized such that again $J[\mathbf{u},\boldsymbol{\phi}]=0$. For example, if $\mathbf{u}$ is replaced by $(\mathbf{u}-\bar{\mathbf{u}})$ in the cost functional, where $\bar{\mathbf{u}}(t)$ is the target distribution of the state variable, then minimization of the cost functional would produce $\mathbf{u}(t)=\bar{\mathbf{u}}(t)$ exactly, and the state equation (10.23) could be solved explicitly to determine the required control input $\boldsymbol{\phi}(t)$ necessary for the state variable to exactly replicate the target distribution. Thus, it is only when γ^2 is finite and non-zero, that is, both the performance measure and penalty function are present in the cost functional, that a coupled state/control problem results from the following optimization procedure. Control problems of this type are referred to as LQ problems as they involve a *linear* governing (or state) equation (10.23) with a *quadratic* cost functional involving the state variables and control input (10.25).

In order to minimize the cost functional (10.25) subject to the constraint that the state equations (10.23) are satisfied, we form the augmented functional

$$\tilde{J}[\mathbf{u},\boldsymbol{\phi},\boldsymbol{\Lambda}]=\int_0^{t_f}\left\{\frac{1}{2}\left[\gamma^2\mathbf{u}^T\mathbf{Q}\mathbf{u}+\boldsymbol{\phi}^T\mathbf{R}\boldsymbol{\phi}\right]+\boldsymbol{\Lambda}^T\left(\dot{\mathbf{u}}-\mathbf{A}\mathbf{u}-\mathbf{B}\boldsymbol{\phi}\right)\right\}dt,$$

where $\boldsymbol{\Lambda}^T(t)=[\Lambda_1(t),\Lambda_2(t),\ldots,\Lambda_n(t)]$ are the Lagrange multipliers. Note that the augmented term corresponds to taking the dot product of the vector of Lagrange multipliers with the state equations, and that there is one Lagrange multiplier for each state equation. The Lagrange multipliers are also referred to as the *adjoint variables* as they "adjoin" the differential constraint to the cost functional. As is the case here, when the differential constraint governs a dynamic system, it is sometimes called a *dynamic constraint.* Taking the variation and setting it equal to zero leads to

$$\delta\tilde{J}=\int_0^{t_f}\left[\gamma^2\mathbf{u}^T\mathbf{Q}\delta\mathbf{u}+\boldsymbol{\phi}^T\mathbf{R}\delta\boldsymbol{\phi}+\boldsymbol{\Lambda}^T\left(\delta\dot{\mathbf{u}}-\mathbf{A}\delta\mathbf{u}-\mathbf{B}\delta\boldsymbol{\phi}\right)\right]dt=0. \tag{10.26}$$

In order to see how the first two terms arise, consider the following. Because $\mathbf{Q}$ and $\mathbf{R}$ are symmetric, $\mathbf{u}^T\mathbf{Q}\mathbf{u}$ and $\boldsymbol{\phi}^T\mathbf{R}\boldsymbol{\phi}$ are quadratic forms. For example, the first term may be written in the form

$$\mathbf{u}^T\mathbf{Q}\mathbf{u}=\begin{bmatrix}u_1 & u_2 & \cdots & u_n\end{bmatrix}\begin{bmatrix}Q_{11} & Q_{12} & \cdots & Q_{1n}\\ Q_{12} & Q_{22} & \cdots & Q_{2n}\\ \vdots & \vdots & \ddots & \vdots\\ Q_{1n} & Q_{2n} & \cdots & Q_{nn}\end{bmatrix}\begin{bmatrix}u_1\\ u_2\\ \vdots\\ u_n\end{bmatrix}$$

$$\mathbf{u}^T\mathbf{Q}\mathbf{u}=Q_{11}u_1^2+Q_{22}u_2^2+\ldots+Q_{nn}u_n^2$$

$$+2Q_{12}u_1u_2+2Q_{13}u_1u_3+\ldots+2Q_{n-1,n}u_{n-1}u_n,$$

where each term is quadratic in u_i. Taking the variation of this quadratic yields

$$\begin{aligned}\delta\left(\frac{1}{2}\mathbf{u}^T\mathbf{Q}\mathbf{u}\right) &= Q_{11}u_1\delta u_1 + Q_{22}u_2\delta u_2 + \ldots + Q_{nn}u_n\delta u_n \\ &\quad + Q_{12}u_1\delta u_2 + Q_{12}u_2\delta u_1 + Q_{13}u_1\delta u_3 + Q_{13}u_3\delta u_1 + \ldots \\ &= \begin{bmatrix} u_1 & u_2 & \cdots & u_n \end{bmatrix}\begin{bmatrix} Q_{11} & Q_{12} & \cdots & Q_{1n} \\ Q_{12} & Q_{22} & \cdots & Q_{2n} \\ \vdots & \vdots & \ddots & \vdots \\ Q_{1n} & Q_{2n} & \cdots & Q_{nn} \end{bmatrix}\begin{bmatrix} \delta u_1 \\ \delta u_2 \\ \vdots \\ \delta u_n \end{bmatrix} \\ \delta\left(\frac{1}{2}\mathbf{u}^T\mathbf{Q}\mathbf{u}\right) &= \mathbf{u}^T\mathbf{Q}\delta\mathbf{u}.\end{aligned}$$

Similarly, the second term leads to

$$\delta\left(\frac{1}{2}\boldsymbol{\phi}^T\mathbf{R}\boldsymbol{\phi}\right) = \boldsymbol{\phi}^T\mathbf{R}\delta\boldsymbol{\phi}.$$

Returning to equation (10.26), we apply integration by parts to the third term as follows:

$$\int_0^{t_f} \boldsymbol{\Lambda}^T\delta\dot{\mathbf{u}}dt = \boldsymbol{\Lambda}^T(t_f)\delta\mathbf{u}(t_f) - \boldsymbol{\Lambda}^T(0)\delta\mathbf{u}(0) - \int_0^{t_f} \dot{\boldsymbol{\Lambda}}^T\delta\mathbf{u}dt.$$

Owing to the specified initial condition at $t = 0$, we have $\delta\mathbf{u}(0) = 0$. Therefore, after grouping terms, equation (10.26) becomes

$$\boldsymbol{\Lambda}^T(t_f)\delta\mathbf{u}(t_f) + \int_0^{t_f}\left[\left(\gamma^2\mathbf{u}^T\mathbf{Q} - \dot{\boldsymbol{\Lambda}}^T - \boldsymbol{\Lambda}^T\mathbf{A}\right)\delta\mathbf{u} + \left(\boldsymbol{\phi}^T\mathbf{R} - \boldsymbol{\Lambda}^T\mathbf{B}\right)\delta\boldsymbol{\phi}\right]dt = 0.$$

The expressions multiplied by each variation must vanish leading to the following Euler equations and initial condition:

$$\dot{\boldsymbol{\Lambda}}^T = -\boldsymbol{\Lambda}^T\mathbf{A} + \gamma^2\mathbf{u}^T\mathbf{Q}, \tag{10.27}$$

$$\boldsymbol{\Lambda}^T\mathbf{B} = \boldsymbol{\phi}^T\mathbf{R}, \tag{10.28}$$

$$\boldsymbol{\Lambda}(t_f) = \mathbf{0}. \tag{10.29}$$

Equation (10.27) is the differential *adjoint equation* for the Lagrange multipliers $\boldsymbol{\Lambda}(t)$, also referred to as the adjoint variables. The algebraic *optimality condition* (10.28) relates the Lagrange multipliers (adjoint variables) to the control variables. Taking the transpose of both sides of equations (10.27) and (10.28) gives

$$\dot{\boldsymbol{\Lambda}} = -\mathbf{A}^T\boldsymbol{\Lambda} + \gamma^2\mathbf{Q}\mathbf{u} \tag{10.30}$$

for the adjoint equation, and

$$\mathbf{B}^T\boldsymbol{\Lambda} = \mathbf{R}\boldsymbol{\phi},$$

where we have used the fact that $\mathbf{Q}$ and $\mathbf{R}$ are symmetric, in which case $\mathbf{Q}^T = \mathbf{Q}$ and $\mathbf{R}^T = \mathbf{R}$. Premultiplying both sides of this last equation by $\mathbf{R}^{-1}$ leads to the optimality condition in the form

$$\boldsymbol{\phi} = \mathbf{R}^{-1}\mathbf{B}^T\boldsymbol{\Lambda}, \tag{10.31}$$

where we note that the inverse always exists because $\mathbf{R}$ is positive definite. Given a solution to the adjoint equation for $\mathbf{\Lambda}(t)$, equation (10.31) allows one to obtain the control variable $\boldsymbol{\phi}(t)$.

We can substitute equation (10.31) into the state equation (10.23) to eliminate the control variable, and we are left with

$$\dot{\mathbf{u}} = \mathbf{A}\mathbf{u} + \mathbf{B}\mathbf{R}^{-1}\mathbf{B}^T\mathbf{\Lambda}, \tag{10.32}$$

along with equation (10.30) as two coupled equations for the state variables $\mathbf{u}$ and the Lagrange multipliers (adjoint variables) $\mathbf{\Lambda}$. The initial conditions for the state and adjoint variables are given by equations (10.24) and (10.29), respectively. Again, we see that the adjoint equation (10.30) must be integrated backward in time from $t = t_f$ to $t = 0$.

One could imagine integrating the state equation (10.32) with initial condition (10.24) forward in time and the adjoint equation (10.30) with initial condition (10.29) backward in time and iterating until converged solutions for $\mathbf{u}(t)$ and $\mathbf{\Lambda}(t)$ are obtained. This is the general procedure outlined in Section 10.4.1. However, for the LQ problem being considered here, it is possible to recast these coupled equations in a form that leads to a more direct procedure for obtaining the solution without the need for iteration.

Riccati Equation

For the linear system of equations (10.30), we know that a solution exists for the adjoint variables $\mathbf{\Lambda}(t)$ in terms of the state variables $\mathbf{u}(t)$ in the form

$$\mathbf{\Lambda} = -\mathbf{P}\mathbf{u}, \tag{10.33}$$

where $\mathbf{P}(t)$ is an $n \times n$ matrix. We seek a system of equations that can be solved to obtain $\mathbf{P}(t)$. Differentiating (10.33) yields

$$\dot{\mathbf{\Lambda}} = -\mathbf{P}\dot{\mathbf{u}} - \dot{\mathbf{P}}\mathbf{u}.$$

Substituting this and equation (10.33) into equations (10.30) and (10.32) gives

$$\mathbf{P}\dot{\mathbf{u}} + \dot{\mathbf{P}}\mathbf{u} + \mathbf{A}^T\mathbf{P}\mathbf{u} + \gamma^2\mathbf{Q}\mathbf{u} = \mathbf{0}, \tag{10.34}$$

and

$$\dot{\mathbf{u}} = \mathbf{A}\mathbf{u} - \mathbf{B}\mathbf{R}^{-1}\mathbf{B}^T\mathbf{P}\mathbf{u}, \tag{10.35}$$

respectively. Substituting equation (10.35) into (10.34) yields

$$\left(\mathbf{P}\mathbf{A} - \mathbf{P}\mathbf{B}\mathbf{R}^{-1}\mathbf{B}^T\mathbf{P} + \dot{\mathbf{P}} + \mathbf{A}^T\mathbf{P} + \gamma^2\mathbf{Q}\right)\mathbf{u} = \mathbf{0}.$$

Because the solution $\mathbf{u}$ to the state equations is in general nontrivial, its coefficient must vanish. This leads to the first-order, nonlinear *Riccati equation* for $\mathbf{P}(t)$, which is

$$\dot{\mathbf{P}} = -\mathbf{P}\mathbf{A} - \mathbf{A}^T\mathbf{P} + \mathbf{P}\mathbf{B}\mathbf{R}^{-1}\mathbf{B}^T\mathbf{P} - \gamma^2\mathbf{Q}. \tag{10.36}$$

From equation (10.33), the terminal-time initial condition (10.29) becomes

$$\mathbf{P}(t_f) = \mathbf{0}. \tag{10.37}$$

It can be shown that the matrix $\mathbf{P}(t)$ is symmetric; therefore, the n^2 first-order, nonlinear ordinary differential equations represented by (10.36) reduce to $\frac{1}{2}n(n+1)$ equations.

Given the coefficient matrices $\mathbf{A}$ and $\mathbf{B}$ from the state equations (10.23) and $\mathbf{Q}$ and $\mathbf{R}$ from the cost functional (10.25), the first-order system of equations (10.36) and (10.37) are solved backward in time for matrix $\mathbf{P}(t)$. Substituting equation (10.33) into (10.31) yields the *control law*

$$\phi = -\mathbf{R}^{-1}\mathbf{B}^T\mathbf{P}\mathbf{u}, \tag{10.38}$$

which relates the control directly to the state variables, thereby indicating how the control reacts to changes in the state variables, and providing a control law that is independent of the initial condition $\mathbf{u}_0$.

Substituting (10.38) for the control variable into the state equation (10.23) provides the system of equations (10.35):

$$\dot{\mathbf{u}} = \left(\mathbf{A} - \mathbf{B}\mathbf{R}^{-1}\mathbf{B}^T\mathbf{P}\right)\mathbf{u}. \tag{10.39}$$

Given the solution of the Riccati equation (10.36), with initial conditions (10.37), backward in time for $P(t)$, equation (10.39) is solved forward in time subject to the initial condition (10.24) for the state variable $\mathbf{u}(t)$. In this manner, the solution to the state equations (10.23) that minimize the cost functional (10.25) is obtained directly without iteration.

REMARKS:

1. The advantage of this approach is that only one integration is required for the Riccati equation (10.36) and the system (10.39). The trade-off, however, is that the Riccati equation is nonlinear despite the fact that the underlying system is linear.
2. It can be proven that the control specified by equation (10.38) produces a global minimum and hence is ensured of being optimal for the LQ problem considered here (see, for example, MacCluer 2005, Section 8.4).
3. Changing the sign of the performance measure results in it being maximized rather than minimized.
4. The backward-time integration makes sense physically as it would be necessary to take into account the future behavior of the system in order to determine the *optimal control* that should be implemented at any point in time t to meet the objective. Stengel (1986) likens this to solving a maze in which it is typically easiest to find the solution by working backward from the final destination to the starting location.
5. The performance measures considered thus far seek to use control to drive the system toward the origin of $(u_1, u_2, \ldots, u_n)$-space. We often desire to have the system move toward another target state, say $\bar{\mathbf{u}}$, in which case we simply replace $\mathbf{u}$ with $(\mathbf{u} - \bar{\mathbf{u}})$ in the performance measure. This is known as *tracking*. See Section 10.5.3 for an example of this.
6. As described in Chapters 4–9, physical systems "optimize" themselves based on prescribed *physical* laws, for example, minimization of energy, whereas control

is performed according to ad hoc optimization principles. In the case of physical laws, therefore, we might say that *God* chose the functional, whereas in optimal control, *we* choose the functional. As a result, there is room for debate as to what the appropriate cost functional should be for a given control scenario, and this choice can have a marked influence on the resulting solution.

EXAMPLE 10.3 In order to illustrate use of the Riccati equation to solve LQ problems, let us reconsider control of the undamped harmonic oscillator from Section 10.4.1. Recall that the state equation is

$$\ddot{u}+u=\phi, \tag{10.40}$$

with the initial conditions

$$u(0)=1, \quad \dot{u}(0)=0. \tag{10.41}$$

The cost functional (10.20) to be minimized is

$$J[u,\phi]=\frac{1}{2}\int_0^{t_f}\left(\gamma^2u^2+\phi^2\right)dt.$$

We will use $\gamma^2=1$ for the weight coefficient in the cost functional and $t_f=5\pi$ for the terminal time.

Solution: First, we convert the state equation to a system of first-order, linear equations ($n=2, m=1$) using the transformation

$$u_1=u, \quad u_2=\dot{u}.$$

Differentiating and writing in terms of u_1 and u_2, the first-order system of equations is

$$\dot{u}_1=u_2, \quad \dot{u}_2=-u_1+\phi.$$

Hence, in matrix form, the state equation (10.40) becomes

$$\dot{\mathbf{u}}=\mathbf{A}\mathbf{u}+\mathbf{B}\boldsymbol{\phi}, \tag{10.42}$$

where

$$\mathbf{u}=\begin{bmatrix}u_1\\u_2\end{bmatrix}, \quad \boldsymbol{\phi}=[\phi], \quad \mathbf{A}=\begin{bmatrix}0&1\\-1&0\end{bmatrix}, \quad \mathbf{B}=\begin{bmatrix}0\\1\end{bmatrix}.$$

The cost functional written in matrix form is (with $u=u_1$)

$$J[\mathbf{u},\boldsymbol{\phi}]=\frac{1}{2}\int_0^{t_f}\left(\gamma^2\mathbf{u}^T\mathbf{Q}\mathbf{u}+\boldsymbol{\phi}^T\mathbf{R}\boldsymbol{\phi}\right)dt, \tag{10.43}$$

where

$$\mathbf{Q}=\begin{bmatrix}1&0\\0&0\end{bmatrix}, \quad \mathbf{R}=[1].$$

From equations (10.32) and (10.30), the control problem consists of the state equation

$$\dot{\mathbf{u}}=\mathbf{A}\mathbf{u}+\mathbf{B}\mathbf{R}^{-1}\mathbf{B}^T\boldsymbol{\Lambda}, \tag{10.44}$$

and the adjoint equation

$$\dot{\mathbf{\Lambda}} = -\mathbf{A}^T\mathbf{\Lambda} + \gamma^2\mathbf{Q}\mathbf{u}, \tag{10.45}$$

where the adjoint variables are $\mathbf{\Lambda} = [\Lambda_1, \ \Lambda_2]^T$. We seek a solution of the form (10.33)

$$\mathbf{\Lambda} = -\mathbf{P}\mathbf{u}, \tag{10.46}$$

where the $n \times n$ matrix $\mathbf{P}(t)$ satisfies the Riccati equation (10.36) repeated here:

$$\dot{\mathbf{P}} = -\mathbf{P}\mathbf{A} - \mathbf{A}^T\mathbf{P} + \mathbf{P}\mathbf{B}\mathbf{R}^{-1}\mathbf{B}^T\mathbf{P} - \gamma^2\mathbf{Q}. \tag{10.47}$$

Taking into account that $\mathbf{P}$ is symmetric, for our particular case this becomes

$$\begin{aligned}\begin{bmatrix}\dot{P}_{11} & \dot{P}_{12}\\ \dot{P}_{12} & \dot{P}_{22}\end{bmatrix} &= -\begin{bmatrix}P_{11} & P_{12}\\ P_{12} & P_{22}\end{bmatrix}\begin{bmatrix}0 & 1\\ -1 & 0\end{bmatrix} - \begin{bmatrix}0 & -1\\ 1 & 0\end{bmatrix}\begin{bmatrix}P_{11} & P_{12}\\ P_{12} & P_{22}\end{bmatrix}\\ &\quad + \begin{bmatrix}P_{11} & P_{12}\\ P_{12} & P_{22}\end{bmatrix}\begin{bmatrix}0\\ 1\end{bmatrix}[1]\begin{bmatrix}0 & 1\end{bmatrix}\begin{bmatrix}P_{11} & P_{12}\\ P_{12} & P_{22}\end{bmatrix} - \gamma^2\begin{bmatrix}1 & 0\\ 0 & 0\end{bmatrix}.\end{aligned}$$

Simplifying leads to

$$\begin{bmatrix}\dot{P}_{11} & \dot{P}_{12}\\ \dot{P}_{12} & \dot{P}_{22}\end{bmatrix} = \begin{bmatrix}2P_{12} + P_{12}^2 - \gamma^2 & -P_{11} + P_{22} + P_{12}P_{22}\\ P_{22} - P_{11} + P_{22}P_{12} & -2P_{12} + P_{22}^2\end{bmatrix}.$$

Owing to the symmetry of $\mathbf{P}$, we must solve the following three coupled nonlinear ordinary differential equations for $P_{11}(t), P_{22}(t)$, and $P_{12}(t)$

$$\begin{aligned}\dot{P}_{11} &= P_{12}^2 + 2P_{12} - \gamma^2,\\ \dot{P}_{22} &= P_{22}^2 - 2P_{12},\\ \dot{P}_{12} &= P_{12}P_{22} - P_{11} + P_{22}.\end{aligned} \tag{10.48}$$

These equations are solved backward in time subject to the initial conditions (10.37)

$$P_{11}(t_f) = P_{22}(t_f) = P_{12}(t_f) = 0. \tag{10.49}$$

The nonlinear system (10.48) and (10.49) must be solved numerically, and the solutions are shown in Figure 10.6.

Once $\mathbf{P}(t)$ is obtained, the state equation (10.39), which is

$$\dot{\mathbf{u}} = \left(\mathbf{A} - \mathbf{B}\mathbf{R}^{-1}\mathbf{B}^T\mathbf{P}\right)\mathbf{u}, \tag{10.50}$$

may be solved forward in time for the controlled state variables. In our particular case, the state equations are

$$\begin{bmatrix}\dot{u}_1\\ \dot{u}_2\end{bmatrix} = \left\{\begin{bmatrix}0 & 1\\ -1 & 0\end{bmatrix} - \begin{bmatrix}0\\ 1\end{bmatrix}[1]\begin{bmatrix}0 & 1\end{bmatrix}\begin{bmatrix}P_{11} & P_{12}\\ P_{12} & P_{22}\end{bmatrix}\right\}\begin{bmatrix}u_1\\ u_2\end{bmatrix} = \begin{bmatrix}0 & 1\\ -1 - P_{12} & -P_{22}\end{bmatrix}\begin{bmatrix}u_1\\ u_2\end{bmatrix},$$

or

$$\begin{aligned}\dot{u}_1 &= u_2\\ \dot{u}_2 &= -(1 + P_{12})u_1 - P_{22}u_2\end{aligned}. \tag{10.51}$$

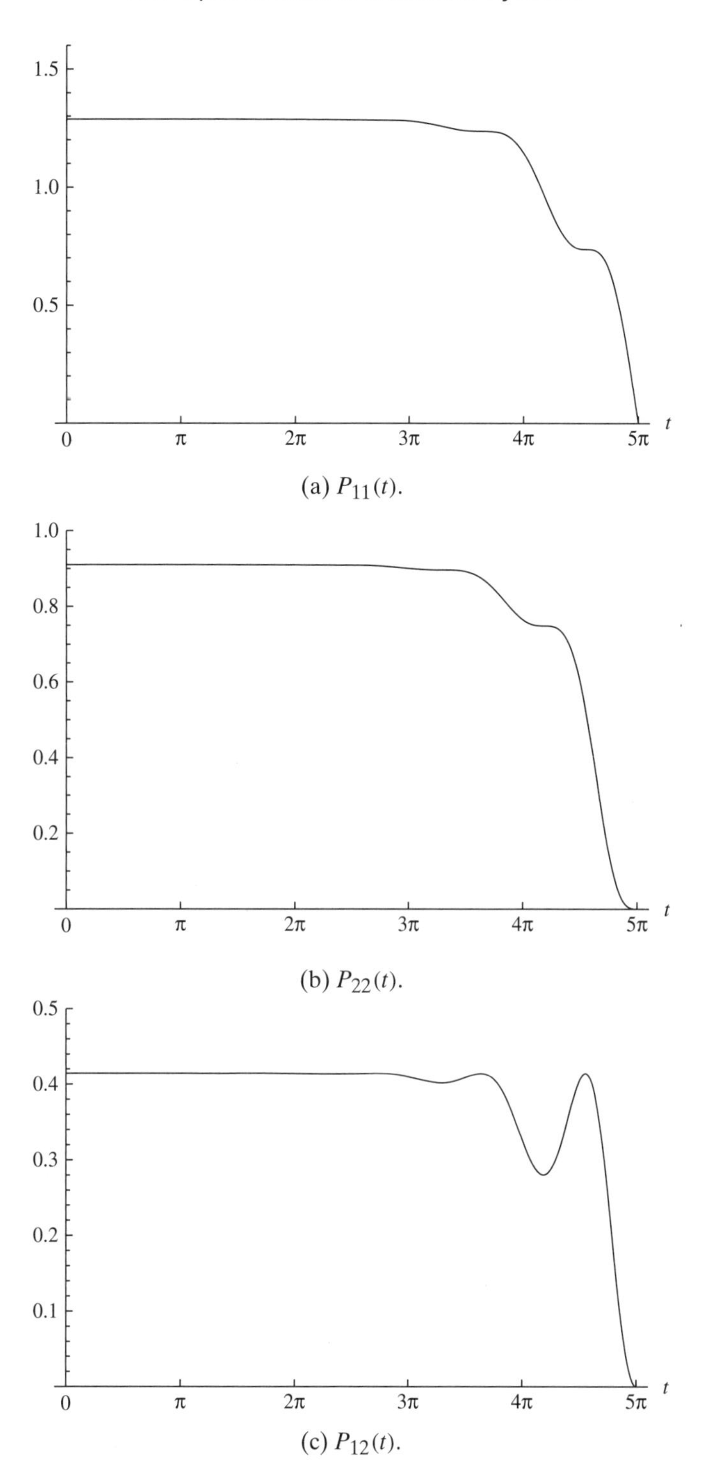

Figure 10.6. Solutions of the Riccati equations for the undamped harmonic oscillator for the case with $t_f = 5\pi$ and weight coefficient $\gamma^2 = 1$.

These equations are solved subject to the initial conditions (10.41)

$$u_1(0) = 1, \quad u_2(0) = 0.$$

Recall that the original state variable is

$$u(t) = u_1(t).$$

Given the solution $\mathbf{u}(t)$ to the state equations, the control variable $\boldsymbol{\phi}(t)$ may be reconstructed using the control law (10.38), which is

$$\boldsymbol{\phi} = -\mathbf{R}^{-1}\mathbf{B}^T\mathbf{P}\mathbf{u}. \tag{10.52}$$

For our particular case

$$\phi = -\begin{bmatrix}1\end{bmatrix}\begin{bmatrix}0 & 1\end{bmatrix}\begin{bmatrix}P_{11} & P_{12}\\ P_{12} & P_{22}\end{bmatrix}\begin{bmatrix}u_1\\ u_2\end{bmatrix},$$

or

$$\phi = -P_{12}u_1 - P_{22}u_2. \tag{10.53}$$

The solutions for the state and control variables are shown in Figure 10.7. These are the same solutions as shown in Figure 10.5 from Section 10.4.1.

Below we list the relative advantages and disadvantages of using the three solution techniques outlined above for solving optimal control problems.

Riccati equation:

- Advantage:
  - Only requires a single backward- and forward-time integration to obtain $\mathbf{P}(t)$ and $\mathbf{u}(t)$, respectively.
- Disadvantages:
  - Only possible for LQ problems.
  - For systems with n state variables, we must solve $\frac{1}{2}n(n+1)$ nonlinear ordinary differential equations to obtain the symmetric matrix function $\mathbf{P}(t)$ in addition to the n linear state equations. Typically, the nonlinear equations must be solved numerically. Such calculations become prohibitively large for large n.
  - The Riccati equation for $\mathbf{P}(t)$ is of a different type than the state equation, for example, nonlinear versus linear. Thus, they generally require different solution techniques.

Iterative solution of the state equation forward in time and control (adjoint) equation backward in time:

- Advantages:
  - The state and control (adjoint) equations are of the same form; therefore, the same solution techniques may be used for both.
  - Because this approach only requires solution of vectors for the state and control variables, it scales readily to systems with very large numbers of degrees of freedom.

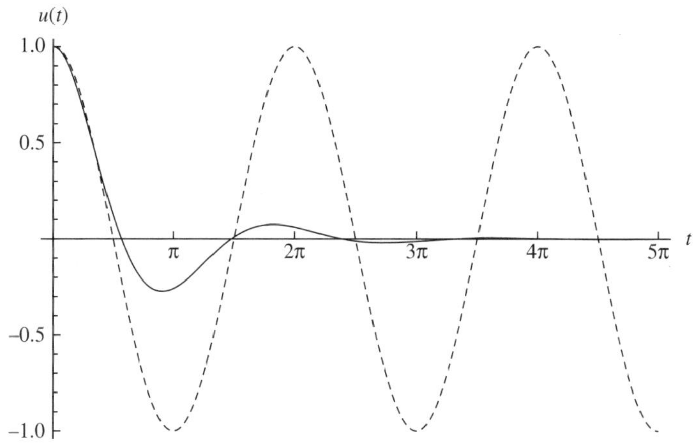

(a) Oscillation amplitude $u(t)$: dashed line represents the uncontrolled solution, $u(t) = \cos(t)$, and the solid line is the controlled solution.

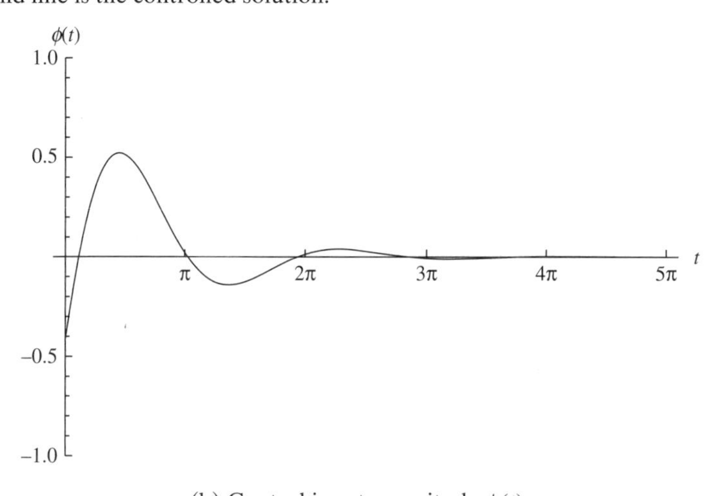

(b) Control input magnitude $\phi(t)$.

Figure 10.7. Optimal control solution for the undamped harmonic oscillator using the Riccati equation for the case with $t_f = 5\pi$ and weight coefficient $\gamma^2 = 1$.

  - The iterative framework can be implemented for more general scenarios, including nonlinear problems.
- Disadvantages:
  - Requires numerous forward- and backward-time integrations of the state and control (adjoint) equations, respectively, until convergence is achieved.
  - Numerical stability of the iterative process is not always ensured.

Combine the state and control (adjoint) equations into a single boundary-value problem over the time domain $0 \leq t \leq t_f$:

- Advantage:
  - Only requires solution of a single coupled boundary-value problem with $2n$ first-order equations (or higher-order equivalent) for $\mathbf{u}$ and $\boldsymbol{\phi}$.

- Disadvantage:
  - It is only possible to recast relatively simple systems into a single boundary-value problem.

10.4.3 Properties of Systems for Control

From a mathematical point of view, there are three desirable properties of dynamical systems that we wish to control. Note that these are properties of the system or its states and are independent of the cost functional.

- **Controllability:** *Controllability* is the ability to transfer a system with an arbitrary initial state to an arbitrary final state in finite time. A system is *controllable* if there is a control history $\boldsymbol{\phi}(t)$ in an interval $t_0 \leq t \leq t_f$ that transfers each element of an initial state $\mathbf{u}(t_0)$ to an arbitrary final state $\mathbf{u}(t_f)$.
  - A *controllable region* is the envelope of initial states $\mathbf{u}(t_0)$ that can be made to reach a particular final state $\mathbf{u}(t_f)$ via control over the interval $t_0 \leq t \leq t_f$.
  - A *reachable region* is the envelope of final states $\mathbf{u}(t_f)$ that can be reached from a given initial state $\mathbf{u}(t_0)$ via control over the interval $t_0 \leq t \leq t_f$.
  - If an n-dimensional linear systems is *controllable*, then all states in n-dimensional space are *reachable* from any initial state. For nonlinear systems, this is not necessarily the case.
  - For linear, autonomous state equations with n state variables of the form

$$\dot{\mathbf{u}} = \mathbf{A}\mathbf{u} + \mathbf{B}\boldsymbol{\phi},$$

  it can be shown that the system is *controllable* if

$$\text{rank}\left[\mathbf{B}\ \mathbf{A}\mathbf{B}\ \mathbf{A}^2\mathbf{B}\ \cdots\ \mathbf{A}^{n-1}\mathbf{B}\right] = n.$$

  Controllability of a system is a necessary, but not sufficient, condition for the existence of a control solution for the system. A sufficient condition must take into account the cost functional defining the control objective.
- **Observability:** *Observability* is the ability to uniquely determine the state of a system from observations of linear combinations of the output of the system in finite time. A state is *observable* at t_0 if an output history $\mathbf{v}(t)$ for $t_0 \leq t \leq t_f$ is sufficient to reconstruct the full state $\mathbf{u}(t_0)$ at $t = t_0$. Note that the output history is the measured state of the system, which may differ from the state $\mathbf{u}(t)$ determined from the state equations (see Section 10.6.3).
  - The *observable region* is the envelope of initial states $\mathbf{u}(t_0)$ that can be determined from the measured output $\mathbf{v}(t)$ given in the interval $t_0 \leq t \leq t_f$.
  - The *constructable region* is the envelope of final states $\mathbf{u}(t_f)$ that can be determined from the measured output $\mathbf{v}(t)$ given in the interval $t_0 \leq t \leq t_f$.
  - If an n-dimensional linear systems is *observable*, then all states in n-dimensional space are *constructable* from any initial state. For nonlinear systems, this is not necessarily the case.
- **Stabilizability:** A system is *stabilizable* if for every initial condition it is possible to find a control input that transfers the state to a stationary point of the system within finite time.

- A controllable system is stabilizable.
- If a system is not controllable, it can still be stabilizable if all uncontrollable states are stable and all unstable states are controllable.

While these properties are desirable from a mathematical point of view, they are often difficult to prove in realistic scenarios. In some cases, we may not want to enforce certain of these conditions. For example, it is often possible to take advantage of instabilities in a system, in which small perturbations have a large impact on the system, in order to accomplish a control objective with dramatically reduced control inputs. For example, this is a significant theme in active flow control (see Section 9.6).

10.4.4 Pontryagin's Principle

Thus far we have framed optimal control in the context of classical calculus of variations as a constrained extremum problem. In 1956, the Russian mathematician Lev Pontryagin reformulated the control problem in terms of a *Hamiltonian* (Pontryagin, Boltyanskii, Gamkrelidze, and Mishchenko 1962). This is not the same Hamiltonian used in classical mechanics (see Chapter 4); however, it is inspired by it and was given the same name by Pontryagin.

Before stating *Pontryagin's principle*, let us reconsider a generalization of the optimal control problem of the previous sections reformulated using Pontryagin's Hamiltonian. The general optimal control problem is to minimize the cost functional

$$J[\mathbf{u},\boldsymbol{\phi}] = \int_0^{t_f} F(t,\mathbf{u},\boldsymbol{\phi})dt \tag{10.54}$$

subject to the state equations

$$\dot{\mathbf{u}} = \mathbf{f}(t,\mathbf{u},\boldsymbol{\phi}) \tag{10.55}$$

and initial conditions

$$\mathbf{u} = \mathbf{u}_0 \quad \text{at} \quad t = 0 \tag{10.56}$$

being satisfied. In the classical variational approach, we form the augmented cost functional as follows

$$\tilde{J}[\mathbf{u},\boldsymbol{\phi},\boldsymbol{\Lambda}] = \int_0^{t_f} \left\{F(t,\mathbf{u},\boldsymbol{\phi}) + \boldsymbol{\Lambda}^T(t)\left[\dot{\mathbf{u}} - \mathbf{f}(t,\mathbf{u},\boldsymbol{\phi})\right]\right\} dt.$$

Pontryagin reformulated the augmented functional in the form

$$\tilde{J}[\mathbf{u},\boldsymbol{\phi},\boldsymbol{\Lambda}] = \int_0^{t_f} \left[H_P(t,\mathbf{u},\boldsymbol{\phi},\boldsymbol{\Lambda}) + \boldsymbol{\Lambda}^T(t)\dot{\mathbf{u}}\right] dt,$$

where Pontryagin's Hamiltonian contains all of the nonderivative terms and is given by

$$H_P(t,\mathbf{u},\boldsymbol{\phi},\boldsymbol{\Lambda}) = F(t,\mathbf{u},\boldsymbol{\phi}) - \boldsymbol{\Lambda}^T(t)\mathbf{f}(t,\mathbf{u},\boldsymbol{\phi}). \tag{10.57}$$

Taking the variation of $\tilde{J}$ and setting equal to zero yields

$$\delta\tilde{J} = \int_0^{t_f} \left[\frac{\partial H_P}{\partial \mathbf{u}}\delta\mathbf{u} + \frac{\partial H_P}{\partial \boldsymbol{\phi}}\delta\boldsymbol{\phi} + \boldsymbol{\Lambda}^T\delta\dot{\mathbf{u}}\right] dt = 0,$$

where the $\delta\boldsymbol{\Lambda}$ term has not been included as it simply leads to the original state equation. Integrating the last term by parts and collecting terms produces

$$\boldsymbol{\Lambda}^T(t_f)\delta\mathbf{u}(t_f) + \int_0^{t_f}\left[\left(\frac{\partial H_P}{\partial \mathbf{u}} - \dot{\boldsymbol{\Lambda}}^T\right)\delta\mathbf{u} + \frac{\partial H_P}{\partial \boldsymbol{\phi}}\delta\boldsymbol{\phi}\right]dt = 0,$$

where we have taken into account that $\mathbf{u}(0)$ is specified from the initial condition. In terms of Pontryagin's Hamiltonian, therefore, the adjoint equation, optimality condition, and terminal-time initial condition are of the form

$$\dot{\boldsymbol{\Lambda}}^T = \frac{\partial H_P}{\partial \mathbf{u}}, \quad \frac{\partial H_P}{\partial \boldsymbol{\phi}} = \mathbf{0}, \quad \boldsymbol{\Lambda}(t_f) = \mathbf{0}, \tag{10.58}$$

respectively. Note that the state equations (10.55) also can be expressed in terms of Pontryagin's Hamiltonian (10.57) as $\dot{\mathbf{u}} = -\partial H_P/\partial\boldsymbol{\Lambda}$.

Applying Pontryagin's approach to the LQ problem of Section 10.4.2, for example, we have

$$F(\mathbf{u},\boldsymbol{\phi}) = \frac{1}{2}\left(\gamma^2\mathbf{u}^T\mathbf{Q}\mathbf{u} + \boldsymbol{\phi}^T\mathbf{R}\boldsymbol{\phi}\right), \quad \mathbf{f}(\mathbf{u},\boldsymbol{\phi}) = \mathbf{A}\mathbf{u} + \mathbf{B}\boldsymbol{\phi}. \tag{10.59}$$

Pontryagin's Hamiltonian (10.57) is then

$$H_P(\mathbf{u},\boldsymbol{\phi},\boldsymbol{\Lambda}) = \frac{1}{2}\left(\gamma^2\mathbf{u}^T\mathbf{Q}\mathbf{u} + \boldsymbol{\phi}^T\mathbf{R}\boldsymbol{\phi}\right) - \boldsymbol{\Lambda}^T(\mathbf{A}\mathbf{u} + \mathbf{B}\boldsymbol{\phi}). \tag{10.60}$$

Evaluating the adjoint equation, optimality condition, and terminal initial condition (10.58) for this Hamiltonian yields

$$\dot{\boldsymbol{\Lambda}}^T = \gamma^2\mathbf{u}^T\mathbf{Q} - \boldsymbol{\Lambda}^T\mathbf{A}, \quad \boldsymbol{\phi}^T\mathbf{R} - \boldsymbol{\Lambda}^T\mathbf{B} = \mathbf{0}, \quad \boldsymbol{\Lambda}(t_f) = \mathbf{0}, \tag{10.61}$$

respectively. Observe that these are the same as equations (10.27), (10.28), and (10.29) derived using classical variational methods. Pontryagin's approach to optimal control is equivalent to the classical variational approach used in the previous section if the control solution is smooth. It allows us to write down the adjoint equations and optimality conditions in a very compact, general, and straightforward form. Note that we have not placed any restrictions on the integrand $F(t,\mathbf{u},\boldsymbol{\phi})$ in the cost functional or the differential equation $\mathbf{f}(t,\mathbf{u},\boldsymbol{\phi})$. Therefore, Pontryagin's approach may be applied to nonlinear state equations and nonquadratic cost functionals as well as nonautonomous systems. The only requirement is that we write the state equation(s) as a system of first-order differential equations, that is, in state-space form.

Not only is Pontryagin's approach equivalent to the classical variational approach emphasized here, it is, in fact, a generalization of it. This is the case because it is not subject to some of the same limitations. Specifically, the classical variational approach requires that all functions, including the state and control variables, must be continuously differentiable functions in order to obtain the Euler equations. In contrast, Pontryagin's approach can allow for piecewise-continuous solutions for the control input by relaxing the optimality condition. Therefore, not only does Pontryagin's principle provide a more straightforward method of obtaining the governing equations for optimal control, it allows for generalization to scenarios in which the optimal control input is not smooth.

Pontryagin's approach is formalized in the form of his *minimum principle*, which asserts that a necessary condition for optimal control is that the control $\boldsymbol{\phi}(t)$ be chosen that minimizes[4] Pontryagin's Hamiltonian $H_P(t,\mathbf{u},\boldsymbol{\phi},\boldsymbol{\Lambda})$ defined by equation (10.57) (see Pontryagin, Boltyanskii, Gamkrelidze, and Mishchenko 1962 for a proof). As shown above, Pontryagin's principle provides a more straightforward method of obtaining the governing equations for optimal control as compared to the classical variational method. Its primary virtue, however, is that it is not subject to the same limitations inherent in the classical variational approach. In particular, if the optimality condition $\partial H_P/\partial\boldsymbol{\phi}=\mathbf{0}$ is not enforced, then there is no need to evaluate derivatives with respect to the control variable $\boldsymbol{\phi}(t)$, and continuity conditions on the control variable can be relaxed. This opens up possible alternative optimal control solutions, such as *bang-bang control* considered in the next section. Some texts, such as Bryson and Ho's (1975), use Pontryagin's formulation for optimal control almost exclusively owing to its ease of use and ability to generalize to nonsmooth optimal solutions. Generally, we will use the optimality condition when the control variables are unconstrained and Pontryagin's principle when they are constrained. See Kirk (1970) for the case when the state variables are subject to inequality constraints. This would be the case, for example, if a maximum altitude is specified for an airplane or the motion of a robotic arm is limited to avoid collisions with other objects.

An important special case of Pontryagin's approach occurs for autonomous systems in which neither the integrand $F(\mathbf{u},\boldsymbol{\phi})$ nor $\mathbf{f}(\mathbf{u},\boldsymbol{\phi})$ in the differential equation depends explicitly on time, and the control input is unconstrained such that the optimality condition is enforced. In this case, Pontryagin's Hamiltonian is of the form

$$H_P(t,\mathbf{u},\boldsymbol{\phi},\boldsymbol{\Lambda})=F(\mathbf{u},\boldsymbol{\phi})-\boldsymbol{\Lambda}^T(t)\mathbf{f}(\mathbf{u},\boldsymbol{\phi}).$$

Let us differentiate with respect to time as follows

$$\begin{aligned}
\frac{dH_P}{dt}&=\frac{\partial F}{\partial\mathbf{u}}\frac{d\mathbf{u}}{dt}+\frac{\partial F}{\partial\boldsymbol{\phi}}\frac{d\boldsymbol{\phi}}{dt}-\boldsymbol{\Lambda}^T\left(\frac{\partial\mathbf{f}}{\partial\mathbf{u}}\frac{d\mathbf{u}}{dt}+\frac{\partial\mathbf{f}}{\partial\boldsymbol{\phi}}\frac{d\boldsymbol{\phi}}{dt}\right)-\dot{\boldsymbol{\Lambda}}^T\mathbf{f}\\
&=\left(\frac{\partial F}{\partial\mathbf{u}}-\boldsymbol{\Lambda}^T\frac{\partial\mathbf{f}}{\partial\mathbf{u}}-\dot{\boldsymbol{\Lambda}}^T\right)\mathbf{f}+\frac{\partial H_P}{\partial\boldsymbol{\phi}}\frac{d\boldsymbol{\phi}}{dt}\\
&=\left(\frac{\partial H_P}{\partial\mathbf{u}}-\dot{\boldsymbol{\Lambda}}^T\right)\mathbf{f}+\frac{\partial H_P}{\partial\boldsymbol{\phi}}\frac{d\boldsymbol{\phi}}{dt}\\
\frac{dH_P}{dt}&=0,
\end{aligned}$$

where the adjoint equation and optimality condition (10.58) have been used to eliminate the two terms on the right-hand side in the last step, respectively. Under these conditions, therefore, Pontryagin's Hamiltonian is such that $H_P=$ constant for the optimal trajectory. Observe that this is analogous to the Hamiltonian in mechanics for conservative, autonomous systems. In addition, if the terminal time t_f is not specified, as in time-optimal control (see the next section), then $H_P=0$ for all $0\le t\le t_f$. It is important to keep in mind that this special case only applies when

[4] In general, we seek to minimize the cost functional, in which case we minimize Pontryagin's Hamiltonian. If we seek to maximize the functional, as in Example 10.5, then we maximize Pontryagin's Hamiltonian.

the optimality condition is enforced, that is, when the control input is unconstrained. If the control input is constrained to a fixed interval, then the optimality condition $\partial H_P/\partial \phi = 0$ does not apply in general, and dH_P/dt cannot be evaluated owing to discontinuities in H_P.

EXAMPLE 10.4 In order to illustrate Pontryagin's approach, let us reconsider control of the undamped harmonic oscillator from Sections 10.4.1 and 10.4.2. We borrow the same cost functional with a $0 \leq t \leq t_f$ performance measure as before according to

$$J[u,\phi] = \frac{1}{2}\int_0^{t_f} \left(\gamma^2 u^2 + \phi^2\right) dt. \tag{10.62}$$

Recall that the state equation is

$$\ddot{u} + u = \phi, \tag{10.63}$$

now with the initial conditions

$$u(0) = a, \quad \dot{u}(0) = b. \tag{10.64}$$

The only difference from the previous case is that we restrict the control authority to the range $-1 \leq \phi \leq 1$.

Solution: As in Section 10.4.2, we convert the state equation to a system of first-order, linear equations using the transformation

$$u_1 = u, \quad u_2 = \dot{u}.$$

Differentiating and writing in terms of u_1 and u_2, the first-order system of state equations is

$$\dot{u}_1 = u_2, \quad \dot{u}_2 = -u_1 + \phi. \tag{10.65}$$

Hence, in matrix form, the state equations become

$$\dot{\mathbf{u}} = \mathbf{f}(\mathbf{u}, \boldsymbol{\phi}), \tag{10.66}$$

where

$$\mathbf{f} = \mathbf{A}\mathbf{u} + \mathbf{B}\boldsymbol{\phi},$$

and

$$\mathbf{u} = \begin{bmatrix} u_1 \\ u_2 \end{bmatrix}, \quad \boldsymbol{\phi} = [\phi], \quad \mathbf{A} = \begin{bmatrix} 0 & 1 \\ -1 & 0 \end{bmatrix}, \quad \mathbf{B} = \begin{bmatrix} 0 \\ 1 \end{bmatrix}.$$

With respect to the new variables, the initial conditions are

$$u_1(0) = a, \quad u_2(0) = b. \tag{10.67}$$

Pontryagin's Hamiltonian for this case is

$$H_P = F(\mathbf{u}, \boldsymbol{\phi}) - \boldsymbol{\Lambda}^T \mathbf{f}(\mathbf{u}, \boldsymbol{\phi}) = \gamma^2 u_1^2 + \phi^2 - \Lambda_1 u_2 - \Lambda_2(-u_1 + \phi). \tag{10.68}$$

Thus, the adjoint equations (10.58) are

$$\dot{\Lambda}_1 = \frac{\partial H_P}{\partial u_1}, \quad \dot{\Lambda}_2 = \frac{\partial H_P}{\partial u_2},$$

which yields

$$\dot{\Lambda}_1 = \Lambda_2 + \gamma^2 u_1, \quad \dot{\Lambda}_2 = -\Lambda_1. \tag{10.69}$$

The optimality condition

$$\frac{\partial H_P}{\partial \phi} = 0$$

requires that

$$\Lambda_2 = \phi. \tag{10.70}$$

The optimality condition only applies if the control remains within its defined bounds $-1 \le \phi \le 1$, in which case we also have $-1 < \Lambda_2 \le 1$. When Λ_2 exceeds these bounds, that is, when the control saturates, we determine the control such that Pontryagin's Hamiltonian (10.68) is a minimum. If $\Lambda_2 \ge 1$, then H_P is a minimum if $\phi = 1$. Similarly, if $\Lambda_2 \le 1$, then H_P is a minimum if $\phi = -1$. Thus, the control input is

$$\phi(t) = \begin{cases} -1, & \Lambda_2 < -1 \\ \Lambda_2, & -1 \le \Lambda_2 \le 1 \\ 1, & \Lambda_2 > 1 \end{cases}, \tag{10.71}$$

which takes the place of the optimality condition (10.70). Observe that it is the adjoint variable Λ_2 that determines the control input.

Summarizing, we solve the state equations (10.65) forward in time and the adjoint equations (10.69) backward in time subject to the appropriate initial conditions at $t = 0$ and $t = t_f$, respectively. Note that if the control does not saturate, in which case Λ_2 (and thus ϕ) remains between the predefined bounds of the control, this gives the same formulation as that obtained in Sections 10.4.1 and 10.4.2 using the classical variational approach. Once again, not only does Pontryagin's principle provide an alternative approach to addressing optimal control problems, it also allows for dealing with nonsmooth inputs, here because of the bounds placed on the control authority.

Armed with Pontryagin's principle, let us revisit the profit reinvestment example considered in Section 10.3. Recall that the classical variational approach, which requires all dependent variables to be continuously differentiable, leads to a contradiction, signaling that no such solution exists. Instead, the optimal solution turns out to be *bang-bang control*, which is considered in more detail in the next section.

EXAMPLE 10.5 In the profit reinvestment example from Example 10.2, we seek to maximize the objective functional

$$J[u(t), \phi(t)] = \int_0^{t_f} (1 - \phi) u dt \tag{10.72}$$

subject to the differential constraint that the state equation

$$\dot{u} = \alpha \phi u, \quad \alpha > 0 \tag{10.73}$$

is satisfied with the inequality constraint $0 \le \phi \le 1$. Recall that u is the profit accumulated per time period, and the control variable ϕ is the percentage of profit to be reinvested.

Solution: Pontryagin's Hamiltonian is given by

$$H_P = (1-\phi)u - \Lambda\alpha\phi u = [1-(1+\alpha\Lambda)\phi]u. \tag{10.74}$$

The adjoint equation $\dot{\Lambda} = \partial H_P/\partial u$ gives

$$\dot{\Lambda} = 1-(1+\alpha\Lambda)\phi, \quad \Lambda(t_f) = 0, \tag{10.75}$$

where the adjoint equation is the same as that in equation (10.4). The reason for the failure of the classical variational approach arises from imposition of the optimality condition. Therefore, we relax the optimality condition and instead appeal to Pontryagin's principle. Because we seek to maximize the objective functional (10.72), Pontryagin's Hamiltonian is to be maximized, rather than minimized as before. Recall that $u \ge 0$; therefore, maximizing H_P given by (10.74) requires that $(1+\alpha\Lambda)\phi$ be minimized. With $0 \le \phi \le 1$, this occurs for

$$\phi(t) = \begin{cases} 0, & \Lambda > -1/\alpha \\ 1, & \Lambda < -1/\alpha \end{cases}, \tag{10.76}$$

where $\alpha > 0$. Hence, we see that the Lagrange multiplier (adjoint variable) is the *switching function* that determines when the control variable switches from one extreme to the other. Thus, the optimal control solution indicates that we should either fully invest all profits back into the business ($\phi = 1$) or invest none of the profits ($\phi = 0$). This is known as *bang-bang control*. We might expect intuitively that we should initially invest all profits early in the time interval $0 \le t \le t_f$ and then switch to zero reinvestment later in the interval in order to maximize total profits over the entire time interval. Let us continue the analysis to confirm this and determine when the switch should occur.

We have two sets of solutions for the state and adjoint equations. When $\phi(t) = 0$, the solution to the state and adjoint equations are

$$u(t) = c_1, \quad \Lambda(t) = t + c_2, \tag{10.77}$$

respectively. When $\phi(t) = 1$, the solutions are

$$u(t) = c_3 e^{\alpha t} \quad \Lambda(t) = c_4 e^{-\alpha t}. \tag{10.78}$$

To determine the optimal solution, let us begin at the terminal time $t = t_f$ and work our way back across the time interval of interest to piece together the solution for the adjoint variable. The terminal-time initial condition is $\Lambda(t_f) = 0$, in which case we have from (10.76) that $\phi(t_f) = 0$, that is, there is no reinvestment of profits at the terminal time as we intuitively expected. Applying the initial condition at the terminal time to the solution to the adjoint equation (10.77) gives $c_2 = -t_f$, and

$$\Lambda(t) = t - t_f \quad \text{for} \quad t_s \le t \le t_f \tag{10.79}$$

over the interval for which $\phi = 0$ beginning at the switch-over time t_s. The switch in control input (profit reinvestment) occurs when $\Lambda = -1/\alpha$, which occurs at the switch-over time of $t_s = t_f - 1/\alpha$. At this time the optimal control switches to the $\phi(t) = 1$ solution given by equations (10.78). Because the adjoint variable is continuous, the constant c_4 can be determined by matching the adjoint variable at $t = t_f - 1/\alpha$, where $\Lambda = -1/\alpha$. Substituting into equation (10.78) for the adjoint variable yields $c_4 = -1/(\alpha e^{1-\alpha t_f})$. Then

$$\Lambda(t) = -\frac{1}{\alpha} e^{\alpha(t_f - t) - 1} \quad \text{for} \quad 0 \le t \le t_f - \frac{1}{\alpha}. \tag{10.80}$$

Because $\Lambda(t)$ remains less than $-1/\alpha$ for the remainder of the interval proceeding toward the initial time $t = 0$, there are no additional switches in control input. The adjoint variable, which is exponential for $0 \le t \le t_s = t_f - 1/\alpha$ and linear thereafter, is plotted in Figure 10.8(a). The Lagrange multiplier (adjoint variable) is sometimes called an *influence*, or *sensitivity*, *function* as it quantifies how sensitive the functional J is to changes in the control input. In this case, we see that the total profit accumulated over the time interval $0 \le t \le t_f$ is most sensitive to changes in reinvestment of profits toward the beginning of the time interval as compared to the end. This makes sense intuitively as profit reinvested near the beginning of the time interval has more time to influence future profits.

Similarly, the state variable is then integrated forward in time from the initial condition $u(0) = u_0$ to give

$$u(t) = \begin{cases} u_0 e^{\alpha t}, & 0 \le t \le t_f - \dfrac{1}{\alpha} \\ u_0 e^{\alpha t_f - 1}, & t_f - \dfrac{1}{\alpha} \le t \le t_f \end{cases}. \tag{10.81}$$

The state variable is shown in Figure 10.8(b) for the case with initial condition $u_0 = 1$. Observe that the profit per period increases exponentially before becoming constant once no profit is reinvested. The control input on profit reinvestment rate, (10.76) is

$$\phi(t) = \begin{cases} 1, & 0 \le t \le t_f - 1/\alpha \\ 0, & t_f - 1/\alpha \le t \le t_f \end{cases}, \tag{10.82}$$

which is shown in Figure 10.8(c). Consequently, we see that in order to maximize total profits over the time period $0 \le t \le t_f$, the optimal solution is to reinvest all profits until the switch-over time $t_s = t_f - 1/\alpha$, after which none of the profits are reinvested. Control that switches suddenly between the extremes of its bounds is called *bang-bang control*, which is considered in more detail in the next section.

10.4.5 Time-Optimal Control

We have focused most of our attention on the LQ problem, which for the linear case has a number of advantages that lend itself to applications. However, there are several variations on the LQ problem that give rise to interesting and useful control behavior. Here, we consider the case of *time-optimal control*, in which the control objective is to be accomplished in the minimum time subject to an inequality constraint on the control magnitude. If our objective is to accomplish a task in a short

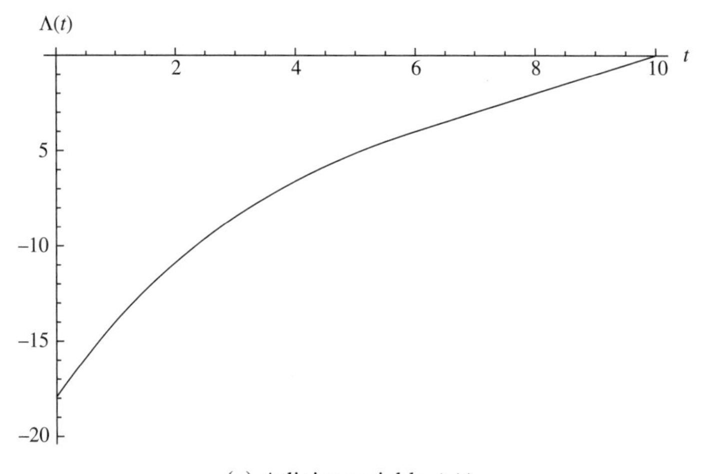

(a) Adjoint variable $\Lambda(t)$.

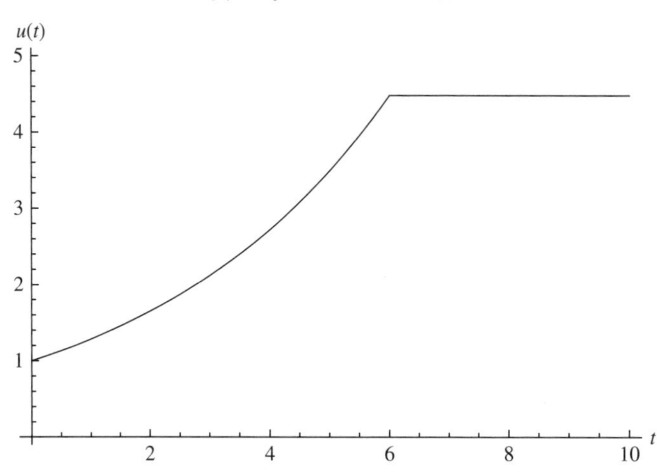

(b) State variable $u(t)$ showing the profit per time period, for example quarterly.

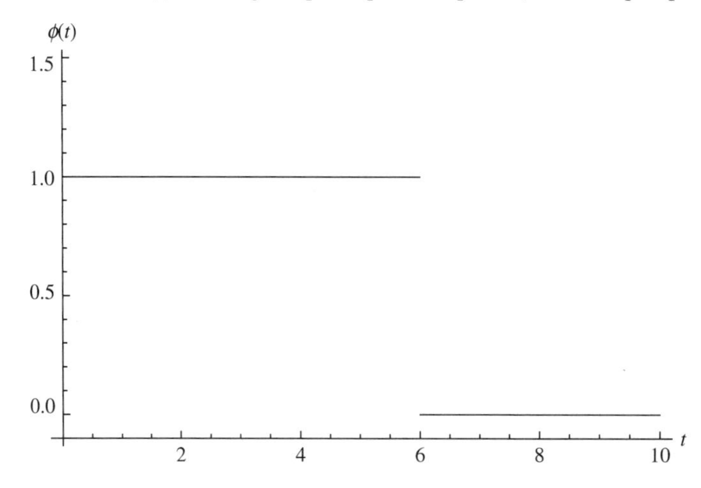

(c) Control input $\phi(t)$ showing the percentage of profit to be reinvested each period.

Figure 10.8. Solution of profit reinvestment Example 10.5 for the case with $u_0 = 1, \alpha = \frac{1}{4}$, and $t_f = 10$ ($t_s = 6$).

time period, then naturally this would typically result in very high control inputs. For example, a rocket could achieve a particular orbit in the shortest time by requiring more thrust. Therefore, it is typically necessary in time-optimal control scenarios to include limits on the control input, often referred to as *control authority*. This would take into account the maximum thrust of the rocket's engines, for example. Hence, time-optimal control is generally associated with inequality constraints. The inequality constraint on the magnitude of the control input replaces the penalty function used in the classical variational approach, which serves to regulate the magnitude of the control input but does not restrict it to predefined bounds.

As we will see, time-optimal control with inequality constraints often leads to *bang-bang* behavior for which the control operates only at its extreme values and switches suddenly (discontinuously) between the minimum and maximum control inputs as in Example 10.5. Because the classical variational approach does not admit piecewise continuous solutions, such scenarios must be treated using Pontryagin's principle, thereby illustrating its generality relative to the classical variational approach.

In time-optimal control, the objective is to transfer the system from one state to another in minimum time, that is, to minimize the cost functional

$$J[\mathbf{u},\boldsymbol{\phi}] = \int_0^{t_f} dt = t_f, \tag{10.83}$$

where again, we replace the penalty function with inequality constraints on the control input, say $-1 \le \phi_i \le 1$, such that each control input is confined to a specified range. Hence, for time-optimal control, the integrand of the cost functional is $F = 1$, and Pontryagin's Hamiltonian is

$$H_P(t,\mathbf{u},\boldsymbol{\phi},\boldsymbol{\Lambda}) = 1 - \boldsymbol{\Lambda}^T(t)\mathbf{f}(t,\mathbf{u},\boldsymbol{\phi}). \tag{10.84}$$

EXAMPLE 10.6 In order to illustrate time-optimal control, let us reconsider control of the undamped harmonic oscillator from Section 10.4.1 and Example 10.4. Here, we evaluate time-optimal control with inequality constraints on the control input as opposed to the LQ-type control considered previously. Recall that the state equation is

$$\ddot{u} + u = \phi, \tag{10.85}$$

now with the initial conditions

$$u(0) = a, \quad \dot{u}(0) = b. \tag{10.86}$$

The control authority is restricted to the range $-1 \le \phi \le 1$.

Solution: As in Example 10.4, the state equation (10.85) becomes

$$\dot{\mathbf{u}} = \mathbf{f}(\mathbf{u},\boldsymbol{\phi}), \tag{10.87}$$

where

$$\mathbf{f} = \mathbf{A}\mathbf{u} + \mathbf{B}\boldsymbol{\phi},$$

and

$$\mathbf{u} = \begin{bmatrix} u_1 \\ u_2 \end{bmatrix}, \quad \boldsymbol{\phi} = [\phi], \quad \mathbf{A} = \begin{bmatrix} 0 & 1 \\ -1 & 0 \end{bmatrix}, \quad \mathbf{B} = \begin{bmatrix} 0 \\ 1 \end{bmatrix}.$$

With respect to the new variables, the initial conditions are

$$u_1(0) = a, \quad u_2(0) = b. \tag{10.88}$$

We can define the target state to be anywhere in the (u_1, u_2)-phase space; however, we can designate this target state to be the origin

$$u_1(t_f) = 0, \quad u_2(t_f) = 0 \tag{10.89}$$

without loss of generality (we can simply shift the variables such that this is the case).

Before invoking Pontryagin's principle, let us follow the procedure in Section 10.4.4 to determine what the consequences are. Pontryagin's Hamiltonian for this case is

$$H_P = 1 - \boldsymbol{\Lambda}^T \mathbf{f}(\mathbf{u}, \boldsymbol{\phi}) = 1 - \Lambda_1 u_2 - \Lambda_2(-u_1 + \phi). \tag{10.90}$$

Thus, the adjoint equations (10.58) are

$$\dot{\Lambda}_1 = \frac{\partial H_P}{\partial u_1}, \quad \dot{\Lambda}_2 = \frac{\partial H_P}{\partial u_2},$$

which give

$$\dot{\Lambda}_1 = \Lambda_2, \quad \dot{\Lambda}_2 = -\Lambda_1. \tag{10.91}$$

The optimality condition

$$\frac{\partial H_P}{\partial \phi} = 0$$

requires that

$$\Lambda_2 = 0.$$

However, if $\Lambda_2(t) = 0$ then $\Lambda_1(t) = 0$ as well from the adjoint equations. This would result in $H_P = 1$. However, recall the result at the end of the last section requiring that $H_P = 0$ when the optimality condition is enforced and t_f is not specified. In addition, observe that ϕ does not appear in the optimality condition as usual, suggesting that it is arbitrary. Clearly, this cannot be the case as different control inputs result in different solutions of the state equations (10.87). These contradictions indicate that we must relax the optimality condition. In other words, the optimal control solution does not allow for intermediate values of the control variable between its extreme bounding values. Instead, the control must operate at its extremes $\phi = -1$ or $\phi = 1$. Thus, we have bang-bang control. Observe that bang-bang control is consistent with the fact that $\delta\boldsymbol{\phi} = 0$ if the optimality condition is not enforced; therefore, $\boldsymbol{\phi}$ is specified. This is analogous to driving a car to a destination with the objective of minimizing the time of travel. Intuitively, we know that this would require alternating between full throttle and full braking. This *optimal* solution minimizes the travel time (and maximizes the discomfort of the passengers).

Given that the optimality condition does not apply for bang-bang control, we must appeal directly to Pontryagin's principle, which requires that the control input

ϕ be such that Pontryagin's Hamiltonian (10.90) is a minimum. For example, given the case when $\phi = -1$, Pontryagin's Hamiltonian will be a minimum when Λ_2 is negative. Similarly, when $\phi = 1$, it is a minimum when Λ_2 is positive. Hence, the adjoint variable and switching function Λ_2 determines the control input as follows:

$$\phi(t) = \begin{cases} -1, & \Lambda_2 < 0 \\ 1, & \Lambda_2 > 0 \end{cases},$$

or simply

$$\phi = \operatorname{sgn} \Lambda_2.$$

From the adjoint equations (10.91), Λ_2 satisfies

$$\ddot{\Lambda}_2 + \Lambda_2 = 0,$$

which has the solution

$$\Lambda_2(t) = c_1 \sin(t + c_2),$$

where we require that $c_1 > 0$ without loss of generality. Thus, the switching function is

$$\phi = \operatorname{sgn}\left[\sin(t + c_2)\right],$$

which indicates that the control input changes sign every π time units. Solutions of the state equations (10.87) for the undamped harmonic oscillator with $\phi = \pm 1$ are circles in the phase plane centered at $(u_1, u_2) = (-1, 0)$ when $\phi = -1$ and $(u_1, u_2) = (1, 0)$ when $\phi = 1$ (see, for example, Pontryagin, Boltyanskii, Gamkrelidze, and Mishchenko 1962 or Denn 1969). The optimal trajectory begins at $(u_1, u_2) = (a, b)$ at $t = 0$ and proceeds along successive circular arcs in the phase plane, switching every π time units until it reaches the origin $(u_1, u_2) = (0, 0)$ within the shortest time $t = t_f$; that is, with the least number of switches for the given initial conditions.

Observe that the results for the control obtained using the classical variational method for the undamped harmonic oscillator in Section 10.4.1 essentially produces smoothed (continuous) versions of the bang-bang control solution obtained here. Specifically, the control input is out-of-phase forcing that changes sign every π time units.

Let us close by summarizing Pontryagin's approach to optimal control. If we seek to minimize the cost functional (10.54) subject to the state equations (10.55) expressed as a system of first-order differential equations and initial conditions (10.56), we form Pontryagin's Hamiltonian (10.57). If the control input is unconstrained, then the adjoint equation(s), optimality condition, and terminal-time initial condition(s) are given by (10.58). This formulation is exactly equivalent to the classical variational method of optimal control. If the control input is constrained within a predefined finite range and/or the cost functional is formulated for time-optimal control, then it may be the case that a solution to the optimal control problem is not possible assuming that the control input is continuous, which is required by the optimality condition. In this case, we do not enforce the optimality condition and instead apply Pontryagin's principle, which states that the control input $\phi(t)$ is chosen such that Pontryagin's Hamiltonian H_P is a minimum. Note that if the cost functional is to be maximized, then we select the control input that maximizes Pontryagin's Hamiltonian.

10.5 Optimal Control of Continuous Systems

We now turn our attention to control of continuous systems, which are typically governed by partial differential equations. Optimal control theory applied to partial differential state equations does not lend itself to as direct a formulation as the LQ problem for discrete systems considered previously. That is, there is no generalized Riccati-type equation for such cases. However, we can revert to more general approaches, and the LQ problem with a linear partial differential state equation does have interesting parallels to the adjoints of differential operators considered in Section 1.8. To illustrate, consider a system that is governed by an unsteady, second-order, linear partial differential equation of the form

$$\frac{\partial u}{\partial t} = a_0 \frac{\partial^2 u}{\partial x^2} + a_1 \frac{\partial u}{\partial x} + a_2 u + a_3 + \phi(x,t), \quad 0 \leq x \leq 1, \quad 0 \leq t \leq t_f, \tag{10.92}$$

where $u(x,t)$ is the state variable, and $\phi(x,t)$ is the control variable. Physically, we regard $\phi(x,t)$ as an input, that is, a source term, to the system as in the discrete case. The constant $a_0 > 0$ in order for the equation to be parabolic forward in time. In other words, equation (10.92) is an initial-value problem from $t = 0$ to $t = t_f$ and is to be solved subject to initial and boundary conditions on the state variable as follows:

$$u(x,0) = u_0(x), \quad u(0,t) = u_L(t), \quad u(1,t) = u_R(t), \tag{10.93}$$

where $u_0(x), u_L(t)$ and $u_R(t)$ are specified. A common special case of equation (10.92) governs one-dimensional, unsteady diffusion (for example, heat conduction) with $a_1 = a_2 = a_3 = 0$ and a_0 being the diffusivity.

10.5.1 Variational Approach

We seek the solution to the state equation (10.92) and the associated control input $\phi(x,t)$ that minimizes the following cost, or objective, functional over the time interval $0 \leq t \leq t_f$

$$J[u,\phi] = \frac{1}{2} \int_0^{t_f} \int_0^1 \left(\gamma^2 u^2 + \phi^2 \right) dx dt. \tag{10.94}$$

Note the similarity between this cost functional and those used for discrete systems, such as equations (10.20) and (10.25). In particular, the first term is the performance measure, and the second term is the penalty function. The primary difference is that in the case of continuous systems, the definite integral is over both time and space.

In order to minimize the functional (10.94) subject to the differential constraint that the state (governing) equation (10.92) is satisfied, we introduce the Lagrange multiplier $\Lambda(x,t)$ and integrate to form the augmented functional

$$\begin{aligned} \tilde{J}[u,\phi,\Lambda] = \int_0^{t_f} \int_0^1 \bigg\{ & \frac{1}{2} \left[\gamma^2 u^2 + \phi^2 \right] \\ & + \Lambda(x,t) \left[u_t - a_0 u_{xx} - a_1 u_x - a_2 u - a_3 - \phi \right] \bigg\} dx dt, \end{aligned} \tag{10.95}$$

where subscripts represent partial differentiation with respect to the indicated variable. Taking $\delta \tilde{J} = 0$ gives

$$\int_0^{t_f} \int_0^1 \Big[\gamma^2 u\delta u + \phi\delta\phi + \Lambda\left(\delta u_t - a_0 \delta u_{xx} - a_1 \delta u_x - a_2 \delta u - \delta\phi\right)\Big] dxdt = 0. \qquad (10.96)$$

Next, the terms that involve derivatives are integrated by parts. The third term yields

$$\int_0^1 \int_0^{t_f} \Lambda\delta u_t dtdx = \int_0^1 \Bigg[\Lambda(x,t_f)\delta u(x,t_f) - \Lambda(x,0)\delta u(x,0) \\ \int_0^{t_f} \Lambda_t \delta u dt\Bigg] dx.$$

The fourth term is integrated by parts twice according to

$$\begin{aligned}\int_0^{t_f} \int_0^1 [-a_0 \Lambda \delta u_{xx}] dxdt = & \int_0^{t_f} \Bigg[- a_0 \Lambda(1,t)\delta u_x(1,t) + a_0 \Lambda(0,t)\delta u_x(0,t) \\ & + \int_0^1 a_0 \Lambda_x \delta u_x dx\Bigg] dt \\ = & \int_0^{t_f} \Bigg[- a_0 \Lambda(1,t)\delta u_x(1,t) + a_0 \Lambda(0,t)\delta u_x(0,t) \\ & + a_0 \Lambda_x(1,t)\delta u(1,t) - a_0 \Lambda_x(0,t)\delta u(0,t) \\ & - \int_0^1 a_0 \Lambda_{xx} \delta u dx\Bigg] dt.\end{aligned}$$

Finally, the fifth term produces

$$\begin{aligned}\int_0^{t_f} \int_0^1 [-a_1 \Lambda \delta u_x] dxdt = & \int_0^{t_f} \Bigg[- a_1 \Lambda(1,t)\delta u(1,t) + a_1 \Lambda(0,t)\delta u(0,t) \\ & + \int_0^1 a_1 \Lambda_x \delta u dx\Bigg] dt.\end{aligned}$$

Note that the integration by parts causes the terms corresponding to odd-order derivatives, u_t and u_x, to change sign, whereas the even-order derivative terms, such as u_{xx}, do not (more accurately, they change sign twice). From the initial and boundary conditions (10.93) for the state equation, $\delta u(x,0) = \delta u(0,t) = \delta u(1,t) = 0$ because they are specified and, therefore, do not vary. For the remaining terms evaluated at the boundaries and at $t = t_f$ to vanish requires the following natural boundary conditions to hold

$$\Lambda(x,t_f) = 0, \quad \Lambda(0,t) = 0, \quad \Lambda(1,t) = 0, \qquad (10.97)$$

which are the initial and boundary conditions for the control (adjoint) equation. Substituting the remaining terms from the integration by parts into equation (10.96) and collecting terms leads to

$$\int_0^{t_f} \int_0^1 \Big\{\Big[\gamma^2 u - \Lambda_t - a_0 \Lambda_{xx} + a_1 \Lambda_x - a_2 \Lambda\Big]\delta u + [\phi - \Lambda]\delta\phi\Big\} dxdt = 0.$$

Because $u(x,t)$ and $\phi(x,t)$ vary, each of the expressions in square brackets must vanish independently over the entire domain $0 \le x \le 1, 0 \le t \le t_f$. Therefore, the Euler equations are

$$\Lambda_t = -a_0\Lambda_{xx} + a_1\Lambda_x - a_2\Lambda + \gamma^2 u, \tag{10.98}$$

$$\Lambda = \phi, \tag{10.99}$$

where equation (10.98) is the *adjoint equation*, and equation (10.99) is called the *optimality condition* as it relates the optimal control $\phi(x,t)$ to the Lagrange multiplier $\Lambda(x,t)$, which is also called the *adjoint variable*. Upon substitution of (10.99) into (10.98), we have the control equation for $\phi(x,t)$

$$\frac{\partial\phi}{\partial t} = -a_0\frac{\partial^2\phi}{\partial x^2} + a_1\frac{\partial\phi}{\partial x} - a_2\phi + \gamma^2 u. \tag{10.100}$$

Therefore, we have the coupled state equation (10.92) and control equation (10.100) for $u(x,t)$ and $\phi(x,t)$. These are solved subject to the initial and boundary conditions (10.93) and (10.97), respectively, with Λ replaced by ϕ according to (10.99) in the initial and boundary conditions for the control equation. The solution to this system of equations satisfies the state equation (10.92) and minimizes the cost functional (10.94).

Observe that the control equation (10.100) is of the same form as the governing equation (10.92), being an unsteady, second-order, linear differential equation. However, solving the control equation (10.100) requires integration backward in time from $t = t_f$ to $t = 0$ owing to the signs of the ϕ_t and ϕ_{xx} terms; in other words, the control equation is parabolic backward in time. Note that the terminal-time initial condition (10.97) for the control equation is consistent with such a backward-time integration. Alternatively, one can view this as a boundary-value problem in time with boundary conditions at $t = 0$ and $t = t_f$ instead of two initial-value problems forward and backward in time.

REMARKS:

1. As in the discrete case, because the Lagrange multiplier $\Lambda(x,t)$ varies, taking the variation of equation (10.95) also produces the term

$$\int_0^{t_f}\int_0^1 (u_t - a_0u_{xx} - a_1u_x - a_2u - a_3 - \phi)\,\delta\Lambda\,dx\,dt,$$

but this simply returns the governing (state or constraint) equation (10.92) and has not been included in the above derivation.

2. The general form of the Euler equations for functionals of the form (10.95) are

$$\frac{\partial F}{\partial u} - \frac{\partial}{\partial t}\left(\frac{\partial F}{\partial u_t}\right) - \frac{\partial}{\partial x}\left(\frac{\partial F}{\partial u_x}\right) + \frac{\partial^2}{\partial x^2}\left(\frac{\partial F}{\partial u_{xx}}\right) = 0, \quad \frac{\partial F}{\partial \phi} = 0,$$

where $F(x,t,u,u_x,u_{xx},u_t,\phi)$ is the integrand of the functional (10.95). These produce the same Euler equations (10.98) and (10.99).

There are many possible generalizations and variations on what has been included in this brief introduction to optimal control theory applied to continuous systems. For example:

- As in equation (10.25) for discrete systems, if the state variable is a vector $\mathbf{u}(x,t)$ corresponding to a system of linear equations, then the quadratic cost functional (10.94) would be of the form

$$J[\mathbf{u},\boldsymbol{\phi}] = \frac{1}{2}\int_0^{t_f}\int_0^1 \left(\gamma^2 \mathbf{u}^T \mathbf{Q}\mathbf{u} + \boldsymbol{\phi}^T \mathbf{R}\boldsymbol{\phi}\right) dxdt,$$

  where $\boldsymbol{\phi}(x,t)$ is now a vector, $\mathbf{Q}$ and $\mathbf{R}$ are symmetric matrices, and $\mathbf{R}$ is also positive definite.
- Rather than specifying a performance measure that applies throughout the entire time interval $0 \leq t \leq t_f$, it is sometimes desirable to enforce a condition at the final time $t = t_f$ by replacing the first term in the cost functional (10.94) with

$$\int_0^1 \frac{1}{2}\gamma^2 Q u(t_f)^2 dx.$$

  Note that the integral only involves the performance measure at the final time $t = t_f$, rather than over the entire interval $0 \leq t \leq t_f$ as above. Thus, the optimization problem is to determine the control distribution $\phi(x,t)$ throughout the time interval $0 \leq t \leq t_f$ that minimizes the performance measure at the final time $t = t_f$. A terminal-time performance measure has been illustrated for the discrete case in Section 10.4.1.
- In the continuous optimal control scenario considered here, the control is imagined to act throughout the domain, and the performance measure assumes that state information is available throughout the entire domain. In many practical situations, however, the state is determined using sensors, and the control is implemented via actuators, and it may not be possible to sense and actuate throughout the entire domain. As a result, one can also consider boundary-based control and/or boundary-based performance measures. Any combination of boundary- or domain-based performance measure or control can be considered using the variational framework outlined here (see Section 10.5.3).
- In practice, it is often the case that the magnitude of the control must be confined to a well-defined range dictated by the choice of actuator or other practical constraints, such as the energy available for control. In order to account for this in the above formulation involving the penalty function in the cost functional (10.94), it would be necessary to determine, for example, by trial and error, the value of γ^2 that limits the total control input at each time below the desired threshold. Alternatively, as in Gunzburger and Manservisi (1999), one could account for this actuation authority directly by replacing the penalty function in the cost functional with an additional inequality constraint of the form

$$\left(\int_0^1 R\phi^2 dx\right)^{1/2} \leq K,$$

  where K is a given positive constant that may represent the maximum control energy input allowed. This scenario has been considered for the discrete case in Section 10.4.5 using *Pontryagin's principle*, which generalizes to the continuous case as well.

- An alternative to solving the coupled state-control problem as described here is to utilize an optimization algorithm based on gradients of the functionals, such as the conjugate-gradient method, or a variant of Newton's method. The gradients are determined using *sensitivity* or *influence functions* that provide the gradient of the state variables with respect to the control variable. In this context, the Lagrange multiplier (adjoint variable) is the *influence function* representing the *sensitivity* of the cost functional to changes in the state of the system.

10.5.2 Adjoint Approach

In order to see how the variational control approach is related to the adjoint method introduced in Section 1.8, consider the governing (state) equation (10.92) in operator form

$$\mathcal{L}u(x,t) = f(x,t), \tag{10.101}$$

where

$$\mathcal{L} = \frac{\partial}{\partial t} - a_0\frac{\partial^2}{\partial x^2} - a_1\frac{\partial}{\partial x} - a_2, \quad f(x,t) = a_3 + \phi(x,t).$$

All initial and boundary conditions must be homogeneous. Recall that the *adjoint* operator $\mathcal{L}^*$ of a differential operator $\mathcal{L}$ is such that

$$\langle u, \mathcal{L}^* v\rangle = \langle v, \mathcal{L}u\rangle, \tag{10.102}$$

where in the present context the inner product of two functions $g(x,t)$ and $h(x,t)$ is defined by

$$\langle g,h\rangle = \int_0^{t_f}\int_0^1 g(x,t)h(x,t)dxdt.$$

Therefore, evaluating the right-hand side of equation (10.102) with the operator $\mathcal{L}$ yields

$$\langle u, \mathcal{L}^* v\rangle = \int_0^{t_f}\int_0^1 v\left(u_t - a_0u_{xx} - a_1u_x - a_2u\right)dxdt. \tag{10.103}$$

From integration by parts (with homogeneous initial and boundary conditions),

$$\int_0^{t_f}\int_0^1 vu_t dxdt = \int_0^{t_f}\int_0^1 (-v_t u)\,dxdt$$

$$\int_0^{t_f}\int_0^1 (-a_0 v u_{xx})\,dxdt = \int_0^{t_f}\int_0^1 (-a_0 v_{xx}u)\,dxdt$$

$$\int_0^{t_f}\int_0^1 (-a_1 v u_x)\,dxdt = \int_0^{t_f}\int_0^1 a_1 v_x u dxdt.$$

Substituting into equation (10.103) gives

$$\langle u, \mathcal{L}^* v\rangle = \int_0^{t_f}\int_0^1 u\left(-v_t - a_0v_{xx} + a_1v_x - a_2v\right)dxdt.$$

Therefore, the adjoint operator is

$$\mathcal{L}^* = -\frac{\partial}{\partial t} - a_0\frac{\partial^2}{\partial x^2} + a_1\frac{\partial}{\partial x} - a_2. \tag{10.104}$$

Table 10.2. *Equivalent terminology for the state and adjoint variables and equations.*

	State Problem	Adjoint Problem
Variables	state variables	co-state variables
	direct variables	indirect variables
		Lagrange multipliers
		adjoint variables
		influence functions
		sensitivity functions
Equations	state equations	co-state equations
		dual-state equations
	direct equations	indirect equations
	constraint equations	Euler equations
		adjoint equations
	governing equations	control equations

For the linear case considered here, the adjoint operator (10.104) is of the same form as the differential operator $\mathcal{L}$ in the governing equation (10.101), but with odd-order derivatives changing sign owing to the integration by parts. Furthermore, the adjoint operator is the same as the differential operator in the control equation (10.100) obtained using variational methods, where the control equation is then

$$\mathcal{L}^*\phi(x,t) = -\gamma^2 u. \tag{10.105}$$

Observe the similarities between determining the adjoint operator and the Euler equations in the variational approach, namely the integration by parts. The differential operator in the control equation is the adjoint of that in the state equation for all LQ problems, in which the state equation is linear and the cost functional is quadratic. Thus, the Euler equation in the variational approach is often referred to as the *adjoint equation* in controls theory. Because of this overlap in the approaches, the terminologies summarized in Table 10.2 are equivalent and are used interchangeably in the literature for LQ problems.

The primary virtue of the variational approach is that no assumption must be made about the cost functional or the governing state equations. For example, nonlinear state equations can be accommodated using the variational approach to obtain the control equation. Moreover, the variational approach may be used for more general forms of the cost functional and initial and boundary conditions (recall that the adjoint approach requires homogeneous boundary conditions).

10.5.3 Example: Control of Plane-Poiseuille Flow

As an illustration of optimal control applied to continuous systems, let us revisit the plane-Poiseuille flow example considered throughout Chapter 9. Recall from equation (9.41) that the governing (state) equation for unsteady plane-Poiseuille flow is

$$\frac{\partial u}{\partial t} = \frac{1}{Re}\frac{\partial^2 u}{\partial y^2} + \frac{2}{Re} + X, \quad 0 \le y \le 2, \tag{10.106}$$

where $X(y,t)$, which is the control variable, is the nondimensional body force in the x-direction, and $2/Re$ is the specified pressure gradient. We could consider any of the four combinations of domain- or boundary-based control with domain- or boundary-based performance measures. Let us first consider the scenario involving domain-based control with a domain-based performance measure. We then formulate the optimal control problem for boundary-based control with a domain-based performance measure.

Domain-Based Control with Domain-Based Performance Measure

In order to determine the optimal control strategy, we seek the body force distribution $X(y,t)$ that will most closely produce a desired target velocity distribution, say $\bar{u}(y,t)$, throughout the domain. The body force required to reproduce the target velocity distribution exactly could be determined simply by replacing the actual velocity $u(y,t)$ in the governing equation (10.106) with the specified target velocity $\bar{u}(y,t)$ and solving for the necessary body force $X(y,t)$. This brute force control approach would typically not be very helpful in practice owing to the fact that the total required energy input to exactly reproduce the target velocity may be excessively large. In addition, the target velocity may not be realizable exactly. For example, if the target velocity in the case of plane-Poiseuille flow being considered here were specified to be a uniform velocity distribution, then the target function does not satisfy the no-slip boundary conditions at the surfaces.

In practice, we desire the body force distribution that produces a velocity profile *close* to that of the target profile but that also minimizes the total energy input required. Therefore, we define optimal control to be the velocity distribution $u(y,t)$ and body force distribution $X(y,t)$ that minimizes the following cost functional over a time interval $0 \leq t \leq t_f$:

$$J[u,X] = \frac{1}{2}\int_0^{t_f}\int_0^2 \left[\gamma^2(u-\bar{u})^2 + X^2\right] dydt. \tag{10.107}$$

Gunzburger and Manservisi (1999, 2000) refer to this approach of driving the velocity toward a specified target distribution as *velocity tracking*. As before, the weight coefficient γ^2 sets the relative weights of the two terms in the cost functional. The first term, the performance measure, minimizes the differential kinetic energy per unit mass between the actual and target velocities. The second term, the penalty function, seeks to minimize one half of the square of the energy per unit mass introduced by the body-force input required to accomplish the control. Setting $\gamma^2 = 0$ corresponds to the no-control case, for which $X(y,t) = 0$, and increasing γ^2 increases the importance of the performance measure relative to the penalty function.

The domain-based control with domain-based performance measure given by minimization of the cost functional (10.107) subject to the constraint that the state equation (10.106) is satisfied is a special case of the more general situation considered in Section 10.5.1. Specifically, the state equation (10.106) is equation (10.92) with

$$a_0 = \frac{1}{Re}, \quad a_1 = 0, \quad a_2 = 0, \quad a_3 = \frac{2}{Re}, \quad \phi(y,t) = X(y,t),$$

and the cost functional (10.107) is equation (10.94) with u^2 replaced by $(u-\bar{u})^2$. Also note that we have replaced the coordinate x with y. Therefore, the control (adjoint) equation (10.100) becomes

$$\frac{\partial X}{\partial t} = -\frac{1}{Re}\frac{\partial^2 X}{\partial y^2} + \gamma^2(u - \bar{u}). \tag{10.108}$$

The initial condition and no-slip boundary conditions for the state equation (10.106) are

$$u(y,0) = u_0(y), \quad u(0,t) = 0, \quad u(2,t) = 0. \tag{10.109}$$

The initial condition (applied at the terminal time) and boundary conditions for the control variable (body force) are

$$X(y,t_f) = 0, \quad X(0,t) = 0, \quad X(2,t) = 0. \tag{10.110}$$

The control equation (10.108) for $X(y,t)$ is coupled with the governing equation (10.106) for $u(y,t)$. Note that the control equation is an unsteady diffusion equation of the same form as the governing state equation except that the diffusion term is negative. Again, this signals that the control equation is to be integrated backward in time.

Boundary-Based Control with Domain-Based Performance Measure
In order to further illustrate the versatility of the variational approach to optimal control, let us now turn our attention to boundary-based control with a domain-based performance measure applied to plane-Poiseuille flow. This would correspond to the fact that for practical applications, it is typically only possible to measure flow conditions using sensors placed at the surfaces of the domain rather than throughout the flow field.

The governing (state) equation (10.106) is now (with no body force)

$$\frac{\partial u}{\partial t} = \frac{1}{Re}\frac{\partial^2 u}{\partial y^2} + \frac{2}{Re}. \tag{10.111}$$

In this case we imagine that we can control the tangential velocity of the fluid at the surfaces; therefore, the wall velocities $u(0,t)$ and $u(2,t)$ are the unspecified control variables and will be determined as part of the solution to the optimal control problem. We introduce the cost functional

$$J[u] = \frac{\gamma^2}{2}\int_0^{t_f}\int_0^2 (u - \bar{u})^2 dy dt + \frac{1}{2}\int_0^{t_f}\left[u(0,t)^2 + u(2,t)^2\right] dt. \tag{10.112}$$

The performance measure seeks to minimize the differential kinetic energy per unit mass as in the body-force control case. The penalty function minimizes the kinetic energy per unit mass of the tangentially moving walls. Observe that while the performance measure is integrated over the entire domain $0 \le y \le 2$, the penalty function is only applied at the boundaries.

Incorporating the differential constraint (10.111), taking the variation, and setting it equal to zero, we have

$$\delta\left\{\frac{\gamma^2}{2}\int_0^{t_f}\int_0^2 (u-\bar{u})^2 dydt + \int_0^{t_f}\int_0^2 \Lambda(y,t)\left[u_t - \frac{u_{yy}}{Re} - \frac{2}{Re}\right]dydt \right.$$
$$\left. + \frac{1}{2}\int_0^{t_f}\left[u(0,t)^2 + u(2,t)^2\right]dt\right\} = 0. \tag{10.113}$$

Evaluating the variation yields

$$\int_0^{t_f}\int_0^2\left[\left\{\gamma^2(u-\bar{u})\delta u + \Lambda\left[\delta u_t - \frac{1}{Re}\delta u_{yy}\right]\right\}dy\right]dt$$
$$+\int_0^{t_f}\left[u(0,t)\delta u(0,t) + u(2,t)\delta u(2,t)\right]dt = 0. \tag{10.114}$$

Performing the required integration by parts leads to

$$\int_0^{t_f}\int_0^2\left[\gamma^2(u-\bar{u}) - \Lambda_t - \frac{\Lambda_{yy}}{Re}\right]\delta u dydt + \int_0^2\left[\Lambda(y,t_f)\delta u(y,t_f) - \Lambda(y,0)\delta u(y,0)\right]dy$$
$$+\int_0^{t_f}\left\{-\frac{1}{Re}\left[\Lambda(2,t)\delta u_y(2,t) - \Lambda(0,t)\delta u_y(0,t) - \Lambda_y(2,t)\delta u(2,t) + \Lambda_y(0,t)\delta u(0,t)\right]\right.$$
$$\left. + u(0,t)\delta u(0,t) + u(2,t)\delta u(2,t)\right\}dt = 0.$$

Then from the δu term, the Euler (adjoint) equation is

$$\Lambda_t = -\frac{\Lambda_{yy}}{Re} + \gamma^2(u-\bar{u}), \tag{10.115}$$

which is the same as in the body-force control case [see equation (10.109)]. Owing to the specified initial condition, the $\delta u(y,0)$ term vanishes. Now consider the remaining terms multiplied by each variation

$$\begin{aligned}
&\delta u(y,t_f) : \Lambda(y,t_f) = 0,\\
&\delta u(0,t)\ : -\tfrac{1}{Re}\Lambda_y(0,t) + u(0,t) = 0,\\
&\delta u(2,t)\ : \tfrac{1}{Re}\Lambda_y(2,t) + u(2,t) = 0,\\
&\delta u_y(0,t) : \Lambda(0,t) = 0,\\
&\delta u_y(2,t) : \Lambda(2,t) = 0.
\end{aligned} \tag{10.116}$$

The second and third equations provide natural boundary conditions for the state equation (10.111) in terms of the solution to the adjoint equation

$$u(0,t) = \frac{1}{Re}\frac{\partial\Lambda}{\partial y}\bigg|_{y=0}, \quad u(2,t) = -\frac{1}{Re}\frac{\partial\Lambda}{\partial y}\bigg|_{y=2}. \tag{10.117}$$

The remaining equations in (10.116) give the initial and boundary conditions for a backward-time integration of the adjoint equation (10.115)

$$\Lambda(y,t_f) = 0, \quad \Lambda(0,t) = 0, \quad \Lambda(2,t) = 0. \tag{10.118}$$

Alternatively, in order to have a formulation that is consistent with the body-force control scenario, we may define the control (adjoint) variable as

$$\Lambda = X, \tag{10.119}$$

but where X no longer represents the body force. Therefore, the state equation, which is integrated forward in time, is

$$\begin{gathered}\frac{\partial u}{\partial t} = \frac{1}{Re}\frac{\partial^2 u}{\partial y^2} + \frac{2}{Re}, \\ u(y,0) = u_0(y), \quad u(0,t) = \frac{1}{Re}\frac{\partial X}{\partial y}\bigg|_{y=0}, \quad u(2,t) = -\frac{1}{Re}\frac{\partial X}{\partial y}\bigg|_{y=2},\end{gathered} \tag{10.120}$$

and the control (adjoint) equation, which is integrated backward in time, becomes

$$\begin{gathered}\frac{\partial X}{\partial t} = -\frac{1}{Re}\frac{\partial^2 X}{\partial y^2} + \gamma^2(u - \bar{u}), \\ X(y,t_f) = 0, \quad X(0,t) = 0, \quad X(2,t) = 0.\end{gathered} \tag{10.121}$$

Note that the formulation (10.120) and (10.121) is the same as that for the body-force control case, except that there is no body force term in the state equation and the velocity boundary conditions have been modified. This is due to the fact that the control now influences the flow through the boundary conditions, rather than a body force. However, the control (adjoint) equation is still a field equation owing to the domain-based performance measure. Finally, it is noted that if $\gamma^2 = 0$, then the solution to the control equation (10.121) is $X(y,t) = 0$ owing to the homogeneous initial and boundary conditions, and the solution to the state equation (10.120) is simply standard plane-Poiseuille flow, corresponding to the no-control case.

10.6 Control of Real Systems

Because our focus is on variational methods in this text, we have primarily viewed control from a theoretical point of view that is subject to a number of inherent assumptions. In particular, the theoretical approach assumes that 1) our system model (state equations) provide a perfect, error free representation of the actual system at every point in space and/or time, and 2) no external disturbances act on the system. To varying degrees, neither of these two assumptions hold for control of real systems, and there are a number of practical issues that arise when implementing these methods. Although these are beyond the scope of this text, it is instructive to briefly mention several of them.

Keep in mind that despite the practical issues with implementing optimal control methodologies for real systems, optimal control solutions provide valuable information in two forms: 1) the optimal control solution provides an upper bound on the potential for control by quantifying the control objective to give a target to aim for when designing practical controllers, and 2) it provides insight into the "best" means of controlling a system, thereby providing guidance in the design of practical control

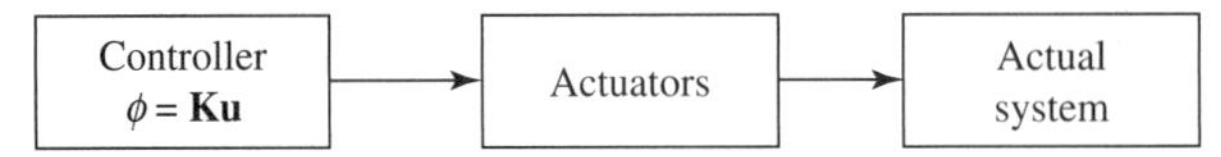

Figure 10.9. Schematic of open-loop control.

techniques for the system. For example, what sensors are required (linked to the performance measure); is domain- or boundary-based control more effective; what actuators most closely provide the optimal control strategy; what control authority (energy input) is required for the intended application; where should the actuators be placed; and when should they be turned on? Therefore, it is often instructive to first carry out an optimal control analysis before undertaking development of a practical control methodology for real systems.

Additional details regarding the variational approach to control theory and optimization are provided in Luenberger (1969), Kirk (1970), Smith (1974), Bryson and Ho (1975), Stengel (1986), and MacCluer (2005). For scenarios involving partial differential equations (continuous systems), see Lions (1971), Li and Yong (1995), and Tröltzsch (2010). In addition, a comprehensive and clear review of the issues related to application of optimal control theory in fluid mechanics is provided by Collis, Joslin, Seifert, and Theofilis (2004).

10.6.1 Open-Loop and Closed-Loop Control

The optimal control methodology described thus far is classified as *open-loop control* and applies to linear and nonlinear, discrete and continuous systems. In open-loop control, the control inputs are applied without regard to the measured (or *estimated*) values of the state variables. In other words, the control can be completely determined prior to operation of the actual system. Open-loop control is appropriate if there are no errors in the model of the state of the system, in which case the state equations are a perfect model of the system, and no external disturbances act on the system that are not accounted for in the state equations. For open-loop control to be effective, the system must be *controllable* (see Section 10.4.3).

A practical implementation of open-loop control is illustrated schematically in Figure 10.9. Solution of the coupled state and control (adjoint) equations would be carried out offline in order to produce the control law, such as equation (10.38), that relates the control output $\boldsymbol{\phi}(t)$ to the state of the system $\mathbf{u}(t)$ in the form

$$\boldsymbol{\phi} = \mathbf{Ku}.$$

In the case of open-loop control, the system state is taken as that predicted by the dynamic model, that is, the state equations, not the actual system. This prescription of the control input is provided to the *controller*, which tells the actuators what to do to the actual system, which is sometimes called the *plant*.

In order to handle the inevitable disturbances and errors, most controllers implemented in practice are *closed-loop*, or *feedback*, *control*. A control methodology that includes feedback of state information into the controller allows for adjustments to the control plan to account for external disturbances acting on the system, errors in control input, and/or differences between the behavior of the actual system as compared to the system model. Because the feedback control will account for the

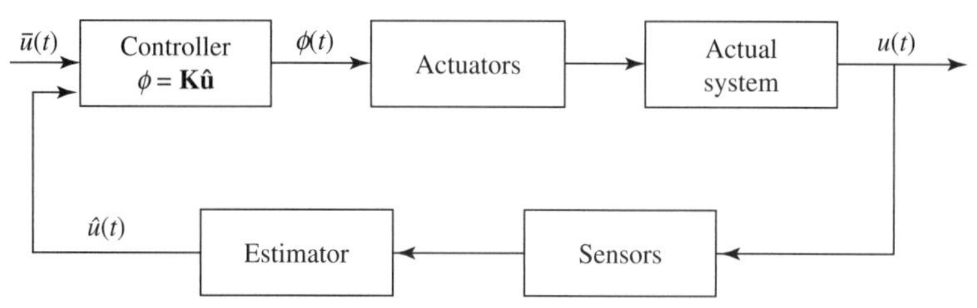

Figure 10.10. Schematic of closed-loop control.

differences between the modeled and actual state of the system, less accurate models can be used as compared to those required for open-loop control. For closed-loop control to be effective, the system must be *controllable* and *observable* (see Section 10.4.3).

Because the control input is adjusted "on the fly" in response to the behavior of the system, it must be determined in real-time during operation of the system. To facilitate this, a *feedback control law* is developed that provides a direct relationship between the control input and the estimated state variables of the form

$$\boldsymbol{\phi} = \mathbf{K}\hat{\mathbf{u}}.$$

It allows for "feedback" of the current state estimate of the system into the controller in order to determine the appropriate control actuation.

A practical implementation of closed-loop control is illustrated schematically in Figure 10.10. Sensors within the actual system, or plant, are used by the estimator to determine an estimate of the current state of the actual system $\hat{\mathbf{u}}$. The estimated state, along with the target state $\bar{\mathbf{u}}$, is used by the controller to determine the control input based on the feedback control law, which is then used to control the actuators to drive the system toward its desired state. Estimation is discussed further in Section 10.6.3.

To see the need for closed-loop control, imagine a chess player. During each move, she plans ahead several moves on her and her opponents part to determine the "optimal" move to make during her current turn. If her opponent makes the move that she anticipated according to her "model," then she can proceed with the next move in her pre-determined sequence based on the previous optimization (open-loop control). However, if her opponent executes a move different from that which she anticipated (because her model of the opponent's strategy is inaccurate or because her opponent simply makes the wrong move), then she must reconsider the optimization problem in response to this new information in order to determine the optimal move (closed-loop control).

10.6.2 Model Predictive Control

A popular means of combining the best features of open- and closed-loop control is *model predictive control*. Employing this approach, an open-loop control methodology can be adapted to provide closed-loop feedback control by embedding the open-loop control within a *receding-time horizon algorithm* according to the following procedure:

1. Based on measurements (estimation) of the current state at time t_i, an open-loop control problem (state and control equations) is solved over the prediction horizon $t_i \leq t \leq t_f$.
2. The resulting control is applied over the interval $t_i \leq t \leq t_i + \Delta t$, where $t_i + \Delta t \leq t_f$, until the next measurement of the system state is available.
3. Because the estimated state of the system may not match that predicted by the model owing to external disturbances and/or modeling errors, steps (1) and (2) are repeated with t_i and t_f incremented by Δt, that is, with the time horizon receding into the future by the increment Δt.

This model predictive control approach allows for the control to adjust, through feedback, to the mismatch between the actual and modeled behavior of the system. This mismatch may be caused by measurement errors, control errors, and/or external disturbances. If no such mismatch occurs, then the receding-time horizon approach is not necessary, and an open-loop approach is sufficient. Receding-time horizon control can be applied to both linear and nonlinear systems. Alternatively, nonlinear systems can be treated by allowing for solution of the nonlinear state equation(s) with a linearized control formulation that consists of the linearized state and control equations; in other words, we linearize about each nonlinear state as time progresses. Although such algorithms provide a great deal of flexibility, they are computationally intensive, particularly for nonlinear systems.

10.6.3 State Estimation and Data Assimilation

State Estimation

In open-loop control, the state of the system is taken to be completely determined from a solution of the state equations, which is assumed to be error free. As such, no measurements of the actual system are required. In closed-loop control, however, the state of the system must be estimated using some combination of the solution of the state equations and measurements of the actual system. One way to deal with this is to augment the predictions of the system's state with actual measurements, or *observations*, using sensors on the system itself. However, these observations are not perfectly accurate, and typically they do not completely fix the state of the entire system as only a finite number of sensors are used. That is, state measurements alone cannot be obtained with sufficient spatial and/or temporal resolution for use in the control algorithm. Although such observations are done to counteract the effect of errors and disturbances, the measurements themselves may introduce additional errors as well. *Optimal state estimation* provides a mathematical means to reconstruct an estimate of the system state from incomplete and imperfect observations. Although possible in an open-loop setting, an estimate of the system state is required for application of closed-loop feedback control laws.

The process of determining the most likely values of the state variables in the past is called "smoothing," in the present is called "filtering," and in the future is called "prediction." The *Kalman-Bucy filter* provides the optimal state estimate for continuous linear systems subject to random Gaussian inputs and errors; the *Kalman filter* is the corresponding optimal filter for discrete data. We seek the optimal state representation of a system over the time interval $0 \leq t \leq t_f$ based on measurements taken over the interval $0 \le t \le t_{est} < t_f$. The estimate is optimal from the point of view that the estimation error, or the difference between our state estimate and the actual state, is minimized in a least-square sense over the relevant time interval. The state equation, which now includes disturbances, is coupled with an *observation model* that relates the observations (measurements) to the state estimates using the measurement errors. If some measurements are known to be more accurate than others, then we could use a weighted least-squares approach. When applied to the linear-quadratic (LQ) problem considered in Section 10.4.2, the optimal linear estimator is the *dual* of the optimal control problem given by equations (10.36) and (10.39). In the estimator, the equation for the state is a Riccati equation very similar to (10.36) except that it is integrated forward in time rather than backward in time as in the control case. As in optimal control, we minimize the cost functional, comprised of the least square error between the model and observation, subject to the constraint that the governing (state) equations are satisfied. See Bryson and Ho (1975) and Stengel (1986) for much more on optimal estimation.

Data Assimilation

Many fields depend upon access to accurate data for making decisions, control, or forecasting. Arguably the largest and most complex state estimation problem is done to facilitate weather prediction, which has advanced considerably from the days of "Red sky at night, sailor's delight. Red sky in morning, sailor's warning." As you read this, some of the largest supercomputers in the world are churning out detailed (and increasingly accurate) weather predictions for the days ahead. The computational models are based on the Navier-Stokes equations derived in Chapter 9 for viscous fluid flow and augmented by additional equations that account for temperature, humidity, and other properties. Rather than being applied to the flow through a channel, however, they are applied to the entire atmosphere surrounding the Earth. The scale of such computations is difficult to imagine, but their accuracy affects us all each and every day.

Accurate numerical weather prediction in the future is predicated on having accurate and complete state information on a fine scale for the entire globe at the present time to use as initial conditions for such predictive models. Ideally, this data would be available throughout the three-dimensional atmosphere and oceans at small increments in time. In reality, however, atmospheric data are primarily based on surface measurements and/or satellite data and are subject to inaccuracies and a severe lack of measured data for certain parts of the atmosphere. Because it is impossible to measure such data at the spatial and temporal scales necessary for prediction and forecasting, the atmospheric and oceanographic communities have developed techniques for *data assimilation*. Data assimilation is based on principles from optimal state estimation with additional features tailored to atmospheric sciences and weather prediction. Essentially, data assimilation is a mathematical approach for assimilating very limited measured (observational) data with numerical model predictions to obtain estimates of the state of a system. Such data can then be used as the basis for future predictions of the states of the system, which is forecasting.

In the past, the initial conditions for numerical weather prediction were typically obtained using a state estimation method known as *three-dimensional variational*

(*3DVar*) data assimilation. Using this approach, a three-dimensional spatial state estimation problem is solved to reconstruct the current state of the atmosphere from observations at the current time. More recently, this is being replaced with *4DVar*, in which observations from previous times are also incorporated into the estimation of the current state in order to provide a more accurate initial condition for numerical prediction models.

For very large dimensional systems, matrix-based Kalman filter methods become computational intractable. Because variational data assimilation methods are vector based, they scale much better to large-dimensional systems. 3DVar is a vector-based equivalent to the Kalman filter. 4DVar is the vector equivalent to the Kalman smoother used to accurately estimate the state over a previous time range based on observational measurements and solution of the state equation(s). The state estimate at the current time (based on an optimization window in the past) resulting from the 4DVar data assimilation is then used as the initial condition for a weather prediction computation carried out using the state equations into the future. For more on data assimilation, see Daley (1991) and Kalnay (2003).

10.6.4 Robust Control

Although optimal estimation produces the "best" representation of the system state, uncertainty persists; therefore, it is necessary to consider the robustness of the control methodology. A *robust controller* maintains stability of the dynamic system and the ability to maintain sufficient control authority to accomplish the control objective despite the presence of uncertainty in the state estimate. Robustness is facilitated through the use of a *compensator*, which combines the *estimator* and *controller*, that compensates for the uncertainties within the system and its control.

10.7 Postscript

An explosion of interest and application of optimization and control theory has occurred in the last half century and parallels closely the increasing availability of computational resources. As we have seen, even for the relatively simple examples considered, very rarely can optimization and control problems be solved in closed form; therefore, their solution had to wait for the arrival of computers of sufficient capability and the development of the necessary algorithms to take advantage of such resources. Today, the size and scope of such problems that can be solved practically is still limited by the computational resources available to the researcher or practitioner. In many fields, such as fluid mechanics, numerical solution of the state equations using computational fluid dynamics (CFD) is a challenging endeavor in and of itself; consequently, wrapping such solution techniques within an optimization or control framework, particularly for unsteady, time-dependent flows, is an extremely challenging problem. Therefore, there will be a delicate interplay between progress in optimization and control theory and techniques, expansion of computer resources, and algorithm development for some time to come. It is both exciting and challenging to imagine the potential applications of this synergy of pursuits in the coming decades.

The variational optimization and control framework outlined here has additional applications in image processing and grid generation that will be explored

Table 10.3. *Control variable for examples of optimization and control.*

Application	Control Variable
Optimal control	System input/forcing
Shape optimization	Domain shape
Image processing	Restored image
Reduced-order models	Optimal basis functions
Grid generation	Grid/node coordinates
Transient stability growth	Initial perturbation

in Chapters 11 and 12, respectively. Along with shape and financial optimization introduced earlier in this chapter, this broad set of applications of the theory highlights the remarkable generality and flexibility that the variational approach furnishes in such a diverse set of applications. Table 10.3 indicates the control variable for each of the optimization and control settings considered. Because most any system, process, or entity can be optimized in some way, this chapter provides the background necessary for the researcher or practitioner to develop a formal framework for their particular area of interest based on variational methods.

One of the primary advantages of the variational framework for optimal control is the flexibility that it provides for incorporating, for example, nonlinear state and control equations, nonhomogeneous boundary conditions, and numerous other special situations that go beyond the traditional linear theory. In fact, this is an important theme for several of the applications that we have or will consider. Optimal control, transient-growth stability analysis, state estimation, data assimilation, and proper-orthogonal decomposition all may be formulated by employing linear systems theory when applied to linear systems, thereby bipassing the variational approach articulated here. For example, transient-growth analysis and proper-orthogonal decomposition can be reduced to that of performing a singular-value decomposition (SVD) under such conditions. More than just providing an alternative framework for these applications, however, variational methods allow us to generalize such techniques to nonlinear settings, which is becoming increasingly essential as research advances in a number of important fields.

EXERCISES

10.1 Given the state equation (10.6) with initial conditions (10.5) and the cost functional (10.20), show that the control equation is (10.21) with the initial conditions (10.22).

10.2 The state $u(t)$ of a system is governed by

$$\frac{d^2u}{dt^2} = \phi,$$

where $\phi(t)$ is the control variable. The initial conditions are

$$u(0) = u_0, \quad \dot{u}(0) = 0.$$

The system is to be controlled such that the cost functional

$$J[\phi] = \frac{1}{2}\int_0^{t_f} \phi^2 dt$$

is minimized. Determine the Riccati equations and the control law relating the control $\phi(t)$ to the solutions $u_i(t)$.

10.3 The states $u_1(t)$ and $u_2(t)$ of a system are governed by

$$\frac{du_1}{dt} = u_2, \quad \frac{du_2}{dt} = 2u_1 - u_2 + \phi,$$

where $\phi(t)$ is the control variable. The initial conditions are

$$u_1(0) = a, \quad u_2(0) = b.$$

The system is to be controlled such that the cost functional

$$J[u_1,u_2,\phi] = \frac{1}{2}\int_0^{t_f} \left[2u_1^2 + u_2^2 + \frac{1}{2}\phi^2\right] dt$$

is minimized. Determine the Riccati equations and the control law relating the control $\phi(t)$ to the solutions $u_i(t)$.

10.4 The state $u(t)$ of a system is governed by

$$\frac{d^2u}{dt^2} + a\frac{du}{dt} + bu = \phi,$$

where a and b are constants, and $\phi(t)$ is the control variable. The initial conditions are

$$u(0) = 1, \quad \dot{u}(0) = 0.$$

The system is to be controlled such that the cost functional

$$J[u,\phi] = \frac{1}{2}\int_0^{t_f} \left(\gamma^2 u^2 + \phi^2\right) dt$$

is minimized. Determine the control equation for $\phi(t)$ and any necessary initial conditions.

10.5 The state $u(t)$ of a system is governed by

$$\frac{d^2u}{dt^2} + \frac{du}{dt} = \phi,$$

where $\phi(t)$ is the control variable. The initial conditions are

$$u(0) = 1, \quad \dot{u}(0) = 0.$$

The system is to be controlled such that the cost functional

$$J[u,\phi] = \frac{1}{2}\int_0^{t_f} \left[\gamma^2\left(u^2 + \dot{u}^2\right) + \phi^2\right] dt$$

is minimized. Determine the control equation for $\phi(t)$ and any necessary initial conditions.

10.6 The state $u(t)$ of a system is governed by

$$\frac{du}{dt} = \phi,$$

where $\phi(t)$ is the control variable. The initial condition is

$$u(0) = u_0.$$

The system is to be controlled such that the cost functional

$$J[u,\phi] = \frac{1}{2}\int_0^{t_f} \left[\gamma^2 u^2 + \phi^2\right] dt$$

is minimized. Determine the solutions to the control and state equations for $\phi(t)$ and $u(t)$, respectively.

10.7 The state $u(t)$ of a system is governed by

$$\frac{du}{dt} + au = b + \phi,$$

where a and b are constants, and $\phi(t)$ is the control variable. The initial condition is

$$u(0) = 0.$$

The system is to be controlled such that the cost functional

$$J[u,\phi] = \frac{1}{2}\int_0^{t_f} \left[\gamma^2 u^2 + (\dot{\phi})^2\right] dt$$

is minimized. Note that this penalty function seeks to minimize the rate of change of the control variable, rather than its magnitude. Determine the solutions to the control and state equations for $\phi(t)$ and $u(t)$, respectively.

10.8 The states $u_1(t)$ and $u_2(t)$ of a system are governed by

$$\frac{du_1}{dt} = u_2, \quad \frac{du_2}{dt} = -u_1 + \left[1 - u_1^2\right] u_2 + \phi,$$

where $\phi(t)$ is the control variable. The initial conditions are

$$u_1(0) = a, \quad u_2(0) = b.$$

The system is to be controlled such that the cost functional

$$J[u_1,u_2,\phi] = \frac{1}{2}\int_0^{t_f} \left[2u_1^2 + u_2^2 + \phi^2\right] dt$$

is minimized. Determine the control equation for $\phi(t)$ in terms of $u_1(t)$.

10.9 Reconsider Example 10.3. Employ Pontryagin's approach to obtain the coupled state, optimality, and adjoint equations along with all necessary initial conditions.

10.10 Reconsider Exercise 10.2. Employ Pontryagin's approach to obtain the coupled state, optimality, and adjoint equations along with all necessary initial conditions.

10.11 Evaluate the time-optimal control solution for the system $\ddot{u} = \phi$, with inequality constraints on the control $-1 \leq \phi \leq 1$. The initial condition is $u(0) = a, \dot{u}(0) = b$, and the terminal state is $u(t_f) = 0, \dot{u}(t_f) = 0$.

10.12 Consider Zermelo's problem introduced in Section 1.3.3. Reformulate the problem as a time-optimal control problem with the state equations given by equation (1.3) (with the heading angle a function of t).

a) Obtain the general form of the control equation for the heading angle $\theta(t)$.

b) Evaluate the special case when the current velocity is constant (uniform) of the form $v(x) = \alpha$.

c) Evaluate the special case when the current velocity distribution is linear of the form $v(x) = \alpha x$, where α is a constant.

10.13 Consider control of the temperature distribution $u(x,t)$ in a metal rod of length ℓ with diffusivity α governed by the one-dimensional unsteady diffusion equation

$$\frac{\partial u}{\partial t} = \alpha \frac{\partial^2 u}{\partial x^2} + \phi,$$

where the heat generation $\phi(x,t)$ is the control variable. The initial and boundary conditions are

$$u(x,0) = 0, \quad u(0,t) = 0, \quad u(\ell,t) = 0.$$

The cost functional is given by

$$J[u,\phi] = \frac{1}{2} \int_0^{t_f} \int_0^{\ell} \left[\gamma^2 (u - \bar{u})^2 + \phi^2 \right] dx dt,$$

such that the differential temperature between the actual temperature $u(x,t)$ and the specified target temperature distribution $\bar{u}(x,t)$ is minimized with the magnitude of the heat generation being penalized. Obtain the control equation along with any boundary or initial conditions to be solved for the control input $\phi(x,t)$.

10.14 Reconsider Exercise 10.13 with the cost functional having a terminal-time performance measure as follows

$$J[u,\phi] = \frac{1}{2}\gamma^2 \int_0^{\ell} \left[u(x,t_f) - \bar{u}(x,t_f) \right]^2 dx + \frac{1}{2} \int_0^{t_f} \int_0^{\ell} \phi^2 dx.$$

11 Image Processing and Data Analysis

> At first [Frodo] could see little. He seemed to be in a world of mist in which there were only shadows: the Ring was upon him. Then here and there the mist gave way and he saw many visions: small and clear as if they were under his eyes upon a table, and yet remote. There was no sound, only bright living images. The world seemed to have shrunk and fallen silent. He was sitting upon the Seat of Seeing, on Amon Hen, the Hill of the Eye of the Men of Númenor.
>
> (J. R. R. Tolkien, *The Lord of the Rings*)

It is hard to imagine a time when our sight was not bombarded with images of far off people and places, when one had to actually be in the presence of someone to see their face without it being filtered through the hands of an artist (or while sitting on a Seat of Seeing). Images allow us to render the very small, the very large, and the very distant; they provide a visual record of time and place; they move us and entertain us; they replace perception with reality. Images are acquired by Earth-orbiting satellites, interplanetary probes, subterranean robots, surveillance cameras in unmanned aircraft and on nearly every street corner, and our mobile phones. They are created using optical principles and magnetic fields, and images display what is "seen" in every corner of the electromagnetic spectrum.

Ever since the discovery in the latter part of the nineteenth century of the ability of X-rays to image the internal anatomy of the body, arguably nothing has done more to revolutionize medical diagnosis and treatment than medical imaging by noninvasively providing unprecedented views inside the human body. The first phase of this revolution has been marked by the development of the imaging technology itself, including X-ray, ultrasound, computed tomography (CT), magnetic resonance imaging (MRI), and positron emission tomography (PET). While these imaging technologies continue to improve, the second phase, which is an active area of current research, involves the processing of those images for diagnostic purposes and as the basis for further analysis. It is now possible to reconstruct detailed patient-specific, three-dimensional anatomical models of blood vessels, organs, and entire bodies based on a series of two-dimensional images. It is also possible to determine the distributed velocity of the blood flowing through the heart based on MRI data.

Other fields have experienced similar revolutions in image acquisition and processing, including astronomy and space exploration, security applications, autonomous navigation, computer graphics, and human and machine vision. For example, military, intelligence, and law enforcement agencies use advanced imaging technologies for target detection and identification, video surveillance, intelligence gathering, and biometric identification. It is now common at popular sporting events, such as the Super Bowl in the United States, for all spectators to have their faces imaged (unbeknownst to them) and compared against a database of individuals who have been identified as security threats. Not only is the image processing technology advancing rapidly, the volume and variety of images to process is growing even more rapidly. It is estimated that there are between two and four million surveillance video cameras in the United Kingdom alone (that's one for every fifteen to thirty people!). Astronomers and atmospheric scientists have learned a great deal about the universe and our home in it by processing the images created across the electromagnetic spectrum from orbiting satellites and space probes throughout the solar system. From forecasting the local weather over the next few days to seeking minute signatures in the cosmos of the origins of the universe, current imaging technology is providing vantage points that only could have been imagined a half-century ago.

Imaging involves *acquisition*, *processing*, and *interpretation*. Each of these endeavors is fascinating in their own right, but it is their integration that holds the most promise for developing still unimagined capabilities. Here, however, we provide an introduction to image processing, as that is where variational methods find their greatest utility at present. Note that image processing relies on the theory of differential geometry, which is beyond the scope of this text but is essential for a deeper understanding of the methods and their applications.

In addition to image data, there are many other types of data that require "optimal" representation. Such data could be the result of experimental measurements or computational simulations (or a combination of both), and it could be discrete or continuous. In Section 10.6.3, we discuss *data assimilation* as implemented in the atmospheric sciences and computational weather prediction. Recall that this entails using optimal-control theory to combine physical measurements with results from computational models to obtain an optimal representation of the state of the atmosphere at a given time. In the present chapter, we discuss two generally applicable and common tasks in data analysis: 1) faithful representation of discrete data with a smooth curve using *splines*, and 2) model reduction of large data sets using *proper-orthogonal decomposition* (POD).

11.1 Variational Image Processing

Variational methods are increasingly being used to construct formal methods for accomplishing tasks that were performed heuristically in the past. This is the case, for example, in image processing and grid generation, along with many other optimization tasks. Whereas traditionally such optimization activities were guided by intuition and executed with considerable trial and error, they are being formalized using mathematical frameworks that allow one to determine the "best" approach, rather than settling for "better" methods and algorithms.

Traditionally, image processing, and its close cousin signal processing, has been based on spectral and Fourier methods. While such methods provide a great deal of flexibility in representing and processing various signals and images, the last two decades have seen a steady rise in the use of variational methods and optimization theory to further advance image processing technologies. Common image processing tasks fall into the following four categories:

1. *Denoising* – remove noise from an image.
2. *Deblurring* – recover a sharp image from a blurred one.
3. *Inpainting* (*image interpolation*) – reconstruct an image from an incomplete one.
4. *Segmentation* – separate discrete objects from the background of an image.

For example, see Figure 11.1 for sample images. Given an observed image, such as the noisy or blurred images in Figures 11.1(b) and (c), the objective of image processing is to reconstruct the original image (a) as closely as possible. Additional topics include *optic flow*, used for motion tracking of objects between two images at successive times, *registration*, in which a source image is transformed to match that of a template image, and *surface* or *geometry reconstruction*.

In order to illustrate the four primary image processing operations, consider a typical scenario in modern biomedical engineering. It is becoming increasingly common in medical applications to use MRI, or other images, as the basis for modeling and analysis of blood vessels and organs. Imagine, for example, computational simulation of the blood flow through a patient's aorta. In order to construct the patient-specific three-dimensional geometry of the aorta for such modeling, it may be necessary to perform denoising, deblurring, inpainting, and segmentation: denoising to remove the random noise that appears in all MRI due to the presence of small-scale nonuniformities in the tissue being imaged, deblurring to remove the effects of the motion of the vessel walls owing to blood pulsations, inpainting to fill in regions where occlusions have been removed from the image, and segmentation to extract the geometry of the aorta to use as the basis for computational modeling of the blood flow.

More often than not today, images are acquired, transmitted, and stored digitally. Digital images are comprised of discrete *pixels*; however, it is helpful to view images as continuous distributions of grayscale or color data. Given an observed two-dimensional image $u_0(x,y)$, which may be noisy or blurry, we seek to reconstruct the (unknown) restored image $u(x,y)$ as faithfully as possible. If the images are grayscale, then $u_0(x,y)$ and $u(x,y)$ are scalar functions representing intensity, and if the images are color, then they are vector functions representing, for example, the RGB colors. The latter case is sometimes called *multichannel* image processing. The two images are related by

$$u_0(x,y) = Ku(x,y) + w(x,y), \tag{11.1}$$

where $w(x,y)$ is random additive noise, and K is a nondifferential linear operator that accounts for blur. When processing video, we have a series of observed images $u_0(x,y)$ at successive times.

The variational framework involves determining the unknown piecewise smooth image $u(x,y)$ that minimizes an objective functional[1] of the general form

$$J[u] = \iint_A \left[\gamma^2 D(u) + S(u)\right] dxdy, \tag{11.2}$$

where $D(u)$ represents optimization of the image data, and $S(u)$ represents smoothing operations (sometimes called the regularization term). Using the optimal control terminology in Chapter 10, we would refer to $D(u)$ as the performance measure and $S(u)$ as the penalty function. The weight coefficient $\gamma^2 > 0$ sets the relative weights of the performance measure and penalty function.

The variational approach to image processing was introduced by Mumford and Shah (1989) originally to perform segmentation. Other image processing tasks follow a similar framework, and a typical form of the objective functional (11.2) is

$$J[u] = \iint_A \left[\gamma^2 (Ku - u_0)^2 + \Phi\left(|\nabla u|\right)\right] dxdy. \tag{11.3}$$

The performance measure $D(u) = (Ku - u_0)^2$ seeks to minimize the difference between the restored image and the observed noisy image, and the penalty function $S(u) = \Phi\left(|\nabla u|\right)$ smooths the result, that is, it penalizes abrupt changes in the image. Note that the *penalty weight function* $\Phi(\cdot)$, often called the *penalizer*, is a function of the absolute value of the gradient of the restored image $u(x,y)$. A number of penalty weight functions have been proposed, but they all have the property that

(a) Original image.

Figure 11.1. Sample images.

[1] Objective functionals in image processing are often referred to as *energy functionals*.

(b) Noisy image.

(c) Blurred image.

Figure 11.1. (Cont.)

they are positive and increasing, such that $\Phi'(\cdot) > 0$, in order to ensure existence and uniqueness of the stationary function $u(x,y)$. Hence, given the observed image data $u_0(x,y)$ and selecting the penalty weight function $\Phi(\cdot)$, we seek the function $u(x,y)$ that minimizes the objective functional for a prescribed weight coefficient γ^2.

To obtain the Euler equation corresponding to the objective functional (11.3), recall from Section 2.5.1 that the general form for an integrand of the form $F = F(x,y,u,\nabla u)$, where $u = u(x,y)$, is

$$\frac{\partial F}{\partial u} - \nabla \cdot \left(\frac{\partial F}{\partial \nabla u} \right) = 0. \tag{11.4}$$

The natural boundary condition for this general case is

$$\mathbf{n}^T \frac{\partial F}{\partial \nabla u} = 0 \quad \text{on} \quad C, \tag{11.5}$$

where C is the boundary of the domain A, and $\mathbf{n}$ is the outward facing normal to curve C. From elementary calculus, the derivative of the absolute value of a function is

$$\frac{d}{dv}|v| = \frac{d}{dv}\sqrt{v^2} = \frac{1}{2}(v^2)^{-1/2}(2v) = \frac{v}{\sqrt{v^2}} = \frac{v}{|v|}.$$

Therefore, the general form of the Euler equation (11.4) for the functional (11.3) is given by the partial differential equation

$$\nabla \cdot \left(\Phi'(|\nabla u|) \frac{\nabla u}{|\nabla u|} \right) - 2\gamma^2 K^*(Ku - u_0) = 0, \tag{11.6}$$

where K^* is the adjoint operator of K (see Section 1.8). To see how the second term comes about, consider the $\partial F / \partial u$ term in (11.4), which is

$$\begin{aligned} \delta \iint_A \gamma^2 (Ku - u_0)^2 \, dxdy &= \iint_A 2\gamma^2 (Ku - u_0)\, \delta (Ku) \, dxdy \\ &= 2\gamma^2 \langle Ku - u_0, K\delta u \rangle \\ &= 2\gamma^2 \langle K^* (Ku - u_0), \delta u \rangle . \end{aligned}$$

In general, the Euler equation (11.6) will be a nonlinear partial differential equation. The natural boundary condition (11.5) requires that normal derivatives of u vanish at the domain boundary according to

$$\frac{\partial u}{\partial \mathbf{n}} = 0 \quad \text{on} \quad C.$$

A rather unique aspect of image processing within the context of variational methods is that we must deal with edges of objects within an image at which $u(x,y)$ can change abruptly. Therefore, it is necessary to consider the nature of such discontinuities within the context of the variational approach. This is done using the *total variation*. For a piecewise smooth image $u(x,y)$, the total variation is the averaged magnitude of the gradient of u over the domain A and is defined by

$$TV[u] = \iint_A |\nabla u| dxdy = \iint_A \sqrt{(\nabla u)^2}\, dxdy = \iint_A \sqrt{u_x^2 + u_y^2}\, dxdy. \tag{11.7}$$

TV is said to be a *bounded variation* if $TV[u] < \infty$. Variational image processing models for denoising, deblurring, and inpainting are typically designed such that they are bounded variations so as to allow for sudden changes within images, such as at object edges, while remaining mathematically well-behaved.

See Chan and Shen (2005) for an integrated treatment of the various methods, including variational, for treating image processing. See also Aubert and Kornprobst (2010) and Vese (2012) for image processing techniques based on partial differential equations and variational methods.

11.1.1 Denoising

Despite the incredible advances in image acquisition technology, dealing with noise in such images is a fact of life. In astronomical applications using ground-based detectors, for example, noise is introduced by property variations throughout the atmosphere that lead to small-scale variations in index of refraction. In medical imaging, inhomogeneity of tissue material leads to noise. Such noise is generally assumed to be random and additive and must be mitigated in order to create clear images for further processing and/or interpretation. This process is referred to as *denoising*.

Traditionally, denoising of signals has been carried out using linear filters, such as the Wiener filter. The Rudin-Osher-Fatemi variational approach to denoising consists of minimizing the objective functional

$$J[u] = \iint_A \left[\gamma^2 (u-u_0)^2 + |\nabla u|\right] dxdy, \tag{11.8}$$

which is (11.3) with $K = 1$ (no blurring) and $\Phi(|\nabla u|) = |\nabla u|$. This is sometimes called *total variation*, or *TV*, *denoising* as the penalty function involves the total variation (11.7). The corresponding nonlinear partial differential Euler equation (11.6) is given by

$$\nabla \cdot \left(\frac{\nabla u}{|\nabla u|}\right) - 2\gamma^2 (u-u_0) = 0, \tag{11.9}$$

where again the boundary conditions require that the normal derivatives vanish at all domain boundaries. Observe that the first term is the expression for the curvature of lines of constant $u(x,y)$ (see Section 9.4.1). In the two-dimensional case considered here, the Euler equation (11.9) becomes

$$\frac{\partial}{\partial x}\left(\frac{u_x}{\sqrt{u_x^2+u_y^2}}\right) + \frac{\partial}{\partial y}\left(\frac{u_y}{\sqrt{u_x^2+u_y^2}}\right) - 2\gamma^2(u-u_0) = 0.$$

The presence of $|\nabla u|$ in the denominator of the Euler equation (11.9) causes computational difficulties when the magnitude of the gradient of the image becomes small. One common remedy to prevent such problems is to modify the gradient definition to be

$$|\nabla u|_\alpha = \sqrt{\alpha^2 + (\nabla u)^2},$$

rather than $|\nabla u| = \sqrt{(\nabla u)^2}$. The parameter α is set by the user in order to regularize the denoising algorithm. One may recognize that if $\alpha = 1$, then this smoothing term becomes the integrand in the minimal surface problem of Section 2.5.2. Recall that the minimal surface equation corresponds to the mean curvature of the surface being zero. Thus, it results in the "smoothest" surface in an averaged sense.

Although the primary purpose of denoising algorithms is to eliminate a uniform level of noise throughout an entire image, which is accomplished through smoothing operators, it is also desired to retain the sharp boundaries of objects in an image. In order to reduce smoothing of sharp edges, that is, reduce the penalty for large gradients, the *Perona-Malik model* suggests a penalty weight function that allows for adaptation of the smoothing operator to the image itself. Such smoothers are sometimes regarded as nonlinear diffusion. Because we want Φ to be inversely proportional to $|\nabla u|$, one possible Perona-Malik model defines the penalty weight function as

$$\Phi\left(|\nabla u|\right) = \frac{1}{1+\frac{1}{\sigma}|\nabla u|},$$

where $\sigma > 0$. Other similar models have also been proposed that successfully denoise an image while retaining the sharp edges of objects.

11.1.2 Deblurring

Blurring of an image can occur because of the imaging device being out of focus, relative motion of the imaging device and the scene being imaged, or variations in optical properties of the intervening medium. In general, blurring can be very complicated; for example, this could be due to the fact that objects may be moving at different speeds during the imaging process, in which case each object effectively has its own K transformation. Despite such complications, blurring is typically modeled with a single K operator that is assumed to be linear and shift-invariant. Shift-invariance implies that if $u_0(\mathbf{x}) = Ku(\mathbf{x})$ as in equation (11.1), then $u_0(\mathbf{x}-\mathbf{a}) = Ku(\mathbf{x}-\mathbf{a})$. In other words, the blurring operator still applies when shifted by a constant amount $\mathbf{a}$. Under such assumptions, the linear operator K is a convolution of the image; therefore, deblurring corresponds to a deconvolution.

The objective functional that accomplishes deblurring is given by

$$J[u] = \iint_A \left[\gamma^2 (k*u - u_0)^2 + |\nabla u|^2\right] dxdy, \tag{11.10}$$

where $k(x,y)$ is the convolution kernel, and $k*u$ is the convolution of the image. This functional is equation (11.3) with the convolution operator $Ku = k*u$ and quadratic penalty weight function $\Phi\left(|\nabla u|\right) = |\nabla u|^2$. From equation (11.6), the Euler equation becomes

$$\nabla^2 u - \gamma^2 K^*(Ku - u_0) = 0, \tag{11.11}$$

where $\nabla^2 = \nabla \cdot \nabla$ is the usual Laplacian operator, and K^* is the adjoint operator of K. The Neumann boundary conditions require homogeneous normal derivatives at the boundaries. In the two-dimensional case considered here, the Euler equation becomes

$$\frac{\partial^2 u}{\partial x^2} + \frac{\partial^2 u}{\partial y^2} - \gamma^2 K^*(Ku - u_0) = 0.$$

11.1.3 Inpainting

Inpainting (interpolation) is used to fill in missing image information on a blank subdomain based on the image information available in the rest of the image. Such blank subdomains may be a result of transmission errors or removal of unwanted objects in an image, for example. Inpainting is the artist's term for what mathematicians and engineers call image interpolation. In its simplest applications, inpainting consists of filling in small blank subdomains by interpolating surrounding local information from the intact image.

A logical extension of the denoising functional (11.8) to the problem of inpainting is to apply the smoothing penalty term to the entire domain A, while only applying the data objective measure to the regions of the image that do not require inpainting. If we denote the subdomains of A that require inpainting as A_i, the objective functional becomes

$$J[u] = \iint_{A-A_i} \gamma^2(u-u_0)^2 dxdy + \iint_A |\nabla u| dxdy. \tag{11.12}$$

Because we once again use the total variation as the penalty function, this is often referred to as *total variation*, or *TV*, *inpainting*. Total variation inpainting is designed to preserve sharp edges while interpolating images across the missing regions.

Total variation inpainting often does not lead to the correct connections between edges of objects as they are interpolated into the regions requiring inpainting. A method based on *Euler's elastica problem*[2] has received considerable attention because it also penalizes curvature of the edges of objects, thereby allowing edge curves to extend properly into inpainting regions. It combines length and curvature measures in the form

$$J_E[u] = \int_\Gamma (a + b\kappa^2) ds,$$

where $a > 0$ and $b > 0$, $\Gamma(s)$ denotes the edges of objects, and ds is the differential arc length along these edges. By squaring the curvature, local positive and negative oscillations that happen to cancel each other are properly penalized. Choosing the coefficients a and b weight the length and curvature terms as desired. Curvature of a one-dimensional function $u(x)$ is given by

$$\kappa(x) = \frac{u''}{[1+(u')^2]^{3/2}}, \tag{11.13}$$

and in two dimensions, the curvature is most easily expressed in the form

$$\kappa(s) = \frac{d\theta}{ds},$$

where $\theta(s)$ is the continuously changing angle from the x-axis to the local tangent to the curve.

The *Mumford-Shah-Euler method* modifies the functional (11.12) to include the Euler elastica term as follows:

$$J[u] = \iint_{A-A_i} \gamma^2(u-u_0)^2 dxdy + \iint_{A-\Gamma} |\nabla u|^2 dxdy + \int_\Gamma (a+b\kappa^2) ds. \tag{11.14}$$

[2] The Euler elastica problem has its origins in elasticity theory applied to large deformation, torsion-free, thin rods (see, for example, Love 1944). The behavior of such thin elastic rods provides the inspiration for splines, which are used extensively in data fitting and geometric modeling and discussed in Section 11.2.

Observe that a quadratic penalty weight function is used, but it is only applied in the interior of objects excluding the edges. Instead, the Euler elastica term applies at the edges. In order to obtain the object edges Γ, segmentation must be used, which is discussed in the next section.

11.1.4 Segmentation

Two-dimensional images are projections of the three-dimensional world comprised of discrete objects. The goal of *segmentation* is to delineate these individual objects from their three-dimensional setting. The celebrated Mumford and Shah (1989) model accomplishes segmentation by removing (blurring) the surface textures of each object, which are typically relatively smooth, to reveal the discrete edges of the individual objects. Thus, the image is segmented by constructing an image with homogeneous regions separated by sharp edges. Segmentation has numerous applications in higher-level image processing and geometry reconstruction. For example, it is important in three-dimensional reconstruction from two-dimensional images, motion analysis, as well as object identification and classification. An example of segmentation of Figure 11.1(a) is shown in Figure 11.2.

Image segmentation is based on the *Mumford and Shah model*, for which the objective functional is given by

$$J[u,\Gamma] = \iint_A \gamma^2 (u-u_0)^2 dxdy + \iint_{A-\Gamma} |\nabla u|^2 dxdy + \beta^2 \int_\Gamma ds, \tag{11.15}$$

where $\Gamma(s)$ comprises the edges between objects, s is the coordinate along the edges Γ, and the penalty weight function is quadratic. Thus, the objective is to

Figure 11.2. Segmentation applied to Figure 11.1(a).

take an observed image $u_0(x,y)$ and determine the reconstructed image $u(x,y)$ that is close to the observed image (first term), is smooth over each object subdomain (second term), and for which the subdomain boundaries Γ minimize the length of the edges between objects (third term). β^2 is the weight coefficient on the edge-length term. The resulting reconstructed image $u(x,y)$ is piecewise smooth, that is, smooth within each object's subdomain and discontinuous on its edges. Observe that in segmentation there are two stationary functions being sought, the reconstructed image $u(x,y)$ and the subdomain boundaries Γ. The solution to the corresponding coupled Euler equations for $u(x,y)$ and $\Gamma(s)$ has been denoised as it is necessary for segmentation to clearly identify edges without the influence of noise. Note the similarities of this model with the Mumford-Shah-Euler model given by equation (11.14). The Mumford-Shah model predates the Mumford-Shah-Euler model and uses a simpler edge length term in the functional.

Observe that a quadratic weight function is used in deblurring, segmentation, and Mumford-Shah-Euler inpainting, and total variation smoothers are used in denoising and classic inpainting. There is a trade-off between the two options as quadratic penalty weight functions lead to diffusion-type terms, in the form of a Laplacian, that smooths out the discontinuous edge intensities. On the other hand, smoothers based on total variations preserve discontinuities at edges while smoothing; however, they require regularization to handle regions where $|\nabla u| \ll 1$.

In segmentation applications where object edges are moving in time, such as in motion tracking, a particularly useful tool is that of the *level set*. The application of level-set methods in variational image segmentation is discussed in detail in Mitichie and Ben Ayed (2011). Level sets, which have many other applications as well, are discussed in Section 9.4.1 in the context of multiphase flow.

11.2 Curve and Surface Optimization Using Splines

The primary distinction between *interpolation* and *curve-fitting* is whether the curve or surface[3] is to pass through all data points (interpolation) or whether the data points are to be used to produce a smooth curve that is close to, but does not pass through, the data points (curve fitting). When a discrete data set is dense with only small errors throughout a domain, straightforward interpolation techniques are adequate to approximate values between data points. When a data set is sparse and/or is subject to large errors, however, more sophisticated methods are called for. One of the most common tools for representing discrete sparse data is the *spline*, of which there are several varieties including natural splines, B-splines, and cubic splines. Splines are piecewise smooth polynomial functions whose primary virtue is their ability to be smoothly joined together. B-splines, in particular, experience widespread use owing to their flexibility and desirable geometric properties. In addition to standard curve fitting, the smoothness properties of splines make them popular in curve and surface design and reconstruction in computer-aided design and manufacturing (CAD/CAM), computer graphics, and path planning.

[3] A *curve* or *surface* need not be a function. That is, while a function must be single valued, having a single value for each point in the domain, a curve or surface may be multi-valued when plotted on a Cartesian coordinate system.

The term spline comes from the draftsmen's use of a thin elastic strip – usually made of wood – placed between movable weights, called "ducks," to aid in drawing smooth curves. Such devices were called splines. The shape of the strip would be such that the strain energy (bending energy in this case) is a minimum along the flexible strip. Over the years, various attempts have been made to develop mathematical models for splines that replicate the behavior of physical splines. Initially, this meant pursuing physically motivated approaches based on emulating the physics of thin elastic rods. More recently, the focus has shifted to methods that display certain aesthetic and/or mathematical properties. After introducing B-splines for comparison, we focus our attention on approaches based on variational methods and illustrate the approaches for curves in a plane; the approaches naturally can be extended to curved surfaces in three-dimensions.

11.2.1 B-Splines

The workhorse of splines is the B-spline, that is, *basis* – or *basic* – *spline*, which is a generalization of Bézier curves. B-splines are defined using a linear combination of basis functions in the form

$$u(s) = \sum_{i=0}^{n} c_i \phi_{i,k}(s), \tag{11.16}$$

where $u(s)$ is the spline defined parametrically in terms of its arc length s, n is the number of terms corresponding to the number of control points, c_i are the control points, and k is the degree of the polynomial basis functions. We define a set of $m+1$ *knots*, $s_0, s_1, \ldots, s_m$, located along the spline curve and $n+1$ *control points*, $c_0, c_1, \ldots, c_n$, that determine the shape of the spline. For polynomial basis functions of order k, the number of knots and control points are related by $k = m - n - 1$. The knots $s_{k+1}, \ldots, s_{m-k-1}$ are called *internal knots*. Note that a Bézier curve is a B-spline with no internal knots. See Figure 11.3 for an illustration with

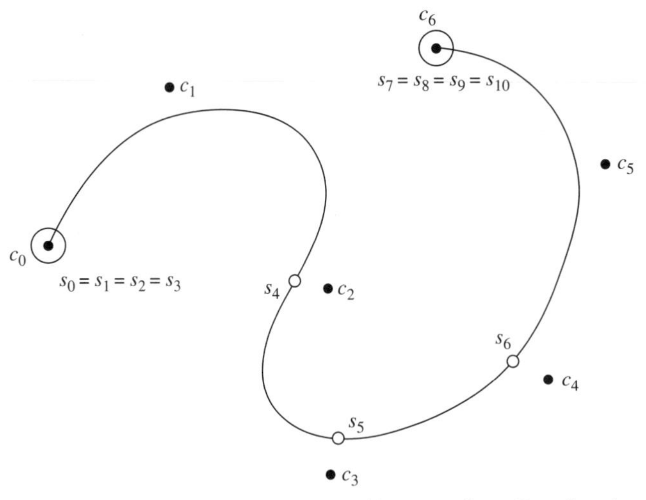

Figure 11.3. Schematic of the control points c_i and knots s_j for a B-spline (s_4, s_5, and s_6 are internal knots).

$m = 10$ and $n = 6$ ($k = 3$). The basis functions are defined through the recursion relation

$$\phi_{i,k}(s) = \frac{s - s_i}{s_{i+j} - s_i}\phi_{i,j-1}(s) + \frac{s_{i+j+1} - s}{s_{i+j+1} - s_{i+1}}\phi_{i+1,j-1}(s),$$

where $j = 1,2,\ldots,k$, and

$$\phi_{i,0}(s) = \begin{cases} 1, & s_i \leq s < s_{i+1} \\ 0, & s < s_i,\ s \geq s_{i+1} \end{cases}.$$

It is common to use *cubic B-splines* ($k = 3$), for which second derivatives are continuous. A common variant of the B-spline used in industry and commercial codes is the *nonuniform rational basis spline* (NURBS). Although B-splines are very versatile and straightforward to calculate, applications in computer graphics and curve or shape representation in CAD/CAM sometimes demand additional aesthetic and/or mathematical properties that can be accommodated using variational methods as shown in the next section.

11.2.2 Spline Functionals

The objective is to determine the optimal curve or surface for representing a data set or a shape. The curve or surface is usually designated to be optimal in terms of its smoothness. For functions $u(x)$ having small slopes, the bending moment and curvature of a thin elastic rod are proportional to the second derivative of the function [see equation (11.13)]. As in Komzsik (2009), therefore, splines may be defined by minimizing the functional

$$J[u] = \int_{s_0}^{s_1} [u''(s)]^2 ds.$$

The integrand is squared because we do not want positive and negative curvature portions of the curve canceling each other. Although it is easy to implement, such a functional does not faithfully mimic the curved shape of a thin elastic strip for cases with non-negligible slopes.

The shape of an elastic strip is such that the bending moments, which are proportional to curvature, is minimized (compare with Euler's elastica in image processing). For the curvature $\kappa(s)$ along the curve $u(s)$ with differential arc length ds, we seek to minimize the *minimum-energy curve* (*MEC*) functional

$$J[\kappa] = \int_{s_0}^{s_1} \kappa(s)^2 ds = \int_{x_0}^{x_1} \frac{(u'')^2}{[1 + (u')^2]^3} dx,$$

where the final expression applies if the curve can be expressed as a function of x. The corresponding Euler equation written in terms of the curvature and arc length is

$$\kappa(s) = 0.$$

A disadvantage of the MEC spline is that it is not scale invariant; that is, zooming in or out and redoing the calculation does not produce the same curve. In addition, it does not produce a circle when the points in fact do lie on a circle.

An alternative to the MEC functional introduced by Moreton (1992) is to minimize a functional involving the square of the *change* of curvature rather than curvature itself. This is known as the *minimum variation curvature* (*MVC*) functional given by

$$J[\kappa] = \int_{s_0}^{s_1} \left(\frac{d\kappa}{ds} \right)^2 ds = \int_{x_0}^{x_1} \frac{[\kappa'(x)]^2}{\sqrt{1+[u'(x)]^2}} dx,$$

where again the final expression applies if the curve is also a function of x. The Euler equation written in terms of the curvature and arc length is

$$\frac{d^2\kappa}{ds^2} = 0,$$

which leads to a linear solution for the curvature

$$\kappa(s) = as + b.$$

Whereas the MEC spline seeks the curve that bends the least, the MVC spline bends as smoothly (uniformly) as possible. In the MVC context, circles are "perfect" shapes from the point of view that their functional is zero; that is, it is a minimum. In addition, the MVC functional is scale invariant.

The variational approach allows for tremendous flexibility in setting how data points will be handled by the spline, for example, whether the spline will pass through the points by including them in a constraint using Lagrange multipliers, or not. Constraints can be placed on total curve length, local tangent slopes, and local curvatures if desired. The variational approach can be extended to curved surfaces in three-dimensional space using, for example, a *minimum variation surface* (*MVS*) functional, which is an extension of the MVC functional to surfaces. Additional material on splines can be found in Dierckx (1993) and de Boor (2001).

11.3 Proper-Orthogonal Decomposition

As discussed in Chapter 10, the computational problems that result from formulation of optimization and control problems can be very large, particularly for time-dependent (unsteady) state equations. Therefore, there has been significant need for, and interest in, developing methods for representing the state of a system in a lower-dimensional form while retaining the essential features of the overall large-dimensional data set. An increasingly popular technique for accomplishing this is *proper-orthogonal decomposition* (*POD*), which is widely applicable in performing *model reduction* of experimental and computational data sets for incorporation in control algorithms or other analysis purposes. This is particularly the case for closed-loop control, in which state estimation and control synthesis must occur in real time during operation of the system. Proper-orthogonal decomposition also goes by the names *Karhunen-Loève decomposition* and *principal-component analysis*.

One possibility for model reduction is to approximate the data set using a linear combination of basis functions. Recall that this is the approach used in the Rayleigh-Ritz, Galerkin, and finite-element methods (see Chapter 3), where we approximate a stationary function using a set of predefined basis functions. Given our discussion of optimization techniques, one may wonder whether there is an "optimal" set of basis functions that should be used for representation of a given function or discrete data set. That is, are there basis functions that contain some of the characteristics of the underlying system? If so, then we would be able to represent the data faithfully with the least number of basis functions, or *POD modes*. Indeed, it turns out that this is the case. Motivated by this, let us consider representing a given continuous data set $u(x,y)$ as an expansion in terms of the basis functions $\phi(x,y)$ as follows

$$u(x,y) = \sum_{i-1}^{\infty} c_i \phi_i(x,y). \tag{11.17}$$

As with image processing considered in Section 11.1, we will first assume that the data are represented continuously for our purposes; the extension to the discrete case takes us out of the realm of variational methods. In addition, we consider the two-dimensional case for illustrative purposes; however, the true power of POD methods is realized in large three-dimensional data sets that may result from experimental measurements or computational simulations.

It is often the case that we would like a *low-dimensional* representation of large data sets in order to 1) extract dominant physics from the data set and/or 2) for use in further analysis, such as optimal control, where it is not practical to work with the entire data set.

Before considering how the optimal basis functions are determined for a given data set, let us summarize the desired properties of such a decomposition:

1. Because of their ease of use, we seek orthonormal basis functions that are orthogonal to one another and normalized to length one (based on some definition of the norm).
2. We would like to be able to represent the original data as faithfully as possible with the fewest number of basis functions.
3. The POD analysis should be completely independent of the source of the data (for example, no knowledge of the governing equations is necessary). Contrast this with data assimilation considered in Section 10.6.3, where a reduced representation of the system is obtained by taking into account the governing equations of the system.

To accomplish these objectives, a set of optimal basis functions is sought that allows us to represent a high-dimensional system with a low-dimensional one containing most of the relevant dynamics of the system.

We could imagine determining the orthogonal basis functions $\phi_i(x,y)$ for a given data set $u(x,y)$ that minimize the functional

$$J_1[\phi] = ||\phi_i(x,y) - u(x,y)||^2 = \iint_A [\phi_i(x,y) - u(x,y)]^2 \, dxdy, \tag{11.18}$$

such that the difference between the orthonormal basis functions and the original data set is minimized in an averaged sense over the entire domain. Observe that the functional seeks the optimal basis functions in a least squares sense as in image processing. It can be shown that this is equivalent to maximizing the functional

$$J_2[\phi] = \langle u(x,y), \phi_i(x,y)\rangle^2 = \left[\iint_A u(x,y)\phi_i(x,y)dxdy\right]^2, \tag{11.19}$$

which can be regarded as determining the optimal basis functions that maximize the average projection of $u(x,y)$ on $\phi(x,y)$. Because orthonormal basis functions are sought, we minimize (11.18) or maximize (11.19) subject to the constraint that the norm of the resulting basis functions are unity; that is,

$$||\phi_i||^2 = \langle \phi_i, \phi_i\rangle = \iint_A \phi_i^2(x,y)dxdy = 1.$$

Thus, the augmented functional is

$$\begin{aligned} \tilde{J}_2[\phi] &= \langle u, \phi_i\rangle^2 - \lambda\left(||\phi_i||^2 - 1\right) \\ &= \langle \phi_i, u\rangle \langle u, \phi_i\rangle - \lambda\left(\langle \phi_i, \phi_i\rangle - 1\right) \\ &= \langle \langle \phi_i, u\rangle u, \phi_i\rangle - \lambda\left(\langle \phi_i, \phi_i\rangle - 1\right) \\ \tilde{J}_2[\phi] &= \langle R\phi_i, \phi_i\rangle - \lambda\left(\langle \phi_i, \phi_i\rangle - 1\right), \end{aligned}$$

where R can be shown to be a self-adjoint operator, and λ is the Lagrange multiplier. Taking the functional derivative of the augmented functional leads to an eigenproblem for which the eigenfunctions are the optimal basis functions that we seek

$$R\phi_i = \lambda_i\phi_i, \quad \lambda_1 \geq \lambda_2 \geq \cdots \geq \lambda_i \geq \cdots \geq 0,$$

and the eigenvalues are the Lagrange multipliers. Because R is self-adjoint and the eigenvalues λ_i are real and nonnegative, the eigenfunctions $\phi_i(x,y)$ are orthogonal and can be normalized to produce an orthonormal set. Each eigenvalue λ_i represents the average energy contained in the corresponding eigenfunction $\phi_i(x,y)$; therefore, the POD modes with the largest eigenvalues represent the most energetic dynamic features in the system (recall that the eigenvalues are arranged in decreasing order).

For a given data set $u(x,y)$ and orthonormal basis functions $\phi_i(x,y)$, the coefficients are obtained from the inner products

$$c_i = \langle u(x,y), \phi_i(x,y)\rangle.$$

Because the basis functions are optimal, however, we can represent most of the information contained in the original data set with a relatively small number of POD modes, say n, such that

$$u(x,y) \approx \sum_{i=1}^{n} c_i\phi_i(x,y). \tag{11.20}$$

In typical applications of POD analysis, $n = O(10)$. This provides the model reduction that we seek, in which the original large data set can be represented by a relatively small number of optimally chosen POD modes.

In time-dependent problems, the basis functions account for the spatial dependence, and the coefficients c_i are functions of time as follows

$$u(x,y,t) \approx \sum_{i=1}^{n} c_i(t)\phi_i(x,y).$$

In this case, we seek the set of spatial orthogonal modes that best represent the time-evolving data set in a time-averaged sense according to

$$J_1[\phi] = \int_0^{t_f} \iint_A [\phi_i(x,y) - u(x,y,t)]^2 \, dxdydt.$$

Because this leads to an extremely large eigenproblem for the basis functions, the usual method for dealing with time-evolving data is to use *snapshots*, which are data sets over the entire spatial domain $u(x,y)$ taken at a set of discrete times. For example, if the data set contains m snapshots, then the functional (11.18) would become

$$J_1[\phi] = \frac{1}{m} \sum_{j=1}^{m} \iint_A \left[\phi_i(x,y) - u_j(x,y)\right]^2 dxdy, \tag{11.21}$$

which is equivalent to maximizing the time-averaged functional

$$J_2[\phi] = \frac{1}{m} \sum_{j=1}^{m} \langle u_j(x,y), \phi_i(x,y) \rangle^2 = \frac{1}{m} \sum_{j=1}^{m} \left[\iint_A u_j(x,y)\phi_i(x,y) dxdy \right]^2. \tag{11.22}$$

This analysis still results in n basis functions; however, they are optimized over the set of m snapshots.

Although we have formulated the POD analysis as a continuous problem, it is more common to analyze discrete experimental or computational data. For example, suppose that we have a data set $u(x,y,t)$ containing k total data points at each time represented as the time-dependent vector $\mathbf{u}(t)$. We then introduce the $k \times k$ matrix

$$\mathbf{R} = \int_0^{t_f} \mathbf{u}(t)\mathbf{u}(t)^T dt,$$

where $\mathbf{R}$ is symmetric and positive semidefinite. Forming the eigenproblem

$$\mathbf{R}\boldsymbol{\phi}_i = \lambda_i \boldsymbol{\phi}_i, \quad \lambda_1 \geq \lambda_2 \geq \cdots \geq \lambda_k \geq 0$$

for $\mathbf{R}$, it can be shown that the eigenvectors are the optimal basis vectors that we seek. Because $\mathbf{R}$ is symmetric and positive semidefinite, the eigenvalues $\lambda_i, i = 1, \ldots, k$ are real and nonnegative, and the eigenvectors $\boldsymbol{\phi}_i$ are orthogonal and can be normalized to produce an orthonormal set. We choose a subset of $n \ll k$ spatial basis vectors $\boldsymbol{\phi}_i$ in order to provide the desired model reduction of the original data set. The original (large) data set at time t is then represented by a linear combination of the n POD modes as follows:

$$\mathbf{u}(t) \approx \sum_{i=1}^{n} c_i \boldsymbol{\phi}_i.$$

For discrete data sets, POD analysis reduces to performing a singular-value decomposition (SVD), where the singular values are related to the eigenvalues.

Once we have obtained the optimal basis functions for a given data set, *Galerkin projection* can be used to obtain the projection of the governing state equation(s) on these optimal basis functions (see Section 3.2, where now the optimal basis functions are used as the weight functions). This is accomplished by taking the inner product of each basis function with the state equation(s), for example, the Navier-Stokes equations of fluid dynamics. In this manner, the partial differential state equations in space and time become a system of ordinary differential equations in time for the $c_i(t)$ coefficients. This system of reduced state equations, called the *reduced-order model*, can then be used in a feedback control setting as it is amenable to rapid computation. Because it is not unusual for an infinite-dimensional system to be adequately represented by $O(10)$ POD modes, this leads to significant economy in terms of the state representation.

The primary advantage of POD analysis is also its primary shortcoming. Because the optimal basis functions are determined from the actual data set, they provide the best representation of the original data with the fewest POD modes. However, this means that they are problem dependent, requiring us to obtain a new set of basis functions each time the data set changes, whether from consideration of a different dynamical model or a different set of data from the same model.

POD and reduced-order modeling are powerful and flexible techniques for model reduction that can be applied to both linear and nonlinear systems. It is being applied in image processing, data compression, signal processing, control, structural dynamics, and is very popular in fluid mechanics. See Holmes, Lumley, and Berkooz (1996) for additional details of the application of POD analysis to fluid mechanics.

12 Numerical Grid Generation

> We have to make every effort to understand for ourselves what the dangers are, and this points up a fundamental thing about computers: They involve more thought and not less thought. They may save certain parts of our efforts, but they do not eliminate the need for intelligence.
>
> (Norbert Wiener, *Computers and the World of the Future*, 1962)

Our final application of optimization theory is in grid generation, which is an important topic in many computational fields in which we desire to obtain numerical solutions of partial differential equations with complex physics and/or on domains with complex shapes. When choosing a numerical grid on which to solve differential equations, it is necessary that the grid both faithfully represent the geometry of the domain and be sufficiently refined in order to maintain the numerical truncation errors at an acceptable level. Traditionally, this has been done in a rather ad hoc manner by choosing a grid that "looks good" in some qualitative sense. Calculus of variations allows us to optimize the choice of a grid in a formal manner by explicitly enforcing certain criteria on the generation of the grid, thereby providing a more intuitive and mathematically formal basis for grid generation. Variational grid generation is generally formulated to produce the "best" grid in a least squares sense, eliminating the trial and error necessary in other grid generation approaches, such as algebraic and elliptic grid generation. We briefly discuss these traditional techniques by way of background for the variational approach which is our primary focus.

12.1 Fundamentals

The basic objective of grid generation is to transform a complex geometry into a simple one and/or to focus grid points in regions where the solution varies rapidly to provide sufficient resolution. The cost of such a transformation from the so called physical to computational domain is that the governing equations typically become more complicated. Thus, we trade solving the equations in their normal form on a complex geometry for solving more complex equations on a simple geometry. Often the "simple" geometry is a rectangular or circular domain on which it is straightforward to implement various numerical algorithms, such as finite-difference or finite-volume methods.

The specific objectives of grid generation are to 1) faithfully represent the domain geometry, 2) provide sufficient resolution to accurately determine the solution throughout the entire domain, 3) implement "smooth" grids in which the grid lines vary smoothly in physical coordinates, and 4) perform the above using the least number of grid points in order to minimize computational resource requirements. Typically, the grid is generated as a *pre-processing* step before the numerical solution is obtained on that grid. This often requires the user to base the grid resolution on an intuitive sense for where and how much the grid should be focused to accurately capture the physics within the domain. Ideally, one would be able to let the solution determine how much resolution is required in an automatic fashion; this is called *solution-adaptive grid generation*.

Many of the tools developed for grid generation originated in the context of solving fluid mechanics problems that often exhibit complex physics and complex geometries in which the flow field is to be solved numerically using computational fluid dynamics (CFD). This is because computational solutions of the nonlinear Navier-Stokes equations are very time consuming, sometimes taking hours, days, or even weeks to complete a single simulation. Broadly, there are two types of grids used in such applications, *structured* and *unstructured*. A structured grid has an ordered arrangement of *cells*, *elements*, or *nodes*, while an unstructured grid does not. Structured grids are typically used for problems in which the geometry of the domain is of a regular shape, whereas unstructured grids provide additional flexibility to handle very complex geometries. Recall that this is one of the primary advantages of finite-element methods considered in Chapter 3. Rather than using unstructured grids, which present a significant challenge in accounting for the irregular orientation and arrangement of cells, an alternative is to use grid generation techniques in which a grid within a complex geometry is transformed to a simple geometry for which a structured grid is straightforward to implement. Finite-difference methods are then commonly used to numerically solve the governing equations on the structured grid within the simple geometry. This is the approach considered here.

We first discuss the traditional grid generation techniques for structured grids based on algebraic transformations and elliptic, or partial differential equation based, grid generation. It is then shown how variational methods can be used as the basis for solution-adaptive grid generation techniques using optimization theory. Ultimately, we desire accuracy of the solution and efficiency of the computational algorithms used to obtain those solutions numerically. To achieve this, the desired properties of a grid transformation are 1) the domain in both physical and computational coordinates is simply connected, 2) the transformation is one-to-one, in which a single image point in the computational domain maps to a unique point in the physical domain, and 3) the transformation is smooth, such that all coordinate functions are continuous and differentiable.

Grid generation is implemented by specifying or determining a transformation $x = x(\xi,\eta), y = y(\xi,\eta)$ from the *physical domain* (x,y) of the original problem to the *computational domain* (ξ,η) on which the equations are to be solved. Traditionally, there are two approaches to determining the grid transformation. In algebraic grid generation, some preselected algebraic function is used to specify the grid transformation. Elliptic grid generation requires numerical solution of a partial differential equation, which is of the elliptic type, in order to obtain the grid transformation. These are discussed briefly in the following two sections in order to motivate the advantages of the variational approach to grid generation, which is introduced in the last section.

We consider transformations of the form

$$(x,y) \Longleftrightarrow (\xi,\eta),$$

where (x,y) are the independent variables in the *physical domain*, and (ξ,η) are those in the *computational domain* such that $\xi = \xi(x,y)$ and $\eta = \eta(x,y)$. To transform back to the physical domain, we have $x = x(\xi,\eta)$ and $y = y(\xi,\eta)$. In order to transform differential equations from one coordinate system to the other, the transformation laws, which arise from the chain rule, are

$$\frac{\partial}{\partial x} = \frac{\partial \xi}{\partial x}\frac{\partial}{\partial \xi} + \frac{\partial \eta}{\partial x}\frac{\partial}{\partial \eta}, \tag{12.1}$$

$$\frac{\partial}{\partial y} = \frac{\partial \xi}{\partial y}\frac{\partial}{\partial \xi} + \frac{\partial \eta}{\partial y}\frac{\partial}{\partial \eta}, \tag{12.2}$$

where $\xi_x = \partial\xi/\partial x$, $\xi_y = \partial\xi/\partial y$, $\eta_x = \partial\eta/\partial x$, and $\eta_y = \partial\eta/\partial y$ are called the *metrics* of the transformation and make up the elements in the Jacobian of the transformation, which is

$$\mathcal{J} = \begin{vmatrix} \xi_x & \eta_x \\ \xi_y & \eta_y \end{vmatrix} = \begin{vmatrix} \xi_x & \xi_y \\ \eta_x & \eta_y \end{vmatrix} = \xi_x\eta_y - \xi_y\eta_x. \tag{12.3}$$

In order to enforce the properties outlined above, the transformation must be such that $\mathcal{J} \neq 0$ in order for the transformation to be one-to-one, and the metrics must be smoothly varying throughout the domain in order to produce a smooth grid distribution. It is also common to seek grids that have orthogonal (or nearly orthogonal) grid lines.

12.2 Algebraic Grid Generation

The simplest grid generation technique, which is sufficient in many circumstances, is algebraic grid generation. In this approach, an algebraic function is preselected for the mapping between physical and computational coordinates. This has the virtue that no additional calculation is required to obtain the grid before seeking the solution of the governing differential equations. Therefore, the computational overhead is relatively small as there are no additional equations to solve. This approach is often used to transform relatively simple geometries, such as a trapezoid, into an even simpler domain, such as a rectangle. It can also be used to concentrate grid points in a fixed, predetermined location within the domain in order to provide additional resolution as needed. The disadvantages of algebraic grid generation are that it is difficult to handle complex geometries, and the algebraic transformation must be selected ahead of time and may not be the most appropriate for the given problem. For example, it is often not known where the grid needs to be focused before the solution has been obtained. As we will see through example, the complexity of the

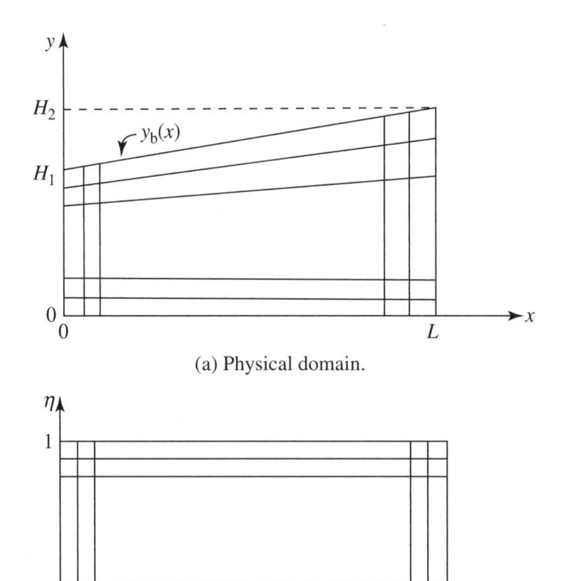

Figure 12.1. Diffuser geometry.

domain boundaries and/or nonuniform grid in the physical domain is essentially moved to the governing equations themselves.

Let us illustrate algebraic grid generation using the example of fluid flow through a diffuser as shown in Figure 12.1(a). Note that only one-half of the domain is considered, and the full domain arises from that shown mirrored around the x-axis. The equation for the upper (sloped) boundary is

$$y_b = f(x) = H_1 + \frac{H_2 - H_1}{L} x.$$

We seek to transform this physical domain into a rectangular computational domain through the algebraic transformation

$$\xi = x, \quad \eta = \frac{y}{f(x)}. \tag{12.4}$$

Thus, we do not apply any transformation in the x-direction, and we use a linear scaling in the y-direction to account for the sloped boundary. We then utilize a uniform grid in the computational domain (ξ, η) as shown in Figure 12.1(b). The inverse transformation from the computational to the physical domain is then

$$x = \xi, \quad y = f(x)\eta = \left(H_1 + \frac{H_2 - H_1}{L} \xi \right) \eta, \tag{12.5}$$

which produces a stretched grid in the y-direction in the physical domain.

Let us check the metrics that comprise the Jacobian of this transformation

$$\frac{\partial \xi}{\partial x} = 1,$$
$$\frac{\partial \xi}{\partial y} = 0,$$
$$\frac{\partial \eta}{\partial x} = -y\frac{f'(x)}{f^2(x)} = -\frac{f'(\xi)}{f(\xi)}\eta = -\frac{\left(\frac{H_2-H_1}{L}\right)\eta}{H_1 + \frac{H_2-H_1}{L}\xi},$$
$$\frac{\partial \eta}{\partial y} = \frac{1}{f(x)} = \frac{1}{f(\xi)} = \frac{1}{H_1 + \frac{H_2-H_1}{L}\xi}.$$

See Figure 12.2 for plots of the latter two metrics showing that they are smooth as desired (the first two are clearly smooth).

The derivatives in the governing equations are transformed using the transformation laws (12.1) and (12.2)

$$\frac{\partial}{\partial x} = \frac{\partial \xi}{\partial x}\frac{\partial}{\partial \xi} + \frac{\partial \eta}{\partial x}\frac{\partial}{\partial \eta} = \frac{\partial}{\partial \xi} - \eta\frac{f'(\xi)}{f(\xi)}\frac{\partial}{\partial \eta},$$
$$\frac{\partial}{\partial y} = \overbrace{\frac{\partial \xi}{\partial y}}^{0}\frac{\partial}{\partial \xi} + \frac{\partial \eta}{\partial y}\frac{\partial}{\partial \eta} = \frac{1}{f(\xi)}\frac{\partial}{\partial \eta}.$$

Having these derivatives operate on themselves leads to the second derivative transformations as follows:

$$\frac{\partial^2}{\partial x^2} = \frac{\partial^2}{\partial \xi^2} - 2G(\xi,\eta)\frac{\partial^2}{\partial \xi \partial \eta} + G^2(\xi,\eta)\frac{\partial^2}{\partial \eta^2} + \left[2\frac{f'(\xi)}{f(\xi)}G(\xi,\eta) - \eta\frac{f''(\xi)}{f(\xi)}\right]\frac{\partial}{\partial \eta},$$
$$\frac{\partial^2}{\partial y^2} = \frac{1}{f^2(\xi)}\frac{\partial^2}{\partial \eta^2},$$

where

$$G(\xi,\eta) = \eta\frac{f'(\xi)}{f(\xi)}.$$

For example, consider one of the convection terms in the Navier-Stokes equation (9.24):

$$u(x,y)\frac{\partial u}{\partial x} = u(\xi,\eta)\left[\frac{\partial u}{\partial \xi} - \eta\frac{f'(\xi)}{f(\xi)}\frac{\partial u}{\partial \eta}\right].$$

The entire set of coupled Navier-Stokes equations would be similarly transformed to produce the equivalent equations written in the computational domain. Similarly, consider the Laplace equation:

$$\frac{\partial^2 \phi}{\partial x^2} + \frac{\partial^2 \phi}{\partial y^2} = 0,$$

which becomes

$$\frac{\partial^2 \phi}{\partial \xi^2} - 2G(\xi,\eta)\frac{\partial^2 \phi}{\partial \xi \partial \eta} + \left[2\frac{f'(\xi)}{f(\xi)}G(\xi,\eta) - \eta\frac{f''(\xi)}{f(\xi)}\right]\frac{\partial \phi}{\partial \eta} + \left[G^2(\xi,\eta) + \frac{1}{f^2(\xi)}\right]\frac{\partial^2 \phi}{\partial \eta^2} = 0.$$

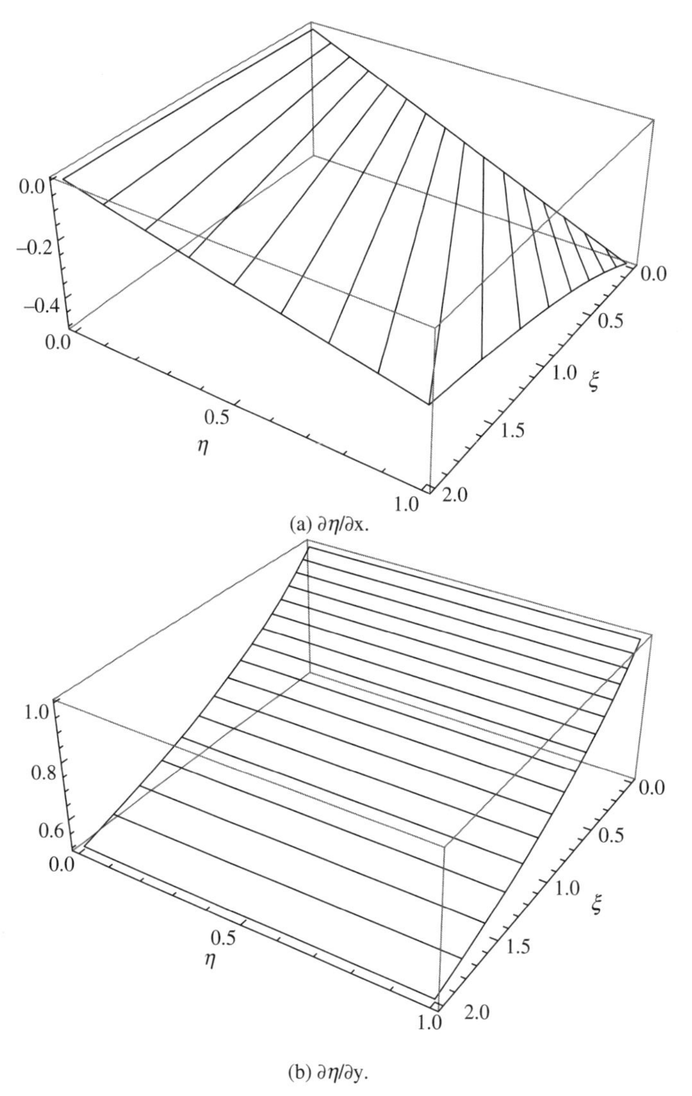

(b) $\partial\eta/\partial y$.

Figure 12.2. Plots of transformation metrics for diffuser with $H_1 = 1$, $H_2 = 2$, and $L = 2$.

Although the coefficients become more complicated in the transformed equations in computational coordinates, the equations are of the same basic form and can be solved using the same algorithms developed for the untransformed Navier-Stokes or Laplace equations. Notice, however, the presence of the mixed derivative term in the Laplace equation.

In the previous illustration, the transformation is used to map the domain from a trapezoid to a rectangle. In order to focus grid points in a particular region of the domain, a useful transformation is

$$\xi = x, \quad \eta = \frac{2}{\pi}\tan^{-1}\left(\frac{y-y_a}{a}\right), \tag{12.6}$$

if applied in the y-direction. Reducing the parameter a concentrates more grid points near the point $y = y_a$ in the physical domain when using a uniform grid in the computational (ξ,η) domain. This transformation also has the benefit that it can be used to transform an infinite (or semi-infinite) domain into a finite one. For example, the transformation (12.6) maps the infinite domain $-\infty \leq y \leq \infty$ to $-1 \leq \eta \leq 1$. The inverse transformation is

$$x = \xi, \quad y = y_a + a\tan\left(\frac{\pi}{2}\eta\right).$$

The transformation laws for the transformation (12.6) are

$$\frac{\partial}{\partial x} = \frac{\partial}{\partial \xi},$$

$$\frac{\partial}{\partial y} = \overset{0}{\cancel{\frac{\partial \xi}{\partial y}}}\frac{\partial}{\partial \xi} + \frac{\partial \eta}{\partial y}\frac{\partial}{\partial \eta} = \Gamma(\eta)\frac{\partial}{\partial \eta},$$

where

$$\Gamma(\eta) = \frac{\partial \eta}{\partial y} = \frac{1}{\pi a}\left[1 + \cos(\pi\eta)\right].$$

Then the second derivative in the y-direction becomes

$$\frac{\partial^2}{\partial y^2} = \Gamma(\eta)\frac{\partial}{\partial \eta}\left[\Gamma(\eta)\frac{\partial}{\partial \eta}\right] = \Gamma^2(\eta)\frac{\partial^2}{\partial \eta^2} + \Gamma(\eta)\Gamma'(\eta)\frac{\partial}{\partial \eta}.$$

When it is sufficient, algebraic grid generation is preferred as it introduces very little computational overhead to implement. Recall, however, that the most significant drawback is the fact that the transformation must be specified in an ad hoc fashion before obtaining the solution. Such transformations are often based on experience and intuition. In addition, the resulting grids are often not orthogonal and may lead to cells with very large aspect ratios, being elongated in one direction, which is typically undesirable.

12.3 Elliptic Grid Generation

In order to treat more complex geometries than possible using algebraic grid generation, elliptic grid generation may be a good choice. It is called elliptic grid generation because it consists of solving an elliptic partial differential equation where the dependent variables are the physical domain coordinates (x,y) and the independent variables are the transformed computational coordinates (ξ,η). Grids generated in this way generally provide smoother grids than using algebraic grid generation even if the grid along the boundary is not smooth.

The motivation behind elliptic grid generation can be illustrated by considering the potential flow (see Section 9.2.3) around a circular cylinder as shown in Figure 12.3, where the streamlines and equipotential lines are plotted. Because such solutions are obtained using Laplace equations, the streamlines and equipotential lines are everywhere orthogonal, which is one of our desired properties of grid transformations. Additionally, the streamlines and equipotential lines are smooth throughout the entire domain. Qualitatively, we might say that the streamlines and equipotential lines would make a good grid on which to base a solution of the full Navier-Stokes equations for this geometry.

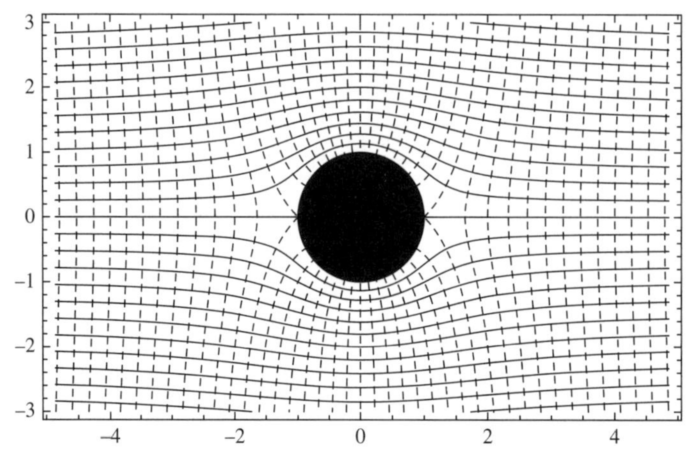

Figure 12.3. Streamlines (solid) and equipotential lines (dashed) for potential flow around a circular cylinder.

The velocity potential $\phi(x,y)$ and streamfunction $\psi(x,y)$ both satisfy the Laplace equation according to

$$\frac{\partial^2 \phi}{\partial x^2} + \frac{\partial^2 \phi}{\partial y^2} = 0, \quad \frac{\partial^2 \psi}{\partial x^2} + \frac{\partial^2 \psi}{\partial y^2} = 0.$$

This suggests that a solution of the equations

$$\frac{\partial^2 \xi}{\partial x^2} + \frac{\partial^2 \xi}{\partial y^2} = 0, \quad \frac{\partial^2 \eta}{\partial x^2} + \frac{\partial^2 \eta}{\partial y^2} = 0 \tag{12.7}$$

would provide a good mapping on which to base the computational grid. As in algebraic grid generation, we would use a uniform grid in the computational domain (ξ, η) that produces the grid shown in Figure 12.3 in physical space. This approach to elliptic grid generation has been pioneered by Winslow (1967).

There are two approaches to obtaining the solution to (12.7). Because the grid transformation is governed by Laplace equations, thereby being harmonic, we could utilize complex variable theory and conformal mapping techniques. This approach is good for two-dimensional flows in certain types of geometries, but it cannot be generalized to three-dimensional flows and can only be used to treat certain classes of geometries. The second, more general, approach is to solve a boundary-value problem to generate the grid, which is what is referred to as *elliptic grid generation*. Owing to its improved versatility over conformal mapping techniques and ability to extend naturally to three-dimensional settings, this is the preferred approach and the basis for many commercial or open-source grid generation tools.

In order to control grid clustering, known functions involving sources and sinks may be added to the right-hand side of equations (12.7) as forcing terms

$$\frac{\partial^2 \xi}{\partial x^2} + \frac{\partial^2 \xi}{\partial y^2} = P(x,y), \quad \frac{\partial^2 \eta}{\partial x^2} + \frac{\partial^2 \eta}{\partial y^2} = Q(x,y), \tag{12.8}$$

where P and Q typically contain exponential functions. Equations (12.8) are Poisson equations. Note that equations (12.8) are for $\xi = \xi(x,y)$ and $\eta = \eta(x,y)$. That is, they must be solved in the physical domain. As with the governing equations, however, we want to solve (12.8) in the computational domain (ξ, η) to obtain the grid transformations $x = x(\xi, \eta)$ and $y = y(\xi, \eta)$. Once again, we seek the transformation laws for moving from the physical to the computational coordinates. Let us consider this process in a more generic manner than performed earlier. For

$$\xi = \xi(x,y), \quad \eta = \eta(x,y),$$

the total differentials are

$$d\xi = \frac{\partial \xi}{\partial x}dx + \frac{\partial \xi}{\partial y}dy, \quad d\eta = \frac{\partial \eta}{\partial x}dx + \frac{\partial \eta}{\partial y}dy,$$

or in matrix form

$$\begin{bmatrix} d\xi \\ d\eta \end{bmatrix} = \begin{bmatrix} \xi_x & \xi_y \\ \eta_x & \eta_y \end{bmatrix} \begin{bmatrix} dx \\ dy \end{bmatrix}. \tag{12.9}$$

Similarly, for the inverse transformation from (ξ, η) to (x,y), we have that

$$dx = \frac{\partial x}{\partial \xi}d\xi + \frac{\partial x}{\partial \eta}d\eta, \quad dy = \frac{\partial y}{\partial \xi}d\xi + \frac{\partial y}{\partial \eta}d\eta,$$

or

$$\begin{bmatrix} dx \\ dy \end{bmatrix} = \begin{bmatrix} x_\xi & x_\eta \\ y_\xi & y_\eta \end{bmatrix} \begin{bmatrix} d\xi \\ d\eta \end{bmatrix}.$$

From the latter expression, we can solve for $[d\xi \ d\eta]^T$ by multiplying by the inverse

$$\begin{bmatrix} d\xi \\ d\eta \end{bmatrix} = \begin{bmatrix} x_\xi & x_\eta \\ y_\xi & y_\eta \end{bmatrix}^{-1} \begin{bmatrix} dx \\ dy \end{bmatrix}$$

$$= \frac{\begin{bmatrix} y_\eta & -x_\eta \\ -y_\xi & x_\xi \end{bmatrix}}{\begin{vmatrix} x_\xi & x_\eta \\ y_\xi & y_\eta \end{vmatrix}} \begin{bmatrix} dx \\ dy \end{bmatrix}$$

$$\begin{bmatrix} d\xi \\ d\eta \end{bmatrix} = \frac{1}{\mathcal{J}} \begin{bmatrix} y_\eta & -x_\eta \\ -y_\xi & x_\xi \end{bmatrix} \begin{bmatrix} dx \\ dy \end{bmatrix},$$

where the Jacobian is

$$\mathcal{J} = \begin{vmatrix} x_\xi & x_\eta \\ y_\xi & y_\eta \end{vmatrix}. \tag{12.10}$$

Comparing with equation (12.9), we see that

$$\begin{bmatrix} \xi_x & \xi_y \\ \eta_x & \eta_y \end{bmatrix} = \frac{1}{\mathcal{J}} \begin{bmatrix} y_\eta & -x_\eta \\ -y_\xi & x_\xi \end{bmatrix}.$$

Consequently, the transformation metrics are

$$\frac{\partial \xi}{\partial x} = \frac{1}{\mathcal{J}}\frac{\partial y}{\partial \eta}, \quad \frac{\partial \xi}{\partial y} = -\frac{1}{\mathcal{J}}\frac{\partial x}{\partial \eta},$$

$$\frac{\partial \eta}{\partial x} = -\frac{1}{\mathcal{J}}\frac{\partial y}{\partial \xi}, \quad \frac{\partial \eta}{\partial y} = \frac{1}{\mathcal{J}}\frac{\partial x}{\partial \xi}.$$

Substituting into the transformation laws (12.1) and (12.2) give

$$\frac{\partial}{\partial x} = \frac{\partial \xi}{\partial x}\frac{\partial}{\partial \xi} + \frac{\partial \eta}{\partial x}\frac{\partial}{\partial \eta} = \frac{1}{\mathcal{J}}\left[\frac{\partial y}{\partial \eta}\frac{\partial}{\partial \xi} - \frac{\partial y}{\partial \xi}\frac{\partial}{\partial \eta}\right],$$

$$\frac{\partial}{\partial y} = \frac{\partial \eta}{\partial y}\frac{\partial}{\partial \eta} + \frac{\partial \xi}{\partial y}\frac{\partial}{\partial \xi} = \frac{1}{\mathcal{J}}\left[\frac{\partial x}{\partial \xi}\frac{\partial}{\partial \eta} - \frac{\partial x}{\partial \eta}\frac{\partial}{\partial \xi}\right].$$

From these relationships, we obtain the second derivative transformation laws as follows

$$\frac{\partial^2}{\partial x^2} = \frac{\partial^2 \xi}{\partial x^2}\frac{\partial}{\partial \xi} + \frac{\partial^2 \eta}{\partial x^2}\frac{\partial}{\partial \eta} + \left(\frac{\partial \xi}{\partial x}\right)^2\frac{\partial^2}{\partial \xi^2} + \left(\frac{\partial \eta}{\partial x}\right)^2\frac{\partial^2}{\partial \eta^2} + 2\frac{\partial \eta}{\partial x}\frac{\partial \xi}{\partial x}\frac{\partial^2}{\partial \eta \partial \xi}, \tag{12.11}$$

and

$$\frac{\partial^2}{\partial y^2} = \frac{\partial^2 \xi}{\partial y^2}\frac{\partial}{\partial \xi} + \frac{\partial^2 \eta}{\partial y^2}\frac{\partial}{\partial \eta} + \left(\frac{\partial \xi}{\partial y}\right)^2\frac{\partial^2}{\partial \xi^2} + \left(\frac{\partial \eta}{\partial y}\right)^2\frac{\partial^2}{\partial \eta^2} + 2\frac{\partial \eta}{\partial y}\frac{\partial \xi}{\partial y}\frac{\partial^2}{\partial \eta \partial \xi}. \tag{12.12}$$

Application of equations (12.11) and (12.12) to the Poisson equations (12.8) gives

$$\begin{aligned} g_{22}\frac{\partial^2 x}{\partial \xi^2} - 2g_{12}\frac{\partial^2 x}{\partial \xi \partial \eta} + g_{11}\frac{\partial^2 x}{\partial \eta^2} &= -\frac{1}{\mathcal{J}^2}\left[P\frac{\partial x}{\partial \xi} + Q\frac{\partial x}{\partial \eta}\right], \\ g_{22}\frac{\partial^2 y}{\partial \xi^2} - 2g_{12}\frac{\partial^2 y}{\partial \xi \partial \eta} + g_{11}\frac{\partial^2 y}{\partial \eta^2} &= -\frac{1}{\mathcal{J}^2}\left[P\frac{\partial y}{\partial \xi} + Q\frac{\partial y}{\partial \eta}\right], \end{aligned} \tag{12.13}$$

where

$$g_{11} = \left(\frac{\partial x}{\partial \xi}\right)^2 + \left(\frac{\partial y}{\partial \xi}\right)^2,$$

$$g_{12} = \frac{\partial x}{\partial \xi}\frac{\partial x}{\partial \eta} + \frac{\partial y}{\partial \xi}\frac{\partial y}{\partial \eta},$$

$$g_{22} = \left(\frac{\partial x}{\partial \eta}\right)^2 + \left(\frac{\partial y}{\partial \eta}\right)^2.$$

Note that the g_{ij} coefficients in the equations are defined such that

$$g = \begin{vmatrix} g_{11} & g_{12} \\ g_{21} & g_{22} \end{vmatrix} = \mathcal{J}^2, \quad (g_{12} = g_{21}).$$

Thus, equations (12.13) would be solved numerically on the computational domain in order to obtain the transformations $x = x(\xi,\eta)$ and $y = y(\xi,\eta)$. The boundary conditions to apply to the transformation equations (12.13) can be either Dirichlet or Neumann. If the grid point locations (x,y) are specified on the boundaries of the physical domain, then Dirichlet conditions are imposed on the solution. Alternatively, one can specify the slopes of the grid lines at the boundaries, for example, to produce grid lines that are orthogonal to the boundary. This gives rise to Neumann (derivative) boundary conditions. Therefore, we have a coupled set of nonlinear elliptic equations with Dirichlet or Neumann boundary conditions that we solve on a uniform grid in (ξ,η) to obtain the inverse transformation $x = x(\xi,\eta), y = y(\xi,\eta)$.

Elliptic grid generation techniques are quite versatile as they can be applied to a variety of physical geometries, and they produce smooth grids owing to solving diffusion-type partial differential equations. It is this diffusion property that leads to discontinuities at the boundaries being smoothed out in the interior. Elliptic grid generation, which can naturally be extended to three dimensional problem, also enforces grid orthogonality and can be used to treat multiply-connected regions. The primary disadvantage is the difficulty in choosing the forcing terms $P(\xi,\eta)$ and $Q(\xi,\eta)$ to produce the desired clustering. Generally, one must use trial and error until something that "looks good" is obtained. In addition, large computational times are required to produce the grid, which can be on the same order as the time required to solve the governing equations themselves. For more details on the traditional approaches using algebraic and elliptic grid generation, see Thompson, Warsi, and Mastin (1985) and Liseikin (2010).

12.4 Variational Grid Adaptation

The fundamental difficulty with the traditional methods discussed thus far is that the grid must be chosen or calculated before the solution is known. This requires that a trial-and-error approach be used to determine an appropriate grid on which an accurate numerical solution is sought based on some subjective criteria. It is expected that grids requiring fewer points while maintaining the desired numerical accuracy could be obtained if the grid is chosen based on the actual behavior of the solution, which is called *solution-adaptive grid generation*, in which the grid "adapts" to local features of the solution as it is computed. In most cases, the "optimal" grid is defined in a least squares sense. In addition, other criteria may be desired, such as smoothness or orthogonality of grid lines, when defining a grid on which to carry out a numerical computation. Brackbill and Saltzman (1982) show that such an optimization fits naturally into a variational framework, for which the goal is to optimize the choice of the grid transformation in order to minimize some objective functional. As with the other applications of optimization theory in this text, grid generation can benefit tremendously from the more intuitive and mathematically formal basis that variational methods provide.

It is expected that we can achieve both solution accuracy and algorithm efficiency by optimizing the grid generated based on the solution itself by supplying more grid points in regions were the solution varies rapidly and fewer grid points elsewhere. Although solution-adaptive grid generation is more computationally intensive than algebraic or elliptic grid generation methods, there is less trial and error required to try different grids to determine what provides sufficient accuracy. In other words, one solution computed numerically with grid adaptation is likely to provide a more "optimal" solution than several computations using trial and error without automatic grid adaptation. The basic approach will be motivated and illustrated using one-dimensional settings followed by a brief discussion of how these approaches are extended to two- and three-dimensional scenarios for which their power really becomes evident.

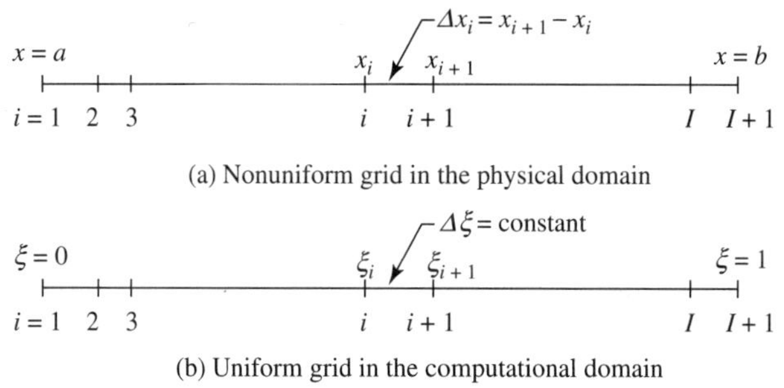

(a) Nonuniform grid in the physical domain

(b) Uniform grid in the computational domain

Figure 12.4. One-dimensional case of variational grid generation.

12.4.1 One-Dimensional Case

We seek to develop functionals for which the stationary function(s) give a transformation in which the grid spacing Δx_i is proportional to a *weight function* $\phi(\xi_{i+1/2})$ evaluated at the midpoint of the interval. This approach is often referred to as *equidistribution* as some quantity is equally distributed among the grid cells. Consider the physical domain $a \le x \le b$ shown in Figure 12.4(a) and the corresponding computational domain[1] $0 \le \xi \le 1$ shown in Figure 12.4(b), where the domain is divided up into I intervals. In the computational domain these are uniform, such that $\Delta\xi$ is a constant, while in the physical domain the grid spacings are not uniform and are to be determined from the stationary function for the functional. Let us begin by considering the following discrete functional

$$S = \sum_{i=1}^{I} \frac{(x_{i+1} - x_i)^2}{2\phi_{i+1/2}} = \sum_{i=1}^{I} \frac{\Delta x_i^2}{2\phi_{i+1/2}},$$

where $x_i = x(\xi_i), x_{i+1} = x(\xi_{i+1})$ and $\phi_{i+1/2} = [\phi(\xi_i) + \phi(\xi_{i+1})]/2$. For a given weight function $\phi(\xi) > 0$, we want to minimize S in the least square sense defined subject to the end conditions $x_1 = a, x_{I+1} = b$. Dividing the above expression by $\Delta\xi$ gives

$$\frac{S}{\Delta\xi} = \sum_{i=1}^{I} \left(\frac{\Delta x_i}{\Delta\xi}\right)^2 \frac{\Delta\xi}{2\phi_{i+1/2}}.$$

Taking $\Delta\xi \to 0$ gives the continuous functional

$$J[x(\xi)] = \frac{1}{2}\int_0^1 \frac{(dx/d\xi)^2}{\phi(\xi)} d\xi = \frac{1}{2}\int_0^1 \frac{x_\xi^2}{\phi} d\xi, \tag{12.14}$$

with the specified end conditions

$$x(0) = a, \quad x(1) = b.$$

This is known as the *weighted-length functional* as the length of each grid interval is proportional to the local weight function ϕ. We seek the grid distribution $x(\xi)$ that minimizes the functional $J[x(\xi)]$.

[1] Some authors refer to the weight functions as *monitor functions* and the computational domain as the *logical domain*.

With the integrand of the functional being

$$F = F(\xi, x, x_\xi) = \frac{1}{2}\frac{x_\xi^2}{\phi},$$

Euler's equation is given by

$$\frac{\partial F}{\partial x} - \frac{d}{d\xi}\left(\frac{\partial F}{\partial x_\xi}\right) = 0$$

$$-\frac{d}{d\xi}\left(\frac{x_\xi}{\phi}\right) = 0 \tag{12.15}$$

$$x_{\xi\xi} - \frac{\phi_\xi}{\phi} x_\xi = 0, \quad (\phi > 0), \tag{12.16}$$

with boundary conditions

$$x(0) = a, \quad x(1) = b.$$

Given a chosen weight function $\phi(\xi)$, we would solve the second-order differential equation (12.16) for the transformation $x(\xi)$ from the uniform grid in the computational domain to the nonuniform grid in the physical domain.

In order to confirm the effect that this transformation has on the grid spacing, let us consider equation (12.15), which is

$$\left(\frac{x_\xi}{\phi}\right)_\xi = 0.$$

Integrating, with integration constant C, and multiplying by $\phi(\xi)$ gives

$$\frac{dx}{d\xi} = C\phi(\xi).$$

Writing this expression in discrete form yields

$$\frac{\Delta x_i}{\Delta \xi} = C\,\phi\left(\frac{\xi_{i+1} + \xi_i}{2}\right).$$

Therefore, we see that this requires that Δx_i be proportional to $\phi_{i+1/2}$, where the proportionality constant is $C\Delta\xi$. In this case, it is x_ξ/ϕ that is being equally distributed throughout the domain.

Note that the weight function $\phi(\xi)$ has been expressed in the computational domain, and it is not necessarily clear how it should be chosen. Conceptually, in solution-adaptive grid generation we prefer to think in terms of *physical weight functions*, say $w(x)$, that express some aspect of the solution in the physical domain x rather than the computational domain ξ. Let us set the grid spacing to be proportional to $[w(x)]^2$ such that

$$\Delta x_i = K\left[w\left(\frac{x_{i+1} + x_i}{2}\right)\right]^2.$$

Therefore, we take $\phi = w^2 > 0$ in equation (12.14) giving the *physical weighted-length functional*

$$J[x(\xi)] = \frac{1}{2}\int_0^1 \frac{x_\xi^2}{w^2(x)}\,d\xi, \tag{12.17}$$

with $x(0) = a, x(1) = b$.

In this case, the integrand of the functional is

$$F = F(\xi, x, x_\xi) = \frac{1}{2}\frac{x_\xi^2}{w^2(x)}.$$

Therefore,

$$\frac{\partial F}{\partial x} = -\frac{w_x}{w^3}x_\xi^2, \quad \frac{\partial F}{\partial x_\xi} = \frac{x_\xi}{w^2},$$

and Euler's equation gives

$$\frac{\partial F}{\partial x} - \frac{d}{d\xi}\left(\frac{\partial F}{\partial x_\xi}\right) = 0$$

$$-\frac{w_x}{w^3}x_\xi^2 - \frac{d}{d\xi}\left(\frac{x_\xi}{w^2}\right) = 0$$

$$\frac{w_x}{w^3}x_\xi^2 + \frac{x_{\xi\xi}}{w^2} - 2\frac{w_x x_\xi}{w^3}x_\xi = 0$$

$$x_{\xi\xi} - \frac{w_x}{w}x_\xi^2 = 0.$$

By the chain rule

$$\frac{d}{dx} = \frac{d\xi}{dx}\frac{d}{d\xi} = \frac{1}{x_\xi}\frac{d}{d\xi} \quad \Rightarrow \quad w_x = \frac{1}{x_\xi}w_\xi.$$

Thus, the Euler equation becomes

$$x_{\xi\xi} - \frac{w_\xi}{w}x_\xi = 0, \quad (w > 0), \tag{12.18}$$

which is of the same form as (12.16), except that now $w(x)$ is a *physical weight function*. The boundary conditions are

$$x(0) = a, \quad x(1) = b.$$

EXAMPLE 12.1 Consider the one-dimensional, steady convection-diffusion equation

$$\begin{gathered} u_{xx} - cu_x = 0, \quad a \leq x \leq b, \\ u(a) = 0, \quad u(b) = 1, \end{gathered} \tag{12.19}$$

where c is the constant convection speed. Increasing c produces an increasingly thin *boundary layer* near $x = b$ in which $u(x)$ changes very rapidly. It is in this boundary layer where we want to focus the grid points in order to resolve the large gradients. Determine the equations to be solved in order to obtain a grid based on a solution-adaptive functional. For comparison, the exact solution to (12.19) is

$$u(x) = \frac{e^{c(x-a)} - 1}{e^{c(b-a)} - 1}. \tag{12.20}$$

Solution: The governing equation (12.19) must be transformed into the computational domain. In one dimension, the transformation laws are

$$\frac{d}{dx} = \frac{d\xi}{dx}\frac{d}{d\xi} = \frac{1}{x_\xi}\frac{d}{d\xi},$$
$$\frac{d^2}{dx^2} = \frac{1}{x_\xi}\frac{d}{d\xi}\left(\frac{1}{x_\xi}\frac{d}{d\xi}\right) = \frac{1}{x_\xi^2}\frac{d^2}{d\xi^2} - \frac{x_{\xi\xi}}{x_\xi^3}\frac{d}{d\xi}.$$

Transforming the governing equation (12.19) to computational coordinates gives

$$\frac{1}{x_\xi^2}\frac{d^2u}{d\xi^2} - \frac{x_{\xi\xi}}{x_\xi^3}\frac{du}{d\xi} - c\frac{1}{x_\xi}\frac{du}{d\xi} = 0,$$

or

$$u_{\xi\xi} - \left(\frac{x_{\xi\xi}}{x_\xi} + cx_\xi\right)u_\xi = 0, \tag{12.21}$$

with boundary conditions

$$u(0) = 0, \quad u(1) = 1.$$

To solve the governing equation (12.21) for the velocity $u(\xi)$ in the computational domain requires the grid function $x(\xi)$. Here we will use the grid equation (12.18) [corresponding to the physical weighted-length functional (12.17)]

$$x_{\xi\xi} - \frac{w_\xi}{w}x_\xi = 0, \tag{12.22}$$

with the boundary conditions $x(0) = a, x(1) = b$.

Now the question arises as to how to choose the weight function $w(x)$. We would like it to be based on the gradient of the dependent variable $u(x)$, such that as the magnitude of the gradient increases, the local grid size decreases to better resolve such regions. For example, consider the possibility

$$w(x) = \frac{1}{\sqrt{u_x^2}},$$

for which the weight function can be in the range $0 < w \leq \infty$. But what if $u_x \to 0$, then $w \to \infty$, which is not acceptable. So let us try

$$w(x) = \frac{1}{\sqrt{1+u_x^2}},$$

in which case $0 < w \leq 1$. Although this works for any value of the gradient u_x, there is no flexibility to adjust the grid resolution as desired. Therefore, it is common to use

$$w(x) = \frac{1}{\sqrt{1+\alpha^2 u_x^2}}, \quad 0 < w \leq 1, \tag{12.23}$$

where α is a parameter chosen to determine how much focusing occurs when the solution gradient becomes large. Observe that when u_x is small, the weight function $w \approx 1$. When u_x is large, on the other hand, $w \approx 1/(\alpha|u_x|)$. Consequently, when the magnitude of the gradient becomes large, the weight function gets smaller in magnitude causing the local grid spacing to reduce as well.

For use in the grid equation (12.22), we need the weight function (12.23) in terms of the computational coordinate ξ, which is

$$w\left(x(\xi)\right) = \frac{1}{\sqrt{1+\frac{\alpha^2}{x_\xi^2}u_\xi^2}}. \tag{12.24}$$

Differentiating yields

$$w_\xi = -\frac{\alpha^2}{2}\frac{\frac{2}{x_\xi^2}u_\xi u_{\xi\xi} - 2\frac{x_{\xi\xi}}{x_\xi^3}u_\xi^2}{\left(1+\frac{\alpha^2}{x_\xi^2}u_\xi^2\right)^{3/2}} = \frac{\alpha^2}{x_\xi^2}\frac{u_\xi\left(\frac{x_{\xi\xi}}{x_\xi}u_\xi - u_{\xi\xi}\right)}{\left(1+\frac{\alpha^2}{x_\xi^2}u_\xi^2\right)^{3/2}}.$$

Therefore, the coefficient in equation (12.22) is

$$\frac{w_\xi}{w} = \alpha^2\frac{u_\xi\left(\frac{x_{\xi\xi}}{x_\xi}u_\xi - u_{\xi\xi}\right)}{x_\xi^2+\alpha^2u_\xi^2}. \tag{12.25}$$

Thus, we have two coupled, second-order, linear ordinary differential equations to solve for the grid $x(\xi)$ and velocity distribution $u(\xi)$. The velocity in the physical domain is then obtained by applying the inverse transformation to obtain $u(x)$. Observe that a uniform grid in the physical domain results if $\alpha = 0$ ($w = 1, w_\xi = 0$), that is, $x(\xi) = (b-a)\xi + a$. Thus, increasing α increases the influence of the transformation in regions with large gradients. Note that all of the grid transformation information is encapsulated in the coefficient of the first-order derivative in the governing equation (12.21). Likewise, all the weight function information is contained in the coefficient of the first-order derivative in the grid equation (12.22).

Let us consider numerical solutions of equations (12.21) and (12.22) with (12.25) for various convection speeds c, weight parameters α, and number of grid intervals I. In all cases, we define the physical domain with $a = 0$ and $b = 10$. For reference, let us first plot the solution for the case with $\alpha = 0$, in which case the grid is uniform in both the computational and physical domains. Because we know the exact solution, we can compare our numerical solutions and compute the error of our numerical solution. Figure 12.5 shows the numerical and exact solutions, the error, and the inverse grid transformation $x(\xi)$. Observe that the maximum error using a uniform grid occurs near $x \approx 9$, and that the transformation $x(\xi)$ for a uniform grid is simply a straight line given by $x(\xi) = (b-a)\xi + a$. That is, the grid sizes in the physical domain are simply ten times those in the computational domain.

Now let us consider the case with $\alpha = 7$ leaving all other parameters the same. Figure 12.6 shows the results for the grid transformation [the solution for the velocity $u(x)$ is the same as shown in Figure 12.5(a)]. Comparing Figure 12.6(a) with Figure 12.5(a), we see that the change in the solution is more gradual with the grid transformation than without. This would be expected to reduce the error of the numerical solution. Indeed, the maximum error in the numerical solution for this case is $E_{max} = 0.0008838$. By comparison, the maximum error using a uniform grid with the same number of points is $E_{max} = 0.001231$. Thus, there is a reduction of approximately thirty percent in the error with $\alpha = 7$. The plot of the inverse transformation $x(\xi)$ in the computational domain shown in Figure 12.6 indicates that

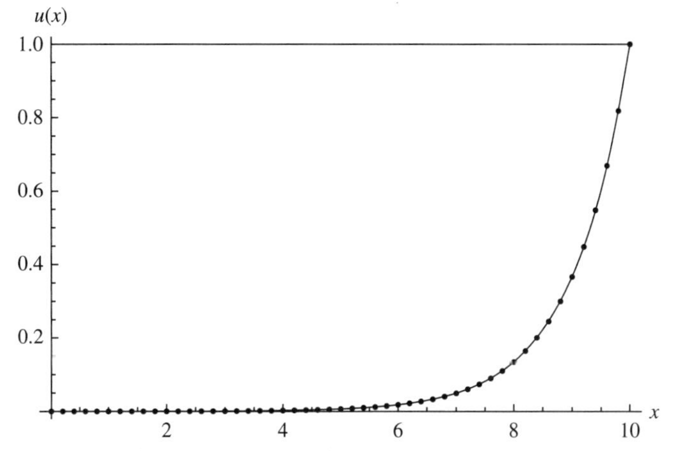

(a) Numerical solution (dots) and exact solution (solid line) for $u(x)$ in physical domain.

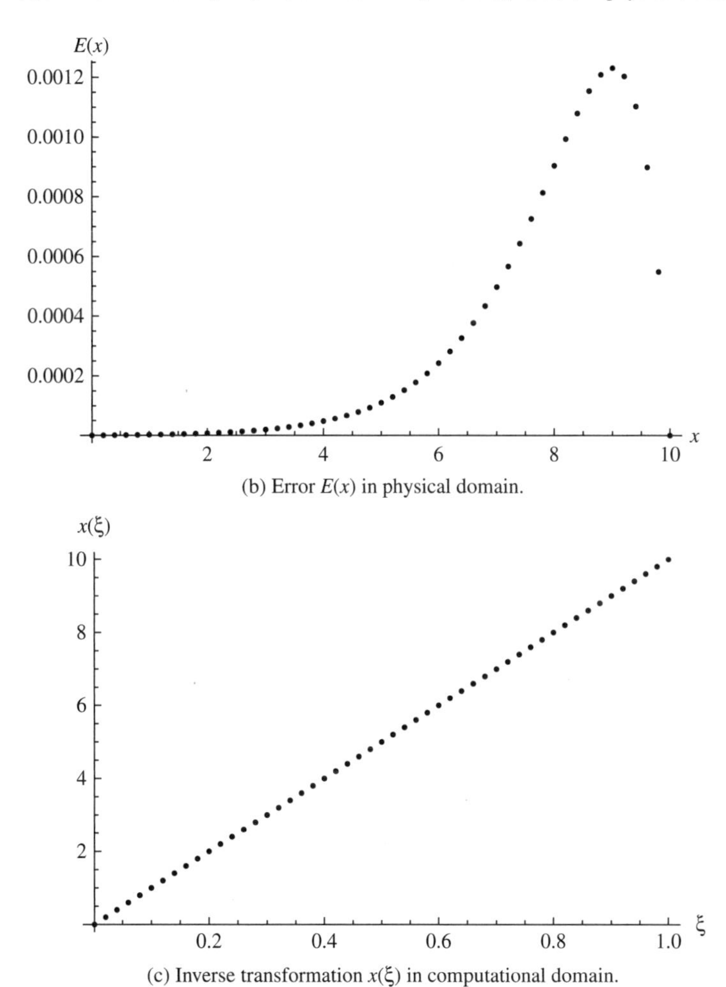

(b) Error $E(x)$ in physical domain.

(c) Inverse transformation $x(\xi)$ in computational domain.

Figure 12.5. Solution of the governing equation and grid equation for the case with $c = 1$, $\alpha = 0, I = 50$.

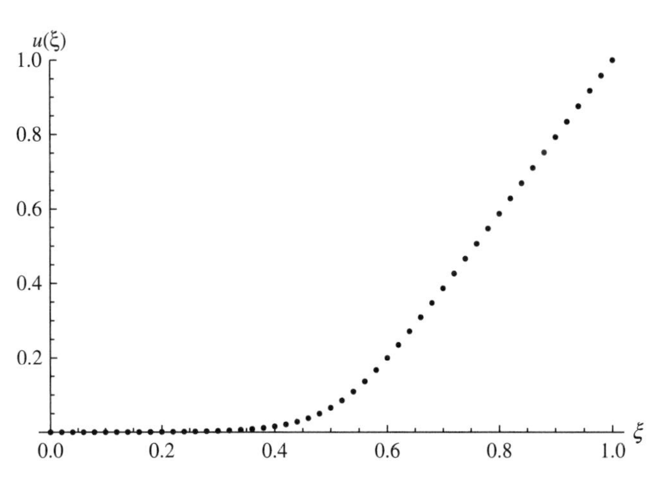

(a) Numerical solution of $u(\xi)$ in computational domain.

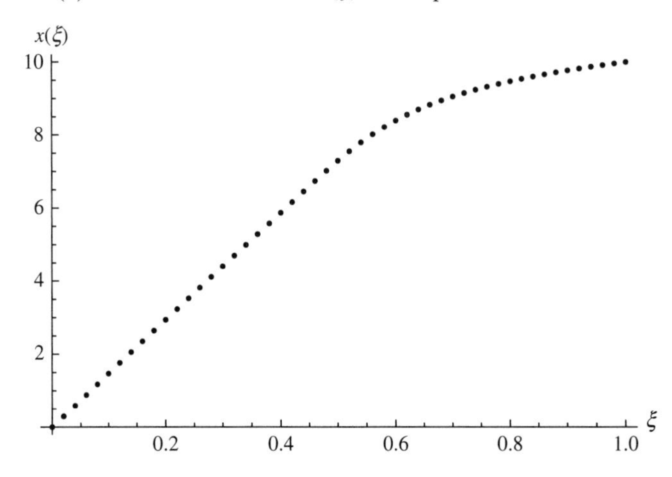

(b) Inverse transformation $x(\xi)$ in computational domain.

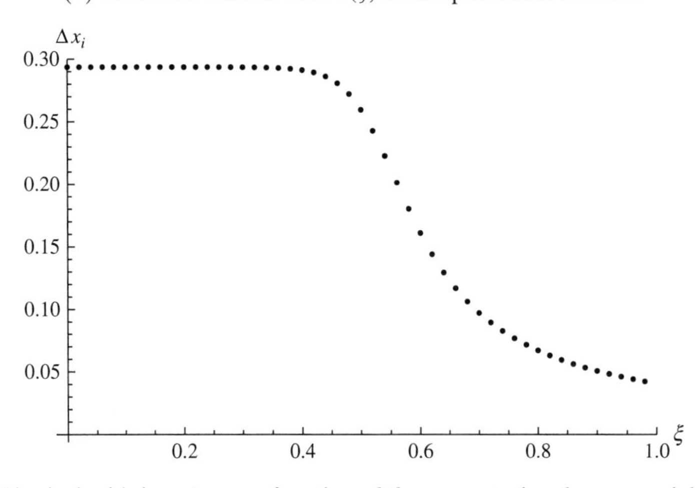

(c) Physical grid sizes Δx_i as a function of the computational space variable ξ.

Figure 12.6. Solution of the governing equation and grid equation for the case with $c = 1$, $\alpha = 7, I = 50$.

Table 12.1. *Grid intervals I required to produce a maximum error of $E_{max} = 10^{-3}$ for a given c and α.*

c	α	I
5	0	280
5	5	150
5	7	145
5	10	150
10	0	560
10	5	225
10	7	220
10	10	225

the uniformly spaced grid points in the computational domain have shifted toward the $x = b$ boundary in the physical domain (compare with the straight line for the case without the transformation). The corresponding grid intervals Δx_i are shown in Figure 12.6(c). Observe that the grid sizes are nearly constant in the left portion of the domain where the solution changes very little. The grid sizes then decrease rapidly toward a minimum at the right boundary.

As the convection speed c is increased, the region over which the solution adjusts from $u = 0$ to $u = 1$ at the right boundary decreases causing the solution gradients to increase. A useful way to envision the effect of such an increase in gradients is to determine the minimum number of grid intervals required to achieve a certain error level. To examine this, we consider two cases with $c = 5$ and $c = 10$ and determine the number of grid intervals required to produce a maximum error of $E_{max} = 10^{-3}$ for a given weight parameter α. We can then determine the "best" value of α for a given c. These results are shown in Table 12.1. Observe that increasing c requires additional resolution (for the same value of α) in order to reduce the maximum error below the designated threshold, and that the minimum resolution to achieve this threshold corresponds to $\alpha = 7$ for both values of c. In addition, the number of points required is not very sensitive to the value of α as the number of points required only differs by 5 for $\alpha = 5$ and 10. Note also that the reduction in grid intervals from the uniform grid ($\alpha = 0$) to the optimal value of $\alpha = 7$ is approximately fifty percent and sixty percent for the $c = 5$ and $c = 10$ cases, respectively. While such a decrease is respectable, remember that to achieve this reduction requires twice as much computation as simply solving the governing equation on a uniform grid.

The approach used here with the weighted-length functional and weight function related to the gradient of the solution through equation (12.23) leaves significant room for improvement. To see this, observe that the minimum grid spacings do not occur at the location in the domain where the maximum errors occur. Whereas the maximum errors occur between $8 < x < 9$, the finest grid resolution occurs adjacent to $x = b = 10$ where the gradient is the largest. Therefore, this approach does not result in an optimal grid. The "best" grid for this problem would result in the solution in the computational domain being a straight line between the two boundary conditions such that the changes in solution are spread evenly across the domain. As it only takes two points to define such a line, the resolution required for essentially zero error (other than that introduced by the discrete mapping) is only one grid interval! That is, we should seek the transformation that leads to the straight-line solution for $u(\xi)$ in the computational domain [compare with Figure 12.6(a)]. Obviously, such an extreme reduction in resolution is not to be expected in problems of a general nature, including multiple dimensions; however, it does hint at a more optimal approach that will be discussed at the end of this chapter.

12.4.2 Two-Dimensional Case

Given the motivation from the one-dimensional case, let us now consider the more applicable two-dimensional case with straightforward extension to three dimensions. The two-dimensional analog to the one-dimensional *weighted-length functional* (12.14) is

$$J_L[x,y] = \frac{1}{2}\iint_A \left[\frac{x_\xi^2 + y_\xi^2}{\phi(\xi,\eta)} + \frac{x_\eta^2 + y_\eta^2}{\psi(\xi,\eta)}\right] d\xi\, d\eta,$$

where we now have two weight functions $\phi(\xi,\eta) > 0$ and $\psi(\xi,\eta) > 0$. This produces a grid for which the lengths of the coordinate lines are proportional to the weight functions, that is,

$$\sqrt{g_{11}} = \sqrt{x_\xi^2 + y_\xi^2} = K_1\phi(\xi,\eta),$$

$$\sqrt{g_{22}} = \sqrt{x_\eta^2 + y_\eta^2} = K_2\psi(\xi,\eta),$$

where K_1 and K_2 are proportionality constants. Recall that g_{11} and g_{22} are defined in Section 12.3. The resulting Euler equations are then

$$\left(\frac{x_\xi}{\phi}\right)_\xi + \left(\frac{x_\eta}{\psi}\right)_\eta = 0, \quad \left(\frac{y_\xi}{\phi}\right)_\xi + \left(\frac{y_\eta}{\psi}\right)_\eta = 0.$$

Thus, we have two coupled, typically nonlinear, elliptic partial differential equations to solve for the grid functions $x(\xi,\eta)$ and $y(\xi,\eta)$. For example, if the weight functions are constant with $\phi = \psi = c$, then the Euler equations are the linear Laplace equations

$$x_{\xi\xi} + x_{\eta\eta} = 0, \quad y_{\xi\xi} + y_{\eta\eta} = 0,$$

which is comparable to elliptic grid generation with no grid focusing. The weight functions ϕ and ψ control the interior grid locations in a much more formal manner than the P and Q forcing functions in elliptic grid generation. That is, the local grid spacing is directly proportional to the specified weight functions, which may be related directly to the physical solution.

Another possible objective functional is the *area functional*, for which the area of each cell, which is proportional to the Jacobian $\mathcal{J} = \sqrt{g}$, is proportional to a weight function $\phi(\xi,\eta) > 0$. The area functional is given by

$$J_A[x,y] = \frac{1}{2}\iint_A \frac{\mathcal{J}^2}{\phi}\, d\xi\, d\eta,$$

and leads to the coupled, nonlinear Euler equations

$$\left(\frac{\mathcal{J}\,x_\eta}{\phi}\right)_\xi - \left(\frac{\mathcal{J}\,x_\xi}{\phi}\right)_\eta = 0, \quad \left(\frac{\mathcal{J}\,y_\eta}{\phi}\right)_\xi - \left(\frac{\mathcal{J}\,y_\xi}{\phi}\right)_\eta = 0.$$

Rather than designating a weight function to which some aspect of the grid is to be proportional, another approach is to minimize some quantity, for example, a measure of the smoothness or orthogonality of the grid lines. For example, we consider the *orthogonality functional*. The grid is orthogonal if $g_{12} = x_\xi x_\eta + y_\xi y_\eta = 0$; therefore, the orthogonality functional is

$$J_O[x,y] = \frac{1}{2}\iint_A g_{12}^2 d\xi\, d\eta,$$

such that the magnitude of g_{12} is minimized. The nonlinear Euler equations are given by

$$\left(g_{12}x_\eta\right)_\xi + \left(g_{12}x_\xi\right)_\eta = 0, \quad \left(g_{12}y_\eta\right)_\xi + \left(g_{12}y_\xi\right)_\eta = 0.$$

Orthogonality is desirable because: 1) the transformed equation in the computational domain is generally less complicated, 2) boundary conditions are applied naturally, and 3) it is believed that orthogonal grids reduce the growth of truncation errors during the numerical solution. The orthogonality functional is difficult to apply in three dimensions.

The use of any one of the above functionals alone often produces unsatisfactory results in practical domains. To improve on this situation, one possibility is to combine multiple functionals to produce *combination functionals* that enforce complementary objectives in some average sense. For example, consider the *area-orthogonality functional*, which is

$$J_{AO}[x,y] = \frac{1}{2}\iint_A \frac{\mathcal{J}^2 + g_{12}^2}{\phi} d\xi\, d\eta = \frac{1}{2}\iint_A \frac{g_{11}g_{22}}{\phi} d\xi\, d\eta,$$

where again the fact that $\mathcal{J}^2 = g = g_{11}g_{22} - g_{12}^2$ has been used. This results in a grid such that the average of the area and orthogonality functionals is minimized. One criterion could be emphasized over the other by including constant weight coefficients in the respective terms in the functional as in Chapter 10. The Euler equations for the area-orthogonality functional are

$$\left(\frac{g_{22}x_\xi}{\phi}\right)_\xi + \left(\frac{g_{11}x_\eta}{\phi}\right)_\eta = 0, \quad \left(\frac{g_{22}y_\xi}{\phi}\right)_\xi + \left(\frac{g_{11}y_\eta}{\phi}\right)_\eta = 0.$$

The area-orthogonality functional generally produces satisfactory grids that are smooth and nearly orthogonal.

All of the two-dimensional functionals above have been written in the form

$$J[x,y] = \frac{1}{2}\iint_A F(\xi,\eta,x,y,x_\xi,y_\xi,x_\eta,y_\eta)d\xi\, d\eta$$

in order to determine the mappings $x(\xi,\eta)$ and $y(\xi,\eta)$. The problem with this so-called *covariant* approach is that generally there is not a straightforward interpretation of the weight functions $\phi(\xi,\eta)$ and $\psi(\xi,\eta)$ in the computational domain. Just as in the one-dimensional case, it is generally preferable to define weight functions in the physical domain, $w(x,y)$, rather than in the computational domain, $\phi(\xi,\eta)$. Using this so called *contravariant* approach, functionals may be written to determine $\xi(x,y)$ and $\eta(x,y)$ in the form

$$J[\xi,\eta] = \frac{1}{2}\iint_A F(x,y,\xi,\eta,\xi_x,\eta_x,\xi_y,\eta_y)dxdy.$$

These are called *contravariant functionals* and are discussed in detail in Knupp and Steinberg (1994).

Contravariant functionals are typically composed in the general form

$$J[\xi,\eta] = \frac{1}{2}\iint_A \left(\nabla\xi^T \mathbf{G}_1^{-1}\nabla\xi + \nabla\eta^T \mathbf{G}_2^{-1}\nabla\eta\right) dxdy, \tag{12.26}$$

where $\mathbf{G}_1(x,y)$ and $\mathbf{G}_2(x,y)$ are symmetric positive definite matrices containing the weight, or monitor, functions. The eigenvalues and eigenvectors of $\mathbf{G}_1$ and $\mathbf{G}_2$ give the amount and direction, respectively, of stretching enforced by the weight functions. Physical weight functions used in the context of contravariant functionals generally result in simpler Euler equations than those that result from the covariant approach. The Euler equations for the general functional (12.26) are of the form

$$\nabla\cdot\left(\mathbf{G}_1^{-1}\nabla\xi\right) = 0, \quad \nabla\cdot\left(\mathbf{G}_2^{-1}\nabla\eta\right) = 0.$$

For example, if $\mathbf{G}_1 = \mathbf{G}_2 = w\mathbf{I}$, where $\mathbf{I}$ is the identity matrix and $w(x,y)$ is the physical weight function, then the functional (12.26) becomes

$$J[\xi,\eta] = \frac{1}{2}\iint_A \frac{1}{w}\left[(\nabla\xi)^2 + (\nabla\eta)^2\right] dxdy = \frac{1}{2}\iint_A \frac{1}{w}\left(\xi_x^2 + \xi_y^2 + \eta_x^2 + \eta_y^2\right) dxdy,$$

and the corresponding Euler equations are

$$\nabla\cdot\left(\frac{1}{w}\nabla\xi\right) = 0, \quad \nabla\cdot\left(\frac{1}{w}\nabla\eta\right) = 0.$$

This corresponds to Winslow's variable diffusion method (Winslow 1967) originally formulated using an elliptic grid generation approach.

REMARKS:

1. The number of coupled Euler equations to be solved for the grid is equal to the number of spatial dimensions of the problem.
2. Variational grid generation requires a great deal of computational overhead to solve the Euler equations for the grid distribution. Therefore, a significant reduction in total grid points over a comparable uniform grid, for example, would be necessary to justify this additional computational cost. In other words, the number of grid points required to achieve a desired level of accuracy must be reduced sufficiently to offset the additional computational cost of computing the grid equations.

3. Additional terms are required in the governing equations for unsteady, time-dependent problems owing to movement of the grid in the physical domain [the computational grid (ξ,η) remains fixed and uniform], that is,

$$\xi = \xi(x,y,t), \quad \eta = \eta(x,y,t).$$

4. For additional details on variational grid generation, see Knupp and Steinberg (1994), Liseikin (2010), and Huang and Russell (2011).

Although the variational framework presented here provides a flexible and formal means to optimize grid generation in a wide variety of circumstances, one issue still remains somewhat unsatisfactory, namely the choice of the physical weight function, which is somewhat arbitrary. Thus far the weight functions have been chosen heuristically based on the notion that we want to focus grid points where the gradients in the solution are large. As such, we have shown how to obtain the optimal grid for a given weight function, but not what the optimal choice for the weight function is. This is an active area of research where the most promising approach appears to be based on linking the grid optimization directly to some measure of the local error in the numerical solution.

By targeting the numerical error directly, one is getting at the true objective of grid generation, which is to devise the grid on which the governing equations can be solved to the desired level of accuracy with the minimum number of grid points. Doing so will reduce the computational cost of obtaining the numerical solution. Therefore, a method that provides for the equidistribution of some measure of the solution error is sought. Various measures of the solution error exist, including the truncation error of the numerical algorithm, the residual, and the interpolation error.

In traditional solution-adaptive approaches, the grid is determined only by the behavior of the solution. By choosing the grid such that the truncation error of the numerical scheme is minimized, the grid would be determined both by the solution and the numerical discretization method. One would expect that this should provide a definitive and quantitative measure of the optimal grid. As such, it could be argued that this approach does indeed provide the true optimum for an adaptive grid generation problem. However, the fact that it requires incorporation of the truncation error of the numerical scheme applied to the differential equation means that any changes to the governing equation and/or numerical discretization will result in a different functional to be minimized and corresponding Euler equations to be solved. Generally, this is not desirable because one would then need to ensure uniqueness of the grid solutions and potentially modify the numerical algorithm to solve the various forms of the differential Euler equations that may arise. For two approaches to grid adaptation based on minimization of the truncation error, see Yamaleev (2001) and Lapenta (2003).

Additional areas of ongoing research in variational grid generation include: 1) application to unsteady, time-dependent problems having moving meshes that are adapting to the evolving solution, 2) treatment of moving-boundary problems, 3) grid generation applied to coupled equations having multiple dependent variables,

4) efficient extension to three-dimensional problems, 5) application to unstructured grids, and 6) constrained minimization approaches using Lagrange multipliers. For an example of the latter, see Delzanno, Chacón, Finn, Chung, and Lapenta (2008), who propose a functional that minimizes some measure of the grid quality subject to the constraint that a local equidistribution principle is enforced.

Bibliography

Abergel, F. and Temam, R. (1990). On some control problems in fluid mechanics. *Theoretical and Computational Fluid Dynamics* **1**, 303–325.
Akhiezer, N. I. (1962). *The calculus of variations*, Blaisdell Publishing Company.
Anderson, D. M., McFadden, G. B., and Wheeler, A. A. (1998). Diffuse-interface methods in fluid mechanics. *Annual Review of Fluid Mechanics* **30**, 139–165.
Anderson, D. M., McFadden, G. B., and Wheeler, A. A. (2000). A phase-field model of solidification with convection. *Physica D* **135**, 175–194.
Arpaci, V. S. (1966). *Conduction heat transfer*, Addison-Wesley Publishing Co.
Aubert, G. and Kornprobst, P. (2010). *Mathematical problems in image processing. Partial differential equations and the calculus of variations*, 2nd edition, Springer Verlag.
Basdevant, J. L. (2007). *Variational principles in physics*, Springer.
Batchelor, G. K. (1967). *Fluid dynamics*, Cambridge University Press.
Berdichevsky, V. L. (2009a). *Variational principles of continuum mechanics 1: fundamentals*, Springer.
Berdichevsky, V. L. (2009b). *Variational principles of continuum mechanics 2: applications*, Springer.
Bewley, T. R. (2001). Flow control: new challenges for a new renaissance. *Progress in Aerospace Sciences* **37**, 21–58.
Bewley, T. R., Moin, P., and Temam, R. (2001). DNS-based predictive control of turbulence: an optimal benchmark for feedback algorithms. *Journal of Fluid Mechanics* **447**, 179–225.
Bliss, G. A. (1946). *Lectures on the calculus of variations*, University of Chicago Press.
Bloch, A. M. with the collaboration of Baillieul, J., Crouch, P., and Marsden, J. E. (2003). *Nonholonomic mechanics and control*, Springer.
Bloch, A. M., Fernandez, O. E., and Mestdag, T. (2009). Hamiltonization of nonholonomic systems and the inverse problem of the calculus of variations. *Reports on Mathematical Physics* **63**, 225–249.
Bolza, O. (1904). *Lectures on the calculus of variations*, University of Chicago Press. (Dover 1961).
Brackbill, J. U. and Saltzman, J. S. (1982). Adaptive zoning for singular problems in two dimensions. *Journal of Computational Physics* **46**, 342–368.
Brebbia, C. A. and Dominguez, J. (1992). *Boundary elements – an introductory course*, McGraw-Hill Book Company.
Bryson, A. E. and Ho, Y. C. (1975). *Applied optimal control*, Hemisphere Publishing.
Burger, M. and Osher, S. J. (2005). A survey on level set methods for inverse problems and optimal design. *European Journal of Applied Mathematics* **16**, 263–301.
Butler, K. M. and Farrell, B. F. (1992). Three-dimensional optimal perturbations in viscous shear flow. *Physics of Fluids A* **4**, 1637–1650.
Butt, H. J., Graf, K., and Kappl, M. (2004). *Physics and chemistry of interfaces*, John Wiley and Sons.

Chan, T. F. and Shen, J. (2005). *Image processing and analysis: variational, pde, wavelet, and stochastic methods*, SIAM.
Chen, L. Q. (2002). Phase-field models for microstructure evolution. *Annual Review of Materials Research* **32**, 113–140.
Chomaz, J. M. (2005). Global instabilities in spatially developing flows: non-normality and nonlinearity. *Annual Review of Fluid Mechanics* **37**, 357–392.
Chung, T. J. (1978). *Finite element analysis in fluid dynamics*, McGraw-Hill, Inc.
Chung, T. J. (2010). *Computational fluid dynamics*, Cambridge University Press.
Citron, S. J. (1969). *Elements of optimal control*, Holt, Rinehart and Winston.
Collis, S. S., Joslin, R. D., Seifert, A., and Theofilis, V. (2004). Issues in active flow control: theory, control, simulation, and experiment. *Progress in Aerospace Sciences* **40**, 237–289.
Courant, R. and Hilbert, D. (1953). *Methods of mathematical physics*, Interscience Publishers.
Dafermos, C. M. and Pokorny, M. (2008). *Handbook of differential equations: evolutionary equations, Volume 4*, North-Holland.
Daley, R. (1991). *Atmospheric data analysis*, Cambridge University Press.
de Boor, C. R. (2001). *A practical guide to splines: revised edition*, Springer.
Delzanno, G. L., Chacón, L., Finn, J. M., Chung, Y., and Lapenta, G. (2008). An optimal robust equidistribution method for two-dimensional grid adaptation based on Monge-Kantorovich optimization. *Journal of Computational Physics* **227**, 9841–9864.
Den Hartog, J. P. (1952). *Advanced strength of materials*, McGraw-Hill. (Dover 1987)
Denn, M. M. (1969). *Optimization by variational methods*, McGraw-Hill.
Dierckx, P. (1993). *Curve and surface fitting with splines*, Oxford University Press.
Dirac, P. A. M. (1958). *The principles of quantum mechanics*, 4th edition, Oxford University Press.
Drazin, P. G. (2002). *Introduction to hydrodynamic stability*, Cambridge University Press.
Drazin, P. G. and Reid, W. H. (1981). *Hydrodynamic stability*, Cambridge University Press.
Drazin, P. G. and Riley, N. (2006). *The Navier-Stokes equations: a classification of flows and exact solutions*, Cambridge University Press.
Dreizler, R. M. and Gross, E. K. U. (1990). *Density functional theory*, Springer-Verlag.
Eisenberg, B., Hyon, Y., and Liu, C. (2010). Energy variational analysis of ions in water and channels: field theory for primitive models of complex ionic fluids. *The Journal of Chemical Physics* **133**, 104104-1–23.
Farrell, B. F. (1988). Optimal excitation of perturbations in viscous shear flow. *Physics of Fluids* **31**, 2093–2102.
Farrell, B. F. and Ioannou, P. J. (1996). Generalized stability theory. Part 1: autonomous operators. *Journal of the Atmospheric Sciences* **53**, 2025–2040.
Ferziger, J. H. and Peric, M. (1996). *Computational methods for fluid dynamics*, Springer-Verlag.
Feynman, R. P. and Hibbs, A. R. (1965). *Quantum mechanics and path integrals*, McGraw-Hill.
Fish, J. and Belytschko, T. (2007). *A first course in finite elements*, John Wiley and Sons.
Fletcher, C. A. J. (1991a). *Computational techniques for fluid dynamics 1: fundamental and general techniques*, Springer-Verlag.
Fletcher, C. A. J. (1991b). *Computational techniques for fluid dynamics 2: specific techniques for different flow categories*, Springer-Verlag.
Forsyth, A. R. (1926). *Calculus of variations*, Cambridge University Press. (Dover 1960).
Gelfand, I. M. and Fomin, S. V. (1963). *Calculus of variations*, Prentice-Hall. (Dover 2000).
Goldstein, H. (1959). *Classical mechanics*, Addison-Wesley Publishing.
Greenberg, M. D. (1998). *Advanced engineering mathematics*, 2nd edition, Prentice Hall.
Gunzburger, M. D. (2003). *Perspectives in flow control and optimization.* SIAM.
Gunzburger, M. D. and Manservisi, S. (1999). The velocity tracking problem for Navier-Stokes flows with bounded distributed controls. *SIAM Journal of Control Optimization* **37**, 1913–1945.
Gunzburger, M. D. and Manservisi, S. (2000). The velocity tracking problem for Navier-Stokes flows with boundary control. *SIAM Journal of Control Optimization* **39**, 594–634.

Hildebrand, F. B. (1965). *Methods of applied mathematics*, 2nd edition, Prentice-Hall. (Dover 1992).
Hildebrand, F. B. (1976). *Advanced calculus for applications*, Prentice-Hall.
Hinch, E. J. (1991). *Perturbation methods*, Cambridge University Press.
Holmes, P., Lumley, J. L., and Berkooz, G. (1996). *Turbulence, coherent structures, dynamical systems, and symmetry*, Cambridge University Press.
Huang, W. and Russell, R. D. (2011). *Adaptive moving mesh methods*, Springer.
Jameson, A., Martinelli, L., and Pierce, N. A. (1998). Optimum aerodynamic design using the Navier-Stokes equations. *Theoretical and Computational Fluid Dynamics* **10**, 213–237.
Jeffrey, A. (2002). *Advanced engineering mathematics*, Harcourt/Academic Press.
José, J. V. and Saletan, E. J. (1998). *Classical dynamics: a contemporary approach*, Cambridge University Press.
Joseph, D. D. (1976). *Stability of fluid motions I*, Springer-Verlag.
Joseph, D. D. and Carmi, S. (1969). Stability of Poiseuille flow in pipes, annuli and channels. *Quarterly of Applied Mathematics* **26**, 575–599.
Kalnay, E. (2003). *Atmospheric modeling, data assimilation and predictability*, Cambridge University Press.
Kim, J. and Bewley, T. R. (2007). A linear systems approach to flow control. *Annual Review of Fluid Mechanics* **39**, 383–417.
Kirk, D. E. (1970). *Optimal control theory: an introduction*, Prentice-Hall, Inc. (Dover 1998).
Knupp, P. and Steinberg, S. (1994). *Fundamentals of grid generation*, CRC Press.
Koch, W. and Holthausen, M. C. (2001). *A chemist's guide to density functional theory*, Wiley.
Komzsik, L. (2009). *Applied calculus of variations for engineers*, CRC Press.
Kreyszig (2011). *Advanced engineering mathematics*, 10th edition, John Wiley and Sons.
Lanczos, C. (1970). *The variational principles of mechanics*, 4th edition, University of Toronto Press. (Dover 1986).
Lapenta, G. (2003). Variational grid adaptation based on the minimization of local truncation error: time-independent problems. *Journal of Computational Physics* **193**, 159–179.
Leitmann, G. (1966). *An introduction to optimal control*, McGraw-Hill.
Li, X. and Yong, J. (1995). Optimal control theory for infinite dimensional systems, Birkhäuser.
Lions, J. L. (1971). *Optimal control of systems governed by partial differential equations*, Springer-Verlag.
Liseikin, V. D. (2010). *Grid generation methods*, 2nd edition, Springer.
Love, A. E. H. (1944). *A treatise on the mathematical theory of elasticity*, 4th edition, Dover.
Luenberger, D. G. (1969). *Optimization by vector space methods*, John Wiley and Sons.
MacCluer, C. R. (2005). *Calculus of variations: mechanics, control, and other applications*, Pearson Education.
Min, C. and Choi, H. (1999). Suboptimal feedback control of vortex shedding at low Reynolds numbers. *Journal of Fluid Mechanics* **401**, 123–156.
Mitichie, A. and Ben Ayed, I. (2011). *Variational and level set methods in image segmentation*, Springer.
Mohammadi, B. and Pironneau, O. (2004). Shape optimization in fluid mechanics. *Annual Review of Fluid Mechanics* **36**, 11.1–11.25.
Moiseiwitsch, B. L. (1966). *Variational principles*, John Wiley and Sons.
Moreton, H. P. (1992). *Minimum curvature variation curves, networks, and surfaces for fair free-form shape design*, PhD Thesis, University of California at Berkeley.
Morse, P. M. and Feshbach, H. (1953). *Methods of theoretical physics*, McGraw-Hill Book Company.
Mumford, D. and Shah, J. (1989). Optimal approximations by piecewise smooth functions and associated variational problems. *Communications on Pure and Applied Mathematics* **42**, 577–685.
Mura, T. and Koya, T. (1992). *Variational methods in mechanics*, Oxford University Press.
Nair, S. (2009). *Introduction to continuum mechanics*, Cambridge University Press.
Nair, S. (2011). *Advanced topics in applied mathematics*, Cambridge University Press.

Nesbet, R. K. (2003). *Variational principles and methods in theoretical physics and chemistry*, Cambridge University Press.

Neuenschwander, D. E. (2011). *Emmy Noether's wonderful theorem*, Johns Hopkins University Press.

Noether, E. (1918). Invariante variationsprobleme. *Nachr. d. König. Gesellsch. d. Wiss. zu Göttingen, Math-phys. Klasse*, 235–257. English translation by Tavel, M. A. (1971). Invariant variation problems. *Transport Theory in Statistical Mechanics* **1**, 183–207.

Oden, J. T. and Reddy, J. N. (1983). *Variational methods in theoretical mechanics*, 2nd edition, Springer-Verlag.

O'Neil, P. V. (2012). *Advanced engineering mathematics*, 7th edition, Cengage Learning.

Orszag, S. A. (1971). Accurate solution of the Orr-Sommerfeld stability equation. *Journal of Fluid Mechanics* **50**, 689–703.

Osher, S. J. and Fedkiw, R. P. (2003). *Level set methods and dynamic implicit surfaces*, Springer.

Osher, S. J. and Sethian, J. A. (1988). Fronts propagating with curvature dependent speed: algorithms based on Hamilton-Jacobi formulations. *Journal of Computational Physics* **79**, 12–49.

Palacios, A. F. (2006). *The Hamiltonian-type principle in fluid dynamics – fundamentals and applications to magnetohydrodynamics, thermodynamics, and astrophysics*, Spring-Verlag.

Panton, R. L. (2005). *Incompressible flow*, 3rd edition, John Wiley and Sons.

Parr, R. G. and Yang, W. (1989). *Density-functional theory of atoms and molecules*, Oxford University Press.

Pironneau, O. (1974). On optimum design in fluid mechanics. *Journal of Fluid Mechanics* **64**, 97–110.

Pironneau, O. (1984). *Optimal shape design for elliptic systems*, Spring-Verlag.

Pontryagin, L. S., Boltyanskii, V. G., Gamkrelidze, R. V., and Mishchenko, E. F. (1962). *The mathematical theory of optimal processes*, John Wiley and Sons.

Pruessner, G. and Sutton, A. P. (2008). Phase-field model of interfaces in single-component systems derived from classical density functional theory. *Physical Review B* **77**, 054101.

Reddy, J. N. (2002). *Energy principles and variational methods in applied mechanics*, 2nd edition, John Wiley and Sons.

Reddy, J. N. (1986). *Applied functional analysis and variational methods*, McGraw-Hill Book Company.

Reddy, S. C. and Henningson, D. S. (1993). Energy growth in viscous channel flows. *Journal of Fluid Mechanics* **252**, 209–238.

Saffman, P. G. (1992). *Vortex dynamics*, Cambridge University Press.

Sagan, H. (1961). *Boundary and eigenvalue problems in mathematical physics*, John Wiley and Sons.

Sagan, H. (1969). *Introduction to the calculus of variations*, McGraw-Hill Book Company. (Dover 1993).

Salmon, R. (1988). Hamiltonian fluid mechanics. *Annual Review of Fluid Mechanics* **20**, 225–256.

Schmid, P. J. (2007). Nonmodal stability theory. *Annual Review of Fluid Mechanics* **39**, 129–162.

Schmid, P. J. and Henningson, D. S. (2001). *Stability and transition in shear flows*, Springer-Verlag.

Schulman, L. S. (1981). *Techniques and applications of path integration*, John Wiley and Sons.

Seierstad, A. and Sydsaeter, K. (1987). *Optimal control theory with economic applications*, Elsevier Science.

Seliger, R. L. and Whitham, G. B. (1968). Variational principles in continuum mechanics. *Proceedings of the Royal Society A* **305**, 1–25.

Serrin, J. (1959). On the stability of viscous fluid motions. *Archive for Rational Mechanics and Analysis* **3**, 1–13.

Sethian, J. A. and Smereka, P. (2003). Level set methods in fluid interfaces. *Annual Review of Fluid Mechanics* **35**, 341–372.

Sethian, J. A. (1999). *Level set methods and fast marching methods: evolving interfaces in computational geometry, fluid mechanics, computer vision, and materials science*, Cambridge University Press.

Singer-Loginova, I. and Singer, H. M. (2008). The phase field technique for modeling multiphase materials. *Reports on Progress in Physics* **71**, 106501.

Smith, D. R. (1974). *Variational methods in optimization*, Prentice-Hall, Inc. (Dover 1998).

Sokolnikoff, I. S. (1956). *Mathematical theory of elasticity*, Krieger Publishing Company.

Stengel, R. F. (1986). *Stochastic optimal control: theory and application*, John Wiley and Sons. (Dover 1994 *Optimal control and estimation*).

Sydsaeter, K. and Hammond, P. (2012). *Essential mathematics for economic analysis*, 4th edition, Prentice Hall.

Sydsaeter, K., Hammond, P., Seierstad, A., and Strom, A. (2008). *Further mathematics for economic analysis*, 2nd edition, Prentice Hall.

Tannehill, J. C., Anderson, D. A., and Pletcher, R. H. (1997). *Computational fluid mechanics and heat transfer*, Taylor and Francis.

Theofilis, V. (2003). Advances in global linear instability analysis of nonparallel and three-dimensional flows. *Progress in Aerospace Sciences* **39**, 249–315.

Theofilis, V. (2011). Global linear instability. *Annual Review of Fluid Mechanics* **43**, 319–352.

Thompson, J. F., Warsi, Z. U. A., and Mastin, C. W. (1985). *Numerical grid generation – foundation and applications*, North Holland.

Timoshenko, S. P. and Goodier, J. N. (1934). *Theory of elasticity*, McGraw-Hill Book Company.

Tröltzsch, F. (2010). *Optimal control of partial differential equations: theory, methods, and applications*, American Mathematical Society.

Tse, F. S., Morse, I. E., and Hinkle, R. T. (1978). *Mechanical vibrations – theory and applications*, Allyn and Bacon.

Van Dyke, M. (1975). *Perturbation methods in fluid mechanics*, Parabolic Press.

Ventsel, E. and Krauthammer, T. (2001). *Thin plates and shells: theory, analysis, and applications*, Marcel Dekker.

Vese, L. A. (2012). *Variational methods in image processing*, CRC Press.

Vujanovic, B. D. and Jones, S. E. (1989). *Variational methods in nonconservative phenomena*. Academic Press.

Weinstock, R. (1952). *Calculus of variations with applications to physics and engineering*, McGraw-Hill. (Dover 1974).

White, F. M. (1974). *Viscous fluid flow*, McGraw-Hill Book Company.

Winslow, A. (1967). Numerical solution of the quasi-linear Poisson equation in a nonuniform triangle mesh. *Journal of Computational Physics* **1**, 149–172.

Yamaleev, N. K. (2001). Minimization of truncation error by grid adaptation. *Journal of Computational Physics* **170**, 459–497.

Yourgrau, W. and Mandelstam, S. (1968). *Variational principles in dynamics and quantum theory*, Sir Isaac Pitman and Sons. (Dover 1979).

Yue, P., Feng, J. J., Liu, C., and Shen, J. (2004). A diffuse-interface method for simulating two-phase flows of complex fluids. *Journal of Fluid Mechanics* **515**, 293–317.

Zill, D. G. and Wright, W. S. (2014). *Advanced engineering mathematics*, 5th edition, Jones and Bartlett Learning.

Index